Contemporary Manufacturing Processes

by

J. Barry DuVall, Ph.D., CMfgE
School of Industry and Technology
Department of Industrial Technology
East Carolina University
Greenville, NC

Publishers
The Goodheart-Willcox Company, Inc.
Tinley Park, Illinois

Copyright 1996
by
THE GOODHEART-WILLCOX COMPANY, INC.

All rights reserved. No part of this book may be reproduced, stored in a retrieval system, or transmitted in any form or by any means, electronic, mechanical, photocopying, recording, or otherwise, without the prior written permission of The Goodheart-Willcox Co., Inc. Manufactured in the United States of America.

Library of Congress Catalog Card Number 94-47298
International Standard Book Number 1-56637-158-9

3 4 5 6 7 8 9 10 96 99 98

Library of Congress Cataloging in Publication Data

DuVall, John Barry
    Contemporary manufacturing processes / John Barry DuVall.
    p. cm.
   Includes index.
   ISBN 1-56637-158-9
    1. Manufacturing processes.   I. Title.
TS183.D88 1996
670—dc20   94-47298
                                              CIP

About the Author

J. Barry DuVall is Director of the Defense Industry Partnership Project, a project entitled *The Factory as a Learning Laboratory*. This federally funded $1.2 million project is designed to provide education and training to Black and Decker (U.S.), Inc., associates and defense industry scientists and engineers in six locations across the country, using interactive television and computer-based E-mail. Dr. DuVall is also a professor in the Department of Industrial Technology at East Carolina University, Greenville, NC, where he enjoys teaching manufacturing processes, industrial materials, and productivity improvement. He has more than twenty-five years experience teaching in both industry and at various universities, and has worked in industry in the fields of electronics, design, and manufacturing. Dr. DuVall has served as Vice President of Strategic Manufacturing Services, Inc., a Greenville-based industrial consulting firm; is the former Director of Graduate Studies in the School of Industry and Technology at East Carolina University, and has been involved with a number of educational/industrial partnerships. One of these, ECU's Graduate Industrial Fellowship Program, has resulted in manufacturing firms providing more than $800,000 in funding for students majoring in industrial technology while on assignment as Graduate Fellows in industry. Dr. DuVall also is a Certified Manufacturing Engineer (CMfgE) with the Society of Manufacturing Engineers.

Using a high-speed machining technique employing ceramic tool inserts, this process removes up to 3 cubic inches of material per minute to quickly finish cast-iron engine blocks with minimal heat distortion. (LeBlond-Makino)

Introduction

Contemporary Manufacturing Processes is intended to serve as a source for evaluating and using manufacturing processes. This book provides a comprehensive survey of manufacturing processes for use in introductory courses in Industrial Technology, Manufacturing Engineering Technology, Industrial Engineering, and similar programs. It will also prove useful as a valuable reference guide in industry, particularly in the areas of industrial, project, and process engineering.

Literally thousands of different manufacturing processes are in use today, and the number continues to grow. Some are used by primary manufacturing firms, which start with raw materials such as trees or minerals and convert them into industrial stock. A far larger number of processes is used by secondary processing firms. These are firms that purchase industrial stock and transform it into products for sale to consumers. Many of these products, such as foods and paper goods, are consumable. Other products, such as automobiles, appliances, or machine tools, are referred to as nonconsumable, hard good products.

The focus is on those manufacturing processes being used by secondary processing firms to produce consumer products classified as nonconsumable hard goods. Even with this narrowed field of study, there remain thousands of processes in use.

To achieve this specific scope, the focus is placed on the major material families: metallics, ceramics, polymerics (plastics and woods), and composites. Most processes used in the manufacture of hard good consumer products involve one or more of these major material families. This text does not include printed, electronic, tobacco, and textile products. While these industries are responsible for many different types of manufactured products, they represent specialized groupings and unique processes that are outside its scope.

The manufacturing processes selected for study in *Contemporary Manufacturing Processes* were identified by systems analysis of 20 major industries. It was first necessary to determine which processes were used most frequently in each of the major material processing families. Some processes, of course, are used primarily with metals, others mostly with plastics or woods, and still others are common to the field of industrial ceramics. Since composites consist of several different materials integrated in a matrix, the processes employed are often hybrid versions of those used with other materials.

As noted, some processes are used with a number of different materials; others can be used efficiently with only one. Throughout, efforts have been made to present each process in conjunction with the material family with which that process is most commonly used.

Two overview chapters open the book. They provide the reader with an introduction to today's manufacturing environment and describe the continuing trend toward greater production efficiency resulting from adoption of design for manufacturing (DFM) principles.

The main portion of the text is organized into sections corresponding to the material families. Each section, in turn, consists of an introductory chapter and separate chapters on forming, separating, fabricating, conditioning, and finishing processes used with that material family. Even though plastics and woods are both polymeric materials, each is given its own section to more adequately cover the differing processes used.

The final chapter, Automated Manufacturing Systems, examines the impact of the computer on virtually every aspect of the manufacturing process. To aid in understanding the specialized vocabulary of manufacturing, the book includes a thorough Glossary in which approximately 1000 terms are defined.

Manufacturing is truly the driving force behind any industrialized nation, bearing a direct relationship to the quality of life and standard of living of the nation's people. By providing a thorough exploration of this vital area of living, I hope that the reader will be stimulated to continue a program of exploration and discovery. What lies ahead is an exciting world in which new materials and processes are being developed each day, so there is always something new to learn.

J. Barry DuVall

Acknowledgments

The author would like to thank the following individuals and companies for their assistance and contributions:

My wife, Jean DuVall, who deserves special thanks for her belief in the importance of this project, and for freely giving up most of the vacation time we might have spent together over the past five years.

All of my friends and Associates at Black and Decker (U.S.), Inc., Tarboro and Fayetteville, NC, and Easton, MD, for the opportunity to work with and learn from you.

My students for tolerating my enthusiasm for this subject, and for patiently field-testing all of this material through various stages of evolution.

Dr. David R. Hillis, Associate Professor, Department of Industrial Technology, East Carolina University, for writing and obtaining many of the photographs and illustrations for Chapter 31. He also field-tested some of the materials in Chapters 3-7.

Dr. John T. Jones, Vice President of R & D, Lenox China Technical Center, Absecon, NJ, for his help in arranging for special photographs, and for providing photographs and technical detail for use in Chapters 14-19.

Roger Kuntz, President, Kuntz Automation Engineering, Santa Ana, CA, who provided a number of very helpful application photographs illustrating turnkey systems and unique applications of automation developed by his company.

The late Dr. Don Cassel, who served as a reader for Chapters 3-7.

Mrs. Deborah Parker, Drafter and Engineer, Ilco Unican Corporation, Rocky Mount, NC, who prepared many of the illustrations used in the first half of the book.

The following companies and agencies generously provided photographs and other illustrations:

3D Systems, Inc.
Abbey Etna Machine Co.
ACIPCO Steel Products
American Plywood Association
American SIP Corporation
AMP Inc.
Anocut Inc.
Apple Rubber Products, Inc.
Arbonite Corporation
Aremco Products, Inc.
Aristech Chemical Corporation
Ball Corporation
Benjamin Moore
Betar Inc.
Black and Decker (U.S.) Inc.
Boston Digital

Branson Ultrasonics Corporation
Cadillac Division, General Motors
Ceradyne Inc.
Ceramic Refractory Corporation
Cincinnati Milacron
Clausing Industrial Inc.
Coherent General
Composite Corporation
Corning Incorporated
Creative Pultrusions, Inc.
Delta International Machinery Corporation
Diamond Star Motors
Dixon Automatic Tool, Inc.
DoAll Company
Duo-Fast Corporation
Eldorado
Elox Corporation, an AGIE Group Company
Ford Motor Company
Forest Products Laboratory, USDA Forest Service
Garrett Engine Division, Allied Signal Aerospace
GE Plastics
Giddings and Lewis, Inc.
Grady White Boats
Hammermill Paper Group
HDE Systems, Inc.
Herman Miller, Inc.
Hexacomb Honeycomb Corporation
Hitachi America
Hoge Lumber Company
Hudson Machinery Cutting Systems, Inc.
Ingersoll-Rand Waterjet Cutting Systems
ISO-Spectrum Inc.
Kennametal, Inc.
Kuntz Automation Engineering
Lake Geneva Spindustries, Inc.
Laramy Products Co., Inc.
Laser Technology, Inc.
Lincoln Electric Co.
Macco Adhesives
MagneTek
Malco Products, Inc.
Manufacturing Technology Inc.
Martec Plastics
Masonite Corporation
Metal Improvement Company
Milwaukee Electric Tool
Mitsubishi Carbide
Motoman, Inc.
Mykroy/Mycalex Ceramics
National Broach & Machine Co.
National Machinery Co.
National Twist Drill Inc.
Pekey Industries, Inc.
Penberthy Electromelt International, Inc.
PHI CONRAC
PMW Products, Inc.
Polygon Company
Polymer Systems
Polymerland, Inc.
Powermatic-Houdaille, Inc.
PPG Industries, Inc.
Prodel Automation Inc.
Racon Equipment Co.
Reed Rolled Thread Co.
Robersonville Products
Rofin-Sinar, Inc.
Rohm and Haas Co.
Roto-Finish Company, Inc.
S. S. White Industrial Products, Inc.
Saturn Corporation
Schuler Incorporated
Sonic-Mill, Albuquerque Division; Rio Grande Albuquerque, Inc.
Soudronic LTD.
South Bend Lathe, Inc.
Stanton Manufacturing Co.
Stapla Ultrasonics Corporation
Step 2 Corporation
Tantec
Technicut
Teksoft
The Cyril Bath Co.
The Minister Machine Co.
Timesavers Inc.
Toyoda Machinery Co.
Tree Machine Tool Company
United States Air Force
U.S. Amada, Ltd.
USG Interiors
Vought Aircraft Company
Workrite, Inc.

CONTENTS

MANUFACTURING TODAY

Competition among world-class manufacturers is growing ever more intense, challenging the positions of superiority in many industrial sectors long held by U.S. firms. The wave of change in manufacturing requires a new way of thinking in terms of lean production and increased emphasis on quality and flexible manufacturing, rather than quantity. The hope for the future of American manufacturing lies with employees who know how to work both harder and smarter.

1 Introduction to Manufacturing 13
2 Planning for Production 33

METALLIC MATERIALS

Despite the many exotic space-age materials available today, and the hundreds of traditional composite materials used over the ages, metals are still used more than any other manufacturing material. Metals are particularly useful in manufacturing because of their molecular composition, which distinguishes them from other engineering materials. Few other materials can be subjected to the same range of hot and cold temperatures and retain their essential properties.

3 Introduction to Metallic Materials 53
4 Forming Metallic Materials 65
5 Separating Metallic Materials 113
6 Fabricating Metallic Materials 167
7 Conditioning and Finishing Metallic Materials 185

PLASTIC MATERIALS

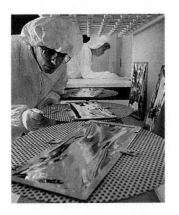

Plastics are synthetic materials that are capable of being formed and molded to produce finished products. They exhibit many characteristics that provide design advantages over other industrial materials. These include light weight, corrosion resistance, strength, durability, thermal and electrical insulation value, and ease of manufacturing.

8	Introduction to Plastic Materials	207
9	Forming Plastic Materials	219
10	Separating Plastic Materials	257
11	Fabricating Plastic Materials	269
12	Conditioning Plastic Materials	283
13	Finishing Plastic Materials	293

CERAMIC MATERIALS

Ceramic products are made of clay; the fusion of silica dioxide produces glass. Clay and glass are major divisions of the ceramics industry. When compared to metallic parts, ceramic parts exhibit many advantages. Their strong chemical bonding makes them hard and able to withstand high temperatures without melting, as well as surviving attacks by corrosive gases, molten metals, and acids. Ceramic materials can be made with densities so low that they will float on water, or can be mixed in such a way that they will be as dense and heavy as lead.

14	Introduction to Ceramic Materials	305
15	Forming Ceramic Materials	317
16	Separating Ceramic Materials	347
17	Fabricating Ceramic Materials	359
18	Conditioning Ceramic Materials	367
19	Finishing Ceramic Materials	379

WOOD MATERIALS

Wood is one of the few natural materials that humans have used throughout history without finding it necessary to drastically change its properties. Wood requires little modification to make it useful for most industrial applications, because it can easily be formed, shaped, and smoothed with a multitude of manufacturing processes. In recent years, wood has even been used in product applications that previously required plastic or metal. The open pores of wood can be impregnated with synthetic polymers to improve stiffness, water repellency, strength, and stability.

20	Introduction to Wood Materials	389
21	Forming Wood Materials	403
22	Separating Wood Materials	413

23	Fabricating Wood Materials	437
24	Conditioning Wood Materials	449
25	Finishing Wood Materials	459

COMPOSITE MATERIALS

A composite is a material in which different substances are combined to create a new material with better attributes than any of those making it up. Using composites usually results in a product that is both stronger and lighter. It also makes possible combining parts to provide a significant cost savings over conventional assembly methods. While composite materials are often more costly than conventional materials, the final product is frequently less expensive. It appears likely that composites will be the material of choice for a majority of sophisticated engineering applications in the future.

26	Introduction to Composite Materials	475
27	Forming Composite Materials	491
28	Separating Composite Materials	501
29	Fabricating Composite Materials	515
30	Conditioning and Finishing Composite Materials	527

AUTOMATING PRODUCTION PROCESSES

The word "automation" was introduced almost 50 years ago to describe the process of applying automatic control devices to production equipment (usually a single machine or process). While far more sophisticated methods are used today, automation still has three basic ingredients: a repeatable manufacturing operation or process, a control system, and a material placement system. Linking several automated processes or operations together with a management control system will provide the information necessary to achieve the level of sophistication referred to computer-integrated manufacturing (CIM).

31	Automated Manufacturing Systems	539
	Glossary of Terms	569
	Index	599

MANUFACTURING TODAY

Competition among world-class manufacturers is growing ever more intense, challenging the positions of superiority in many industrial sectors long held by U.S. firms. The wave of change in manufacturing requires a new way of thinking in terms of lean production and increased emphasis on quality and flexible manufacturing, rather than quantity. The hope for the future of American manufacturing lies with employees who know how to work both harder and smarter.

Mobil Oil Corporation

Introduction to Manufacturing

Key Concepts

△ Strategies for establishing market superiority among international competitors.

△ Development of regional power blocs as the basis for dominating global markets.

△ Strengthening alliances with the *maquiladoras* under NAFTA.

△ Becoming *lean* and *mean*—a contrast between Toyota and GM Framingham.

△ The manufacturing process, using tools, techniques, and technical systems to convert raw materials into products.

△ The distinction between primary processing and secondary processing.

△ Using process actions as the mechanism to classify manufacturing processes.

As we stand on the threshold of the twenty-first century, there are many exciting opportunities facing American manufacturing. Competition among world-class manufacturers is intense, challenging the positions of superiority in many industrial sectors long held by U.S. firms. In areas where complacency has led to stagnation, entire industries (such as steel, shipbuilding, and textiles) have been lost to foreign competition. Aggressive leadership is needed to reverse this trend and regain lost territory. The hope for the future of American manufacturing lies with employees who know how to work both harder and smarter. It is encouraging to observe that most successful companies are beginning to recognize that *people*, not just technology, must be involved in providing solutions to the productivity puzzle.

The Global Challenge

According to most analysts, the United States still holds an overall lead throughout the world in science and technology. However, the gap is closing as many new nations gain experience in specialized fields. Over the next twenty years, our major challenge for world industrial leadership will continue to come from Japan. Let's take a moment to consider what that nation has accomplished.

In the steel industry, Japan is the world's leading exporter. It has accomplished this despite having to contend with importing the basic raw materials for steel production. Japan currently has ten steel plants that are larger than any in the United States.

The automotive industry has a tremendous impact on most of manufacturing. For approximately three decades (since 1961), one of the classic models of **mass production** was the General Motors assembly plant in Framingham, Massachusetts. Although the plant is now closed, it was one of the best automotive plants in the United States during the heyday of mass production. Let's contrast data from this plant against what was once one of its major competitors, Toyota in Takaoka, Japan. According to the International Motor Vehicle Program Assembly Plant Survey conducted in 1986 by the Massachusetts Institute of Technology, the gross assembly time to produce one vehicle at GM Framingham was 40.7 hours. At Toyota, the time to assemble a comparable vehicle was 18.0 hours. There was great disparity in the number of assembly defects per 100 cars: GM had 130; Toyota, only 45. In terms of the inventory of parts kept on hand to support assembly, Framingham had a two-week supply. Takaoka, only two *hours*.

From the preceding, it should be apparent that the Japanese have made very significant advancements in productivity and quality. These advancements in management and technology have an impact reaching far beyond the automotive industry. All types of production are being influenced by the new dynamics of manufacturing, Figure 1-1.

Figure 1-1
The installation of efficient, automated production equipment, like this manufacturing cell, is helping U.S. industries in their efforts to remain competitive in the world marketplace. This cell includes both a CNC machining center and a CNC turning center, allowing "start-to-finish" machining of parts, as well as rapid changeover from one type of part to another. The gantry-mounted robot handles the loading and unloading of parts from the turning center at left and the machining center at right. (Cincinnati Milacron)

Lean Production vs. Mass Production

In their book, *The Machine that Changed the World,* authors James P. Womack, Daniel T. Jones, and Daniel Roos refer to the production concept that has been implemented at Toyota as **lean production.** Lean production incorporates some of the best thinking from continuous (mass) production and craft (one of a kind) production. It also involves major changes in the way we think about product design, parts inventories, quality control, product rework, and reducing **work in process (WIP)** (components that are partially processed, waiting for the time when they will be used).

A major change in thinking relates to the relationships that must be created with parts suppliers. The number of parts warehoused by the manufacturing plant for immediate use are greatly reduced, thus cutting costs that result from maintaining this inventory. But the concept is much more complex than this. It involves a whole new way of *thinking about the process of making quality products.*

Most automotive assembly plants in the 1950s used a moving production line, not much different than the one designed by Henry Ford to produce his Model T automobiles in the 1920s at Highland Park, Michigan. While the **assembly line** has long been the backbone of American industry and productivity, many manufacturing plants today are making major changes in the traditional concept of continuous, or mass, production. Often there is a need for continuous production, but the methodology needs to be refined to provide additional flexibility.

With conventional mass production, many thousands of "look-alike" products might be made in what is referred to as a **batch,** or **job lot.** In the heyday of mass production, consumers were satisfied with lower-cost products that were exactly like those of their neighbors.

The demands of the market have changed. With increasing global competition, manufacturers must be able to provide product variety. To remain competitive, manufacturers today must often respond quickly to demand with smaller batch sizes. In some cases, the size of these batches ("lots") may be as small as *one:* a single unit of one product may be produced, followed by a single unit or small batch of a different product. Production is still *continuous,* but major adjustments must be made to reduce changeover time for each new product. Major changes taking place in continuous production are evident in development of concepts such as **computer-integrated manufacturing (CIM).**

In the future, Japan will continue to be aggressive, attempting to capture many new markets through extensive investments in research and development. Some areas where Japan's efforts will be concentrated include semiconductors, factory automation, telecommunications, lasers, fiber optics, pharmaceuticals, and industrial ceramics.

Legislation has been enacted in the United States in an attempt to restrict and balance competition. This has stimulated the creation of a number of **international joint ventures,** Figure 1-2. NUMMI (New United Motor Manufacturing, Inc.), a Japanese/American automotive assembly plant in Fremont, California, was one of the first of these successful joint ventures. The plant is responsible for assembling GM cars and pickup trucks for the U.S. West Coast. NUMMI represents a successful marriage between the classic mass producer,

Figure 1-2
The highly automated Diamond Star Motors plant in Bloomington, Illinois, opened in 1987, was a joint venture involving Mitsubishi of Japan and Chrysler Corporation of the United States. In the early 1990s, Mitsubishi became the sole owner by purchasing Chrysler's interest in the company. (Diamond Star Motors)

GM, and the classic lean producer, Toyota. The net result is a car or truck that is made faster, cheaper, and with fewer defects.

In other markets, names such as Hitachi, Toshiba, Sony, and Fujitsu are replacing those of American manufacturers as leaders in the production of high-quality products. However, in fields such as space technology, aerospace, microprocessors, medical technology, food technology, and bioengineering, the United States still maintains a competitive advantage.

In other areas, such as fiber optics, composites, and superconductivity, the race is still on. Only time will tell which nation will be able to establish superiority in these fields, which provide tremendous potential for future growth. The *trickledown effect* of success in these areas will have an impact on many different types of manufacturing industries. Any nation that achieves superiority in these fields will have a tremendous manufacturing advantage.

Other Competitors for Manufacturing Superiority

Capturing and maintaining manufacturing superiority would be much simpler if Japan was the United States' only serious competitor. However, this is no longer the case: many countries throughout the world are gaining rapidly in specialized high-growth technology fields.

Automotive assembly plants in countries such as Brazil, Korea, Mexico, and Taiwan are showing extraordinary performance. In terms of manufacturing revitalization in the developing countries, the Ford plant in Hermosillo,

Mexico, has surfaced as a model of innovation for others to follow. This plant is based on conventional mass production, updated to include the new thinking of lean production.

Other countries, particularly those in Western Europe, are now working hard to surpass the United States in terms of annual growth in ***manufacturing productivity.*** However, contrary to popular belief, U.S. productivity in manufacturing is still going up, at a rate that exceeds annual increases in Japanese and German manufacturing. This is in contrast to what is happening in other developed countries, where the productivity revolution in manufacturing is over.

Peter Drucker, in his 1993 book, *Managing for the Future—the 1990s and Beyond,* point out that about one-fifth of the total capital invested by U.S. manufacturing firms is in plants outside the United States. This is interesting to note when you consider that at least one-third of the world trade in manufactured products takes place between plants operated by the same company but located in different countries. Such trade might involve materials transported from a Sony plant in Mexico to a Sony plant just across the border in the U.S. at San Diego, California. Today, Mexico is the United States' third-largest trading partner, ranking behind only Canada and Japan.

One of the great steps in revitalizing the world economy following World War II was the General Agreement on Tariffs and Trade (GATT). More recently, the U.S., Mexico, and Canada negotiated the ***North American Free Trade Agreement (NAFTA).*** This agreement will encourage the development of many new manufacturing partnerships between and among the three member countries.

Many U.S. business experts predict that NAFTA will lead to the creation of a North American Economic Community consisting of Canada, the U.S., and Mexico. Such a community would be more populous than the European Economic Community (EEC), and would be not far behind the EEC in terms of combined gross national product.

Mexico will be an active partner with the U.S. and Canada in the decades ahead. One reason for this is Mexico's greatest economic initiative: its industrial parks or ***maquiladoras.*** There are already hundreds of these parks strung along the Mexico-U.S. border from Tijuana to Reynosa. Factories located in these parks can import parts and supplies from the U.S. duty-free. Many of the factories, or *maquilas,* are owned by corporations with headquarters in the United States.

Plants that are not located in a maquiladora must pay costly import duties and deal with bureaucratic red tape at the border. In contrast, the maquilas, or factories that *are* located in one of the industrial parks, pay no duties on goods imported from the U.S., and when their products are moved back across the border, duty is assessed only on the value that was added to those products in Mexico.

During the 1980s, approximately one-half million new jobs were created in the maquiladoras. The largest of the maquilas resulting from this program is the Ford Motor Co. plant at Hermosillo, about 175 miles south of the U.S. border. It is recognized as one of the most efficient automobile plants in the world.

By United States standards, workers at plants in the maquiladoras are paid very low wages (ranging from .75 to $2.00 per hour). Workers also are provided with one hot meal per day at the plant, and qualify for other benefits. Some observers note that the maquiladoras are taking jobs away from workers in the United States. This is true, in terms of the lowest-skilled assembly jobs. Business spokespeople contend, however, that without the maquiladora program, some U.S. firms would be unable to compete with products produced in other countries that have lower labor costs. They also note that by shifting low-skilled, lower-paying jobs to the maquiladoras, companies have often been able to retain more highly skilled jobs in the United States.

Traditional blue-collar wage costs are becoming less significant as a driver for product costs. This is an important factor to consider when thinking about the relationship of labor costs to the total value of the manufactured product. It is more important to analyze transportation costs, considering the distance the raw materials must move from their source to the manufacturing plant, as well as the distance finished goods must move from factory to market.

A rule of thumb used by many companies with multi-national operations is that offshore production must be five percent to eight percent lower in cost than domestic production to compensate for the higher costs attributed to distance. Offshore production incurs higher costs for communication, financing, travel, transportation of goods, and insurance.

Much work has been completed by such U.S. automakers as Ford and GM to integrate in their plants some of the concepts typified by Toyota and NUMMI. It is likely that Ford is now just about as lean in its North American assembly plants as the average Japanese transplant to the United States. Many automotive component manufacturers are also incorporating the new thinking related to continuous production. These efforts will require time to be implemented and tested. A catch-up game will have to be played: if major changes do not take place in all types of American manufacturing, additional world markets will be lost to foreign competition.

What is important to understand is that competition between industrialized nations has become much more intense, with many countries entering the race for manufacturing superiority. In the decades ahead, survival in the marketplace will force companies to recruit more skilled personnel. These will be people who are aggressive and know how to make things happen.

Technological change is so rapid that there are really few "experts" anymore—there is just too much to know in most specialized fields for anyone to learn it all. A new breed of technical manager is needed. These managers will have to function more like *technical strategists* than managers. They will have to know where to go to find answers, how to solve problems, and what to do to motivate others to be more productive. Companies will call these people various names: engineers, technologists, supervisors, associates, managers, or technicians. None of these traditional job titles really communicates what these people will have to accomplish, however. The bottom line is that they must know how to increase profits for their employers.

These strategists will need to know how to work with both people and technology. Merely understanding sophisticated technical systems will no longer be sufficient; they must out-think the competition. Although technology is becoming increasingly sophisticated, there are still almost no totally automated, workerless factories. Technology is of little value without people.

Competing in a Global Marketplace

Before the 1990s, many small manufacturers were satisfied with maintaining a strong market for their products in one geographical region or sales territory. Only corporate giants were strong enough to compete in larger markets. Now, with firms competing on a world-wide scale, such a limited vision of the market is totally inadequate. In the years ahead, this trend toward world-class manufacturing will continue as an increasing number of mid-sized and small-sized companies begin to actively participate in world-class markets. In order to maintain a leadership position in any one developed country, it soon will be necessary for manufacturing firms to be able to research, design, develop, engineer, and manufacture their products in any part of the developed world. They will have to go *transnational*.

The approach to competing on a global basis that is being used by many companies is what the Germans call a ***"community of interest."*** In the U.S., this arrangement might be called a joint venture or cross-licensing agreement. It can also be thought of in terms of "leveraging knowledge."

No matter what they are called, such partnerships are not limited to small or medium-sized companies. For example, ASEA of Sweden, the world's largest heavy engineering company, has joined forces with equipment manufacturer Brown Boveri of Switzerland. Already world leaders in their own right, the two firms feel that forming a "community of interest" will make them able to be even more competitive in the North American and Far Eastern markets.

The global nature of the market will lead to major changes in the way many manufacturers make their products. In order to meet the needs of consumers in different markets, ***customized products*** will be necessary. In a global marketplace, the traditional concept of mass production, making thousands of identical products for consumers with essentially identical needs, will have limited value. In addition to producing customized products, manufacturing firms will have to produce better-quality products at lower cost to the consumer. They will have to learn how to accomplish more with fewer resources.

A vital consideration today for any company hoping to compete in the world marketplace, and especially, in the European Economic Community, is ***ISO 9000 certification.*** The ISO 9000 standard, developed by the ***International Standards Organization,*** is a recognized and agreed-upon method of determining quality. Certification is a process that involves meeting strict requirements demonstrating the company's ability to closely control all the processes that affect the acceptability of its product to the buyer. Certification can take a year or more, since it requires considerable work and strong commitment at all levels of an organization, from top management down to production, clerical, and support personnel.

It will be difficult for some companies to adjust to all of the challenges and opportunities that now face manufacturing. This is where the technology strategist must make the difference. People working with changing technology will have to be better informed than ever before.

The primary purpose of this book is to serve as a resource tool for evaluating and using manufacturing processes. It is hoped that this material will stimulate continued exploration and discovery. What lies ahead is an exciting world in which new materials and processes are being developed each day, so there is always something new to learn.

What is Manufacturing?

Manufacturing is really the driving force behind any industrialized nation, since it has a direct relationship to the quality of life and standard of living of the nation's people. Nearly two-thirds of all of the wealth-producing activities in the U.S. come from manufacturing.

The ***Standard Industrial Classification (SIC)*** Manual, prepared by the U.S. Office of Management and Budget, is the source for determining what industries and types of firms can be classified as manufacturing. The manual lists the classifications and sub-classifications of industries with their numerical codes, and is the basis for economic decisionmaking related to business and industry.

Manufacturing can be defined as the use of tools, processes, and machines to transform or change materials or substances into new products. Repair or rework of old products is not classified as manufacturing. However, manufacturing does include companies that assemble component parts into a product.

Transforming materials into products can be accomplished through mechanical or chemical means. Manufacturing is usually done in a factory, mill, or plant, but can take place just about anywhere.

In order to understand what manufacturing is, it may be helpful to spend some time clarifying terminology. There are many different types of manufacturing firms, or companies. Each of these produces and/or assembles a specific line of products, such as motorcycles or cosmetics or woodworking tools or computers. Typically, a company will manufacture a variety of models or styles in its particular product line. For example, an automobile manufacturer would offer two-door, four-door, or hatchback models; different "packages" of features, and several power train options.

If you were to consider all of the companies manufacturing a particular product (such as fishhooks), and a significant quantity of that product was sold each year, then chances are that it would be classified as an *"industry"*— in this case, the fishhook industry. Currently, there aren't too many fishhook manufacturers around, nor are vast numbers of fishhooks being sold, so the Standard Industrial Classification Manual doesn't list this as a separate industry. Fishhook production is covered as a subclassification under a major type of manufacturing industry, miscellaneous manufacturing. Other products, such as cars and trucks, have established their identity as a separate industry (the *automotive industry*).

There are many different types of industries classified as "manufacturing." They range from companies that make hard good products, such as earthmoving equipment and laser cutting machinery, to others that blend plastics, resins, and oils. Even textile, tobacco, food, and chewing gum producers are classified under manufacturing.

Major Manufacturing Industries

The twenty major manufacturing industries that are listed in the Standard Industrial Classification Manual are shown in Figure 1-3. However, there is much more to manufacturing than this. According to the SIC, industries involved in the production of liquors and wines, and products of agriculture such as mining, fishing, and quarrying, are classified as manufacturing. Other industries that you might not normally associate with manufacturing are also included: oyster shucking, apparel jobbing, publishing, logging, and ready-mixed concrete production.

Figure 1-3
These twenty groups comprise the manufacturing segment in the Standard Industrial Classification Manual. The groups vary greatly in size, with some made up of a relatively small number of producers, while others number their members in the thousands.

Industrial Groups

Manufacturing Industries Listed in the SIC Manual

Food and Kindred Products	Tobacco Products	Textile Mill Products	Furniture / Fixtures
Paper and Allied Products	Printing, Publishing and Allied Industries	Chemicals and Allied Products	Lumber and Wood Products Except Furniture
Leather and Leather Goods	Rubber and Plastic Products	Miscellaneous Manufacturing	Stone, Clay, Glass, and Concrete Products
Primary Metal Products	Fabricated Metal Products	Transportation Equipment	Electronic Equipment
Measuring, Analyzing, and Controlling Instruments	Industrial and Commercial Machinery and Computer Equipment	Apparel and Other Finished Products Made from Fabrics and Similar Materials	Petroleum Refining and Related Industries

Given the SIC classification, simplicity may not be possible. Mining of copper is classified as manufacturing, while the extraction of coal or other nonmetallic materials is classified as mining. Tobacco products and milk bottling and pasteurizing are manufacturing; milk processing on farms and stemming leaf tobacco are not.

The twenty industrial groups shown in Figure 1-3 should help you to understand the tremendous diversity that exists in the field of manufacturing. Each of these major industrial groups can be subdivided further into more specific product classifications.

With the twenty major manufacturing industries, and four or five sub-classifications of each, there are nearly 100 different types of industries listed in the SIC Manual. Figure 1-4 illustrates product types (material families) that

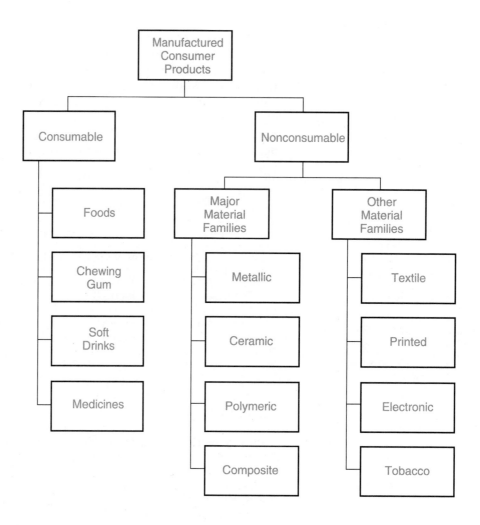

Figure 1-4
Nonconsumable, or hard good, products are produced by industries that can be classified under various material families.

are associated with each of the major manufacturing industries. Each of these industries can be further broken down into even more specific types of industries and products. There are many different types of manufacturing firms associated with each type of material. For example, the glass industry (under the ceramics material family) has many subdivisions, or firms making specific types of glass or glass products: plate glass, fiber optic rods, electrical insulators, and scientific glassware, to list only a few. There are hundreds of manufacturing industries and thousands of individual companies associated with them.

Although there are many different types of products and materials involved, *all* manufacturing firms have this in common: they either make new products from *raw materials,* or assemble parts into new products. Service activities such as repairing, servicing, or rebuilding products are not classified as manufacturing.

Manufacturing processes come into play when raw materials, called **industrial stock,** are changed into products. Figure 1-5 shows what happens

Figure 1-5

Turning is a typical manufacturing process, in which industrial stock is transformed into a product. Turning consists of holding a cutting tool against a rotating workpiece to remove material. In this example, a replaceable cutter made of ceramic carbide is being used to shape hard titanium metal. (Kennametal, Inc.)

in one typical manufacturing process. Here, a process called turning is used to remove metal from titanium stock that is mounted on a rotating machine tool known as a lathe. Since the titanium metal is difficult to cut because of its extreme hardness, a ceramic carbide insert is used as the cutting tool. The photograph provides a close-up view of the titanium chip that is being removed as the tool cuts into the *workpiece* (stock).

Every manufacturing process involves following a planned sequence of operations or steps to create the desired product. If the same sequence of operations is followed each time the process is used, and if the conditions in the work environment do not change, then products of consistent quality can be produced.

Material Processing in Manufacturing

While all manufacturing companies are concerned with transforming materials into products, not all of them make or assemble products that are ready for use or consumption by the consumer. In some cases, a product from a manufacturing firm is ready for use or consumption when it leaves the manufacturing plant. In other cases, the product of one plant becomes the raw material for use by another, thus undergoing successive transformations. Figure 1-6 shows a microwave amplifier package that was hermetically sealed (made airtight) with the heat generated by a process called laser welding. With this product, the cover was welded using a Nd:YAG (neodymium yttrium aluminum garnet) laser moving at a rate of 8 to 10 inches per minute. Individual parts making up the amplifier package might have been produced by several different manufacturers, and then sold to the amplifier manufacturing firm as *components* to be assembled in the product shown. The amplifier, in turn, might become a component in a more complex product.

Figure 1-6
A laser beam was used to weld the cover of this microwave amplifier in place, sealing it airtight. The attached tag provides an idea of the size of this small component. (HDE Systems, Inc.)

The process of *successive transformation* (production in stages) can extend across a number of industries. Consider the manufacture of electric motors: first, the copper smelting plant produces copper for use by electrolytic refineries. The refinery then manufactures a product called refined copper. This is purchased as the raw material for use by the copper wire mill. The copper wire it produces is purchased by the electrical equipment manufacturer as the raw material for use in making electric motors. The electric motor may then be purchased as a subassembly by other manufacturers for use in their products, such as portable electric drills, battery-operated toothbrushes, or coffee grinders.

Material Processing Categories

There are many ways to organize the way we think about the various stages of transformation, or phases of material processing. To simplify the process, we will classify all of the different types of processing into two categories:

△ *Primary processing,* carried out by firms engaged in the preparation of standard industrial stock.

△ *Secondary processing,* done by firms that manufacture hard good consumer products.

These two categories and their relationship are shown in Figure 1-7.

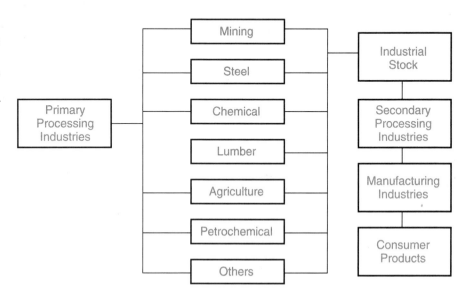

Figure 1-7
Primary processing industries are those that convert raw materials into industrial stock. Secondary processing industries, in turn, transform industrial stock into products. These products may be ready for sale to consumers, or may be used as components by other manufacturers who assemble consumer products.

There are literally thousands of different manufacturing processes used by primary and secondary manufacturing firms. Primary processors start with raw materials, like trees or minerals, and then use manufacturing processes to change these materials into industrial stock: sheets, rods, pellets, marbles, or liquids. As noted earlier, products often go through many different stages in production. For example, some primary processing firms deal with smelting and separating metal from the impurities. Others distill petroleum to manufacture jet fuel or make plastic pellets.

There are many types of industrial stock produced by primary manufacturing firms, ranging from perfumes, ceramic particles, and sheets of composite materials to hunks of clay and containers of lubricating oil. The chemical refining, filtering, mixing, and separating processes used by primary processing firms are outside the scope of this book.

Secondary processing firms are those companies that purchase industrial stock, such as powdered clay or plastic pellets, and then use manufacturing processes to transform the stock into a product for sale to the consumer. By turning stock into useful products, secondary processing adds more value to the basic raw material. Many products, such as foods or soft drinks or paper goods, are consumable. Other products, such as electric mixers, earthmoving equipment, or bathtubs, can be classified as nonconsumable, hard good products.

The focus of this book is on the *manufacturing processes* used by *secondary processing* firms to produce nonconsumable, *hard good consumer products.* Even with this attempt to narrow the field of study, there are thousands of manufacturing processes used by these manufacturing firms. Further delimitation is necessary.

Major Material Families

There are many types of materials used by the industries shown in Figure 1-7. Some of these materials are used less frequently than others by manufacturers concerned with producing hard good consumer products. Figure 1-8 shows the *major material families* (MMFs): **metallics, ceramics, polymerics** (plastics and woods), and **composites.** These four are identified as major material families because most processes used in the manufacture of hard good consumer products involve one or more of them. The four major material families constitute what could be called "the common body of knowledge in manufacturing." Most manufacturing sector employment involves the making of products from these materials.

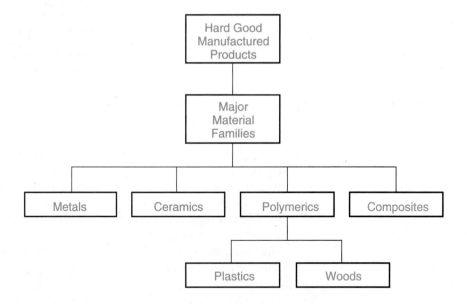

Figure 1-8
Most manufacturing processes are used to transform material from one or more of the major material families. A majority of manufacturing employment is involved in some way with processing materials from these families.

You may recall that, in Figure 1-4, a major grouping labeled as Other Material Families was shown. This grouping included printed, electronic, tobacco, and textile products. While these material families are also responsible for many different types of manufactured products, they are not within the scope of this book. These have been given a secondary priority because the manufacturing processes used in these areas are often associated with other industries. Examples would be electronic products (communications or telecommunications industry), printed products (printing industry), and tobacco products (tobacco industry). While these industries *do* manufacture products, they represent specialized groupings and unique processes that are outside the common body of knowledge in manufacturing.

Refer to Figure 1-4 again. Note that there are two types of manufactured products, consumable and nonconsumable. ***Consumable products*** would include foods, chewing gum, soft drinks, and even some pharmaceutical products (medicines). ***Nonconsumable products*** are hard goods, such as plastic recycling containers, ceramic grinding wheels, and metal lawn furniture.

It should be realized that there is nothing finite about the use of classification schemes to organize thinking. Graphic models can be used to group information according to the best judgment of the designer.

Although we have delimited the scope of our study a great deal, there are still thousands of manufacturing processes that are used to produce hard good consumer products using materials from the four major families (metallics, ceramics, polymerics, and composites). For each of these material families, there are major industries that can be identified. These are:

- Metallic (metals industry).
- Ceramic (clay and glass industries).
- Polymeric (woods and plastics industries).
- Composite (aerospace industry).

Composite materials are used by all of the manufacturing industries. However, the metallic, ceramic, and polymeric material families have their own unique types of products. Let's look more closely at the structure of each of these industries.

The different types of manufactured *metal* products are listed in Figure 1-9. Products in this taxonomy represent the major types of metal industries. There are seven major industries concerned with the manufacture of metal products.

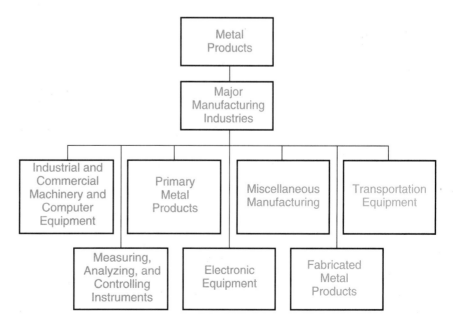

Figure 1-9
A large number of major manufacturing industries produce products that involve metals.

Figure 1-10 illustrates the four major types of industries that manufacture *ceramic* products. The term "ceramics" gives the initial impression that this industry is solely concerned with clay products. However, as shown, products made from glass, stone, and concrete are also included in this major industrial grouping.

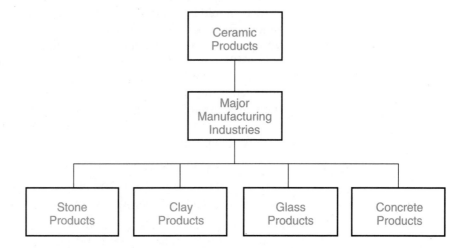

Figure 1-10
In addition to clay products, the ceramics family includes those made from stone, glass, and concrete.

Figure 1-11 shows the six major types of industries that use *polymeric* materials. Processes used to manufacture products in all these classifications can be studied in conjunction with two of the industries, plastics and woods.

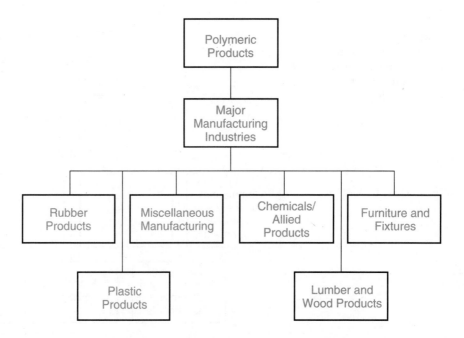

Figure 1-11
Plastics and wood products represent the major elements of the polymeric materials category.

Material Processing Families

Over the past quarter century, new materials in the families of polymers (plastics), ceramics, and composites have brought revolutionary changes in the way we design and manufacture products. Never before have we had the opportunity to not only design and build products, but also to create the materials used in their manufacture. Sophisticated new materials are now available which can withstand the abuse of harsh environments, including the rigors of space exploration. These *engineered materials* can withstand exposure to chemicals, pollution, contamination, and even collision with floating debris in space. Such human-engineered materials offer great promise for manufacturing; they also provide new challenges for the technologist.

Each material has its own set of behavioral characteristics; not all manufacturing processes work equally well with all materials. Important production decisions often have to be made in selecting the correct process for a particular material and unique set of conditions. Manufacturing strategists (manufacturing supervisors, engineers, and technicians) have to be able to make informed judgments to choose the process that will accomplish the most for the least (provide the highest quality with the least expenditure of resources). On the plant floor, good decisionmaking is interpreted in terms of profit. *Profit* results from making the correct number of quality products at a cost that makes them more attractive than those offered by competitors. Poor judgment in selecting and using manufacturing processes will result in poor quality, lost time, and lost profits.

To identify *manufacturing processes* for study in this book, it was first necessary to determine which processes are used most frequently in each of the four major material processing families. Some processes, of course, are used primarily with metals. Others are used mostly with plastics or woods. Still other processes are common to the field of industrial ceramics. However, many of the processes used with ceramics are also used with plastics, metals, ceramics, or woods.

Since composites involve several different materials integrated in a matrix as one material, the processes employed are often hybrid versions of those used with other materials. Sometimes, the process can be used in exactly the same form it is with other materials. In other cases, the process must be modified so that it is more suitable for use with the type of composite that has been created.

Some processes are used with a number of different materials; others can be used efficiently with only one. Throughout this text, efforts will be made to present each process in conjunction with the material family where the process is most frequently used.

Process Action

Because of the many manufacturing processes used with the four material families, it would be difficult to develop perspective on how all of these processes work if they were presented and studied at random. In order to help you understand the similarities and differences between processes, they will be presented according to the type of *process action* that they perform. The major process actions are:

- Forming.
- Separating.
- Fabricating.
- Conditioning.
- Finishing.

A process action is not the same thing as a manufacturing process. A ***process action*** is the description of what happens when a process changes the internal structure or the outward appearance of a material (ceramic, composite, metallic, or polymeric).

Forming

Forming processes are used to change the size and/or shape of industrial stock. With forming processes, pressure is normally employed to squeeze or shape the material. Often, there is no loss of weight or volume when material is changed using a forming process. Bending, stretching, casting, and molding processes are all classified under the heading of forming. These are processes used to form metal, polymeric, ceramic, and composite materials.

Separating

Separating processes are used to remove material or volume. They can be classified according to the type of chip produced as a cutting tool strikes the material. The force involved is also a determining factor in classifying separating processes. There are three different types of separating processes:

- *Mechanical chip-producing separating.* This process involves wedge-type cutting action with loss of material or volume (Example: sawing).
- *Mechanical non-chip-producing separating.* In this process, there is wedge cutting action with no chip being produced and no loss of volume of the material (Example: shearing).
- *Non-mechanical separating.* This process involves cutting action that may or may not generate a chip, with no force involved (Example: photochemical milling).

Fabricating

Fabricating processes are those used to join and fasten materials together. There are three major types of fabricating processes: adhesion, cohesion, and mechanical joining.

Soldering, brazing, and gluing are fabricating processes involving adhesion. Adhesive joining processes normally take place with two materials being joined through use of another material. The material that joins the others together creates a bond at the point of attachment. Adhesion processes do not provide the strength attained through cohesive bonding or mechanical joining.

Fabricating processes based on the principle of cohesion create a molecular bond that unites the atoms of the materials being joined. Fusion welding and resistance welding are examples.

Conditioning

Conditioning processes are used to change the internal or external properties of materials. Conditioning processes often involve heat, shock, electrical impulses, magnetic methods, chemical action, or mechanical means to change the structural characteristics of the material. Many times, the result of conditioning is not visually apparent. Often, conditioning processes change the arrangement of atoms within the material. Conditioning processes are used to influence the hardness, strength, rigidity, resistance to wear, and fatigue (product life) of a material.

Finishing

Finishing processes are often used to improve the outward appearance or protect the exterior surface of a part. In other cases, finishing processes are used to prepare the surface of a workpiece prior to the application of a coating.

There are hundreds of coating processes used in manufacturing. Many of these are unique, ranging from fluidized bed coating of ceramics to spray metallizing of plastic and curtain coating of wood products.

Important Terms

assembly line
batch
ceramics
community of interest
components
composites
computer-integrated manufacturing (CIM)
conditioning
consumable products
customized products
engineered materials
fabricating
finishing
forming
hard good consumer products
industrial stock
industry
international joint ventures
International Standards Organization
ISO 9000 Certification
job lot
lean production
major material families
manufacturing
manufacturing processes
manufacturing productivity
maquiladoras
mass production
metallics
nonconsumable products
North American Free Trade Agreement (NAFTA)
polymerics
primary processing
process action
raw materials
secondary processing
separating
Standard Industrial Classification (SIC)
successive transformation
technical strategists
trickledown effect
work in process (WIP)
workpiece

Questions for Review and Discussion

1. With the passage of NAFTA and the rapid growth of maquiladoras just across the U.S.-Mexico border, what do you expect will be the impact on low-skilled workers who are presently assembling household products in U.S. plants?

2. From which direction would you expect the real challenge to United States manufacturing to emerge in the next decade: Japan, China, or the

European Economic Community? Provide a rationale to support your answer.

3. List the U.S. manufacturing industries that you believe could be labeled "major growth industries." Explain why you feel these industries are growing faster than others.

4. In as much detail as necessary to differentiate the two, discuss the difference between a manufacturing process and a process action.

Planning for Production

Key Concepts

△ The third wave of change in manufacturing requires a new way of thinking in terms of lean production and increased emphasis on quality and flexible manufacturing, rather than quantity.

△ Anything in the manufacturing process that does not add value to the product should be eliminated.

△ Typical sources of waste in manufacturing are overproduction, waiting, excess inventory, lost motion, and rework.

△ Design is a major driver of the cost of manufactured products.

△ Value analysis and design for assembly are techniques that can be used to accomplish continuous improvement.

Change, defined as "departure from the ordinary," seems to occur in cycles that are later recognized and labeled by historians as they attempt to describe the course of human thought and action. The expression of progress in terms of successive (and sometimes colliding) *waves of change* was popularized by historians who described the settlement of the American West. The first wave of western settlement was the movement of pioneers. They were followed by a wave of farmers, and finally, by the entrepreneurs (businessmen and industrialists).

Waves of Change in Manufacturing

The concept of change taking place in waves has relevance, in a manufacturing context, to any discussion of planning for production. An effective planner must consider what has come before, or time will be wasted in creating a desirable future.

We are now riding the crest of a new wave in American manufacturing. Historians of technology will probably one day describe this as the *third wave* of change influencing productivity in the United States, Figure 2-1. It is impossible to tell, at this point, exactly what phrase will be used to describe this

Figure 2-1
Third-wave manufacturing, with its emphasis on quality and just-in-time material management, relies heavily upon efficient and highly adaptable equipment like this flexible manufacturing cell. The cell can operate as a "stand-alone" unit, or can be integrated with other equipment in a Flexible Manufacturing System (FMS). (Cincinnati Milacron)

third wave of change. It will probably be something like "world-class manufacturing," or "lean production." We will have to wait and see.

What is important to understand is that energies directed toward second wave thinking will not be productive in a third wave world. Before effective planning for production can occur in world-,class manufacturing, a new way of thinking is required. Failure to make the transition in thinking from the second to third wave is, in the opinion of this author, the major cause of waste, stagnation, and deteriorating profits in many industries that are struggling for survival. Loss of jobs cannot be blamed on automation, or foreign competition. The company that is able to produce the best product at a fair price, with the shortest lead time, will be successful in the third wave.

First-Wave Manufacturing

The *first wave* of change began in the mid-1920s, with the advent of mass production. Many people contributed to the development, refinement, and diffusion of mass production, but certainly much of the credit must be awarded to such pioneers as Henry Ford and General Motors' Alfred Sloan.

The model of *mass production* for all the world to follow was the gigantic Ford Rouge plant in Detroit, Michigan, which covered an area one and one-half miles long and three-fourths of a mile wide. The plant had 75,000 employees, 93 buildings, 93 miles of railroad track, and 27 miles of conveyor belts. It even had its own steel mill and glass plant on site. When the Rouge switched from the Model T to the Model A in 1928, it was the most integrated plant ever developed. Lead time necessary to finish one complete vehicle was reduced from 21 days to only four.

One of the distinguishing characteristics of first-wave manufacturers was their emphasis on the *product*. Quantity was paramount, with most products being continuously manufactured on an assembly line. Quality inspections were normally conducted on products when they reached the end of the production cycle, just after final assembly. Such an approach resulted in a great deal of wasted time spent reworking defective parts (many of which even had to be partially disassembled to locate defects). By the end of the decade of the 1940s, most of the large manufacturing industries—such as automotive, textile, steel—were heavily immersed in first-wave thinking.

Today, first-wave thinking continues to be evident in many industries, stretching far beyond steel and automotive plants. Even fast-food restaurants owe much of their success to the proven techniques of mass production.

There are many limiting features associated with mass production, however. Many businesses that fail do so not because of foreign competition, but because they were still guided by first-wave thinking with its emphasis on quantity rather than quality.

For example, in 1985, one photocopier manufacturer averaged around 50,000 defective parts per million parts produced. In the steel industry, the situation became so bad that the Ford Motor Company reported, in 1983, that it was rejecting as much as eight percent of all of the steel delivered from domestic plants.

Second-Wave Manufacturing

After World War II, many firms began to change their way of thinking; a new wave, a new philosophy of manufacturing, was born. The transformation from the first wave to the second did not take place overnight. Even today, with many firms clearly following what can be described as third-wave thinking, others continue to operate with a second-wave (or even first-wave) mentality.

As the *second wave* developed, a totally new way of thinking emerged, with emphasis shifting from a focus on quantity to one on *quality*. This involved far more than an enhanced emphasis on quality inspection at the end of the production line.

The second wave really began when an American, Dr. W. Edwards Deming, began to teach his ***"Deming Plan for Total Quality Control"*** to Japanese manufacturers. Eiji Toyoda and Taiichi Ohno found his message was so significant that they immediately adopted the Deming principles at the Toyota Motor Company. Within a decade, the Japanese had completely turned around public opinion toward the quality of their products, not only in the automotive field but in nearly all of their industries.

Dr. Deming's principles went far beyond changing employee attitudes toward scrap and rework. Deming's belief was that quality control should check the ***process***, not the product. He placed emphasis on finding the cause of a problem, so that it could be eliminated, rather than fixing the problem after it occurred. The diffusion of the Deming approach (first in Japan, and then more recently in the United States), has resulted in the development of a completely new attitude toward responsibility for quality.

An important tool of the second wave was the computer, which made it possible to process information at speeds which were unheard of in the first wave. (The first computer used in manufacturing was installed in 1954 at the General Electric appliance plant in Louisville, Kentucky.) Without the computer, many refinements in second-wave management technology would have been stifled.

In the 1950s, 1960s, and 1970s, many refinements were made to second-wave thinking in the manufacturing field. One of the major focuses of research and writing at this time was a vigorous attack on the problem of waste. Toyota defined *waste* as "anything other than the minimum amount of equipment, materials, parts, space, and worker's time, which are absolutely essential to add value to the product."

The emphasis on *value* being *added* to a product is important. It does not take long to realize that most of the time spent in production, or in planning for production, is not spent adding value to the product. Instead, much of the time spent by employees is spent adding *cost* to the product. For example, much less time is spent on processing materials than is devoted to *waiting*—in storage, in transit to another operation, or in a batch being held for inspection. Many firms have realized the need for changes in this area.

Koyoshi Suzaki, in his book *The New Manufacturing Challenge*, summarizes what many writers list as the major areas of waste by second-wave manufacturing firms. Suzaki lists seven different types of waste that are common in manufacturing firms exhibiting second-wave thinking:

- Overproduction.
- Waiting.
- Transporting.
- Process inefficiency.
- Excess inventory.
- Lost motion.
- Rework.

Overproduction commonly occurs when market demand slows and there is no control over manufacturing. This results in the company carrying unsold products as excess inventory. Money is spent unnecessarily for raw materials, and wages have to be paid to employees for unneeded services. All of the activity expended on production, recordskeeping, and planning for production gets in the way of priority issues that should be addressed. Only enough product should be produced to meet the needs of the next stage of production. Waste from overproduction does not add value to the product.

Waiting as a form of waste is not difficult to observe in many manufacturing firms. When visiting plants, you will often find operators passively watching machines run. If the operator must wait until something goes *wrong* to take corrective action, what purpose does this serve? Control systems and instrumentation should be used to detect impending problems, such as broken or dulled tools, before they become acute. Wasting time waiting for a problem to occur does not add value to the product.

A great deal of time is typically wasted in ***transporting*** parts from one operation to another, or waiting for materials to be brought to a workstation from storage. Often, many hours are spent trying to track material, due largely to poor layout of a work area, or to sloppy housekeeping. Some companies create "warehouses on wheels" by loading parts in carts, and push these from one location to another in an attempt to reduce time lost in waiting. Most often, however, the carts merely serve as obstacles and safety hazards in the plant's aisles. All of these factors create confusion and loss of time. Wasting time waiting for material movement does not add value to the product.

One of the most significant causes of wasted time is *inefficiency* in manufacturing processes. For example, if injection molding of a nylon case results in excessive flash (excess material on the part at the point where the mold halves close) being generated, then a secondary process may be required. Such a process adds unnecessarily to the cost of producing the part. A better solution would be to change the stock used, or modify the die set, to eliminate the problem of excessive flash. Wasting time and money for inefficient processing does not add value to the product.

Many companies seem obsessed with the need to develop large inventories. Such *excess inventory* increases the cost of the part, simply because of additional handling, facility costs, and paperwork. With a low inventory, problems with the product can be detected much more quickly. Thus, there is a direct relationship between low inventory and high quality. In many manufacturing companies, investments in inventories and facilities comprise more than two-thirds of the investment in the company. Excessive inventories do not add value to the product.

Many movements performed by employees are nothing more than *lost motion*, expending energy without adding any value to the product. It is becoming more and more important to consider the amount of time spent on various operations and activities: "staying busy" is not the same as "being productive." Companies have become increasingly concerned with ergonomics, or human factors engineering. The layout of the work area or *workcell* (environment where several machines are operated by one employee), must be supportive of efficient movement of employees throughout the work environment. Moving for movement's sake does not add value to the product.

Rework of parts, assemblies, or entire products as a result of defects discovered by inspection adds to costs because of additional handling and use of production resources. Components that cannot be reworked must be scrapped, adding further to operating costs. Rework of defective items does not add value to the product.

Third-Wave Manufacturing

According to many observers, we are now in the midst of the *third wave* in manufacturing, which first became noticeable in the early 1980s. By that time, a significant number of U.S. manufacturing firms were aggressively implementing a new set of management techniques.

When first implemented, these techniques centered upon two major emphases. The first of these was what many companies later referred to as a *Total Quality Approach* or *Total Quality Management*. While virtually every firm was interested in quality improvement, some also chose to pursue a second emphasis: reducing inventories through the application of a just-in-time (JIT) supply and production system.

Just-in-time manufacturing, also referred to by the Japanese term *kanban,* eliminates excess inventories by placing the burden directly on suppliers. In JIT at its most ideal, the manufacturer tells the supplier the quantity and time of delivery, and the supplier delivers the required quantity at the prescribed time. JIT works well when the supplier is close by and when the relationship between the company and supplier is strong. Many firms have *captive suppliers* (those for whom the firm is the principal, or sometimes only, customer).

In some cases, suppliers or the company itself will "cheat" a bit to guarantee JIT performance, using a buffer storage area with parts stored in adjacent warehouses or parked semitrailers. One advantage of JIT for the manufacturer is that the firm doesn't have to pay for parts until they use them.

General Electric was an early adopter of the JIT approach, starting with two projects in 1980. A Kawasaki plant in Nebraska, and a Toyota facility in California started with JIT at about the same time. Several months later, IBM Corporation and Nashua Corporation announced that they were initiating JIT programs. Hundreds of other companies followed suit through the mid-1980s.

While these efforts were certainly important, they were not comprehensive enough to suggest an oncoming transformation (an actual "third wave"). A more integrative approach was needed. Many have referred to this as *world-class manufacturing (WCM).*

A world-class manufacturing firm is what was described in Chapter 1 as a "lean producer," a company that has gone through continual improvement to the extent necessary to assure that better quality products are being consistently produced at less cost. World-class manufacturers are experts at producing quality products. Efficiency in performance has been sharpened to the point where lead times are almost nonexistent. WCM firms can be effectively viewed as the *integrators*: they know how to utilize all of the principles of effective technology management, and they can do it faster and better. In most cases this involves JIT, but it may also involve technologies such as computer-integrated manufacturing (CIM). Computer-integrated manufacturing is discussed in more detail in Chapter 31.

World-class manufacturers are the innovative companies that have taken advantage of what has been learned during the first and second waves. These firms are anticipatory: they don't wait for a problem to occur before they take steps to eliminate it.

An understanding of all of the principles that are important to WCM firms would take many volumes. What is important to realize is that these techniques are all directed toward the reduction of waste. But it cannot end with that.

When the major sources of waste are eliminated, and the best thinking related to information management, quality assurance, and efficiency in production have been incorporated into the system, it still would be premature to engage in planning for production.

The major sources of waste have been presented as factors that disturb productivity and efficiency in a production system. While they can be viewed as drivers of manufacturing, control of these factors will not necessarily result in survival in the marketplace as a WCM producer.

What is left to be considered is the *product* itself. It is becoming increasingly difficult to produce high quality products and sell them at competitive prices when they are not designed to facilitate ease of assembly and simplicity in manufacture. If the product is poorly designed, it may be impossible to shorten lead times or to cut waste related to processing.

Design in Manufacturing

The *design* of the product is not something which can be taken lightly, since design is often the primary driver of a product's cost. It is the design that determines whether the product can be manufactured efficiently.

Design determines the product's structure, method of assembly, number of parts, design of components, type of materials, tolerances, and surface finishes. The product gradually assumes its final shape as machined parts, components, and subassemblies are joined through the process called *assembly*.

Many firms find that as much as 80 percent of a product's manufacturing cost is driven by its design. Most firms allocate at least 40 to 60 percent of manufacturing cost to assembly. False hopes have often been raised by the belief that this cost could be significantly reduced by automating assembly and other production operations. In most instances, however, this approach is unsuccessful. A poorly designed product that is difficult to assemble *manually* will be just as difficult to assemble using robots or other automated devices and machines.

Because of the rate of change in the marketplace, it is almost impossible for a manufacturer to stay current without expediting the process of design improvement. Once the basic design for a product has been exposed to the market, it is carefully studied by aggressive competitors. Design improvement does not take place in a vacuum.

Value Analysis

One of the primary methods that firms use to gather information on the design of competitive products is called *value analysis*. Value analysis is a continuing process, not a one-shot objective, and continues as long as the part is being produced.

There are two different approaches to value analysis. The first involves analyzing the costs involved in each step of production. This introspective view is directed toward *continuous improvement*, which Japanese WCM firms refer to as *"kaizen"* or incremental improvement.

A second type of value analysis involves analyzing the competitor's product. The approach involves obtaining a sample of the product, then using an in-house design team to analyze it. Some firms find it effective to develop a review team consisting of a designer, manufacturing engineer, marketing specialist, and value engineer. Other specialists from manufacturing engineering can also be brought into the team as needed.

The value analysis normally includes a critique of the product, a determination of how it works, and an assessment of the importance (value) of each part. Considerations such as production cost, quality, reliability, and maintainability are evaluated. The competitor's product is compared with the firm's own product to determine strengths and weaknesses.

There are three major factors that are emphasized in most value analysis programs. The first is the *function* of the product in terms of its cost. This includes a careful analysis of all parts and components.

The second factor is *creativity*. How well did the competitor address factors such as ease of use, aftermarket service, and maintenance requirements?

The third major consideration is *design simplicity.* Are parts included on the product that are not critical? What about the material used? Could less expensive materials accomplish the same objective, and still maintain the life of the tool or device? How easy is the product to assemble and service? Is the product overdesigned?

Simultaneous Engineering

Value analysis was refined into a related concept, called *simultaneous engineering,* that really described what was taking place in terms of linking production and design. Simultaneous engineering brings the manufacturing engineer into the design team to determine how the new product will be manufactured. By introducing this dimension at an early stage in the design cycle, many problems that might later create waste in manufacturing processes can be eliminated. Examples of such problems are a need for exotic fixturing, unnecessary secondary processing, or difficult assembly operations.

Value analysis, and simultaneous engineering, have been operationalized by WCM firms through what most companies call Design for Assembly (DFA), or Design for Manufacturability.

Design for Assembly (DFA)

With a very simple product, it might be possible to eliminate all of the concern over assembly, since a one-piece product doesn't require any assembly operations. In most cases, however, products are much more complicated, so assembly is necessary.

A product that is specifically designed to emphasize its manufacturability will produce significant cost savings. These savings will result from a reduction in monies expended for materials, labor, equipment, and material handling systems.

While many manufacturing firms have spent millions of dollars on automation, it was the world-class manufacturers that really began to emphasize design for manufacturability. A slogan used by General Electric says it well: "If we can't be simple, we can't be fast. If we can't be fast, we can't win." Being *fast* is critical to survival in the third wave.

The operational philosophy behind *design for assembly (DFA)* requires that products be evaluated in terms of their suitability to modern methods of assembly. DFA is a process that simulates the actual assembly of a product, assigning penalties to parts that are difficult to assemble because of their design. Data developed from DFA sessions is critical to redesigning products to facilitate their assembly and manufacture.

The most extensive use of DFA to date has been in the computer industry, but firms in other fields (such as Ford Motor Company and Xerox) have also been pioneers. Each of these firms has saved millions of dollars by implementing DFA programs. An example of a successful DFA project was one completed by Ford: a door trim panel that was redesigned to reduce the part count and decrease the cost of labor for assembly. Figure 2-2 illustrates the original door panel; the redesigned panel in shown in Figure 2-3. The door panel

Figure 2-2
A typical automobile door trim panel before DFA analysis. Note the number of separate parts and subassemblies. (Ford Motor Co.)

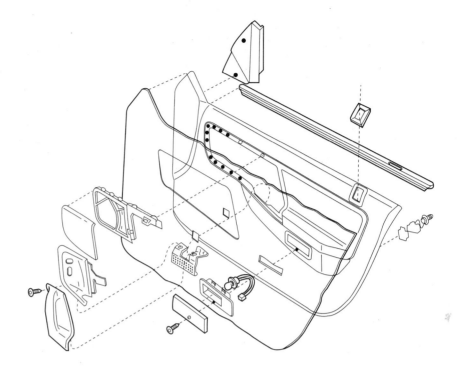

redesign, made possible through DFA task force meetings, resulted in a 79 percent reduction in the number of parts. Assembly costs were reduced by 94 percent, and material costs by 27 percent.

Figure 2-3
The same door panel shown in Figure 2-2, after being redesigned using DFA methods. Note the dramatic decrease in separate parts that must be attached. (Ford Motor Co.)

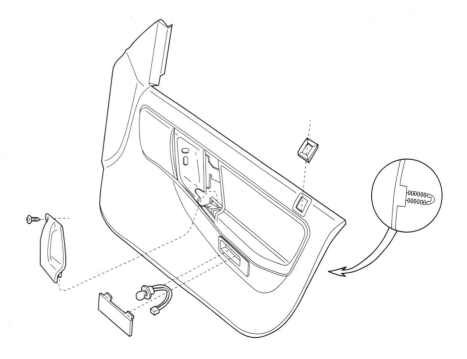

Companies large and small report DFA savings in the part count as great as 20 percent and in labor cost, savings of up to 40 percent. These savings are common without automation. However, DFA is also useful when assembling products using robots or hard technology (nonprogrammable automation). Philips International, a worldwide manufacturer of products ranging from household appliances to digital recording systems, has established success with DFA using automation. The company notes that costs for automated feeding and assembly devices are often reduced by as much as 40 percent by incorporating simple DFA design changes.

Parts can be assembled by many different techniques. These can be classified as follows:

- Δ Physical contact (enclosing, inlaying, inserting, hanging).
- Δ Filling (filling a cavity with liquids or gases).
- Δ Interference fits (squeezing, shrinking, forcing, wedging, or using fasteners such as screws or nails).
- Δ Phase changes (casting, forging).
- Δ Form changes (folding, bending, rolling, crimping).
- Δ Joining (welding, soldering, gluing).
- Δ Mechanical attachment (braiding, sewing, weaving).

Often the shift from manual to automatic assembly approaches can be accomplished with simple design changes, as shown in Figure 2-4. Such simple design changes often can produce significant savings in handling costs.

Boothroyd and Dewhurst Method

The most widely accepted approach for implementing design for assembly was developed and extensively implemented in industry by consultants Geoffrey Boothroyd and Peter Dewhurst. The method can be used to analyze and rethink the configuration of products that are to be assembled manually, or those that are assembled using robotics or hard automation devices.

The *Boothroyd and Dewhurst method* yields numerical data that can be used to assess design efficiency, and to evaluate the need for redesign of the product under study. Methods used to obtain the data will vary, depending on the type of assembly (manual or automatic) that is desired. For simplicity, we will concentrate only on the approach applied to manual assembly.

Before the design for analysis process is begun, a design team meeting is conducted to review items such as:

- Δ Product background.
- Δ Marketing plan.
- Δ Expected sales.
- Δ Product design.
- Δ Manufacturing processes.
- Δ DFA procedure.
- Δ Preliminary assessment.

Once the background on the product has been established and the proposed strategy for manufacturing outlined, the design group conducts a pre-

Figure 2-4
Simple design changes in parts, like those shown, will make them suitable for automated assembly

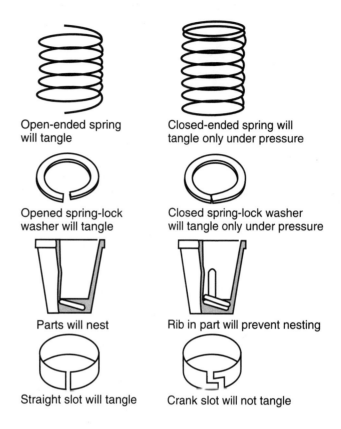

liminary assessment of the product under review. An exploded view drawing is helpful at this point, as is a design prototype (simulation of the actual product) or the actual product. The design for analysis process begins with the group disassembling the actual product, if available. If the product has not yet been built, group members envision how the proposed product would be disassembled. Each of the parts on the exploded view drawing is assigned an identification number. In cases where the product includes a subassembly, the major parts are considered first, and the subassemblies later.

At this point, the design team reassembles the product, beginning with the part that has the highest identification number. Parts are added to the assembly, one at a time. No parts are to be assembled together, using both hands, before placing them in the completed assembly.

There are two different types of assembly operations: *handling and orienting*, and *inserting and securing*. Data must be obtained on each type of operations. The design team analyzes the features of each part and estimates assembly times. This enables the user to obtain accurate assembly times according to different types of assembly processes, shapes of products, and types and sizes of stock.

The ideal efficiency for manual assembly of the product can be generated by using the model shown below. The model for calculating the ideal manual assembly efficiency of each product shown here is a simplification of the Boothroyd and Dewhurst method, and was developed by Bart Huthwaite, President of Troy Engineering, in Rochester, Michigan.

Huthwaite's model is as follows:

$$EM = \frac{3 \times NM}{TM}$$

where:

EM = the ideal manual assembly efficiency for the product being reviewed.

3 = estimated ideal assembly time of 3 seconds. This is based on an ideal orienting and handling time of 1.5 seconds and an ideal insertion and securing time of 1.5 seconds.

NM = the theoretical minimum number of parts. This is calculated by analyzing all parts in the product, in terms of the questions below. If any of these questions can be answered, "no," then the part is unnecessary and should be eliminated or integrated with another part.

TM = the estimated total manual assembly time required to assemble the present product.

1. Does the part move in terms of its relationship to parts already assembled?
2. Is the part made of a material different from the parts already assembled?
3. Must the part be isolated from the parts already assembled?
4. Does the part have to remain separate from the parts already assembled because integrating it would make the required assembly or disassembly of other parts impossible?

An ideal manual assembly efficiency can be calculated for each product. This is useful when comparing different products in terms of design for manufacturability. It can also be useful to compare existing products with redesigned products to facilitate ease of assembly and producibility.

We can apply the model presented above in terms of a hypothetical product, which we will call a "wingbatt." Data for the original and redesigned wingbatt is shown in Figure 2-5.

Figure 2-5
By applying design for assembly principles to the hypothetical "wingbatt," 18 of 23 parts were eliminated. This resulted in a 43 percent cost reduction.

Traditional Design		DFA Design
1	Platen	1
2	Flex Posts	2
2	Clamp Levers	2
2	Paper Clamps	0
12	Screws	0
4	Washers	0
23	Parts Count	5

The ideal manual assembly efficiency (EM) for the wingbatt, after value analysis, would be calculated as follows:

$$EM = 3 \times NM \text{ (theoretical minimum number of parts)}/TM$$

In this case, the theoretical minimum number of parts required is five. Eighteen parts have been eliminated from the original wingbatt because they did not pass the test of the four questions presented earlier, so:

$$EM = 3 \times 5 = 15/TM$$

To keep things simple, we will assume that each of the assembly operations for the wingbatt takes three seconds. With 23 parts in the original product, 69 seconds would be needed to assemble the wingbatt. So, TM equals 69. Let us continue:

$$EM = 15/69, \text{ or } 0.21$$

This means that the original product had a design efficiency of 21 percent. Contrast this with the redesigned wingbatt. The new product has only 5 parts, and thus, a TM of 15. Placing this figure in the formula, we would have:

$$EM = 15/5 = 300 \text{ percent}$$

The difference between 21 percent and 300 percent efficiency represents a significant savings.

Thus far in this chapter, the major techniques for controlling the management of technology have been presented. It has also been established that, in order to survive in our highly competitive society, manufacturers must to be able to quickly respond to a changing market with quality products. The components of world-class manufacturing were discussed, as were techniques for analyzing products for their ease of assembly in manufacturing. Now we are ready to consider planning for production.

Planning for Production

The process of organizing for production often begins with study of the design package, complete with an engineering drawing, assembly drawing, parts listing (bill of material), and sequence of manufacturing operations. Unfortunately, in some cases, simultaneous engineering has not taken place, resulting in a great deal of wasted time in getting ready for production. World-class manufacturing firms do not have this problem, since they have learned how to integrate design and manufacturing.

In most cases, the first step in planning for production is to use the list of manufacturing operations, created by the design for assembly team, as a basis for developing a graphic model. This model will show the sequence of operations (processes, transportation, delays, storage, and inspection activities) that will be required in the manufacture of the product. These pictures of manufacturing are presented through what is called a ***process flow model.*** One of the most common types of process flow models is PERT/CPM.

PERT (Program Evaluation and Review Technique), was first used in the late 1950s by DuPont and Remington Rand for work under Department of Defense and NASA contracts. PERT is used in conjunction with ***CPM*** (the Critical Path Method). PERT/CPM is widely used today in many areas of planning, particularly in the manufacturing and construction industries.

The first step in constructing a process flow model is to complete a process flow chart (PFC). See Figure 2-6. The PFC is developed directly from the list of manufacturing operations, and specifies each of the activities necessary to manufacture the product. This includes not only manufacturing operations, but also transportation, assembly, delays, and inspection. Times for each of these activities are normally indicated on the chart.

Figure 2-6
A process flow chart (PFC) is used to specify each activity involved in production of the product. Time estimates for each operation are entered in the four right-hand columns.

Process Flow Chart
Product ______________ Part # _________
Department ___________

Process Actions	**#**
Transportations	
Inspections	
Delays	
Storages	
Operations	
Total Steps	

	transport delay storage operation inspect		OT	PT	MT	ET
1	➤ ◗ ▼ ● ■					
2	➤ ◗ ▼ ● ■					
3	➤ ◗ ▼ ● ■					
4	➤ ◗ ▼ ● ■					
5	➤ ◗ ▼ ● ■					
6	➤ ◗ ▼ ● ■					
7	➤ ◗ ▼ ● ■					
8	➤ ◗ ▼ ● ■					
9	➤ ◗ ▼ ● ■					
10	➤ ◗ ▼ ● ■					
11	➤ ◗ ▼ ● ■					
12	➤ ◗ ▼ ● ■					
13	➤ ◗ ▼ ● ■					
14	➤ ◗ ▼ ● ■					
15	➤ ◗ ▼ ● ■					
16	➤ ◗ ▼ ● ■					
17	➤ ◗ ▼ ● ■					
18	➤ ◗ ▼ ● ■					
19	➤ ◗ ▼ ● ■					
20	➤ ◗ ▼ ● ■					

Once the process flow chart is completed, the foundation for establishing a process flow model has been created. The sequencing order found on the process flow chart is used to construct the model.

The process flow model consists of arrows (representing activities) and circles (indicating events). An event is the completion of one or more activities. Arrows starting at a circle indicate activities that can begin only after all activities terminating at that circle have been completed. Refer to the network conventions shown in Figure 2-7.

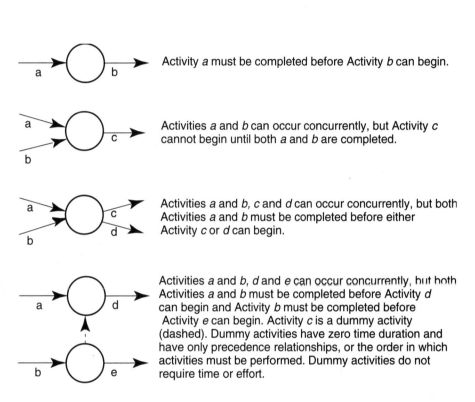

Figure 2-7
These PERT network conventions are useful when attempting to interpret the activities and events shown on a network chart. Arrows represent activities; circles are events.

On a PERT/CPM model, events are often shown by numbers or letters in the event circles; each succeeding event has a higher number or letter than its preceding event. Usually, activities are represented by arrows, which are later replaced by word descriptions.

After a flow process chart has been developed to specify all of the activities and time estimates, a process flow *model* is drawn. Then, the planner can use the data shown on the model to determine the **critical path.** The critical path is the longest route. It is said to be critical because, if any task or operation is delayed in this path, there is a corresponding delay in the time needed to build the product. See Figure 2-8.

In the model shown in Figure 2-8, three different time estimates are presented above each of the activity lines (arrows). The first number is the estimate of the most optimistic time (OT) required for completion of this activity. The middle number is the most likely time (MT). The third number shows the pessimistic time (PT), an estimate of the longest time that is likely to be required to complete the activity. Most likely time (MT) is the average of OT and PT.

Figure 2-8
A typical PERT model, with letters used in the circles to show sequence of the events represented. The three numbers near each arrow are differing time estimates for completion of the activity the arrow represents. The critical path is the longest path from A to H.

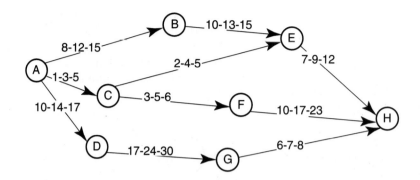

Normally, the planner would use numbers in the circles indicating a progression of events from low to high. In Figures 2-7 and 2-8, however, letters are used in the circles to avoid confusing these events with numbers shown on the activity arrows.

Since neither the optimistic time nor pessimistic time are likely to occur in practice, an estimated time (ET) for accomplishing each activity must be determined. The ET is calculated for each activity by using the following formula:

$$ET = \frac{OT + 4(MT) + PT}{6}$$

Calculating the Critical Path

When creating the process flow model using the PERT/CPM technique, the most important component is the critical path. As noted earlier, the critical path is the longest sequence of events and activities in the network. In the network model, Figure 2-9, four paths are apparent: ABEJK, ACFHJK, ADGHJK, and ADGIJK. The OT, MT, and PT times are shown above each activity arrow, with the ET shown in bold face below the arrow.

Figure 2-9
In this network model created with the PERT/CPM techniques, the boldface numbers below each arrow represent the estimated time (ET) for completion of that activity. Critical path on this model is the sequence of activities labeled ADGHJK.

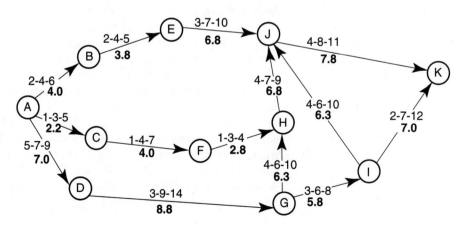

Total time along each path in the network model, using the data shown in Figure 2-9, is as follows:

Path	Estimated Times (ET)	Total Times
ABEJK	4.0+3.8+6.8+7.8	22.4
ACFHJK	2.2+4.0+2.8+6.8+7.8	23.6
ADGHJK	7.0+8.8+6.3+6.8+7.8	36.7
ADGIJK	7.0+8.8+5.8+6.3+7.8	35.7
ADGJK	7.0+8.8+5.8+7.0	28.6

With the model shown in Figure 2-9, the sequence of activities in the ADGHJK path takes longer than any other path. It is, therefore, the critical path. If the target completion time of the project is the same as the critical path, or 36.7 units (seconds or minutes, usually), then all paths will have slack time. This means that they will be completed before the target for completion of the product, and might be idle.

Slack time can refer to the difference between the critical path time and the target completion time, if the target completion time is different from the critical path time. Slack time can also be referred to as the difference between the target time and the time associated with each path. With critical path time equal to target time, the slack for each path using the above data would be as follows:

Path	Target	Total Time	Slack Time
ABEJK	36.7	22.4	14.3
ACFHJK	36.7	23.6	13.1
ADGHJK	36.7	36.7	0.0
ADGIJK	36.7	35.7	1.0
ADGIK	36.7	28.6	8.1

How to Reduce Slack Time

When a process flow model has been completed and the critical path has been identified, it is important to consider techniques for eliminating slack time. This is called *line balancing.*

When the planner develops the first process flow model for a product, more than likely there will be a great amount of slack time between the lines that produce and assemble the various parts of the product. Efficiency can be improved by line balancing. Techniques for line balancing include:

- ∆ Remove activities and events that are not essential.
- ∆ Reconfigure the model. With the possibility of added risk, time advantages can be gained by reorganizing serial activities into parallel activities. Refer to Figure 2-7.
- ∆ Reallocate resources. The redistribution of resources (time, labor, equipment) may shorten paths.
- ∆ Reduce specificity. Breaking down gross descriptions of activities into fine detail may help to better configure the network.

Once the planner is satisfied with the process flow model and input from all critical areas has been included, then it is time to proceed with planning for production. By this time, the necessary materials have been located by Purchasing, and are ready for JIT delivery. The target date for the pilot run of a limited number of test products, often referred to as an "engineering lot," has been established, and the plant floor is being readied for production. Often, training at this stage is completed by Engineering. Once the pilot run has been completed, and refinements made, then a more formal training program may be instituted. Shortly after training is completed, actual production begins.

Important Terms

assembly
Boothroyd and Dewhurst method
captive suppliers
continuous improvement
CPM
creativity
critical path
Deming Plan for Total Quality Control
design
design for assembly (DFA)
design simplicity
excess inventory
first wave
function
inefficiency
integrators
just-in-time manufacturing
kaizen
kanban
line balancing
lost motion
mass production
overproduction
PERT
process
process flow model
rework
second wave
simultaneous engineering
slack time
third wave
Total Quality Approach
transporting
value analysis
waiting
waste
workcell
world-class manufacturing (WCM)

Questions for Review and Discussion

1. Select a product and conduct a design for assembly study. Use the formula given in this chapter for calculating design efficiency. Did your DFA work contribute to improved efficiency? How realistic is it to think that the changes you suggested for this product could be carried out?

2. Disassemble a manufactured product, then develop a process flow chart that shows how you think this product could be manufactured. Be sure to list estimated activity times that will be needed for each operation (process). Which operation requires the greatest amount of time? Could you eliminate this operation and replace it with another that would require less time to perform? What are the disadvantages (if any) of substituting the new process?

3. Discuss how you would go about conducting a value analysis study of a product, using the Japanese process called *kaizen*. What major tasks would be involved in doing the study with the *kaizen* process?

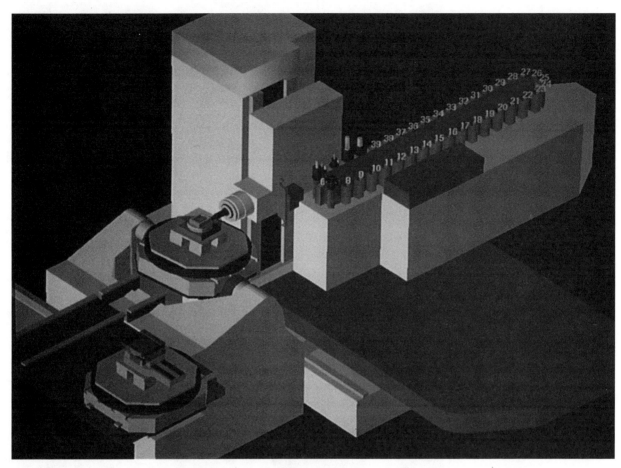

Simulation software is now being used in a number of industries to verify computer numerical control (CNC) programming to identify potential problems and assess efficiency of the process without tying up production equipment for trials. This simulation is being done with a computer-modeled 5-axis machining center with dual pallet loaders and a tool changer. (Silma Incorporated)

METALLIC MATERIALS

Despite the many exotic space-age materials available today, and the hundreds of traditional composite materials used over the ages, metals are still used more than any other manufacturing material. Metals are particularly useful in manufacturing because of their molecular composition, which distinguishes them from other engineering materials. Few other materials can be subjected to the same range of hot and cold temperatures and retain their essential properties.

Gleason Corporation

Introduction to Metallic Materials

Key Concepts

- △ There are a number of basic properties that are important to classifying a metal for use with a particular manufacturing process and product application. These properties are the material's weight, color, electrical conductivity, and reaction when exposed to heat.
- △ There are four major types or classifications of metals: ferrous, nonferrous, high-temperature superalloy, and refractory.
- △ The two major types of ferrous metals are iron and steel.
- △ Since they are resistant to corrosion, nonferrous metals are more durable than iron and steel.
- △ Superalloys are classified according to the base metal used (typically iron, nickel, or cobalt).
- △ There are three major methods used to identify types of steel: AISI numbering, grinding, and color-coding.

Despite the development of the many exotic space-age materials available today, and the hundreds of traditional composite materials that have been used over the ages, *metals* are still used more than any other manufacturing material. If you attempt to determine the significance of the metals industry by analyzing it in terms of employment, you will find that the top four employment sectors emphasize products made of metal. These largest manufacturing employers are machinery, electrical/electronic equipment, transportation equipment, and fabricated metal products. See Figure 3-1. Not only are metals used in the products made by these industries, they are also used to make the machines that *produce* these products. Of the known chemical elements, more than 50 (almost half the total) are classified as metals.

Unique Characteristics of Metals

Metals are particularly useful in manufacturing because of their molecular composition, which distinguishes them from other engineering materials. Unlike plastics, ceramics, and woods, metals can survive drastic changes

Figure 3-1
Automotive manufacturing is a major user of metals. This unique materials handling and storage system is used to transport engine bore liners between machining operations at a large manufacturing plant. (Cadillac Division, General Motors)

in the environment in which they are used. At one extreme, they remain strong and rigid enough to support heavy loads while being subjected to heat. At the opposite extreme, in frigid environments, they retain flexibility and can still be easily formed. Few other materials can be subjected to the same range of hot and cold temperatures and retain their essential properties.

Metals can accomplish these feats because of the way that their atoms are clustered together when the material is in a solid state. The atomic structure of a *pure metal* is relatively simple. *Alloys* have structures that are more complex.

The major metals are aluminum, copper, iron, lead, tin, magnesium, nickel, titanium, and zinc. Other metals such as steel, brass, bronze, and titanium are alloys of these base metals. In manufacturing, alloys are used more frequently than pure metals. Alloys consist of several different metals that are blended together, sometimes with the addition of other nonmetallic elements. Brass, for example, is a simple alloying of copper and zinc. Steel is far more complex: a mixture of iron, carbon, magnesium, vanadium, nickel, and chromium. Through the principle of alloying, new metals can be created to "showcase" the best attributes of each metal. For example, an alloy of tin and copper would be stronger than either tin or copper alone.

Because of the many different types of alloys, a metal can be created to withstand exposure to just about any environment. There are more than 25,000 different types of steel, and more than 200 standard copper alloys, not counting the many different types of brass, bronze, and nickel silver. Each of these is identified by its own code number.

Properties of Materials

Every material has its own unique physical and mechanical properties. *Physical properties* are used to distinguish one material from another. There are basically four physical properties used to classify metals: weight, color, electrical conductivity, and the reaction of the material when it is exposed to heat.

Metals that are more *dense*, or have tighter molecular structures, are heavier. Other metals can easily be distinguished by their color, such as the bright silver of aluminum or dull gray of lead.

Metals have different properties when they are subjected to temperature extremes. Each type of metal has a different thermal conductivity and its own coefficient of thermal expansion, depending on its molecular structure.

Physical Properties

Often, the physical property of a metal is more important than the actual mechanical property. A property such as the specific heat of the material is important to know when a particular metal is to be cast. **Specific heat** is the amount of energy necessary to produce a one-degree change in the temperature of the material.

A physical property such as thermal conductivity, thermal expansion, and the reaction of a particular metal to magnetic fields, may be critical in a specialized application. **Thermal conductivity** relates to the amount of heat that is carried or stored in a material. Good electrical conductors exhibit high thermal conductivity.

The property of **thermal expansion** is particularly important to consider when metals are being cast and cooled. Like ceramics, metals shrink during the curing or cooling cycle. This amount of shrinkage must be taken into consideration by the manufacturing technologist, or products that are created will be undersized.

Electrical parts are naturally more sensitive to properties such as electrical conductivity and response to magnetic fields. Temperature changes affect the response of the material to electrical conductivity.

The suitability of a given metal to a particular design application can be further complicated when you understand that different metals respond differently to a magnetic field. Magnetic properties are particularly important in such manufacturing processes as electrical discharge machining and electrochemical grinding, which cannot be used with materials which do not conduct electricity. See Figure 3-2.

Some metals change simply by being placed in an electrical field (the *piezoelectric effect*). Others expand and contract when subjected to a magnetic field. Nickel, in particular, exhibits this type of behavior.

Mechanical Properties

While physical properties often are critical to the selection of a metallic material for a particular end use, there are many times when the mechanical properties are even more important. It is the mechanical properties that determine how the part or product will survive continued use or abuse. Mechanical properties are evaluated by conducting laboratory tests on the reaction of the particular metal to applied forces. Typical tests used to assess mechanical properties deal with hardness (Rockwell test, Brinell test, or Charpy impact test), tensile strength (tensile test), ductility (tensile test), toughness (tensile test), elasticity (Young's Modulus of Elasticity), and fatigue (fatigue failure test).

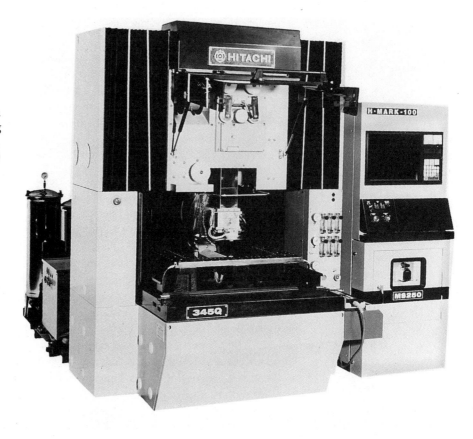

Figure 3-2
Electrical discharge machining equipment, such as this wire EDM, is assuming increasing importance as a production tool. Physical properties of a material often are important in determining which processes can be used effectively. (Hitachi America)

Behavioral characteristics related to stress and fatigue must be taken into consideration when selecting a metallic material. Stresses are created when a load is applied to a metal structure, part, or component. Stress occurs when the material attempts to respond to deformation created by an applied force.

Let's consider an example. If a heavy enough weight is placed on the middle of a length of bar stock that is suspended by its ends, the weight will cause deformation of the bar, Figure 3-3. This deformation will produce some lengthening of the bar stock. In this instance, the stress produced results from tensile strain, and is called *tensile stress.* The amount of stress resulting from a particular load depends on the characteristics of the metal. What is particularly important to know is the plasticity of the material. *Plasticity* refers to the ability of the material to change shape or size as a result of force being applied. Information related to plasticity is useful to know when shaping and forming metal.

A similar concept, called ductility, also has to be considered. *Ductility* is the ability of the material to be formed plastically, without breaking. This is an important characteristic when selecting a manufacturing process. If the material is easily fractured (if it is not *ductile*), it is more difficult to form.

Another characteristic of metals that must be considered is called fatigue. *Fatigue* results in the breaking of a piece of metal after the metal has

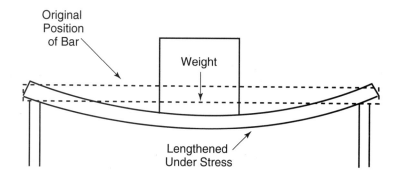

Figure 3-3. Deformation of the bar stock results from stress created by application of the load (weight). The bar lengthened slightly and sagged downward under tensile stress.

been bent back and forth repeatedly. With metal parts, most breakage is caused by fatigue.

Another term that is related to the behavior of metal, creep, is also important to the manufacturing technologist. It is something that must be addressed when manufacturing durable products of metal or plastic. **Creep** is the elongation of material that occurs when it is exposed to elevated temperatures while under stress. If a material is exposed to heat over time, and this exposure continues to produce elongation, the material will eventually rupture. This is called *creep failure*. The creep rate of steel can be reduced by adding elements such as nickel and chromium to the alloy.

Classifications of Metals

There are essentially four major classifications, or types, of metals. These are: ferrous metals, nonferrous metals, high-temperature superalloys, and refractory metals.

Most of the metal products manufactured today are made from either ferrous or nonferrous metals. However, tremendous growth is taking place in high-temperature superalloys. They can survive exposure to extreme heat, cold, and unusual forces that were not even imaginable ten years ago.

There are metals that are produced from iron ore and others that contain no iron at all. **Ferrous metals** are produced from iron ore. The word ferrous comes from the Latin word *ferrum* which means iron. Figure 3-4 illustrates the two types of ferrous metals, iron and steel.

Figure 3-4 There are two types of ferrous metals, iron and steel. Iron can be further divided in wrought iron and cast iron; steel into hot-rolled and cold-rolled.

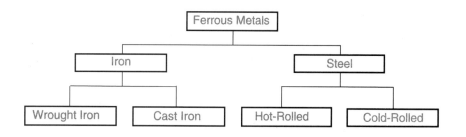

Ferrous Metals

Iron is produced in two basic types, wrought iron and cast iron. **Wrought iron** is tough, because it contains very little carbon. Wrought iron is easy to bend, even without heating, which makes it a prime candidate for use in ornamental ironwork.

Cast iron is a very useful material in manufacturing. Cast iron is made by pouring into a mold molten iron that has been mixed with between 1.7 percent and 4 percent carbon. It is used when strength and durability are critical.

Cast iron is hard. This makes it brittle and easy to crack, but gives it tremendous wear and abrasion resistance. Cast iron cannot be bent, stretched, or formed by forging. Typical uses of cast iron include engine blocks, machine frames, machine parts, and other castings.

There are several different types of cast iron. The most popular type is called *gray iron*. It is easily cast and is less expensive than the other types.

Another type is called *white cast iron*. White cast iron is very hard and is used for making parts that must combat fatigue from extreme wear and abrasion conditions.

A third type of cast iron, called *malleable iron*, is made by heating white cast iron, and then cooling it slowly (a softening process called *annealing*). Annealing is an important process in the metal manufacturing industries, and will be discussed in greater detail later in this section.

Another major type is called *ductile cast iron*. It is also known as nodular, or spheroidal grade (S.G.), cast iron. Ductile cast iron is heat-treatable and is used for making parts such as crankshafts, camshafts, and connecting rods for both gasoline and diesel engines. This is one of the newest cast iron products, and has displaced much of the malleable and white cast iron production in recent years.

Steel is an alloy of iron and carbon, or of iron and other alloying elements. When carbon is not used, elements such as tungsten, molybdenum, or vanadium are often added to make the steel harder and tougher. Stainless steel is an example of a class of alloy steel made by blending chromium and iron. Alloy steels are hard and resistant to corrosion.

When steel includes carbon as an alloying element, it is called **carbon steel**. There are three types of carbon steel: low-carbon, medium-carbon, and high-carbon.

There are several methods that are used to classify steel. One method classifies by number, according to the percentage of carbon content in the sample. Steel with one percent carbon content is classified as *100 carbon steel*.

Steels are also classified according to how they are made. **Hot-rolled steel** is squeezed between rollers while it is hot, and can be identified by its bluish surface coating. **Cold-rolled steel** is rolled while cold, and has a shinier surface finish than hot-rolled material.

Steel is highly refined iron with a carbon content of less than 1.7 percent. Low-carbon steel, or mild steel, usually has from between 0.05 and 0.30 percent carbon. Since it is very soft, it can be easily formed and machined. The major disadvantage of mild steel is that it can't be annealed (softened) or hardened.

Medium-carbon steel has between 0.30 percent and 0.60 percent carbon. It is more difficult to bend and shape, but it can be hardened through heat treatment. Hardening causes the material to become more brittle.

High-carbon steel is often referred to as *tool steel.* It has from 0.60 percent to 1.50 percent carbon. Tool steel is hard and difficult to bend. It can be made even harder through heat treating. Tool steel is used to make tools, such as forging dies, screwdriver shafts, chisels, and milling cutters.

Next, we will look more closely at how ferrous metals are used. Machine tools, machine bases, support frames, and similar products that require strength and rigidity are typically made of ferrous metals. Cast iron is extremely durable, and is often the preferred choice for castings, gears, and crankshafts. One of the limitations of cast iron parts is that machining is difficult because of the presence of silicon.

In making cast iron, pig iron (refined wrought iron) is mixed with scrap iron or steel to help control the carbon content. Most cast iron production is accomplished in a cupola, or small blast furnace.

Ferrous metals resist the tendency to bend and flex, but they will rust. Cast and wrought iron are ferrous metals which do not rust as quickly as other ferrous metals.

Wrought iron is tough and ductile, because it is made from almost pure iron. When wrought iron is heated almost to its melting point it becomes soft. At this stage, two pieces can be hammered together to create a union between them. This is called forging.

Wrought iron is easy to bend, so it is useful for making products requiring ornamental bends and tight circles. A limiting feature of wrought iron is that it is expensive. Low-carbon steel is often chosen as a cheaper alternative.

Metals with no iron content—copper, aluminum, lead, nickel, zinc, tin, and brass—are called *nonferrous metals.* See Figure 3-5. Nonferrous metals are softer and easier to form. In spite of the fact that they are softer, these metals are usually more durable than iron and steel, since they are resistant to corrosion.

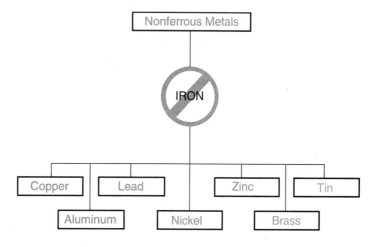

Figure 3-5
Nonferrous metals are those that do not contain iron. In effect, all metals other than iron and steel are nonferrous.

Copper and copper alloys are classified in a Unified Numbering System specified by the Copper Development Association. The numbers C10100 through C15500 are reserved for copper alloys with 99.5 percent or greater purity. Higher numbers represent alloys with a smaller percentage of copper. Copper and its alloys are useful in product applications that require superior electrical and thermal conductivity. These materials are soft, easy to form, and resistant to corrosion.

Aluminum is another nonferrous metal that is soft and easy to form. But its major advantage over most other metals is light weight. Like copper, it is resistant to corrosion and is a good conductor. Aluminum and aluminum alloy stock are classified by their own numbering system. Numbers consist of four digits followed by a hardness classification number. The first four digits refer to the purity of the metal and the nature of the alloying. An aluminum alloy with the number 1000-H10 could be interpreted as follows: "1000" means that the principal ingredient is aluminum of 99 percent or greater purity. The "H10" describes the type and hardness of the material.

High-Temperature Superalloys

We have discussed both ferrous and nonferrous metals. Now we can turn to the third major classification of metals, *high-temperature superalloys.* The superalloys were created by the defense and space-related industries in response to the need for new materials that could withstand abuse from extreme forces and severely oxidizing high temperature environments. Superalloys are used in the manufacture of rocket engines, turbines, jet engines, and space vehicles. See Figure 3-6. They are also being used in the chemical processing and nuclear fuels industries.

Superalloys are classified according to the base metal in the alloy. In most cases, this is iron, nickel, or cobalt. Thus, superalloys can be iron-based, nickel-based, or cobalt-based. Other alloys such as chromium, titanium, aluminum, or tungsten are then added to the base metal.

What all these superalloys have in common is their ability to survive, without degradation, a temperature of 2200°F (1200°C) for reasonable periods of time when used in a nonloadbearing structure. Under load, most superalloys can be used at 1800°F (1000°C).

Refractory Metals

Refractory metals, the fourth major category, are high-temperature metals. They can withstand heat and maintain their strength at temperatures ranging from 4474°F (2468°C) for niobium to 6170°F (3410°C) for tungsten.

Refractory metals include niobium (Nb), tungsten (W), and molybdenum (Mo). All of these metals have very strong interatomic bonding, which is the reason why they can survive such high melting temperatures. Tungsten has the highest known melting point of any metal.

Refractory metals are used in many ways. When molybdenum is alloyed with stainless steel, it gives the metal greatly improved resistance to corrosion. Often tantalum is also added, creating an alloy that is impervious to chemical attack in environments below 302°F (150°C).

These metals are used in products such as incandescent light filaments and welding electrodes. They are also used for tools and dies, rocket engines, gas turbines, and containers for holding and dispensing molten metal.

Metals are used for so many different product applications that sometimes the industry is described in terms of specific product lines: automobiles, appliances, containers, pipe and tube, bicycles, fasteners. One is quick to remember that other major industries such as foundry, shipbuilding, and transportation equipment are also dependent on metals.

Figure 3-6
Superalloys are used for many aircraft and aerospace parts. This turbine component for a jet engine is being checked for machining accuracy by a coordinate measuring machine. (Garrett Engine Division, Allied Signal Aerospace)

Analysis of an industry might logically follow product lines. However, the diversity of the "metals" industry might better be visualized by considering the thousands of individual (often small) machine shops throughout the world that produce metal products in support of other manufacturing industries. Manufacturing is almost totally dependent on metals as a basic manufacturing material.

Nature of Industrial Stock

As noted earlier, the primary emphasis of this book is on the secondary manufacturing industries—those that start with industrial stock and transform it into a useful consumer product. Industrial stock is the product of primary manufacturing industries. In the case of metals, primary industries are concerned with mining, refining, and production of metals.

Steel is a very important form of industrial stock. The steelmaking industry shapes molten steel into solid ingots. Ingots are rolled or extruded into various shapes for use by industry to make bars, hot-rolled strip ("band iron"), plates, round or hexagonal rod, tubing, wire, or angle stock. These forms of stock are continuously produced in standard lengths, and are normally sold by the pound.

There are other types of stock which are widely used. Metal stock in the form of powder or of billets (for example, aluminum bricks for casting) is also available. Stock in powder or billet form is normally sold by the pound.

Stock is available in sheet form, as well. Steel plate or sheet stock is sized in thickness by *gage number*. The smaller the gage number, the thicker the stock. However, care is needed when purchasing different types of sheet stock, since not all plate has the same numbering system.

Tin plate (mild steel coated with tin), has its own unique gage system. With tin, the greater the number of "Xs," the thicker the plate. *Nonferrous metals* are usually sold according to their weight per square foot, their thickness in decimal equivalents, or their gage number.

Determining the Type of Steel

Most of the time, it is impossible to look at stock and determine what type of steel it is. There are three basic techniques that you can use to determine the type of steel. The most accurate is the numbering system devised by the American Iron and Steel Institute (AISI). The Society of Automotive Engineers (SAE) uses the same system. See Figure 3-7.

Figure 3-7
The numbering system devised by the AISI for classifying steels is widely used in industry.

Here's how the system works: the first number stands for the type of metal. The second number indicates the percentage of alloy, and the last two tell how much carbon is in the metal. For example, the number 1020 for a steel would be interpreted as follows: 1 means that it is carbon steel, 0 that it contains no alloy, and 20 that it has 20 percent carbon.

A second method that can be used to identify the type of steel is to grind a small section of stock and study the pattern of the sparks. As shown in Figure 3-8, mild (low-carbon) steel creates a long and consistently distributed pattern of sparks. High-carbon steel produces a wider and more-tightly clustered pattern. Wrought iron produces a smaller cluster of sparks. High-speed steel creates a more-scattered long and thin pattern of chrome yellow sparks.

A third method for determining the type of steel in a bar or rod is by a color code. Most manufacturers paint the end of the stock with one or more

Figure 3-8
The spark color and pattern that results from grinding is one of several methods used to identify different types of steels.

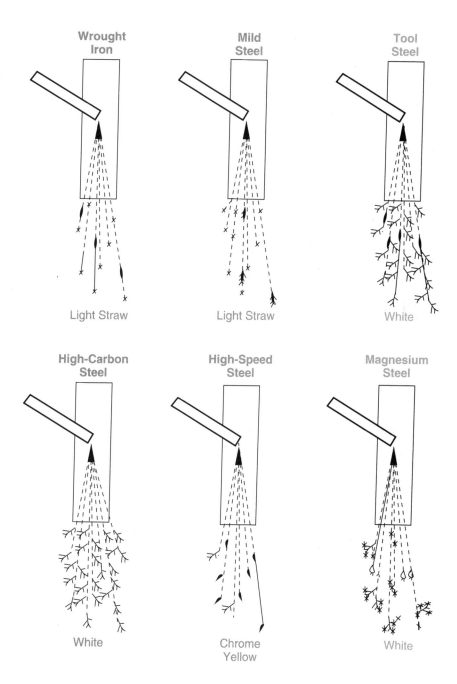

colors to indicate the particular type of steel. The end of a 1020 carbon steel rod would be brown. Other types of carbon steel are painted white, red, blue, green, orange, bronze, or aluminum. Alloy steel normally is painted with two colors. Nickel steel, for example, is painted red and brown, red and bronze, red and blue, red and white, etc., depending upon its chemical composition.

Important Terms

alloys
carbon steel
cast iron
cold-rolled steel
creep
ductility
fatigue
ferrous metals
gage number
high-temperature superalloys
hot-rolled steel
nonferrous metals
physical properties
plasticity
pure metal
refractory metals
specific heat
steel
tensile stress
thermal conductivity
thermal expansion
tin plate
tool steel
wrought iron

Questions for Review and Discussion

1. In manufacturing, alloys are used more often than pure metals. Alloys are usually more costly than pure metals. Why are they so popular?
2. Discuss the concept of plasticity in terms of a material's ability to withstand changes in its shape when it is being formed under pressure.
3. How do the material qualities of *plasticity* and *ductility* differ?
4. List some products made with refractory metals. How are the metals made?
5. Explain how the AISI numbering system is used in classifying types of steel.

Forming Metallic Materials

Key Concepts

- Δ Some manufacturing processes are considered foundational technical developments, since they have played a major role in shaping the development of civilization. The process of forging is one such foundational technical development.

- Δ Forged parts may have surface cracks as a result of the hammer action of the forging dies. These cracks can cause premature part failure.

- Δ Swaging is a process that is used for the forming of products (such as ballpoint pens), as well as the assembly of parts without need for additional fasteners.

- Δ The term "pressworking" includes a number of forming processes, such as stamping, embossing, drawing, expanding, and bulging.

- Δ Necking is a process that can be used to reduce the diameter of a part while improving its tensile strength.

There are many manufacturing processes that are used to form metals. Some of these, such as swaging, extrusion, and drawing, require considerable pressure to stretch and constrict metal as it is being formed in die orifices. Other forming processes, such as roll-forming and thread rolling, stretch or force the metal between rolls or cylindrical dies. There are also forming processes that compact powdered metal in a mold or die (powder metallurgy). In processes like permanent mold casting, metal is heated to a liquid state, so that it can be poured into a mold. Other forming processes physically form the material through pressworking operations, such as stamping and blanking.

There are literally hundreds of processes that can be used to shape or form metallic materials. Many of these will be omitted from this chapter, however, since they are seldom used in contemporary industry.

Forging

Hand forging is one of the most ancient of all of the metal forming processes, dating back to about 4000 BC. In those times, forging was used to shape primitive metal tools; it was later refined for use in making fine jewelry and coins. Forging is considered a *foundational technical development.* That is, it played a major role in shaping the development of our technological culture, and led to the development of many other forming processes.

Schoolchildren often learn about hand forging by watching a blacksmith forming metal. The blacksmith forms metal, such as a horseshoe, by heating it until it is cherry red and then hammering it against an anvil. Today, the most extensive use of hand forging is by crafters working with ornamental or wrought iron. The process is similar to that used by the blacksmith. Forging is considered to be a hot forming process, but it can be used to form cold metal as well.

Forging is done in industry using heavy presses and drop forging hammers that force heated metal into shape with heavy pounding or in dies. All forging processes involve successive hammering, resulting in shaping hot or cold metal by compressive force. With forging, little material is lost in the forming process. The shape of the raw material is changed by repositioning metal, rather than removing it.

One of the advantages of forging is that the flow of the metal and structure of the grain in the workpiece can be controlled. This results in a stronger and tougher part than could be produced with most other forming processes. Today, forging is often used to produce bolts and rivets, connecting rods, gears, and structural members for equipment.

Open-Die Forging

Open-die forging, also called Smith forging, is illustrated in Figure 4-1. Smith forging is the simplest of all of the forging processes. In this method of forging, the workpiece is formed between flat dies that compact, but do not completely enclose, the heated metal part. Open-die forging could be used in the case where a round part is needed, and flat bar stock is available to make the parts.

The open-die method is also used to produce rectangular shapes from round stock. In this type of application, heated round bar stock is placed between the flat dies. The dies close, creating a rod with two flat sides. The rod is then turned over in the die by hand and the flat dies press the part again. This forms two more flat sides, resulting in a square produced from round rod. This process would be used to compact metal along the entire length of the rod. Open-die forging is time-consuming and the quality is dependent on the skill of the smith performing the work. Because of this, it has limited application in high-volume production. It is sometimes used to produce simple shapes by replacing the flat open dies with special dies needed to produce v-shapes and rounds.

Figure 4-1
In open-die forging, the workpiece is compacted and formed between two die halves that do not completely enclose it.

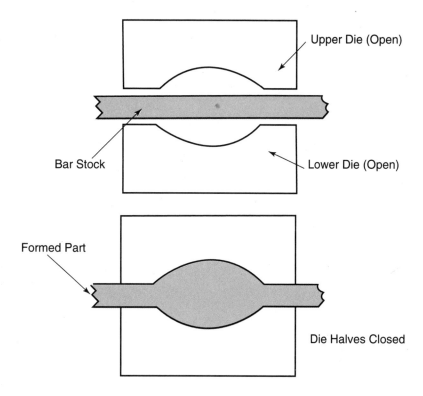

Closed-Die Forging

Closed-die forging is also called impression-die forging. In this type of forging, a preheated metal billet (a "preform") is shaped by being placed between dies that completely encompass the billet and restrict its flow. Closed-die forging is usually done with horizontally opposed die sets. These opposing dies simultaneously hammer against the billet until the forming process is complete. See Figure 4-2.

Since the two-part dies used for closed-die forging must endure tremendous abuse from hammering and contact with heated stock, they are made of hardened steel. Note, as shown in Figure 4-2, that the male and female die halves fit together perfectly. This enables the two-part die to completely surround the part. In some closed-die forging applications, more stock is used than is needed to ensure a completely filled and compacted part. This helps to produce a strong product with relatively uniform density. Some dies are made with a groove (called a gutter) cut into both mold halves. When the top and bottom die halves compress, the excess material (*flash*) is forced out into the gutter. Some forging dies do not have gutters. Instead, a thin wing or seam of flash is produced on the outside of the part where the dies close. The seam where the dies close and flash is generated is called the *parting line.*

The flash that is produced around the edge of the part must be removed by grinding or some other material removal process. The time lost in removing flash adds to production cost. For this reason, many firms carefully meter the amount of material supplied, using only what is absolutely necessary to fill the dies and produce a quality part. Flashless forging is possible using closed-die forging with properly constructed dies and careful control of material.

Figure 4-2
Closed-die forging is a process that completely encloses the workpiece between the two die halves.

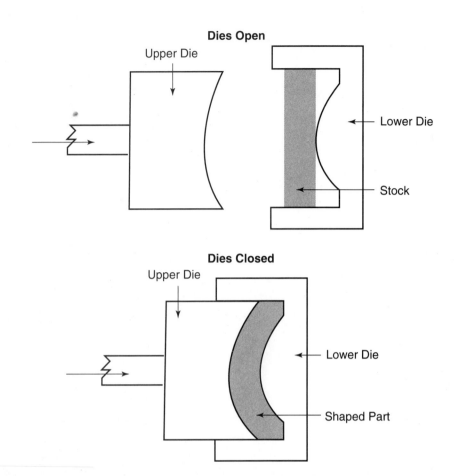

Some metals are easier to forge than others. The softer alloys, such as aluminum, magnesium, and copper can be easily forged at relatively low temperatures 750-1650°F (400-900°C). The harder metals, such as titanium and tungsten, require temperatures in excess of 1740°F (950°C) for forging. Other characteristics, such as the ductility, strength, and frictional behavior of a particular metal, also influence its forgeability.

The major weakness of forged products is that they may have surface cracks caused by the forging process. Overlapping of material, as well as complex radii in a particular pattern, can cause internal and external cracking. Cracks and other defects can be produced if the process is not carefully controlled. These can lead to early part failure resulting from fatigue, particularly if the part is used where it must withstand heavy wear.

Coining

Coining is a high-pressure, precision closed-die forging process that is used to produce such items as coins and fine medallions. Coining causes metal to move, as the die halves are closing, from thinner to thicker areas. For this reason, coining is often the process chosen when a part requires built-up corners or edges.

To achieve required detail, coining may be accomplished in one or more operations. No lubricant is used, since it would be trapped in the die and prevent reproduction of fine detail.

Figure 4-3 illustrates vertical coining. This method of coining is similar to closed-die forging, but requires much more pressure to ensure production of fine detail. Coining can also be done using round dies turning against round stock (rotary coining), or with flat dies contacting round stock (roll-on marking).

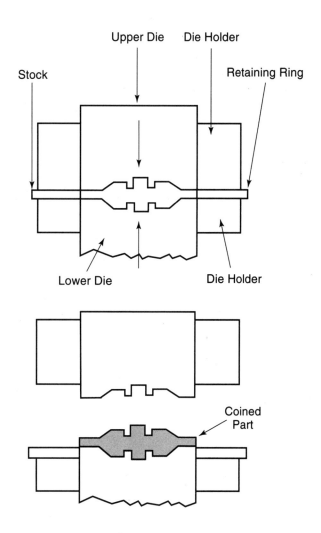

Figure 4-3
Coining is a form of closed-die forging, but is done under great pressure to better reproduce fine detail.

Coining is also used with forged parts to improve surface finish. When coining is conducted in this manner, it is called *sizing*. Forged parts are sized if more accurately rendered detail is needed on particular surfaces, or if improved dimensional accuracy is desired. Sizing can be used to squeeze material into difficult shapes, or to create fine detail. Sizing is not used to move *masses* of material, as is the case in vertical or rotary coining.

Rotary Forming Processes

There are many rotary processes used for forming metals. Some of these involve progressive forming of long lengths of stock using rollers. Others involve rotary bending.

Roller Forming

Roller forming is also called *contour roll forming* and *cold-roll forming*. Any ductile metal can be roller-formed without heating. Roller forming is used to form straight lengths into nearly any imaginable shape. No dies are needed: rollers progressively squeeze the continuous strips of metal into the desired shape.

Several sets of rolls are used to grip and then form the strip or rod as it passes through the rolling machine. Stock is usually purchased in large rolls (also called "coils"), which permits continuous feeding through the machine. Roller forming machines consume a lot of floor space, because the forming process uses many sets of rollers.

The thickness of the metal that is roller-formed remains fairly constant throughout the process; only its *shape* is changed. Each set of rollers produces a small change (a little more bend) in the stock. When the part reaches the end of the rolls, it is in the desired form.

Roller forming is a fast process that is suited to continuous production. Sometimes, it is combined with other processes to reduce the length of time necessary to produce a final product. An example would be in the production of welded tubing. Many firms turn flat stock into tubes using roller forming. The edges are then automatically joined in a continuous welding operation.

One of the advantages of roller forming is that bent parts have *springback,* Figure 4-4. Springback is often used as a method for joining material or applying pressure between parts. This makes the process particularly useful for manufacturing aluminum window frames, door trim, angles, channels, trim parts, and lightweight metal furniture.

Roll Bending

Roll bending is a process that is used to bend circular, curved, and cylindrical shapes from bar, rod, tube, angle, and channel stock. Roll bending is a variation of roller forming.

There are various bending roll arrangements used on roller machines (called *roll benders*). The pyramid type has one roll on top of a pair of side-by-side rolls. The position of the top roll is adjustable; the other two rolls are fixed. Stock is fed between the top and two bottom rolls. As the top roll turns, the material is moved through the rolls and bent.

A widely used roll bending configuration is referred to as the *two-roll machine.* The configuration is unique because it can produce bends of any diameter using only two rolls and a slip-on tube. The two rolls are aligned with a slight gap between them. A tube is inserted over the top roll, as shown in Figure 4-5. The upper roll applies pressure, pushing the stock down against the flexible lower urethane roll. The tube controls the bending diameter of the workpiece: the larger the tube, the bigger the radius that can be bent. Stock is fed between the rolls and the top roll turns, pushing the material through and bending it.

Figure 4-4
Springback is a characteristic of metals that are shaped by the roller forming process. Springback is useful in such products as an aluminum door frame. The frame might be compressed to permit it to fit into the door opening; when released from the compression, the frame would spring back to fit tightly against the opening.

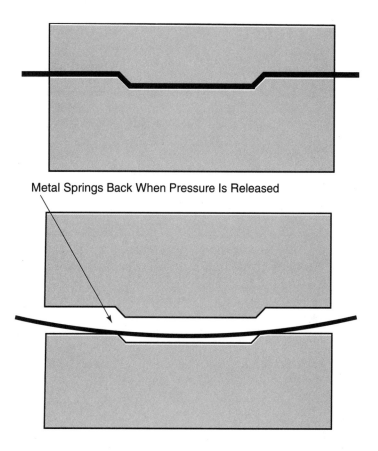

Figure 4-5
The diameter of the slip-on tube used in two-roll bending controls the diameter of the bend produced in the stock.

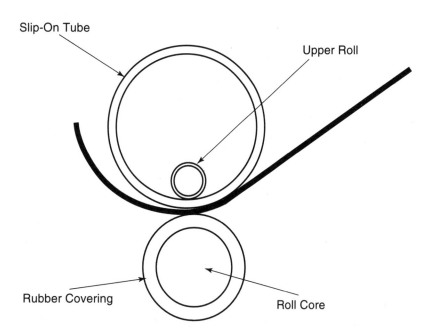

Another roll arrangement, called the *three-roll double-pinch machine*, is the most popular. The top roll of the three-roll machine turns freely, and the bottom two rolls are powered and adjustable. Rolls on the three-roll machine can be arranged in pyramid fashion for symmetrical bending of products such as barrel hoops, as shown in Figure 4-6. Irregular bends can be done by repositioning the bottom rollers so the distance between the top and left roller is different from the difference between the top and right roller.

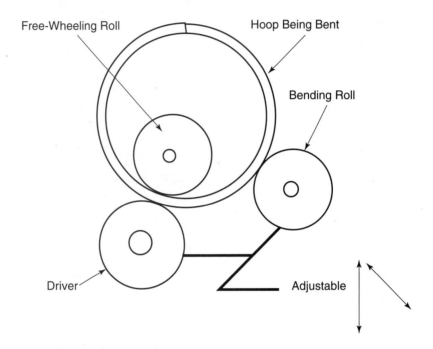

Figure 4-6
With a three-roll double-pinch machine, stock can easily be bent symmetrically to form a hoop.

Metal can also be bent without the use of rollers. Figure 4-7 illustrates a computer-controlled workcell being used to produce tension on a heavy steel bar. Information pertaining to the metal under stress is recorded and displayed on a monitor while the bar is being bent to the desired angle. Another bending machine, used to conduct what is referred to as bullnose forming, is shown in Figure 4-8. In this process, the part is draped over a forming mandrel, and the die table is raised to stretch the metal sheet into the desired angle.

Thread Rolling

Threads can be produced on the outside of a rod (forming a *bolt* or *screw*), or on the inside of a part (forming a *nut*). One of the processes often used to accomplish this is called thread rolling.

Thread rolling is a chipless cold-forming process that can be used to produce either straight or tapered threads.

When internal threads are rolled, the process that is used is similar to *tapping*. In tapping, threads are turned into a hole and chips are produced. When threads are *rolled*, a special fluteless forming tap is used; no chips are produced. Chipless swaging taps are good for rolling threads in deep, small-

Chapter 4   Forming Metallic Materials

Figure 4-7
This workcell is stretch-forming metal bar stock. The equipment is computer-controlled. (The Cyril Bath Co.)

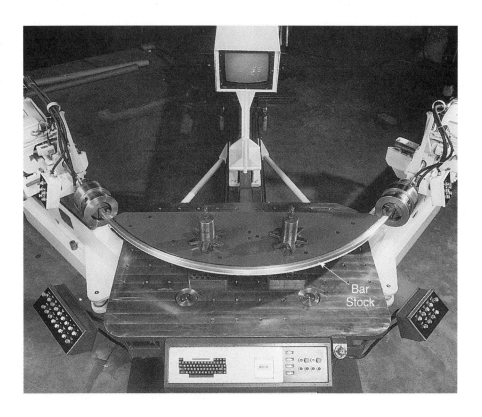

Figure 4-8
In bullnose forming, the material to be shaped is draped over a forming mandrel (center) and clamped in place. The die table is raised to stretch the metal. (The Cyril Bath Co.)

diameter holes. If conventional taps were used in this situation, tap breakage would normally be a problem. The chipless swaging tap allows the threads to be rolled continuously. There is no need to periodically back the tap out.

Dies used to roll threads can take the form of either flat plates or cylindrical rolls. Threads are rolled with one pass of the dies across the blank.

The photograph in Figure 4-9 shows a typical flat die used to roll threads on round stock. The stock, called a *blank*, is placed between the dies and the threads are rolled on. This is how bolts are made. One of the dies moves and the other is stationary. Many different types of dies are used on thread-rolling machines. See Figure 4-10.

Figure 4-9
These flat grooved dies are used for rolling threads. In the thread-rolling machine, one die is stationary and the other moves, rotating the round stock as the threads are formed.

Threads also can be rolled using three dies positioned in a pyramid arrangement. The equipment used to hold the dies varies, depending on the size and type of part. Most cylindrical thread rolling is accomplished using thread-rolling machines, automatic screw machines, or automatic lathes. Flat dies are used in thread-rolling machines (boltmaking machines).

When many parts must be produced, thread rolling is the fastest way to form threads. Dies used on thread-rolling machines can also be made for *knurling*, a rough-textured pattern usually used as a gripping surface on tool handles. See Figure 4-11.

Automatic thread-rolling machines, similar to the one in Figure 4-12, are highly cost-effective for producing large quantities of production parts. Some firms are producing in excess of 90 parts per second using rotary dies.

Rolled threads have several advantages over conventional threads. First, rolled threads are stronger, because the stock does not undergo drastic shifts in the direction of the metal. With conventional threading, material is removed by cutting through the grain-flow lines of the metal. Rolled threads also have superior surface finishes.

Second, they have a larger major diameter than conventional threads. This is a result of the accuracy of the process, which produces a product that is uniform throughout.

Chapter 4 Forming Metallic Materials 75

Figure 4-10
Some of the many different types of dies and attachments available for use in thread-rolling machines. (Reed Rolled Thread Co.)

Figure 4-11
Knurling dies produce a pattern of raised lines or diamonds on a surface, usually to allow better gripping of a tool or similar device. (Reed Rolled Thread Co.)

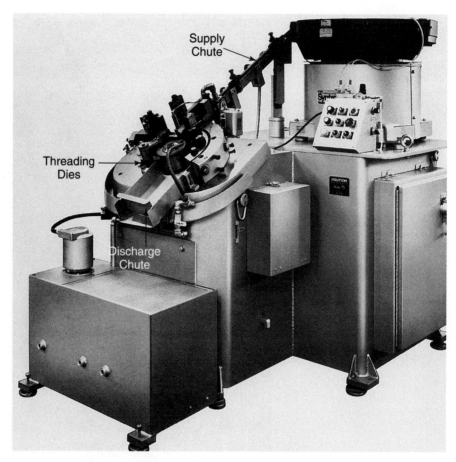

Figure 4-12
This automatic thread-rolling machine is highly productive. Parts to be threaded are stored in the hopper at upper right, and fed into a supply chute by vibratory action. After passing through the threading dies, they are ejected into a discharge chute. A container for quantities of finished threaded items would be placed below the chute, at left. (Reed Rolled Thread Co.)

Swaging

Rotary swaging, also referred to as *radial forging,* is a process that takes a solid rod, wire, or tube and progressively reduces its cross-sectional shape through repeated impacts from two or four opposing dies. Swaging is a noisy process that gradually squeezes away the desired amount of material. Operators wear ear protection to keep from damaging their hearing. Figure 4-13 illustrates the process of rotary swaging. Dies rotate around the workpiece, opening and closing rapidly to generate a hammering action against the rod or bar stock. The blank is fed continuously into the die opening in the swaging machine, Figure 4-14. There, a revolving spindle creates centrifugal force, throwing the opposing dies outward. The dies hit against rollers carried in a circle around the spindle. This impact causes the dies to rebound against the workpiece. Some swaging operations create as many as 5000 impacts per minute.

Swaging is normally used to produce round parts up to 2 in. (51 mm) in diameter. Special machines have been constructed permitting swaging of tubes to 6.5 in. (165 mm) and bars to 4 in. (102 mm). The swaging machine shown in Figure 4-15 is equipped with an automatic feeder, although some machines are fed by an operator.

Chapter 4 Forming Metallic Materials

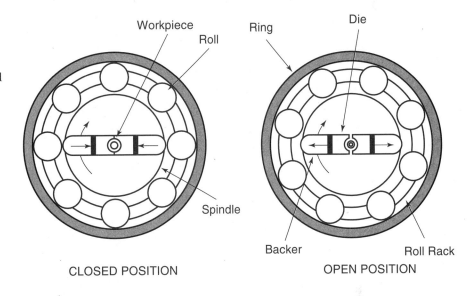

Figure 4-13
In rotary swaging, the dies rotate around the cylindrical or tubular workpiece, reducing its diameter with repeated blows. Stock is fed continuously through the swaging machine.

Figure 4-14
Rod or other material to be processed is fed into the round opening of this rotary swaging machine. (Abbey Etna Machine Co.)

Figure 4-15
A copper tube being swaged to smaller diameter in an automatic-feed rotary swaging machine. Note diameter of tube in workholder at left, and partial reduction (from earlier pass). The workpiece has been reduced to still-smaller diameter at right, where it can be seen emerging from the machine. (Abbey Etna Machine Co.)

The rifling in gun barrels is done by swaging. This is accomplished by deforming a metal tube placed over a mandrel with spiral grooves. Swaging is used for a great variety of products ranging from screwdriver tips and fasteners to ballpoint pen caps and metal chair legs. Swaging can also be used for assembly, such as swaging a bushing to a shaft or fittings to a cable.

Extrusion

Extrusion is a manufacturing process that compresses metal beyond its elastic limit before it is forced through a die opening. Openings in extrusion dies can be round, or any of a number of unique shapes, depending on the cross-section of the continuous length that is desired. Metals such as aluminum, copper, magnesium, and stainless steel are easily extruded.

Extrusion is a continuous-pressure forming process. In concept, it works much like squeezing toothpaste out of its tube. Figure 4-16 illustrates two of the most popular forms of extrusion: direct extrusion and the stationary mandrel method. There are several variations of each of these methods.

In direct extrusion, the ram or punch moves down on the molten metal billet, and pressure from the ram forces the stock to extrude out of the die opening. This method is sometimes referred to as the Hooker process.

The stationary mandrel method is used to make hollow tubing. With this method, the ram is attached to a stationary mandrel. As the ram travels down into the metal billet, the mandrel contacts the stock first and guides it through the die opening. The mandrel also keeps the inside of the tube open.

Figure 4-16
Two popular types of extrusion. In direct extrusion, the ram applies pressure to force the hot metal out through the die opening. In the stationary mandrel method, the mandrel keeps the inside of the tube open as the ram applies pressure to the metal being extruded.

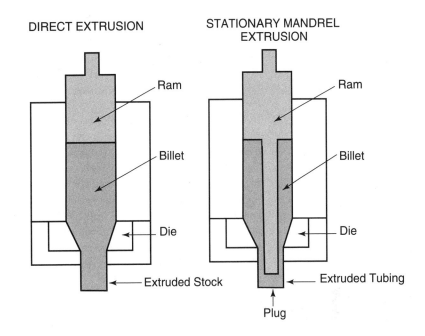

Another form of extrusion, called hydrostatic extrusion, uses water to increase the forming pressure. In this process, the billet is placed in a container filled with fluid, and a hydraulic ram applies pressure to the metal billet.

Hot extrusion is used to apply a plastic insulation coating to copper electrical wire. This application is covered in more detail in the section on polymeric/plastic processes. In the wire-coating process, both the wire and plastic material are fed through the die orifice.

In the hot extrusion process, lubricants are used. When hard metals such as stainless steel are being extruded, glass is often used as a lubricant. Glass is applied to the opening of the extruder by placing a circular glass pad at the die opening. The glass is kept in a molten state, so it serves as both lubricant and as a coating for the metal that is forced through the die opening. Powdered glass is also applied to the inner liner of the chamber of the extruder before the metal billet is introduced.

Metal products that are normally extruded include lipstick cases, soft drink cans, fire extinguishers, and vacuum bottle cases. Long lengths of aluminum and stainless steel channel are also produced by hot extrusion. Extrusion is even gaining some popularity with larger products such as refrigerators, washing machines, and air conditioners.

Upsetting

The process called *upsetting* (also known as *cold forming* or *cold heading*), is a type of forging that thickens or bulges the workpiece while also shortening it by compression. Upsetting is actually a combination of forging and extrusion. It is one of the fastest processes used by industry for high-volume production of heads on nails, bolts, and rivets.

Cold heading forms a head on the end of an unheated metal rod by compressing its length in a die cavity. Machines used to perform this action are called "headers." There are many different types of machines and shapes of dies that can be used for cold heading. Some machines utilize a two-part die. Others have a punch and die. Some shear the stock to length and compress it inside closed dies. Others use an open die, with forming done by a punch hammering against the die.

A typical closed-die arrangement is shown in the Figure 4-17. A closed die consists of a hardened cylinder with a hole running through its center. This is called the *die block*. The second die half, called a *punch*, automatically pushes wire stock in precut lengths through the hole. With the desired amount of blank projecting out of the hole, a punch hammers down on the stock. One or two blows are used to form the head. Then the punch opens and an ejector pin pushes the part out of the die.

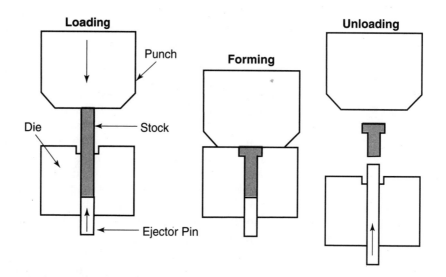

Figure 4-17
Upsetting, or cold heading, is the process of forming a head on a bolt, nail, or other fastener through a combination of forging and extruding. As shown, the punch forces the stock downward into the die, so that the metal bulges out to form the head. When the punch retracts, the ejector pin pushes the workpiece out of the die.

There are many different variations of upsetting and heading machines. Production boltmakers combine several operations—forming, head trimming, and thread rolling—into one continuous process. A finished part is generated with each stroke of the machine. The forming center with robot-assisted tool changer shown in Figure 4-18 is an example of such a state-of-the-art cold-forming machine. This equipment can take preforms, coiled wire, or bar stock and produce a multitude of complex, odd-shaped parts requiring no surface finishing. See Figure 4-19.

Cold forming is a process which automatically feeds coiled wire into the machine, transfers cutoff blanks through a series of progressive forming dies, and creates a finished part. Cold forming has been widely adopted for use in the high-volume production of small- to medium-sized parts from low alloy

Chapter 4   Forming Metallic Materials 81

Figure 4-18
This automated forming center, equipped with a gantry-type robotic tool changer, is an example of a state-of-the-art cold-forming machine. (National Machinery Co.)

Figure 4-19
Some of the varied types of parts that can be formed at high production rates by automated cold-heading machines. Such parts seldom need surface finishing. (National Machinery Co.)

steels. One of the best candidates for cold forming is the production of spark plug shells, Figure 4-20. Traditionally, these shells were produced using screw machines. Now, in addition to the spark plug shell, many manufacturers are cold forming electrodes and center posts as well.

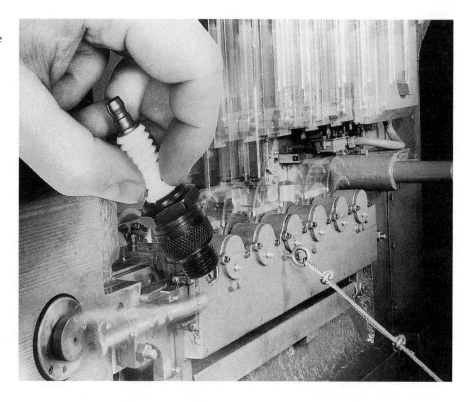

Figure 4-20
Metal spark plug shells can be produced at the rate of two per second by automated cold-heading equipment. (National Machinery Co.)

Cold Forming Advantages

There are several advantages of cold forming over traditional machining and forging methods. These include higher production rates, little or no scrap, production to size with close tolerances, and increased part strength. One manufacturer, whose six-die cold former has a capacity of 120 pieces per minute, claims that cold forming results in a reduction of waste from 74 percent to 6 percent and increases part strength by 18 percent. An overall increase in production by ten times or more is possible with cold forming, the manufacturer says.

The basic cold forming machinery can be modified for use with warm and hot forming processes, as well. The press shown in Figure 4-21 is used primarily for warm and hot forming.

Parts that are frequently produced by warm forming include automatic transmission gears, planetary pinion gears, lifter rollers, and outer bearing races. The automotive industry is a primary user for parts that are automatically hot formed. Typical hot formed products are connecting rods, track links for off-the-road vehicles, and large gear blanks.

Stamping

Stamping is sometimes referred to as pressworking. However, pressworking is a very general title—it covers a number of processes that are conducted in a heavy stamping press.

Figure 4-21
This large transfer press is used for warm and hot forming. The automotive industry is a major user of such equipment, especially for hot forming. (Schuler Incorporated)

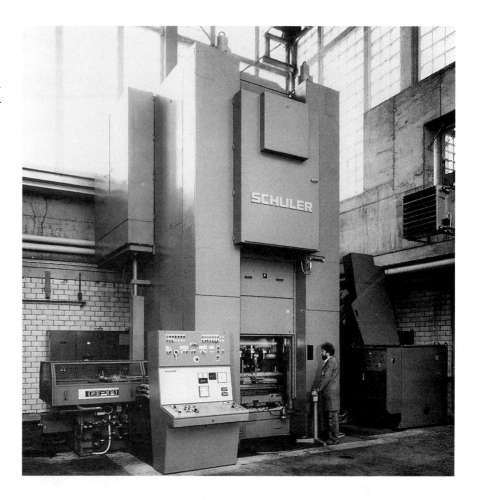

Stamping is a chipless process which produces a sheet metal part with one downward stroke of the ram in a stamping press (usually referred to as a *punch press*). Parts are typically stamped cold. One of the largest markets for stamped products is the automotive industry, which uses many thousands of medium-sized and large body panels each year. See Figure 4-22.

In addition to performing stamping operations, these presses are used to punch, notch, pierce, or trim sheet metal parts. This somewhat complicates the meaning of the term *stamping*. In actual industrial usage, "stamping" generally refers to all of the presswork operations related to sheet metal processing. A stamped part may be cut to size at the same time that slots, holes, and notches are generated. All of these operations can be done with a single downward stroke of the ram.

Stamping presses are large (often several stories in height). The large automated transfer press shown in Figure 4-23 is used to produce body panels for the automotive industry from sheet stock. A press can be designed to incorporate as many die stations as desired.

In most stamping presses, the ram is at the top and travels down into the work. The ram is often equipped with several punches. As the ram travels

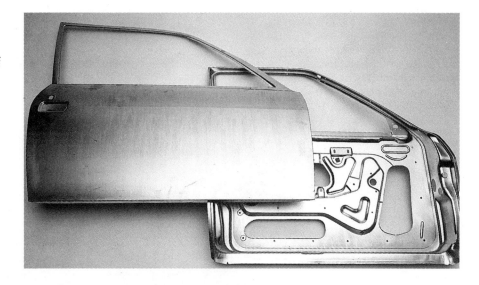

Figure 4-22
Automotive body panels, such as these door components, are typical products of large stamping presses. (Schuler Incorporated)

downward, the desired punch moves into a die block attached to the rigid press bed, acting upon the workpiece placed on top of the die. The punches and die block assembly make up what is called the *die set*. A typical die block assembly is shown in Figure 4-24.

Figure 4-23
Highly automated equipment, such as this transfer press for stamping and forming large body panels, is widely used in the automotive industry. (Schuler Incorporated)

Figure 4-24

The die block assembly consists of a ram that forces punches down into a die block mounted on the press bed. Note that stock moves through this press as a continuous ribbon of sheet metal, advancing each time the ram rises after stamping. (The Minster Machine Co.).

Stamping Press Classifications

Stamping presses are classified in different ways. Some are classified by the types of parts that are to be stamped. The press shown in Figure 4-25 is set up for drawing oil filter canisters for use in the automotive industry. Consequently, it can be classified as a drawing press. Presses also may be classified according to the type of power, the structure of the press, or the type of ram which they employ.

Figure 4-25

This stamping press is used to draw metal shells for oil filters, so it would be classified as a "drawing press." (The Minster Machine Co.)

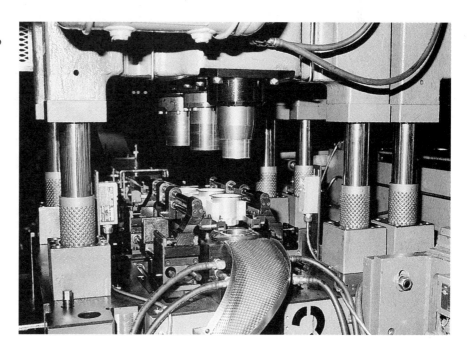

The Joint Industry Conference (JIC) system has been established to aid in classifying stamping presses. With this system, presses are given an identifying letter and number code. A typical number might be D-700-90-72. This could be interpreted as: double action, 700-ton capacity, with a bed width (left to right) of 90 inches and a depth (front to back) of 72 inches. The heavy stamping press shown in Figure 4-26 is numbered E2-400-96-42. This is a two-point, eccentric-shaft 400-ton press that is 96 inches wide and 42 inches deep.

Stamping is a high-volume production process, capable of producing thousands of parts per hour. This made the process particularly useful at a time when large job lots were the rule in industry. With the trend today toward *decreasing* the size of lots, however, new technology must be developed to shorten the time necessary for changing die sets. Some progress is being made in this area. Conventional die changes used to take several hours. The trend now is to the production of smaller presses with quick-change dies capable of being replaced in minutes.

Figure 4-26
A numbering system describes each press. The identifying number of this large stamping press (E2-400-96-42) is shown next to the manufacturer's name on the top front, above the ram.

Embossing

Embossing is a process that is similar to stamping; both can produce impressions or markings on metal surfaces. A common product application would be placing names or numbers on such items as military dog tags or machine identification plates. One side of the tag would have raised printing, while the other would be depressed. This is where the similarity between stamping and embossing ends. The two processes use different types of dies, and produce differing levels of quality in the impression that results.

In simplest form, stamping of data on metal could be done by striking the workpiece with a punch. Embossing requires a two-part matched die. The male die half has a raised impression surface; while the female die half has sunken (depressed) detail that conforms to the raised areas of the male punch. While stamping punches *deform* or stretch the metal, embossing *compacts* the metal in the area of contact. This produces finer detail and improved strength in the part. Embossing can be performed in a punch press or stamping press. It can also be done using rotary dies to continuously emboss patterns on thin metal foil and sheet stock.

Drawing

A great variety of sheet metal products, ranging from kitchen sinks and automobile fenders to pots and pans and various dome-shaped parts, are made by using the manufacturing process called drawing. What is unusual about drawing is that it can be used to produce a three-dimensional part from flat sheet or metal plate. *Drawing* involves both stretching and compressing.

There are several variations of the basic process of drawing. The most common is called *deep drawing* or shell drawing. Drawing was once used to make large artillery shells. A typical drawn part is shown in Figure 4-27.

Figure 4-27
An example of parts made by the drawing process. These are shells for oil automobile filters. (The Minster Machine Co.)

In *deep drawing,* a female die is pressed into thin sheet stock, stretching the metal over a male forming punch. Usually the sheet stock is placed on a pressure ring surrounding a bottom punch. An overhead ram then moves the female die down and compresses the stock against the bottom punch. The process is completed when the female die retracts and the pressure ring pushes the part back up to the unload position.

Figure 4-28 illustrates the *stretch draw forming* process, which uses mating dies that must match perfectly. Such dies are costly to construct and maintain. These are the types of dies used in stamping processes. Not all draw forming processes involve mating dies, however. One flexible draw forming process that is often a cost-efficient alternative to stretch draw forming is called *marforming*. In the marform, no expensive dies are involved, so there will be no die marks on the product. Both shallow and deep-drawn parts can be produced.

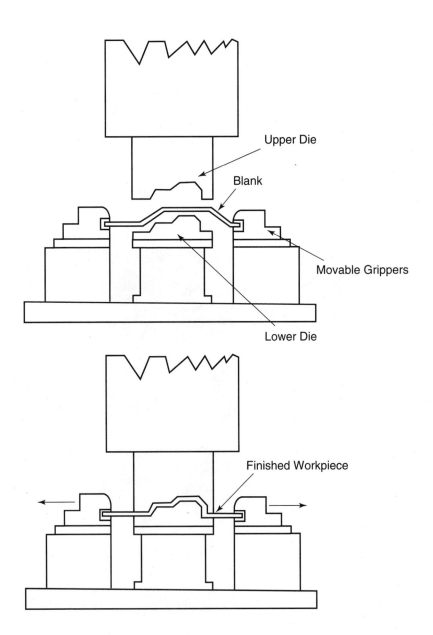

Figure 4-28
Mating dies used in the stretch draw forming process must match perfectly. The workpiece, or blank, is drawn over the lower dies, then stretched tight as the two dies come together to shape the part.

As shown in Figure 4-29, the workpiece is held on a stationary lower forming punch. Drawing begins when the upper platen, which holds a rubber pad, moves down against the forming punch. The workpiece is pressed between the thick rubber pad and the forming punch. Marforming enables parts to be produced with greater depth and with fewer wrinkles than is possible with many other drawing processes.

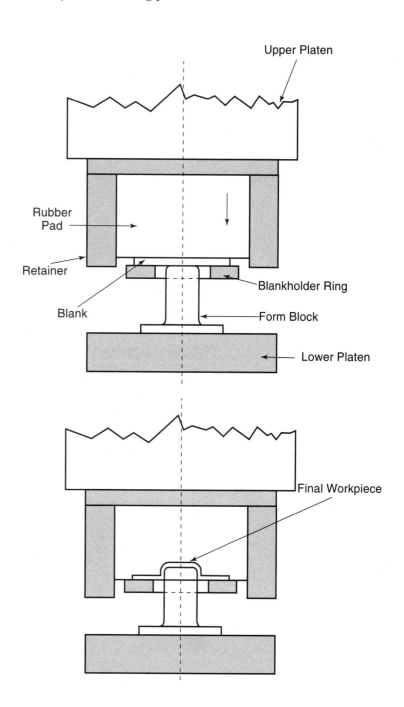

Figure 4-29
In the process called marforming, no dies are used. A thick rubber pad in the upper platen forces the blank down over a lower forming punch.

Die forming can be done using many different types of machines. Normally, a stamping press or punch press is used. Die forming can also be done in a press brake that would usually be used to cut metal. Forming can also be done on specially designed presses that are adapted to a particular type of forming process.

Another draw forming process, called *hydroforming,* employs hydraulic fluid in place of the rubber pad used in marforming. Hydroforming incorporates the use of a rubber diaphragm that seals the hydraulic fluid in the pressure-forming chamber, Figure 4-30. The workpiece is positioned on the bottom die block, and the pressure forming chamber is lowered into place. Forming begins when the punch is forced upward into the workpiece, increasing the pressure in the fluid chamber. This increase in pressure causes the workpiece to be compressed around the punch. When forming is completed, pressure is released and the workpiece is stripped from the punch.

Hydroforming is done using high-speed presses capable of producing parts as large as 25 in. (635 mm) in diameter, with a draw depth of 12 in. (305 mm). Some companies have hydroformed steel parts as thick as 3/8 in. (9.5 mm) and aluminum up to 1 in. (25.4 mm) in thickness.

Expanding

Expanding is a process that can be used to increase the size of tubular parts that have been formed by drawing. Figure 4-31 shows how a drawn part can be expanded by a forming punch. The punch travels down through a stripper ring, pressing the top of the drawn part against the walls of the die block. When the punch is retracted, the stripper ring helps to remove the expanded part from the forming punch.

A typical application for expanding would be to form a bell on the end of a tube. Expanding can be used to enlarge other portions of drawn workpieces, as well. Specialty machines, called expanders, carry movable shoes on a circular face plate. The shoes are moved outward to expand the work. Expanding can be done on other machines as well. All that is necessary is the capability of compressing the expansion die.

Before the workpiece is expanded it should be softened, or annealed. Methods for annealing metals will be discussed in Chapter 7. If the part is to be expanded no more than 30 percent of the diameter of the workpiece, then the process can be conducted with one stroke. If it is more than 30 percent, several strokes are required, with the part being annealed before each stroke.

Bulging

Another process that can be used to expand tubular shapes such as corrugated tubes, bellows, teapots, or musical instruments is called *bulging.*

There are two basic methods of bulging: the rubber core method and two-piece die method. The rubber core method is shown in Figure 4-32. Here's how it works:

The tubular workpiece is expanded by applying compressive force to the end of a rubber die core placed inside it. Under high pressure, rubber pro-

Figure 4-30
Hydroforming is very similar to marforming, but substitutes a forming chamber filled with fluid under pressure for the rubber pad. In this process, the forming chamber is lowered over the workpiece, then the punch moves upward to shape the blank.

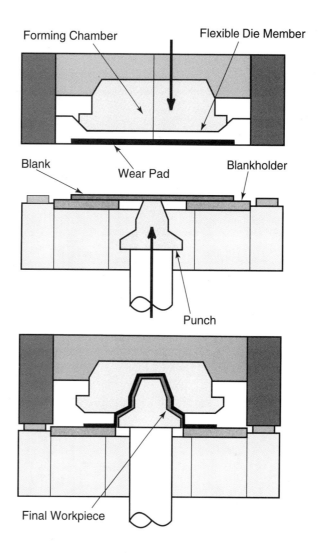

vides equalized hydraulic pressure in all directions. When the platens on the press close, a metal punch compresses the rubber core and the metal is shifted to the middle section of the tube, forming a bulge. The rubber core method creates a shorter tube with narrow top and bottom sections and an expanded middle.

The two-part die method compresses the part without using the rubber core. In this method, the punch is forced against the end of the tube, producing a workpiece that is shorter in length with a bulge in the middle.

Necking

The forming process called *necking* is used to reduce the diameter of the end of a tubular part. Figure 4-33 shows a typical necked part, a small canister used for carbon dioxide gas. With one operation, the end of the canister is reduced to about 20 percent of the diameter of the tube.

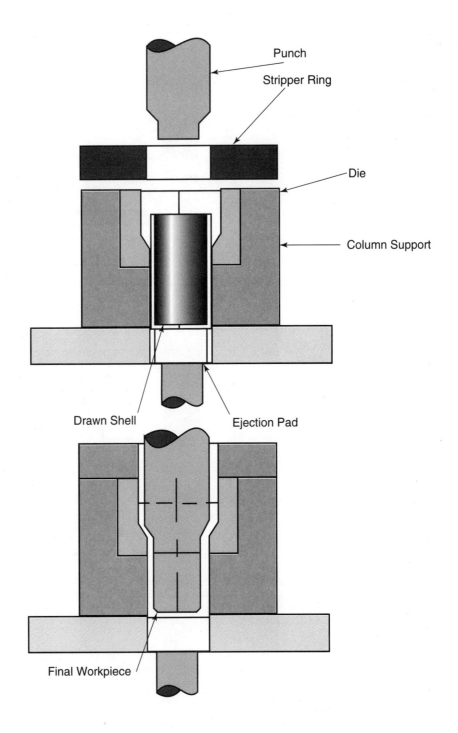

Figure 4-31
Expanding involves driving a punch into the opening of a drawn part to increase its diameter.

Necking is a die-reduction process that stretches the relatively soft and ductile metal part at the same time that it reduces its cross-sectional area. At first glance, it might appear that the process is used only to reduce the diameter of the workpiece, but there is more to it than simple reduction. As they stretch or elongate the workpiece, the dies *turn*. The diameter of the stock is

Figure 4-32
A rubber core is placed inside the part to be shaped by bulging, then a plunger presses down on the rubber. This causes it to exert outward force.

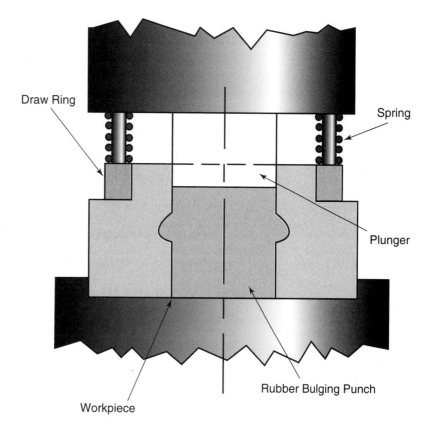

reduced in the area where the dies are working. The process results in some inconsistency in wall thickness. When the cross-sectional area is reduced, the stress and strain on the material becomes greater than the overall loadbearing capacity of the material. This is referred to as the ***ultimate tensile strength*** (UTS) of the material. The actual point where this occurs depends upon the ductility and hardness of the metal.

An important point to remember about necking is that the ***tensile strength*** of the part is increased when the specimen is stretched or elongated. This increase in tensile strength makes the part better able to support heavy loads.

The necking process must be used with care, however—there is a point at which increasing strain on the part will result in it breaking. This is called the *breaking point* or *fracture strength* of the material.

Figure 4-33
Necking is a die-reduction process that achieves progressive reduction of the part's diameter.

Nosing

Nosing is similar to necking, but is used to partially close the end of a tube, rather than just reduce the diameter in one section. Nosing can be used to taper or round the end of tubing. This is a process that is particularly useful in making rifle and pistol cartridges, Figure 4-34.

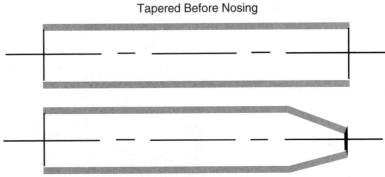

Figure 4-34. Tubing is first slightly tapered, top. Then additional pressure while rotating causes the diameter to be reduced, bottom.

Nosing is normally done by using dies to close the end of the tubing. In principle, it is somewhat like the process of rotary swaging. Nosing can be extended to its limits, stretching the metal in order to cause it to close. This type of nosing, called spinning, will be discussed later in this chapter.

Electromagnetic Forming

Electromagnetic forming, sometimes called magnetic pulse forming, is a process that forms a workpiece by using intense pulsating magnetic forces. Energy is stored in capacitors, which discharge short (measured in microseconds) pulses to a forming coil, Figure 4-35. The coil produces a magnetic field and passes it to the conductor (workpiece). Eddy currents are produced in the workpiece. This current creates its own magnetic field. When the two magnetic fields oppose each other, a repelling force is created between the coil and workpiece. This force pushes the workpiece against a forming die. Since the metal is stressed beyond its yield strength, permanent deformation occurs.

Magnetic pulse forming is capable of generating pressures as high as 50,000 psi (344 750 kPa) on the workpiece. The process is used extensively in industry as an assembly technique to join tubular parts to other components.

Magnetic pulse forming can be used to reduce the diameter of a section of a tube or to join tubes to rods. In such an application, the tube is inserted into the coil, then a rod is inserted in the tube. The electrical/magnetic force created around the coil compresses the tube, pressing it forcefully against the rod.

In the automotive industry, magnetic pulse forming is used to assemble steering gears, ball joints, and shock absorbers. The electrical equipment man-

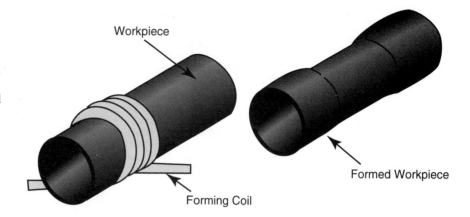

Figure 4-35 A forming coil wrapped around a tube generates opposing magnetic fields, with a resulting reduction in diameter of the workpiece. The process can also be used to expand a tube by placing the coil inside it.

ufacturing industry uses the process for assembling coaxial cable, electric motors, potentiometers, and various parts that would be difficult to assemble by other methods.

Joints can be assembled using magnetic pulse forming, instead of welding or brazing. Metal rings can even be assembled to ceramic, thermoset plastic, or phenolic parts, since the magnetic field will pass through materials that are electrical nonconductors.

The use of magnetic pulse forming for external forming of tubes was shown in Figure 4-35. The process can also be used for internal forming by placing the coil inside a tube. When the pulsating electrical current travels through the coil, the force causes the tube to expand.

Contouring

Magnetic pulse forming also can be used for shaping flat parts, a process called *contouring*. In this application, the forming coil looks much like a burner on an electric stove. The coil is placed against the sheet, and presses the flat stock into a forming die.

It is easiest to use magnetic pulse forming with conductive metals. Nonconductive metals can be formed, but a conductive material (called a "driver") is placed between the forming coil and the workpiece. An aluminum driver is used to form stainless steel parts, for example.

This process cannot be used to form complicated shapes. The forming rate is consistent around the coil, so you cannot produce high pressure in one area and low pressure in an adjacent area. Another limitation is that the process does not work well with parts that have many holes, notches, or slots. These voids interfere with the flow of current across the part.

Peen Forming

Shot peening is a cold-working process that is accomplished by bombarding the surface of a part with small spheres called *shot*. Shot can be made of cast steel, glass, or ceramic particles. Figure 4-36 shows a part being shot peened.

Figure 4-36
The many thousands of tiny impacts caused by shot striking the metal surface strengthens the part and makes it less subject to surface cracking. (Metal Improvement Company)

Each piece of shot that strikes the metal acts like a tiny peening hammer, creating a pattern of small overlapping surface indentions. Peening causes plastic deformation of the surface of the metal. As such, it can really be compared to the process of forging.

Since surface deformation is not consistent through the thickness of the part, compressive residual stresses are created on the surface. Cracks will not occur in a compressively stressed zone. Most fatigue failures originate on the surface of a part, so the compressive stresses produced by shot peening actually result in considerable increases in part life.

Shot peening is extensively used on parts such as oil-well drilling equipment, turbine and compressor blades, shafts, and gears. Compression coil springs are probably the best-known and most widely used parts that are shot peened. All automobile engines use valve springs that are shot peened. Springs as small as 0.005 in. (0.13 mm) in diameter and as large as 3 in. (76 mm) in diameter have been shot peened to increase their resistance to fatigue.

Shot peening can also used to produce contours on parts. Figure 4-37 shows an aircraft wing skin being formed on a peening machine by shot peening.

Figure 4-37
Shot peening is being used to form the contour on this aircraft wing skin. (Metal Improvement Company)

Often parts are too large or difficult to transport from the field to the plant for shot preening. Units such as chemical storage tanks and pressure vessels, gas and steam turbines, tubing in heat exchangers and steam generators, and wood processing digesters have been shot peened on site using portable equipment. Many different types of machines for spraying shot are commercially available.

Explosive Forming

Some forming processes, such as stretch draw forming, matched die method, and coining, require both a male and female die to perform the forming operation. Others, such as flexible die forming and hydroforming, use only one die and a rubber forming block. All of these forming processes are used to shape relatively small parts.

With *explosive forming,* only a female die form is used and there is no limit to the size of the workpiece. In explosive forming, the sheet metal workpiece is clamped to the die and the assembly is placed in the bottom of a water-filled tank. A vacuum line attached to the die cavity removes air from the void between the workpiece and the die. Next, an explosive charge is suspended in the water above the workpiece. When the charge is detonated, a shock wave forces the sheet stock against the forming die. As shown in Figure 4-38, the same principle is used to expand a tubular workpiece, except that the dies enclose the workpiece and the explosive charge is placed inside it.

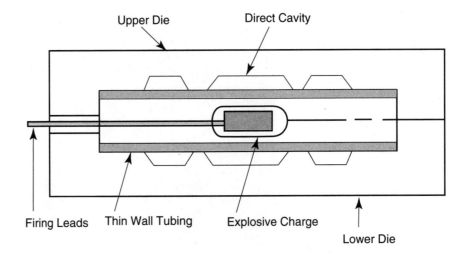

Figure 4-38
Arrangement for explosive forming used to expand a tubular workpiece.

HERF

A variation of the explosive forming process, called electrohydraulic, electrospark, or *high-energy-rate forming (HERF)*, uses a spark-generated shock wave, rather than a chemical explosion. See Figure 4-39.

With HERF, a capacitor bank stores electricity until a switch closes a circuit to the forming tank. The tank is normally filled with water, glycerin, or a light oil. When the energy is released, an aluminum or magnesium bridge wire connecting the electrodes is vaporized. This creates a plasma channel in the liquid for the spark to cross. The spark generates a shock wave that propagates radially. The concussion forces the stock into a forming die. Sometimes a bridge wire is not used. In such a case, the capacitor voltage is increased from about 4500 to 20,000 volts. With high energy rate forming, pressures approaching 175,000 psi (1 206 625 kPa) can be created.

Spinning

Spinning is a process that involves stretching sheet stock over a rotating male or female mold. The mold is attached to a lathe or spinning machine. Spinning can be either a cold-working or hot-working process, but it is most often done without heating the workpiece.

The beginning stage in the production of a microwave reflector by the spinning process is shown in Figure 4-40. A flat disk is snugged up against a male *forming block* (also called a mandrel or mold) that is attached to the headstock of the spinning lathe.

Let's take a closer look at what is happening in the photo. The mandrel is fastened to a faceplate that is screwed onto the headstock spindle of the

Figure 4-39
The energy released in HERF causes a shock wave through the transfer medium, forcing the workpiece into the cavity of the forming die.

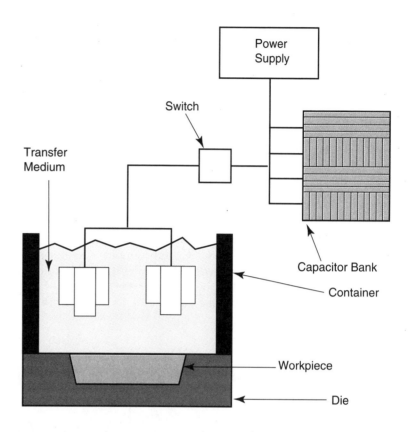

Figure 4-40
Using spinning to stretch metal over a rotating mold to form a microwave reflector. (Lake Geneva Spindustries, Inc.)

lathe. The headstock is located to the left in the photograph. The tailstock, located on the right end of the lathe in the photo, can be loosened and slid on horizontal ways toward the headstock.

To begin spinning, the operator would normally place a *follower* (a free-wheeling spindle) in the tailstock of the lathe. The tailstock would then be slid over to snug up against the stock to be spun. After the tailstock is secured again to the ways of the lathe, the pressure between the forming block and follower would be sufficient to hold the stock in place.

However, there is more to the photograph than this. First, the machine shown is a lathe operating under *computer numerical control (CNC)*. This enables the operator to run a program stored in the controller. The program is responsible for positioning the machine properly for setup and for conducting the actual spinning operations.

To continue with the spinning, the lathe would be turned on and the program would be initiated. This would cause the forming block, workpiece, and follower to rotate. The CNC program would then transmit information to a *forming tool*, and the tool would automatically apply pressure against the face of the disk. This would stretch the metal over the forming block.

High-production-rate machines are normally computer-assisted, reducing dependence on the operator. When spinning is accomplished without the aid of CNC, considerable operator skill is required to properly stretch the metal without making it too thin or causing the workpiece to buckle. Figure 4-41 shows an operator spinning a large aluminum canister by hand. Roller tools like those in the foreground are used to stretch the metal over the form.

Figure 4-41
Large canisters and other products may be spun by hand, using roller tools to stretch the metal over the forming block. Some typical roller tools are visible in the foreground. (Lake Geneva Spindustries, Inc.)

Bowl and dome-shaped products made from soft metals, such as copper and aluminum, are often spun. Spinning incorporates low-cost tooling, so production of limited volume products is practical. Parts such as flood lamp reflectors, bowls, and bells for musical instruments are normally spun using CNC equipment.

Casting and Molding

The foundry industry is directly concerned with the casting and molding of metals. In many instances, firms in this industry can be classified as primary manufacturers. They are concerned with producing stock for use by secondary manufacturing establishments that, in turn, make cast metal products. Cast products are made from steel, iron, and other metals or alloys.

Cast Iron

Since cast iron is an important material, a review of the various types and their characteristics might be useful.

Cast iron is produced in a high-temperature furnace called a *cupola*. It is made from pig iron (molten iron poured from the blast furnace) combined with scraps of high-carbon-content steel. Cast iron contains from two to four percent carbon. Cast iron parts cannot be bent, but they can be easily machined. The high graphite content in cast iron provides a lubricant for cutting tools.

There are many types of cast iron, including ductile, gray, high-alloy, malleable, and white. Gray and ductile cast irons are most common. Of all of the cast ferrous alloys, gray cast iron has the lowest melting temperature, and the best overall castability. It also has tremendous compressive strength. For this reason, machine beds and frames are usually made of gray cast iron. The major limitation of gray cast iron is that it is brittle and difficult to weld.

Ductile, or nodular, iron is used to cast axles, brackets, and crankshafts. Ductile iron is made by adding elements such as magnesium, nickel, or silicon in small amounts to the molten alloy. These elements improve the strength, castability, and machinability of cast iron.

Some foundries are set up as small job shops that contract with customers to produce limited quantities of cast products. Many large companies have foundries in their own facilities.

The raw material to be cast is often purchased as an **ingot** (brick of stock), then turned into a liquid for casting at the foundry. Firms that produce large quantities of cast iron products usually have sophisticated facilities for melting castable steel or iron alloys. Molten metal is produced in electric furnaces under very carefully monitored temperatures. Figure 4-42 shows the molten metal being released from a furnace into a heavy metal container, or ladle. This process is called *tapping a heat*. Furnaces typically produce heats ranging from 500 lbs. to 35,000 lbs. (227 kg to 15 876 kg).

Casting Processes

It is common to think about casting as the process of pouring molten metals. However, not all casting involves pouring *liquids*. Powders are also cast through compacting and heat-setting processes.

Casting processes can be studied according to the type of mold used in the process. Molds are either permanent (reusable) or consumable (expendable). The most significant metal casting processes that use a permanent mold (in terms of industrial use) are powder metallurgy, die casting, and permanent mold casting. Major casting processes that use consumable molds are sand casting and investment casting.

Figure 4-42
Molten steel pours from an electric furnace into a large ladle that will be used to transport the metal to an area where it will be poured into molds to form cast products. (ACIPCO Steel Products)

Powder metallurgy

Powder metallurgy (P/M) is a process that involves compacting metallic powders in a permanent (reusable) mold. It can be used to cast both ferrous and nonferrous parts. For many product applications, it is a highly cost-effective process. Powder metallurgy typically uses more than 97 percent of the initially supplied raw material in the finished part. Therefore, it conserves energy and materials.

Powder metallurgy is generally a high-volume production process, capable of producing thousands of parts per hour. In low-volume production (fewer than 1000 pieces, for example), powder metallurgy is not generally a cost-effective process. The cost of the tooling required would be too expensive to justify its use. A typical powder metallurgy die set is shown in Figure 4-43.

Powder metallurgy is a *near-net-shape manufacturing process.* This means that parts are produced in the desired final size. Little shrinkage occurs during firing and virtually no machining is necessary. Machining is needed only if the part has threads, holes, or unusual design features. Machinability of P/M parts is similar to that of poured castings. An advantage of this process is that additives such as lead, copper, or graphite can be mixed with the powder prior to casting to improve machinability. When P/M parts are machined, carbide-tipped cutting tools are recommended.

Powder metallurgy parts are particularly interesting. Their ability to withstand unusual conditions can be greatly influenced by the way that the powders are mixed, and the various processes which they undergo once they are pressed into shape.

Figure 4-43
A typical powder metallurgy die set consists of the die itself, an upper punch, a lower punch, and a core.

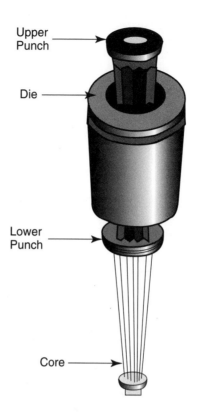

There are three steps involved in manufacturing a powdered metal part: *powder mixing, compacting,* and *sintering.* After sintering, the P/M part is normally ready to use. However, like other metal-formed parts, secondary heat treating and processing operations may be necessary to improve product life in specific environments. Like other metal parts, P/M parts can be annealed, hardened, and tempered.

Very fine powders are used to make P/M parts. *Mixing* of alloys can be done by the powder producer or the parts manufacturer. The powders from which parts are pressed can be made to assume the characteristics of other metals through a process called infiltration. Self-lubricating bearing surfaces can even be created through the process of impregnation. These and other unusual processes will be discussed below.

Once the powders are mixed as desired, a metered amount of powder is automatically gravity-fed into a precision die, Figure 4-44. After the die is filled, the upper and lower punches close, *compacting* the powder in the die. The densely compacted powder shape is then ejected and the die is refilled so that another part can be pressed.

The pressure required to press a P/M part varies from 10 to 60 tons per square inch. The higher the pressure, the greater the density of the part. Pressing of the powder is done cold, at room temperature.

When the compressed part is ejected from the die, it is referred to as a **green compact.** At this point, the part can be handled, but is extremely fragile—it will fracture if dropped or carelessly handled.

Figure 4-44
To make a part in the powder metallurgy process, the die is filled with blended powder. The upper and lower punches then close, compacting the powder tightly. The upper punch retracts, and the lower punch ejects the compact from the die. Finally, the lower die returns to its original position, ready for the next cycle.

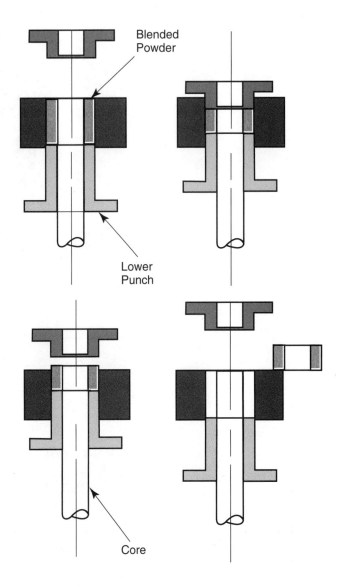

In most production environments, the green compact is transported by conveyor to the sintering furnace, where it travels slowly through a temperature-controlled environment.

Sintering takes place at a temperature just below the melting point of the base metal. Parts are held at this temperature for a short length of time, then cooled. The purpose of sintering is to create an internal metallurgical bond between the particles of powder. After sintering, the parts can be pressed again to improve detail (coining), or can be plated, heat-treated, impregnated, or machined as desired.

Impregnation is the process of immersing porous P/M parts in heated oil or resin after they are removed from the sintering oven. The parts are allowed to soak long enough so that the pores are saturated by oil. The process has been used by the automobile industry since the 1920s. Parts can also be impregnated with oil using a process called vacuum impregnation.

Care must be taken in handling oil-impregnated parts. Such parts should never be allowed to rest for any length of time on cardboard or uncoated paper, or oil will be drawn off by a process called *wicking*. To avoid wicking, place parts on coated paper or a nonporous surface.

What does impregnation do for the P/M part? If the part is a bearing, for example, it must withstand continued friction. Friction creates heat, which causes the oil to expand and flow to the surface of the bearing. The oil at the surface reduces the friction, and the bearing cools. As the bearing is cooled, the oil is drawn back into the pores of the metal by capillary action.

It is possible to improve the *mechanical* properties of P/M parts through another process called **infiltration.** Infiltration is accomplished by placing pieces of the infiltrant metal on or below the green compact before sintering.

The infiltrant metal must have a lower melting point than the porous P/M part. When the part is sintered, the infiltrant is drawn into the part's pores by capillary action. This enables the formation of a composite metal. Not only does the composite provide improved mechanical properties, but the infiltrant also seals the pores of the part. This is particularly helpful when the part is to be electroplated or machined.

When *maximum density* of P/M parts is desired, a process called **hot isostatic pressing** can be used. This compacting process is discussed in detail in the section on Ceramic Material processing. Commercial products, such as cylindrical billets up to 9 feet in length and hollow cylinders with diameters as large as 24 inches, have been produced by hot isostatic pressing.

Powder metallurgy is also used in conjunction with the molding process that is commonly used for making plastic parts. In **P/M injection molding,** very finely ground powders are coated with thermoplastic resin. The mixture is then used in the conventional manner in an injection molding press.

Modified thermoplastic injection molding machines are used at approximately 300°F and 1000 psi (149°C and 6895 kPa). After molding, the part is sintered. This burns off the polymer and improves the density of the metal.

Powdered metal parts are found in many products, ranging from automobiles, home appliances, and lawn mowers to farm equipment and sporting goods. Gears and cams, support mounts, racks, and an infinite array of specialty products are being produced each day using powder metallurgy.

Die casting

Like powder metallurgy, die casting is a process that also uses a permanent, or reusable, mold. In **die casting,** molten metal is forced under high pressure into a cavity in the die. Low-melting-point alloys such as zinc, tin, or lead are frequently die cast. Die casting has been used in industry since the early 1900s, and is still widely used because it is one of the fastest of all of the casting processes.

Dies used in die casting are constructed of steel alloys. Normally, dies are in two parts: the cover half and the ejector half. The die set, with its die halves, forms a block that is locked into the die casting machine. Dies may be constructed with single or multiple cavities. Dies are expensive, often costing from $10,000 to $50,000 or more to design and machine.

There are two types of die casting machines, *hot chamber* and *cold chamber*. Both types operate by forcing melted metal into a forming chamber. The hot chamber machine is the fastest. It is capable of introducing more than 900 good-sized shots of metal into the chamber per hour, using zinc. With small parts like zipper teeth, 18,000 shots have been reported per hour.

Molten metal is normally forced into the die cavity of the hot chamber die casting machine at about 2000 psi (13 800 kPa). This pressure is maintained until the metal solidifies. The dies are then cooled by water or oil circulating through water jackets. Cold chamber machines also inject molten metal into the dies, but solidification is done in a cold chamber.

One of the disadvantages of die casting is the time needed to develop and produce dies sets for a new product. Before a new product can be produced, the casting dies must be made by the tool and die maker. To shorten this waiting time, and to reduce changeover time for removing and replacing dies, a number of suppliers offer quick-change die systems. One proprietary system, called the ***master unit die method,*** was developed by Master Unit Die Products, Inc. of Greenville, Michigan. That system uses quick change inserts in a permanent frame or fixture. The master unit frame stays in the machine. To change a die, the operator loosens four clamps, disconnects heating or cooling lines, then removes one insert and slides another into place. With this die casting method, tooling costs are reduced because only the insert must be replaced, not the entire mold base. Additional cost savings are achieved through an increase in productivity resulting from less time spent changing dies. Master unit dies are available for both die casting and injection molding.

Die casting is an efficient process where high production rates and high strength products are desired. Products such as components for appliances, carburetors, and metal hand tools are often die cast. The process is also effective for combining parts to eliminate a later assembly step, saving time and cost. Visualize a product shaped like an ice cream bar, with two components, a body and a stick. A die could be designed to permit casting the body and stick together as a single unit.

Permanent mold casting

Permanent mold casting, also known as gravity die casting, is used with molten metal. Permanent molds are filled much like the die is filled in powder metallurgy. The major difference is in terms of the material that is used—in permanent mold casting, molten metal is poured into the mold, reaching the cavity through a gating system. When the part solidifies, the mold is opened and the part is removed. Often molds are hinged to permit easy removal of the cast part.

Since the molds employed for permanent mold casting are used again and again, they must be constructed of materials that can withstand continuous heating and cooling. Molds are normally made of cast iron, steel, bronze, graphite, ceramic materials, or refractory metal alloys.

A simplified mold of the type used for permanent mold casting is shown in Figure 4-45. The metal is poured into the mold through the pouring basin. The molten metal flows on into the mold cavity. When the cavity is filled, extra metal flows back out of the riser. This helps to ensure that the metal completely fills the cavity.

When permanent mold casting is used in industry, it is normally an automated process. This makes it cost-efficient to use as many mold cavities

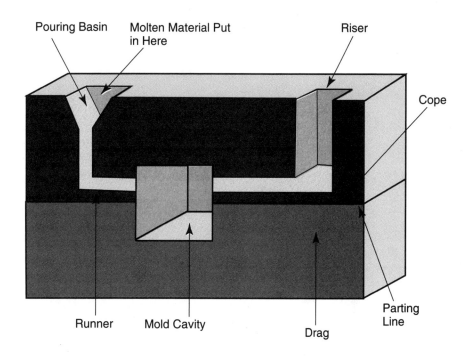

Figure 4-45
A mold used for permanent mold casting. This is a simplified cutaway example to emphasize the major parts of such a mold.

as possible. A typical production setup for permanent mold casting might consist of a circular turntable with ten or twelve individual stations. The casting machine would automatically blow out each mold to clean it, then pour in molten metal. The machine would automatically cool each mold, unlock the dies, and eject the casting.

Permanent mold casting is often used to cast parts from iron and nonferrous alloys. The process produces a good surface finish and maintains close tolerances. It is also fast and is suitable to high-volume production.

Permanent mold casting can also be used to make hollow castings. This adaptation of the permanent mold casting process is called slush casting. *Slush casting* is done by removing the casting from the mold when it has just begun to solidify. The part is inverted to pour out the metal that is still molten. The only metal that remains is the outer shell. This process is often used to make ornamental parts and decorative items.

Sand casting

Although sand casting dates back to antiquity, it is still used more frequently than any other casting process. In the United States alone, sand casting accounts for more that 15 million tons of metal poured each year. Typical products manufactured using sand casting include engine blocks and cylinder heads.

Sand casting uses a mold that is *expendable*, rather than permanent. The mold that is created can be used only once, because it is destroyed when the part is removed after casting. With sand casting, a pattern slightly larger than the final product is placed in a rectangular box called a *flask*. The wood or metal pattern is made slightly oversize, since the casting will shrink during cooling. Moist molding sand is then packed into the flask around the pattern to create the shape to be cast. With wood or metal patterns, the pattern is removed before the mold is used. If an opening through the part is needed (for a shaft, for example), a molded sand *core* is inserted in the cavity. The core will be broken out after the casting cools. Before the mold is closed, entry and exit holes for the molten metal and pathways to carry it to the cavity are cut in the sand.

A typical mold configuration for sand casting is shown in Figure 4-46. The flask consists of two frames that fit together, called a *cope* and a *drag.* For many of the parts cast using this method, a *split pattern* makes mold preparation easier.

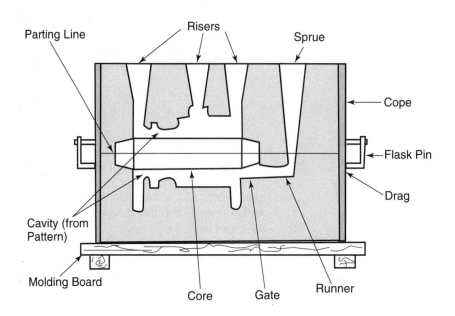

Figure 4-46
After the cope and drag mold halves have been prepared, the pattern is removed. When the mold is reassembled, its center contains a cavity in the shape of the pattern. If the part is pierced by a hole or other opening, a molded sand core is inserted. Molten metal poured in through the risers will fill the mold. When metal appears in the sprue, the cavity has been filled. Once the metal cools, the mold halves can be opened and the metal part removed. The sand mold is destroyed in the process; a new one must be made for each part that is cast. If a core is used, it is broken out after the part is removed from the mold.

Here's how a typical sand-cast part is made:

First, the lower half of the split pattern is placed flat side down on a molding board. The drag is placed over the pattern, with the pins pointing downward. A fine powdered parting compound would then be dusted onto the pattern to aid in its removal from the sand. A screened sifter, called a *riddle,* is used to shake sand over the pattern until it is well-covered.

The drag is filled completely by pouring sand to the top, and the sand is firmly packed with a wooden rammer. A bottom board is then placed on top of the drag. The entire sandwich—bottom board, drag, and molding board—is turned completely over so that the molding board is on top.

Once the molding board is removed, the upper half of the pattern is set on top of the lower half that is already embedded in the drag. Pins or other devices are often used to align the two pattern halves. Then, the cope is stacked on top of the drag. Parting compound is dusted on the pattern and sand.

The next step is to insert sprue and riser pins in the sand about one inch from each end of the pattern. These pins are tubes that will make holes in the sand filling the cope to serve as pathways for the metal to enter and exit the mold cavity. Metal will be poured in through the sprue and the excess will flow out of the mold cavity through the riser.

Sand is next riddled into the cope to cover the top half of the pattern. Additional sand is used to fill the cope, then compacted with the rammer. The sprue and riser pins are then removed. The cope is lifted off the drag and turned pattern-side-up. The pattern is carefully removed. On the surface of the drag, a one-inch wide gate, or pathway, is cut in the sand from the sprue to the cavity left when the pattern was removed. A similar gate is cut from the riser to the pattern cavity. The gates permit molten metal to flow into and out of the mold cavity.

At this point, the cope is once again placed atop the drag and the two clamped together so that the mold can be poured. Molten metal is poured down the sprue until it begins to exit the mold at the riser. This assures complete filling of the mold cavity. After cooling, the sand mold is broken away and the casting removed.

There are many different types of patterns that can be used for sand casting. The split (two-part) pattern already described is popular. An interesting variation is the one-piece polystyrene pattern. This type of pattern can be left in the mold sand, because it is burned up when the molten metal enters the mold cavity.

Investment casting

Like sand casting, investment casting also uses a consumable mold. *Investment casting* is a precision casting method used with materials that are difficult to work or expensive to machine.

In investment casting, the cavity created by the pattern is destroyed when the casting is removed. Investment casting is also known as the *lost wax process*. The process was first used around 3000 BC, and is still in use today as a means of producing intricate jewelry, artwork, and ornate metal products.

A pattern is typically carved from wax or expanded polystyrene. Often a number of patterns are attached to each other, much like limbs on a tree. The finished pattern is then dipped in a ceramic refractory slurry.

The refractory coating is allowed to dry, and the process is repeated until the desired wall thickness is achieved. The entire assembly can now be viewed as a mold.

Next, the mold is placed in an oven to melt the wax out of it. The refractory shell remains intact after the wax is melted. Now the mold is ready to be poured. The metal is poured and allowed to solidify, then the refractory mold materials are broken off and the casting removed.

Dentists are one of the most extensive users of investment casting. They use the process to cast silver, gold, stainless steel, and other alloys for making dental bridges and crowns. The process is also used by the medical profession to repair fractures, bones, and joints.

Centrifugal casting

Centrifugal casting is a process that is used to mold cylindrical parts from plastics and metals. Centrifugal casting is basically the same process when used with either of these materials.

This casting method relies on centrifugal forces created by rotating a cylindrical part to distribute molten metal against the walls of the mold and into the mold cavities. The process is popular for producing tubular parts such as pipes, tanks, and bearing rings.

The process of centrifugally casting a long hollow tube that is partly closed at each end would begin by lining the tube with a slurry of ceramic refractory material. The refractory coating would be dried and baked. This would be much like what is done in investment casting.

The lined mold (tube) would then be spun rapidly and the molten metal would be poured into it. The spinning process would continue until the metal solidified.

Figure 4-47 illustrates centrifugal casting of tubular parts using a process developed by American Cast Iron Pipe Company of Birmingham, Alabama. The process begins by coating the inside of the tube with ceramic slurry. When the slurry has dried, the coated tube is fed into the oven. In the oven, the refractory coating expands to apply pressure to the metal that is being distributed inside of the liner by centrifugal forces. When the material cools, the ceramic shell is removed and the casting is cleaned. The photograph illustrates the casting of a steel tube with close coordination of metal temperature, pouring rate, and speed of rotation.

Figure 4-47
Centrifugal casting is used to create tubular parts. (ACIPCO Steel Products)

Important Terms

- blank
- bulging
- centrifugal casting
- closed-die forging
- coining
- compacts
- computer numerical control (CNC)
- cope
- core
- cupola
- deep drawing
- deform
- die block
- die casting
- die set
- drag
- drawing
- electromagnetic forming
- embossing
- expanding
- explosive forming
- extrusion
- flash
- flask
- follower
- forging
- forming block
- forming tool
- foundational technical development
- green compact
- hand forging
- high-energy-rate forming (HERF)
- hot isostatic pressing
- hydroforming
- impregnation
- infiltration
- ingot
- investment casting
- knurling
- marforming
- master unit die method
- near-net-shape manufacturing process
- necking
- nosing
- open-die forging
- parting line
- permanent mold casting
- P/M injection molding
- powder metallurgy (P/M)
- punch
- punch press
- riddle
- roll bending
- roller forming
- rotary swaging
- sand casting
- shot peening
- sintering
- sizing
- slush casting
- spinning
- split pattern
- stamping
- stretch draw forming
- tapping a heat
- tensile strength
- thread rolling
- ultimate tensile strength
- upsetting

Questions for Review and Discussion

1. What are some of the advantages of rolled threads versus machined threads?
2. Explain the process of upsetting, as used to make rivets.
3. Describe the major pressworking processes. What do they all have in common?
4. What are the parts of a die set? Describe the purpose of each part.
5. Why is it unlikely that the process of necking would be used to improve the tensile strength of a hard metal?

This single-roller spinforming machine uses computer numerical control for precise repeatability. It can be programmed for spindle speeds ranging from 100 rpm to 5000 rpm and can be used with a variety of metals, including steel, aluminum, copper, and titanium. (Electrologic, Inc.)

Separating Metallic Materials

Key Concepts

△ Some separating processes shear, cut, or press parts from flat stock.

△ Processes that reduce the diameter or change the profile of round stock are considered separating activities.

△ Various workholding devices are available to mount different types and shapes of workpieces on the lathe.

△ There are four major types of planing machines used with metallic materials: planers, shapers, slotters, and broaches.

△ The lathe and the milling machine are basic equipment vital to the metalworking industry.

△ For efficient removal of material, you must determine the proper cutting speed for a drill or other cutting tool, as well as the correct feed rate for the machine.

△ Different grinding wheel compositions are available to meet specific material removal requirements.

△ Although conventional EDM and travelling wire EDM use the same operating principle, they are used for different applications.

Types of Separating Processes

There are numerous manufacturing processes that can be used to remove material or volume from a metal workpiece. Many of these are mechanical processes that produce chips, using a wedging action from a cutting tool. Examples would be cutting and grinding processes. These types of processes generate *scrap* (excess material) during the material removal process. Scrap that isn't recycled for reuse in the production process becomes *waste.* Good planning and the proper selection of manufacturing processes will reduce waste.

Other types of separating processes don't produce chips, but they *may* generate scrap. These are processes that are used to punch, cut, or shear the stock cleanly into pieces without any loss of material.

Shearing

A third category of metal separation processes uses no mechanical cutting action. Instead, material removal is accomplished through chemical or electronic means.

Shearing is a mechanical separating process that is often used to cut flat stock from sheet or plate. Shearing processes can be classified according to the type of blade or cutter in the machine. There are three types of shearing—straight shearing, punch-and-die shearing, and rotary shearing.

All shearing processes involve the use of opposing surfaces. The way that the opposing surfaces produce a cutting action is shown in Figure 5-1. The upper blade moves down into a workpiece that is held securely by the bottom fixed blade. It pushes the stock down into the opposing bottom blade, severing the metal.

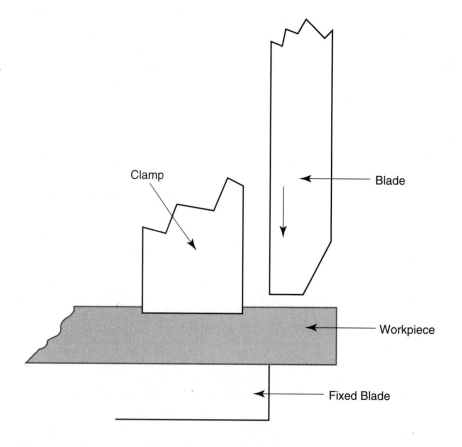

Figure 5-1
Shearing is a process that is accomplished by the use of opposing surfaces: the material is "pinched off" by the action of a moving blade against a fixed blade.

Straight shearing, or cutting of sheet metal to size, is usually done with a machine called a squaring shear. Sheet metal is placed on the horizontal machine table of the shear, where it is held down by a clamping device. The operator then either steps on a foot treadle (if the machine isn't motorized), or engages a lever to power the blade.

Opposing surfaces can also function as a punch and die. The action is like that of the familiar paper hole punch. The downward motion of the punch shears the sheet metal stock by pushing it into and through the fixed die. Figure 5-2 illustrates *punch-and-die shearing.*

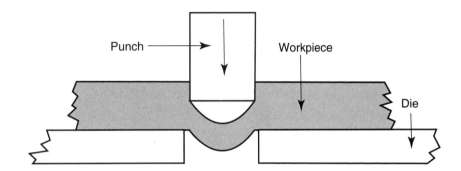

Figure 5-2
In punch-and-die shearing, a moving punch pushes the severed material through the opening in a fixed die.

Rotary shearing is done in a manner similar to straight shearing, except that the blades are rotary wheels. See Figure 5-3. Rotary shearing can be used to cut either straight or circular shapes, and permits cutting small radii and small irregular pieces. It is basically a hand process that is not practical for high-volume production applications.

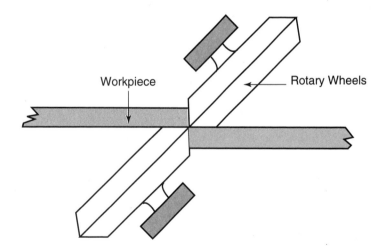

Figure 5-3
Rotary shearing uses opposed wheels to make either straight or curved cuts.

Other shearing processes

In addition to the three major types of shearing described, there are specialty processes that are based on the principle of a shearing or punching action. These include lancing, slitting, notching, perforating, punching, and nibbling. Each of these processes uses a shearing action, but is done with a different type of shearing machine.

Perforating is a punching process used to produce a number of closely and regularly-spaced holes in a straight line across a sheet metal section. This is often done to facilitate bending of metal in a particular area, or to create holes for assembly or heat transfer. See Figure 5-4. Very small holes are difficult to punch and seldom justified in terms of being cost-effective.

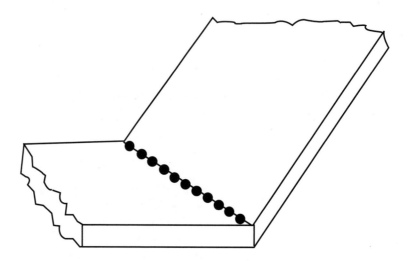

Figure 5-4
The process of perforating results in a line of evenly-spaced holes across a surface.

Both lancing and slitting are accomplished without removing any material. *Lancing* is a punching process that is used to make a tab without removing any material, Figure 5-5. The tab that is punched is normally bent upward to provide an area for gripping or moving a part. *Slitting*, Figure 5-6, is done to create an area where another part or device can be inserted. Both lancing and slitting are variations of the basic process of shearing.

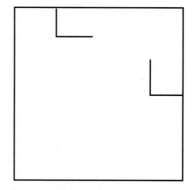

Figure 5-5
Lancing shears metal to make an L-shaped tab for assembly or other purposes. No metal is removed.

Figure 5-6
Slitting is a punching process often used to prepare sheet metal parts for assembly. As shown in the cross-section, the material between the parallel slits is pressed upward or downward to allow a tab to be inserted.

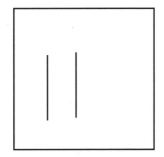

Blanking

The process called **blanking** can be used to punch a flat metal part ("blank") from sheet metal. This process produces no chips and often generates very little waste. Blanks are produced with each stroke of a punch press. Special punches and dies are used.

Blanking is often confused with punching. In the process of **punching** the punch moves down into a mating die. The stock to be punched is fed between the punch and die. As shown in Figure 5-7, the stock that is stamped out with the punch is generally waste, and the *part* is what is left. In blanking, the part consists of the area that is *stamped out*. The stock that is outside the stamped part is generally waste. See Figure 5-8.

Figure 5-7
When punching is used to produce a part, the area that is punched out is usually waste.

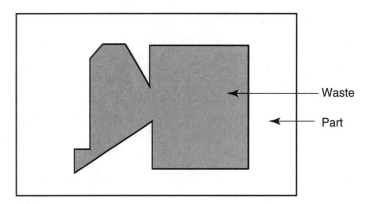

Another difference between punching and blanking is that, in blanking, the punch only travels about halfway through the thickness of the stock. Usually, several blanks are stamped at the same time. These are carefully positioned so that multiple parts can be cut from one workpiece with a minimum amount of waste. Positioning of parts in this manner is called **nesting**. Care must be taken in designing parts to make them nest effectively in the work blank. See Figure 5-9. The purpose of nesting parts, in addition to minimizing waste, is to simplify handling and subsequent processing. Several parts can be handled or processed at one time. After all operations have been performed on a part, it is broken off the sheet.

Figure 5-8
When blanking is used to produce parts, the areas that are punched out are usually the parts, while the surrounding material is waste.

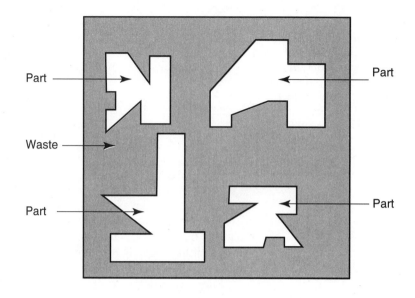

Figure 5-9
To minimize waste, parts to be stamped from a single piece of stock must be carefully positioned. Computer programs are now used in industry to help achieve the most effective nesting of parts on a workpiece. (U.S. Amada)

Nibbling

When individual parts are needed in low volume, other process options may be more practical than blanking. Nibbling and notching are two shearing/punching processes that are used to cut or punch out individual parts. Both of these processes differ from blanking in that they generate chips.

Nibbling is a simple process that requires no special tools or unique fixturing. It is a practical method for cutting out limited numbers of flat parts with complex shapes. The process is done on a *nibbling machine*, which looks something like a foot-powered stapler. The machine supports a round or triangular punch that moves up and down in a mating die at a rate of up to 900 strokes per minute. The process may be cost-effective when limited numbers of unique parts must be cut from sheet metal, but it is slow and time-consuming.

Nibbling is done by inserting sheet metal into the opening between the punch and die. Metal can be manually held or automatically guided into the machine, using a template. The sheet that is nibbled is held down in the machine by a stripper. This helps in producing a part which is not bent or distorted.

Nibbling can cut metal with many overlapping strokes of a punch used to punch out a line in the blank. The material on the inside of the punched line is the finished part, which has a fairly smooth edge. What is located on the outside of the line could be classified as scrap, or as material that can be used to make more nibbled parts. What is unique about the process is that no chips are produced.

A process that is similar to nibbling is called **notching** or piercing. See Figure 5-10. Notching is also a punching process, and is typically done using a punch press. The material that is removed by the punch is scrap.

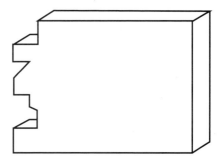

Figure 5-10
Notching to produce irregular contours on the edge of a part is typically done on a punch press.

All shearing, punching, and stamping processes must be carefully selected for use, depending on the type and gage of metal that must be processed. When stamping processes are involved, heavy transfer or punch presses are used. Stock is generally classified according to gage. **Light gage stock** covers thicknesses up to 0.031 in. (0.79 mm). **Medium gage stock** consists of sheet from 0.031 in. to 0.109 in. (0.79 mm to 2.77 mm). **Heavy gage stock** covers anything from 0.125 in. (3.17 mm) and up.

Turning

All the processes discussed so far in this chapter were used to shear, cut, or press parts from flat stock. It is often necessary to reduce the diameter, or change the profile, of *round* stock as well. ***Turning processes*** are used to machine rotating parts. The most common machine used to perform machining operations on round parts is the ***lathe.***

Although metal lathes differ from woodworking lathes, the principle is the same: the machine holds and turns a workpiece against a cutting tool, in order to produce a cylindrical form or reduce its diameter. Today, the lathe is probably the most heavily used machine in the machine tool industry. Many different types of turning operations can be conducted on the lathe. Some of the most common of these are:

- ***Boring,*** used to improve the accuracy of a drilled hole through the use of a boring bar.
- ***Centerdrilling,*** used to produce holes in each end of the stock so the workpiece can be secured in the lathe between centers.
- ***Drilling,*** used to produce a hole in the end of the workpiece. This process uses a drill held in a commercial drill holder or a Jacobs chuck.
- ***Facing,*** used to square the end of the workpiece and reduce it to the desired length.
- ***Knurling,*** used to press diamond-shaped indentations on the circumference of the workpiece. It is used to provide a gripping surface on a part, or to expand the diameter of a part (such as a piston).
- ***Parting,*** or necking, used to cut off a section of the end of a workpiece or to produce a groove around the circumference of the part.
- ***Threading,*** used to cut internal or external threads on a part, using a boring bar and/or cutting tool.
- ***Turning,*** used to reduce the diameter of a workpiece, change its profile, or produce a taper on a length of round stock.

A multitude of other operations can be performed on the lathe, depending on the accessories and workholding devices that are available. Sometimes, it is even possible to complete several operations at the same time.

Types of lathes

There are a number of different types of lathes used for machining metal. However, the most common types of lathes used in manufacturing are the engine lathe, toolroom lathe, turret lathe, and the NC (numerically control) or CNC (computer numerical control) lathe. All are variations of the basic engine lathe. A CNC (computer-operated) lathe is shown in Figure 5-11. This lathe receives information from an IBM PC (personal computer), displays programs on a color monitor, and automatically selects one of eight tool stations to perform the desired operation.

Lathes normally are classified according to two factors. The first is the maximum diameter of workpiece that can be turned, which is referred to as *swing*. The second is the maximum *length* that can be turned between centers.

Figure 5-11
This CNC lathe is controlled by a computer program, allowing exactly repeatable steps in producing each part. (South Bend Lathe, Inc.)

Figure 5-12 shows a 30 x 160 engine lathe. This means that this large lathe has a 30 in. (76.2 cm) swing and a maximum distance between centers of 160 in. (406 cm).

Workholding devices and methods

The method of securing work in the machine must be considered before machining can begin. The type of workholding device used depends on the stock to be turned. Workholding devices and arrangements that are normally used on the lathe are described in the following paragraphs.

∆   A short piece of stock is held in the *headstock* (driven end of the lathe), using a 3-jaw universal chuck, a 4-jaw independent chuck, or a 3-jaw Jacobs chuck. The jaws of a ***universal chuck*** are self-centering, so it is usually used for round stock. The ***Jacobs chuck*** is the type normally used in a drill press. In the lathe, its primary use is to hold smaller workpieces. The ***independent chuck*** is used for irregularly-shaped workpieces—its jaws are turned in toward the work one at a time. Power chucks are used on automated machining centers.

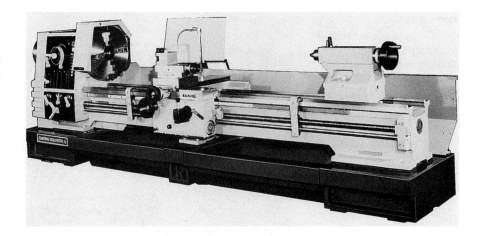

Figure 5-12
This large engine lathe can handle a workpiece up to 30 inches in diameter and more than 13 feet in length. (Clausing Industrial, Inc.)

△ Longer pieces of stock are held between centers. One end is secured using the combination of a *faceplate*, a lathe dog, and a *live center* in place of a chuck. The opposite end of the workpiece is held with either a *dead center* or a live ball-bearing center in the *tailstock*. A *live* center rotates with the work; a *dead* center is stationary (the work rotates against it).
Both ends of the work are centerdrilled to hold the points of the centers.

△ *Collets* are used to automatically center stock that is round, square, octagonal, or hexagonal. Collets are used when accuracy is needed over an extended length of time. Collets have threads on one end and split jaws on the other. The work is either placed inside the jaws (a *draw-in* collet), or over the jaws (a *push-out* collet). When a short length of material is to extend out of the workholding device for machining, it is usually necessary to secure the stock in the lathe using either a chuck or a collet. Both methods clamp around the stock, so that it can be held for operations such as facing, parting, chamfering, drilling, or centerdrilling.

When a longer piece of stock must be machined, it is usually supported at both ends so that it will not wobble as the headstock rotates. Securing the workpiece can be done in a variety of ways, but is usually accomplished by using a tailstock with a dead center.

Turning tools

Many different types of cutting tools are used for turning. Tools that are made of high-speed steel are ground to the desired angle, according to the way that they approach the work. See Figure 5-13. **Right-hand tools** cut from right to left. **Left-hand tools** cut from left to right. A round-nosed tool is used for light turning, and is ground to permit cutting from either direction.

Cemented carbide cutting tools are available for cutting harder material. These tools are also used when greater cutting speeds are desired. The carbide tips are brazed onto the tool blank, Figure 5-14.

Figure 5-13
Typical shapes of tools used for turning. Note that most types are made in two versions, for use in cutting left-to-right or right-to-left. The round-nose tool can be used to make light cuts in either direction.

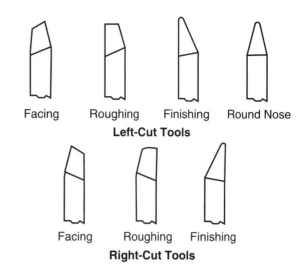

Figure 5-14
Cemented carbide tips brazed onto drills used for high-speed cutting. (Mitsubishi Carbide)

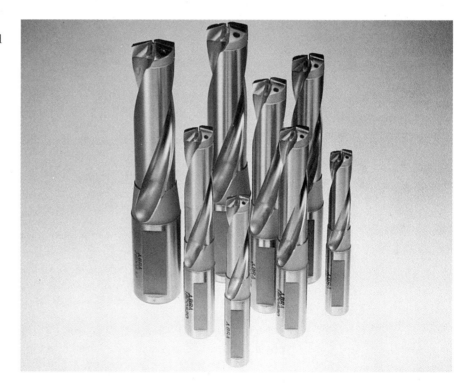

Tools can also be purchased with disposable ceramic cutting tip inserts called *cermets*. The inserts are made from carbide and sintered oxides and are screwed to the tool blank, Figure 5-15. When one side becomes dull, the insert is turned around so another cutting edge can be used. When all sides have been used, the insert is disposed of and a new one is inserted.

Figure 5-15
Cermet inserts and tools used for turning, milling, and drilling. The inserts are screwed into place on the tools, and can be replaced easily when worn. (Mitsubishi Carbide)

Depth of cut

Several factors influence the *depth of cut* that can be taken. If the lathe is running properly at the correct speed (in revolutions per minute, or *rpm*) and the tool is sharp and properly ground, relatively deep *roughing cuts* can be made. The depth of the cut depends upon the material being machined, the cutting tool, the rpm of the headstock, and the rate of feed into the workpiece. Roughing cuts usually are in excess of 0.020 in. (0.50 mm) depth and are done at slower speeds.

The type of finish that is desired is a consideration in setting the depth of *finishing cuts.* These cuts are made at higher speeds, use slower feeds, and involve removal of smaller amounts of material. A finish cut will range from several thousandths of an inch up to 0.020 in. (0.50 mm) deep. The finish obtained depends upon the performance of the lathe, the type and sharpness of the cutting tool, and the hardness of the material being machined.

Lathe components

To better understand the operation of a lathe, it is helpful to study the physical parts of the machine, as shown in Figure 5-16.

Turning force is provided by a belt-driven or gear-driven headstock. Attached to the headstock is a threaded *spindle* to which a chuck or a faceplate can be screwed. *Ways* are the guides that are machined on the top of the heavy cast iron bed. A *carriage* slides left and right along these ways, and supports a *cross slide, compound rest,* and a *tool post.*

The simplest turning operation is called *peripheral turning* or *straight turning.* In this method, the tool is fed into the stock, then moved to the left or right to remove material, reducing the diameter of the workpiece. For straight turning, the compound rest is set at an angle of 30° to the work. A cutting tool

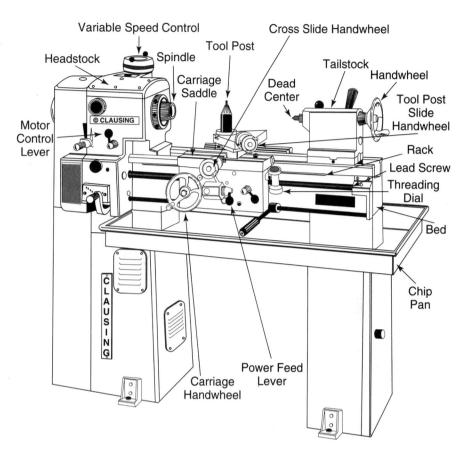

Figure 5-16
The basic parts of a lathe used for metalworking.

is clamped into a ***tool holder*** and the holder is secured in position in the tool post. Normally, the cutting tool would be positioned about 1/64 in. (0.016 in. or 0.4 mm) above the center of the stock to be turned. The angle of the tool against the workpiece depends on the type of tool used. The spindle speed is set according to the type of cut and material being turned. Softer metal, for example, can be turned at a higher speed.

If the length of the stock is *less than three times* its diameter, it may be desirable to use only a chuck, without any additional support. If the stock is *longer than three times* its diameter, it should be secured by placing it in a chuck on one end and using a dead center in the tailstock to provide additional security.

Another mounting method is to screw a faceplate onto the spindle in place of the chuck. The workpiece is then inserted in a ***lathe dog*** (holding device that engages the faceplate to rotate the stock), and the dog screwed to the stock. A lathe center is inserted in the spindle, and the tail of the dog engaged in a slot in the faceplate. The lathe center rests in a hole drilled in the end of the stock, allowing the dog to turn the stock with the faceplate. The tailstock is then moved up to the free end of the stock, and the center is placed in another drilled hole. A lubricant is often squirted into the drilled holes to reduce friction.

Turning operations

In addition to straight turning, there are a number of other ways in which the lathe is used.

To *true* (make perpendicular) both ends of the stock, an operation called *facing* is done. The facing tool is positioned at a slight angle to the center of the working face of the workpiece, which is clamped in the chuck. The carriage is moved into the workpiece slightly, so that the tool removes a chip. Then, the crossfeed is engaged so that stock can automatically be removed from the center of the workpiece to its circumference. Sometimes, facing is done by hand. After one end of the stock has been faced, the lathe is stopped and the workpiece reversed to permit facing the other end.

After both ends have been faced, it may be necessary to perform another operation, called *centerdrilling*. Centerdrilling is necessary if the method of securing the workpiece to the lathe is the use of a faceplate and dog. Even if the workpiece is secured in a chuck, the free end may have to be stabilized with a lathe center placed in the *tailstock* of the lathe. This requires a centerdrilled hole in the stock to accept the tip of the center.

Centerdrilling is done by securing a combination drill and countersink in a chuck and inserting the chuck in the tailstock of the lathe. The tailstock is then moved toward the workpiece until the drill nearly touches it. Next, the lathe is turned on and the hand wheel on the tailstock used to move the drill into the rotating workpiece. The finished depth of the centerdrilled hole should be about half the drill body diameter. When both ends have been centerdrilled, the workpiece is ready for turning between centers.

Drilling and boring of holes into the center of round stock or parts is done in the same manner as centerdrilling, but uses a drill or boring bar instead of a combination drill and countersink.

Operations such as straight turning and knurling are normally done using an *automatic feed*. When the carriage feed is set and engaged, the tool is fed automatically from the right end of the workpiece toward the chuck or lathe dog. Knurling is a process that scores diamond-shaped or straight-lined patterns into the surface of a metal workpiece to create a textured gripping surface. This operation is done using a knurling tool that is locked in the tool post. The tool resembles a tool holder with one or more sets of rollers. When the tool is turned into the work, the rollers create the knurl pattern.

A variety of *screw threads* can be cut on the lathe including American National, Unified, Acme, square, sharp V, metric, and worm threads. Levers on the quick-change gearbox located below the headstock are arranged to select the desired feed for cutting the thread. A feed dial attached to the carriage of most lathes is used as a gage for cutting the threads. The lathe tool for cutting threads is flat, giving it a negative rake angle on top.

External *tapers* can also be produced on the lathe. This is done by using a taper attachment or by offsetting the position of the tailstock. Many other specialty operations can be performed on the lathe as well, since it is a truly versatile machine tool.

Planing

Planing is an operation used to remove large amounts of material from horizontal, vertical, or angular flat surfaces. In some applications, the material is removed to reduce the size of the workpiece. In other instances, the planing process is done to produce slots or angular grooves in the material.

There are four types of planing machines that might be found in industry. These are referred to as planers, shapers, slotters, and broaches. The first two types, the *planer* and the *shaper,* are seldom used any more. These machines are slow and have been replaced by other, more specialized machines such as slotters, broaches, and mills.

Planing processes generate chips while separating material over the length of the workpiece. In the planer, work is moved back and forth into a fixed, overhead cutter. Planers are huge machines useful for handling workpieces as large as 40 feet wide by 80 feet long.

Like the planer, the shaper is seldom used any more in industry. However, a description of how it works should prove useful in understanding the operation of many other planing processes. The shaper operates differently from the planer. It is smaller in size and is used to remove material from smaller workpieces.

In shaping, the work is held stationary in a vise as an overhead tool removes stock in forward passes over the workpiece. Shapers are able to index the work sideways to set up a different tool path for the next forward cutting stroke. Shapers have an overhead ram that moves forward and backward, carrying the cutting tool.

Slotting

Many smaller job shops use the type of planer called a *slotter.* Much of the work once completed by slotters is now being done by another process, called milling.

Slotters are vertical shapers that are sometimes used to cut both internal and external slots and keyways. To visualize how the process is done, consider a workpiece three inches in diameter with a one-inch hole drilled through the center. The slotting machine looks somewhat like a large drill press. The cutting tool would be powered down inside of the hole, removing material and creating a slot or keyway the length of the hole. Keyways are used to keep pulleys from turning on rotating shafts.

Broaching is a process that is faster than slotting and produces a finer surface finish. Like slotting, broaching can be used to do internal or external planing.

Broaching

Broaching is a process that is usually done using a hydraulic-powered broaching machine, Figure 5-17. The process is ideal for internal machining of keyways, splines, and irregularly shaped openings. (Note the shape of the hole shown on the bottom of Figure 5-18.) Broaching can also be used to remove surface material on metal workpieces.

Broaching is a rapid process that can remove as much as 0.250 in. (6.35 mm) of material in one pass of the cutting tool. The final finish that is achieved with broaching is comparable with other fine machining processes. Cutting speeds for broaching depend on the material being broached, but can be as high as 50 fpm (feet per minute) for aluminum and soft alloys. When performing a broaching operation, both the cutting tool and the workpiece are held in fixtures.

The cutting tool used in broaching is called a *broach.* Broaches create a planing action when they are pushed or pulled across or through a workpiece.

Figure 5-17
A broaching machine, with two broaches ready to be pulled down through workpieces. (National Broach & Machine Co.)

The high-speed steel broach is a long rod or bar with many individual teeth. Each tooth is just a few thousandths of an inch longer than the one before. When the tool first enters the work only a little bit of stock is removed. As the broach continues along the workpiece each tooth cuts a little deeper. By the time it is at the end of its pass, the maximum depth is achieved. Broaching is normally done using cutting fluid to lubricate and cool the tool.

There are several different types of broaches. Figure 5-18 shows the tooth configuration that would be found on a *pull broach* used for internal holes. The broach is tapered, with the first teeth being the smallest. The broach starts with chip breakers, expands with roughing teeth, and then ends with finishing teeth.

On an actual broach, the pull end would extend beyond the pilot area. This is the end that is actually gripped and pulled by the broaching machine. Next comes a section with roughing teeth. These teeth remove the largest amount of material. The semi-finishing and finishing teeth then follow to produce a finer finish.

Figure 5-19 shows what an actual pull broach looks like. The pull end is located on the left side of the illustration. On the right side is the end that is last to leave the workpiece. This is called the rear pilot end.

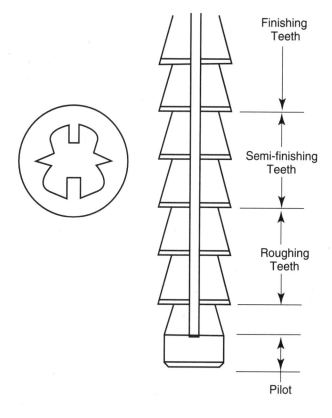

Figure 5-18
Configuration of cutting teeth on a pull broach. This broach would produce the irregularly shaped hole shown below it.

Figure 5-19
A pull broach being assembled. The ring-like cutting tooth units (note different tooth sizes) would be assembled in order on the mandrel in the foreground. The pull end, which enters the workpiece first, is at left in the photo. (National Broach & Machine Co.)

The pull broach is ideal for internal openings through a workpiece. The two pull broaches shown in Figure 5-20 are about to be pulled down through two steel workpieces.

Figure 5-20
Two large pull broaches positioned above the clamping fixtures used to hold workpieces in the machine. (National Broach & Machine Co.)

When material is to be removed on the *outside* of a workpiece another type of broach, called a slab broach, is needed. The **slab broach** consists of a steel blank that may be up to several inches wide and is usually about the same thickness. Slab broaches can be just about any length. This type of broach has cutting teeth arranged on the cutting face. Slab broaches are often constructed in sections that are joined end-to-end. Several slab broaches are shown at left in Figure 5-21. The broaches are installed in slots in the fixture at center. After the fixture is assembled and locked together, a workpiece can be pulled through the opening in the center of the fixture. The slab broaches will then cut slots on the outside of the workpiece.

Another type of internal broach is called the **shell broach.** This broach is constructed in three parts: the broach body, an arbor section holding the shell, and the finishing section. Helical splines, such as the rifling in gun barrels or the teeth of gears, are often formed by rotating these broaches as they are pulled through the part.

Figure 5-21
Slab broaches, like those at lower left, are held in a fixture and used to cut slots on the outside of a workpiece. (National Broach & Machine Co.)

Machining of such parts as crankcases, cylinder heads, and transmission components is often done by broaching. The gears shown in Figure 5-22 were produced with a process called *pot broaching*. This type of broaching is also considered an economical method for producing precision external spur gears and automotive front-wheel-drive transmission gears. In some situations, gears can be machined two at a time, achieving production speeds of more than 500 pieces per hour. A pot broaching machine is shown in Figure 5-23.

In the pot broaching process, the toolholder (called a *pot*) is held stationary, and the parts to be machined are pushed or pulled upward through the broach. Broaching can be done at speeds up to 25 fpm. An important advantage of the pot broaching process is chip removal. Since the broach is inverted, chips fall away from the tool and part while it is being machined. This keeps the teeth clean and aids in producing an improved product.

There are three basic types of broaching tools used with this process: stick, ring, and ring-and-stick. Each has similarities to the slab broach. The *stick broach* has a split two-piece broach holder that supports broach bushings. The bushings mount high-speed steel stick slab broaches.

The *ring broach* has a split two-piece holder that mounts a series of high-speed steel rings. Parts with an outside diameter (OD) in excess of 10 inches can be pot broached with this type of tool.

Figure 5-22
The internal teeth of these ring gears were cut by the pot-broaching process. (National Broach & Machine Co.)

Figure 5-23
Large pot-broaching machines like this one are used extensively in the automotive manufacturing field, and can produce hundreds of finished parts per hour. (National Broach & Machine Co.)

Combination *ring-and-stick broaches* are used to broach teeth on gear blanks at the same time that other operations are being performed. This type of tool is used when complex precision gear forms are needed.

Milling

Milling is a process that uses a multitoothed cutter to produce slots, grooves, contoured surfaces, threads, spirals, and many other configurations. Next to the lathe, the milling machine is the most important machine tool in many manufacturing facilities. Eli Whitney developed the first true milling machine in 1818, while the first universal milling machine was introduced by Joseph Brown in 1861.

Milling machines are normally classified according to their structure. The most popular are the ***column-and-knee-type machine,*** and the ***bed-type machine.*** In the column-and-knee-type machine, a work table is mounted on a knee that moves up and down the column. The bed-type machine doesn't have a knee. Instead, it has a work table that is mounted directly onto the machine bed.

Of the two machine configurations, the column-and-knee-type is the most versatile. Both types can be purchased with either vertical or horizontal spindles. Thus, a milling machine is often referred to as a horizontal machine or a vertical machine.

A ***horizontal milling machine*** has the cutting tool carried on an ***arbor*** (spindle) that travels along an axis parallel to the work table. In a ***vertical milling machine,*** the cutter is positioned perpendicular to the work table, Figure 5-24.

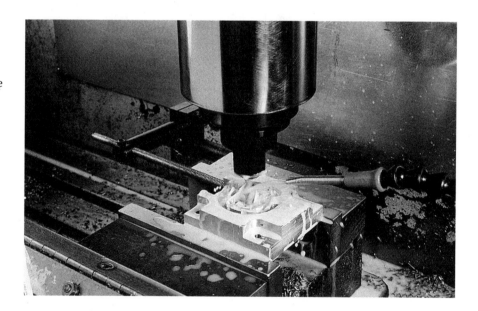

Figure 5-24
A prototype part being machined on a CNC vertical mill. Cutting fluid is being used to cool and lubricate the tool and workpiece. (Tree Machine Tool Co.)

Horizontal mills are used for peripheral milling, removing material along an axis that is parallel to the outside of the cutter. In peripheral milling, the teeth of the cutter are arranged on the *periphery* (outer edge) of the cutter, Figure 5-25. Vertical machines are used for end-milling keyways and slots.

Milling machines are sized according to the distance that the table travels left to right. Number 1 machines have a table travel of 22 inches; number 6 machines have a 60-inch travel.

Cutters of many different sizes and shapes are used to machine vertical, horizontal, and angular surfaces. Milling can be done on both the outside and inside of flat or cylindrical objects.

Figure 5-25
Cermet cutting inserts installed on the periphery of a milling cutter being used to machine a steel workpiece. (Kennametal, Inc.)

In horizontal milling, cutters can either turn *into* or *away* from the work. Since the horizontal machine can cut in both the horizontal (left and right) or vertical (up and down) positions, more horizontal machines than vertical machines are found in metal-machining facilities.

There are two types of operations for horizontal milling—conventional ("up") milling and climb ("down") milling. In both up and down milling, the cutter is positioned on top of the workpiece. With ***up milling,*** the workpiece is fed in the direction opposite the rotation of the cutter. In ***climb milling,*** the work moves in the same direction as the cutter's rotation. See Figure 5-26.

Up milling is the method used most frequently in manufacturing. In this method, the teeth of the milling cutter approach the work with a gradual sliding action that eventually leads to the tooth biting into the work. This produces thin chips and creates milling marks on the workpiece when each tooth begins its cut. Up milling is the preferred method for machining castings or parts with hard surfaces.

Down milling is sometimes called "climb milling," because the cutter rides down on top of the workpiece. In this process, the cutter turns in the same direction that the workpiece is travelling. This means that full downward pressure is placed on each tooth as it contacts the workpiece. There is no gradual slivering as in up milling—the cutter bites into the work with a thick cut that tapers off at the end of the stroke. The downward pressure has the added benefit of helping to hold the stock down under the cutter. Down milling doesn't produce mill marks from the cutter like up milling.

Down milling is a more dangerous approach than up milling and should not be attempted with worn, dated machines that have a great deal of play in the table.

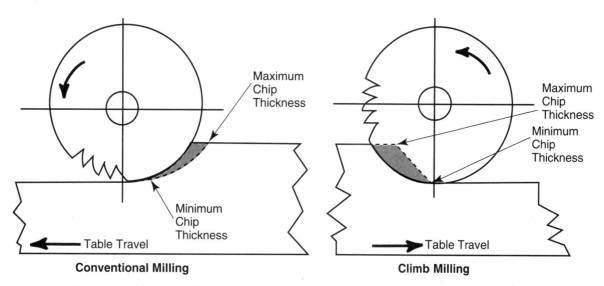

Figure 5-26 Conventional ("up") milling compared to climb ("down") milling. Note opposite direction of table travel in the two systems.

Milling cutters

There are many different shapes and sizes of milling cutters. *Arbor-type cutters* are common. These are mounted directly on the machine arbor (spindle). Other cutters, called **shank-type cutters,** are fastened directly *into* the spindle on the milling machine. Still other cutters, called *facing-type cutters,* are bolted directly on the nose of the machine spindle.

With the vertical mill, the cutter is mounted on a vertical spindle, in a manner similar to the mounting of a drill in a drill press. The cutter normally used in vertical milling is called an **end mill,** or face mill.

Face milling cutters move down onto the workpiece, so the teeth on both the end and outer periphery are used to do the milling. Vertical milling cutters look something like a large-fluted drill with a flat end.

Cutting speed

The speed at which a milling machine cuts depends on the type of cutter, the feed rate (depth of cut), the rotational speed in revolutions per minute (rpm) of the cutter, the condition of the machine, and the type of material being machined. An approximate method for calculating the speed of a cutter is described in the following paragraphs.

The diameter (d) of a drill or cutting tool is expressed in terms of inches. It is possible to determine the cutting speed (cs) of a drill or cutting tool, or the feed rate in feet per minute (fpm) at which a material can efficiently be machined. This is done by multiplying the rpm of the machine times π times d, then dividing by 12. The formula reads as:

$$cs \text{ or } fpm = \frac{\text{rpm} \times \pi \times d}{12}$$

To determine the rpm of a cutter, when the *cs* of the metal to be machined has been determined, use the following formula:

$$\text{rpm} = \frac{cs \times 4}{d}$$

Here's how these formulas could be used: suppose that you had to determine the correct rpm setting for a 3 in. diameter end mill that is to be used to machine a part made of SAE 1095 (tool steel). Determine the rpm by multiplying the cutting speed by 4, then dividing the result by the cutter diameter in inches. If the recommended cutting speed for SAE 1095 is 70 *fpm*, here's how you would substitute into the formula:

$$\text{rpm} = \frac{70 \times 4}{3} = \frac{280}{3} = 93.33$$

This tells us that the spindle speed should be set as close to 93 revolutions per minute as possible to obtain maximum efficiency. It is important to know both the best speed for the cutter (measured in rpm) and the best feed rate for a material. The feed rate is expressed in *feed per tooth during each revolution*, (ftr). The effective selection of the best rpm and ftr results in removal of the greatest amount of stock and the best finish.

Drilling

Drilling is an efficient machining process for producing holes in manufactured products. Drilling can be done on just about any type of manufacturing material. The basic process is always the same, but the speeds and designs of the drills used will differ.

When *drilling*, a hole is created by turning a drill in the spindle of a machine tool (both the drill press and the lathe are used for drilling). The drilling operation produces chips that are carried out of the hole through special channels (*flutes*) in the drill. When multiple-spindle drill presses are used in production, as many holes can be drilled as there are spindles on the drilling machine. This process is called *gang drilling*. Drill presses can also be used for other operations, such as reaming and counterboring.

The most common type of drill used for metals is called a **twist drill**. It is made by forging the basic form, and then twisting the blank to achieve a spiral shape. Twist drills are normally made of high-speed steel or carbon steel. The life of a drill can be extended by coating it with titanium nitride. There are many different types of drills, reamers, and other cutting devices used to make holes. Some of these are shown in Figure 5-27.

For use, the shank of a twist drill is held in the drill press, lathe, or mill. Drill presses equipped with a chuck will hold straight shank drills. Some larger drill presses don't have a chuck; instead, they have a spindle with a tapered hole that accepts a tapered-shank drill.

Figure 5-27
Twist drills and other rotational cutting devices. (National Twist Drill)

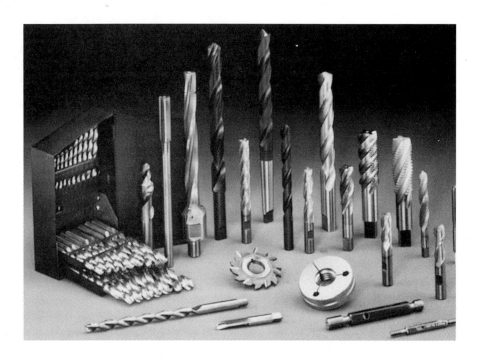

The main body of the drill usually has two grooves, called flutes. The purpose of these flutes is to create spiral channels that run the length of a drill body to carry chips from the cutting area.

The shank of the drill may be tapered or straight. The angle at the tip of the drill (the *tip angle*) varies, depending on the material drilled. The tip angle is the included angle formed when measuring from one side of the tip of a drill to the other. Most metals require an included angle of from 90-118°. Plastics require a 60-90° angle.

When a drill becomes dull through excessive use or abuse, it must be sharpened. While there are several different angles on the drill that are produced, only the lip, point angle, and chisel edge angle are important when sharpening.

Drills can be used to make holes as small as 0.005 in. (0.127 mm) and as large as 4 in. (102 mm) in diameter. Holes larger than this are seldom drilled. Drills can be purchased in many different lengths. Long drills are used in a process referred to as *gun drilling* or deep-hole drilling. Figure 5-28 shows a series of deep holes drilled through a thick steel plate.

Gun drilling, as the name indicates, was originally used to drill gun barrels. The ratio of hole depth to diameter in deep-hole drilling is often greater than 300:1. See Figure 5-29. As drilling proceeds, the long drill shank is supported by pads that slide along the inside of the hole. This enables the gun drill to be self-centering. Cutting fluid is forced under pressure through the center of the drill to aid in cooling and to flush chips from the hole.

Drill speeds vary according to drill size—smaller drills can be run faster than larger drills. However, drills are easily broken, so care must be taken in powering the drill into the workpiece.

Figure 5-28
The series of holes running the length of this steel plate were made with the process known as gun drilling, or deep hole drilling. (Betar, Inc.)

Figure 5-29
This gun drilling machine is capable of drilling holes with a diameter as small as 0.055 in. (1.4 mm) or as large as 2 in. (50.8 mm) to a depth of as much as 60 in. (1524 mm). The finish quality of the holes eliminates need for deburring or reaming. (Eldorado)

Drill sizing systems

Twist drill sizes are designated by one of four systems: numbers, letters, fractions, or millimeters. The largest number-sized drill is number 1, which is 0.228 in. (5.8 mm) in diameter. The smallest number sized drill, number 80, is 0.0135 in. (0.342 mm) in diameter. Letter sized drills are classified from A to Z. The smallest letter sized drill is the A drill, which is 0.234 in. (6 mm) in diameter. The Z drill is 0.413 in. (10.4 mm) in diameter. Fractional drills are available in sizes from 1/64 in. to 3 1/2 in. (0.397 mm to 89 mm). Metric drills are available in sizes from 3.0 mm to 76.0 mm (0.118 in. to 2.992 in.).

When drilling metal, it is important to remember that the chips must flow out of the flutes in the direction *opposite* that in which the drill is turning. The drill must turn in a clockwise direction, so that the lips of the drill cut into the metal. If the drill is turning in the wrong direction, the operator will be *burning a hole* in the stock, rather than drilling it. Cutting fluids are often used in drilling to assist the removal of chips from the cutting area, to provide lubrication to the cutting surface, and to aid in cooling.

When drilling metals, it is often necessary to make an indentation with a center punch to keep the drill from wandering off the point where the hole is to be located. For soft materials, such as wood or plastic, a center-punched indentation is seldom needed. Some drills are ground with a small bullet-type tip that enables the drill to be self-locating. Another point to remember when drilling metal is that it is often good manufacturing practice to drill a hole in two stages. This is particularly true if the hole is large and the metal is hard. The first stage is to drill a small pilot hole, perhaps half the size of the final hole. This will help to guide the larger drill, making the cutting process much easier.

In addition to the common drill press and the multi-spindle gang drill, there are several other drilling machines found in industry. Each of these is sized according to the largest workpiece diameter that can be placed on the table of the press. Radial drills are used for drilling parts that are often as large as 10 feet in diameter. Other units, called **universal drilling machines**, have adjustable heads for drilling holes at an angle.

A sophisticated, ultrasensitive microscopic precision drilling machine has been used in industry to drill holes as small as 0.0001 in. (0.0025 mm). With such a machine, it would theoretically be possible to drill as many as 25 holes in the diameter of one human hair.

Reaming

Reaming is a final finishing process that improves the dimensional accuracy and surface finish of a drilled hole. Most drilled holes have a relatively rough surface finish. Reaming produces the smoother finish required for the insertion of bushings or bearings.

Reamers are long multiple-edged cutting tools made of carbon tool steel or high-speed steel. There are many types of reamers. Some are turned into the drilled hole by hand. Others can be inserted into the hole using a drill press, milling machine, or tailstock spindle of a lathe.

The *hand reamer* has a tip that is flat, rather than being angled like a drill, and has straight or helical flutes that run several inches up the shank. Hand reamers cut on their peripheral edge. They are turned with a T-handle tap wrench.

Machine reamers have either tapered shanks or straight shanks. They are used when greater cutting action is desired. Machine reamers are constructed with a beveled end that cuts somewhat like a drill when entering a hole.

The speed of reaming depends on the type of reamer, the material, and the method (whether it is being done by machine or hand). Machine reaming is normally done at about two-thirds the cutting speed (rpm) that would be used with a drill of the same diameter. The *feed rate* for machine reaming is about three times the rate used in drilling. Cutting fluids should be applied when reaming holes. Holes that are to be reamed must be drilled slightly undersize.

Tapping

Tapping is the process used to cut threads inside a hole (*internal threads*). Tapping is done with a *tap,* a shanked tool with rows of cutting teeth separated by flutes. See Figure 5-30. Tapping can be done by hand, with the tool held in a T-handle tap wrench, or with a machine such as a drill press, automatic screw machine, or lathe. Automated tapping work cells can produce as many as 500 threaded nuts per hour. Hand taps are made of carbon steels. Production taps are made of high-speed steels.

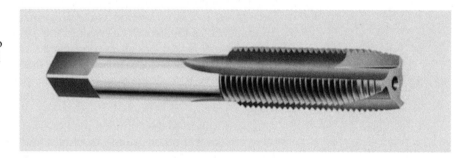

Figure 5-30
A tap can be used manually or on automated machinery to cut internal threads. (National Twist Drill)

To produce a tapped hole, you would first drill an *undersized* hole, according to the dimension specified on a tap drill chart. Such a chart specifies the drill size that is necessary for a particular type of thread.

There are three common types of hand taps: taper, plug, and bottoming. The three are the same, except that the taper tap has the longest chamfer. This makes it easier to start cutting the thread using this tap. Bottoming taps have no chamfer, and are used for tapping **blind holes** (holes that don't run *through* the workpiece). Bottoming taps are normally used when machine tapping.

Often the three taps are used consecutively: taper, then plug, then bottoming. The three differ in the degree of taper on the end of the tap. The bottoming tap has the least amount of taper.

Tapping is done by starting the tap into the hole carefully, taking care to hold it perpendicular to the workpiece. The tap is turned for several revolutions, then backed out to the top of the hole to remove the chips that have been produced. Most taps move the chips back out to the top of the hole using their flutes. A *rake angle* on the end of each row of teeth provides wider flutes at the bottom of the tap. This helps to remove chips when the tap is backed out of the hole. If the chips are not removed, it is easy to break a tap. Figure 5-31 shows the rake angle of two different taps.

Grinding

Grinding is a cutting process that uses abrasive particles to perform the cutting action. Often, these particles are bonded together and shaped into cylindrical grinding wheels. Abrasives also may be bonded to a cloth backing to make sheets, cylinders, disks, or belts. Grinding can even be accomplished by using a paste compound with a binder that clings to the separate grains. Abrasives are also used in such processes as abrasive jet machining and waterjet machining.

Figure 5-31
The end views show two different rake angles used for taps. (National Twist Drill)

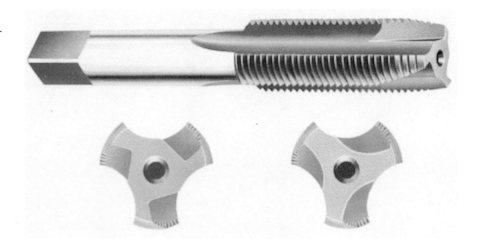

There are two types of grinding, rough and precision. ***Rough grinding,*** also called offhand grinding or snag grinding, is used for rapid material removal on castings, forgings, and welded parts. It is used to remove undesired areas such as parting lines or flash on castings, but is not used for accurate finish work. Rough grinding is normally done with a bench or tool grinder.

Precision grinding is a process that is often used on materials that are too hard to cut with conventional tools. Precision grinding machines are able to grind flat surfaces, cylinders, threads, or cutting tool blades. The gears shown in Figure 5-32 were precision-ground.

Precision grinding requires working to close tolerances and is an effective process for producing a fine finish on the workpiece. The major forms of precision grinding used in industry today are *surface grinding, internal cylindrical grinding,* and *external cylindrical grinding.* Figure 5-33 shows a CNC gear grinder used for external grinding, while Figure 5-34 shows a finished gear that has cycled out of the machine. The 17-inch-diameter aircraft gear in Figure 5-35 is being ground using a 6-axis CNC gear grinding machine that is capable of grinding up to 20-inch-diameter gears.

Grinding wheels

Grinding wheels are used on both rough and precision grinders. Coarse wheels are used for roughing, or rapid material removal. The wheels used for precision grinding are more dense, which makes them stronger and more durable. Abrasives used to manufacture the wheels can be either natural or synthetic.

Natural abrasives include sandstone or quartz, emery, corundum, garnet, and diamond. Natural abrasives are seldom used in production grinding operations. Synthetic abrasives include silicon carbide, aluminum oxide, diamond, and boron nitride.

Grinding wheels are made from abrasive grains held together with a binding agent. There are several major types of binders that are used today: vitrified, silicate, resinoid, metal, rubber, and shellac.

Figure 5-32
Precision grinding provided the fine finish on these two gears, which were ground to very close tolerances. (National Broach & Machine Co.)

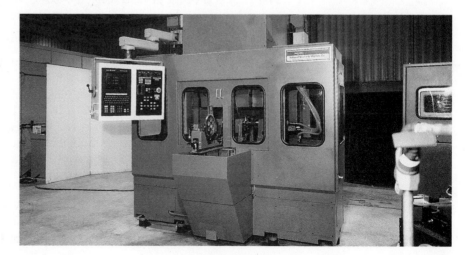

Figure 5-33
This CNC external grinding machine is used to precision-grind the teeth on large gears. (National Broach & Machine Co.)

Figure 5-34
At right is a finished gear that has cycled out of the grinder for unloading. Note the gear blank positioned inside the machine, ready for grinding. (National Broach & Machine Co.)

Figure 5-35
This large aircraft gear is being ground to close tolerances on a 6-axis CNC machine. Note the thin grinding wheel removing material from teeth at the top of the gear. (National Broach & Machine Co.)

Vitrified-bonded wheels are the type most frequently used. They are strong and rigid, and are not affected by water, oils, or acids. Vitrified bonded wheels are designed to operate at speeds approaching 6500 sfpm (surface feet per minute).

Silicate-bonded wheels use sodium silicate as a binder. The mix of binder and abrasives is compressed in a metal mold, then heated for a prolonged length of time. Silicate wheels wear faster than vitrified wheels, because their abrasive grains are released more quickly. Silicate-bonded wheels are recommended for use in such applications as sharpening cutting tools.

Resinoid-bonded wheels use Bakelite™ or Resino™ as a binder. These wheels are very strong and tough, and can operate at high speeds.

Wheels with metal or rubber binders are used for specialty applications. Metal binders are used with diamond abrasive, for grinding hard materials, or for electrolytic grinding. The rubber binder with embedded abrasive is rolled out into sheets, then the wheels are die-cut and vulcanized. Rubber-bound wheels can be made as thin as 0.005 in. (0.127 mm). These wheels are designed for use at high speeds (often exceeding 15,000 sfpm). High-speed rubber wheels are designed primarily for use with materials such as plastic, porcelain, and tile. They are also used on cutoff saws for cutting metal.

When a fine finish is desired, shellac-bonded wheels are recommended. Shellac wheels are made by compacting shellac and abrasives in steel molds. Shellac wheels are thin and flexible.

All metal-grinding processes generate chips and produce heat. Grinding fluids are used to cool the workpiece, and to carry the chips away from the cutting area. This helps to keep the wheel clean so that the open pores between the abrasives can continue to perform their cutting action.

Surface grinding

Surface grinding is a form of precision grinding that is done on flat workpieces. It is a good process for producing fine finishes to close tolerances. Surface grinding is also used for cutting hard materials that are not machinable by conventional methods. Examples would be ceramics, tile, glass, and composite materials.

This process is done using a machine called a surface grinder, which is constructed somewhat like a milling machine. Surface grinders consist of a machine base, worktable, and arbor, and are made with spindles arranged in either a vertical or horizontal position. They are sized by the length of workpiece that can be ground. Horizontal machines, which position the grinding wheel above and parallel to the worktable, are common. See Figure 5-36.

The grinding operation usually begins by securing the workpiece to the worktable of the machine with a magnetic chuck. The worktable moves longitudinally (from left to right), while the grinding wheel removes stock the length of the workpiece. The table position is adjusted laterally (front to back) after each stroke. Grinding is done with cutting fluids to keep the wheel cool and clean, Figure 5-37. The coolant helps to extend the life of the wheel and keeps the working surface cool. Water-soluble coolants and water-soluble oils are preferred by most manufacturers.

Surface grinders can also be used to grind workpieces with irregular shapes, Figure 5-38. The workpiece shown in the photograph was cast with a

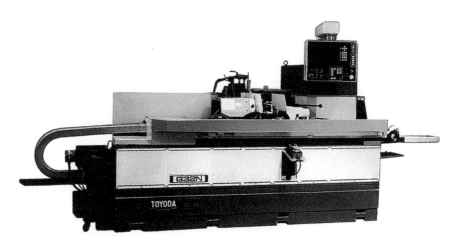

Figure 5-36
This CNC cylindrical grinder can be programmed with up to nine different grinding cycle selections. It can be used effectively for single-part grinding or continuous runs. (Toyoda Machinery Co.)

Figure 5-37
Cutting fluids are used in grinding to cool the wheel and wash away particles of ground-away stock. (DoAll Company)

flat base. It was easy to clamp to the worktable, using the magnetic chuck. Sometimes parts have irregular bases, which can make necessary the use of a special workholding device, called a *fixture*. The fixture is then clamped to the table.

Some applications require the grinding of non-ferrous metals, such as brass or aluminum, or nonmetallics like plastic. Sometimes this is done using fixtures or steel braces to prevent movement. In other instances, particularly when the material is very thin, an operator may use double-faced tape to secure the workpiece to the table.

Figure 5-38
A surface grinder can be used on workpieces with irregular shapes. This workpiece is clamped in a magnetic chuck that holds it tightly in place on the worktable. (DoAll Company)

Cylindrical grinding

Cylindrical grinding is used on workpieces with curved surfaces (cylindrical shapes). Cylindrical grinding can be done on the outside of a workpiece (external), or on the inside (internal). Figure 5-39 illustrates some of the different types of cylindrical grinding applications.

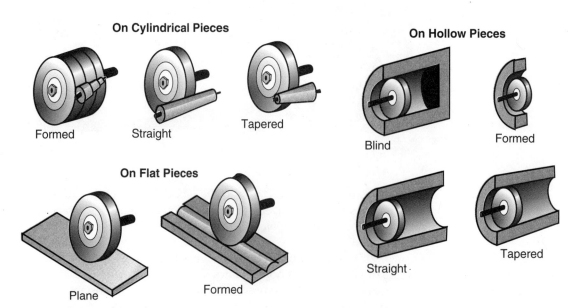

Figure 5-39   Various grinding applications can be performed on cylindrical, flat, or hollow workpieces.

External cylindrical grinding uses a grinding wheel which runs over the outside surface of a workpiece. The work is secured between centers or in a chuck.

Internal cylindrical grinding is done when the surface of a hole must be accurately smoothed. This can be accomplished with a tool post grinder attached to a lathe or with a specialty grinding machine.

Another type of cylindrical grinding is called **centerless grinding.** This form of grinding can be used for external or internal applications. Figure 5-40 illustrates the basic method used for external centerless grinding: the workpiece is held on a work rest blade that is positioned between a grinding wheel and a regulating wheel. The regulating wheel presses against the workpiece, causing it to rotate. The grinding wheel grinds the surface. Products such as lathe centers and roller bearings are often ground using this process.

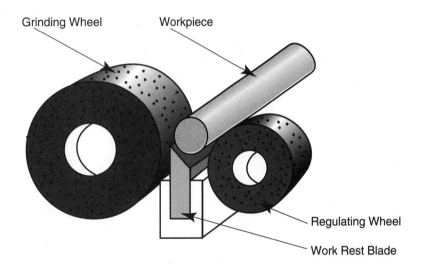

Figure 5-40
The basic arrangement used for external centerless grinding.

Internal centerless grinding is used to finish the inside of a hole in a cylindrical part. There are several variations of this process. The most common method involves positioning the tubular workpiece between three rolls: a pressure roll, a support roll, and a larger regulating roll. The regulating roll turns the stock, while the other two rolls hold it tightly against the regulating roll. The grinding wheel is attached to a shaft that travels inside the tubular part. Figure 5-41 shows how the three rolls work together to hold the tubular part.

Abrasive Jet Machining

Abrasive jet machining is a process that appears, at first glance, to be much like sandblasting. While there are some similarities, there are also major differences.

Sandblasting is a process that uses air pressure to blast sand particles against the surface of a workpiece to clean it or dull its finish. **Abrasive jet machining** is a grinding process that suspends tiny particles of abrasive material

Figure 5-41
Three rolls—pressure, support, and regulating—hold the tubular workpiece for internal cylindrical grinding.

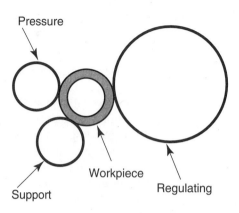

in a low-pressure stream of gas (dry air, carbon dioxide, or nitrogen) sprayed through a sapphire nozzle. This process is used to etch, scribe, groove, and cut holes or slots in hard metals and in such nonmetallic materials as ceramics and glass. See Figure 5-42.

Figure 5-42
Abrasive jet machining is a versatile process with many uses. Here, it is being used to etch and clean the surface of a part. (S. S. White Industrial Products, Inc.)

Many different types of abrasives can be used. When very hard materials are to be machined, typical abrasive choices would be silicon carbide or aluminum oxide. For machining a soft material, a soft abrasive such as bicarbonate of soda would be used. Since the abrasive particles lose their sharpness during use, they cannot be reused.

The pressure that is used for abrasive jet machining is low enough that an operator can safely pass a hand through the abrasive stream. Velocity of the gas stream ranges from nearly nothing up to 1000 fps. The abrasive stream is directed toward the workpiece by guiding the spray nozzle.

Abrasive jet machining is ideal for materials that are heat-sensitive. Conventional grinding processes create heat, so they cannot be used with such parts. Because of the low pressure employed, however, abrasive jet machining is a relatively slow process. Cutting rates are about 0.001 in. (0.025 mm) per minute.

Applications of abrasive jet machining include drilling and slicing thin wafers of hardened metal, etching numbers on parts, removing broken taps from holes, and deburring.

Waterjet Machining

Waterjet machining is a process that uses a high-velocity stream of water to cut materials ranging from paper to stone or metals. The pressurized water stream, by itself, can cut and slit porous materials such as wood, paper, leather, brick, and foam. When abrasives are introduced to the stream of water, the process is effective for cutting hard materials. Abrasive waterjet systems have been used commercially since 1985 for cutting rocks, metals, glass, ceramics, and composites with great speed. See Figure 5-43. A waterjet system can cut hard materials in thicknesses from 0.003 in. (0.08 mm) to 1 in. (25 mm) or more, and softer solids up to 10 in. (25.4 cm) in thickness.

Typically, the process uses focused jets of water 0.006 in. to 0.008 in. (0.15 mm to 0.20 mm) in diameter, pressurized to as high as 55,000 psi (379 225 kPa). In the automotive industry, materials such as carpet, fiber-reinforced plastics, sheet molding compound, reaction injection molded parts, thermoplastics, and fiberglass are being cut with waterjet equipment. Microprocessor-controlled waterjet cutting systems are being used all over the world to cut printed circuit boards.

Waterjet machining is being used on metals for deburring, descaling, and degreasing. It is also used to automatically strip coatings from wire at low pressure, which avoids damaging the wire. For large-diameter cables, stripping takes about 5 to 10 seconds per cable section. This is about five times faster than traditional thermal or mechanical stripping techniques.

The principle of waterjet machining—removing material through erosion from a high-velocity, small-diameter jet of water—was discovered by accident: workers in steam plants observed that pinhole leaks in pressure steam lines had enough power to cut quickly and cleanly through a wooden broomstick. The technology was perfected and patented as waterjet machining in 1960, and was first used in industry in the early 1970s.

An advantage of waterjet machining is that the cutting tool never breaks and never needs sharpening; however, the abrasive action causes nozzles to wear quickly. The sapphire nozzles used by one manufacturer are rated at a useful life of between 50 and 200 hours. Another manufacturer has introduced a new design with a rated useful life of 500 hours, Figure 5-44.

Waterjet machining produces little or no dust and generates no heat during cutting, making it ideal for certain types of applications. An interesting feature of this process is that the cut can be begun at any location in the part. Not many cutting processes can accomplish this.

The equipment required for waterjet machining consists of a large electric motor driving an oil pump. An intensifier is used to increase the water

Figure 5-43
Abrasive waterjet machining can rapidly cut through hard materials, such as this steel plate being cut for use as a saw blade. The tooth configuration being cut is shown on the blade segment in the background. (Ingersoll-Rand Waterjet Cutting Systems)

pressure about forty-fold. Additional parts of the system consist of an accumulator and the waterjet nozzle. The water is adjusted by regulating low-pressure oil actuators. Often, the nozzle is used as an *end effector* (tool on the end of the arm) of an industrial robot. For safety reasons, human operators must be kept out of the work area. A production setup for waterjet machining is shown in Figure 5-45.

Major industries that use waterjet cutting include:

- Δ Aerospace.
- Δ Automotive.
- Δ Electronics.
- Δ Nonwoven textiles.
- Δ Food products.
- Δ Paper and corrugated board.
- Δ Shoe and garment products.
- Δ Building products.

Figure 5-44
This abrasive waterjet nozzle design has a rated life of 500 hours, a far longer period than those previously available. (Ingersoll-Rand Waterjet Cutting Systems)

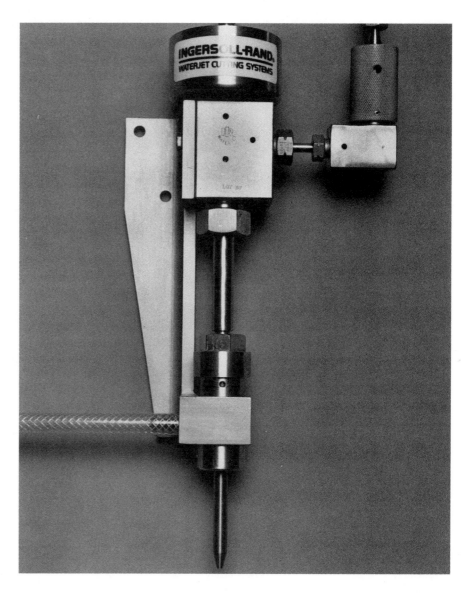

Figure 5-45
A production setup for waterjet cutting from large workpieces. (Ingersoll-Rand Waterjet Cutting Systems)

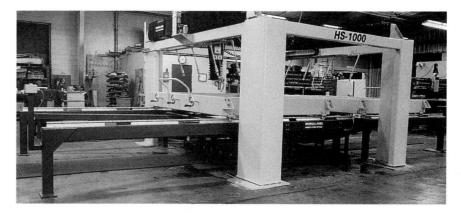

Laser Cutting

The *laser* is used for a variety of processes, including cutting, drilling, welding, heat treating, soldering, and wire stripping. Metal cutting is the single largest application of the laser in manufacturing, and is normally done using a carbon dioxide (CO_2) gas laser, Figure 5-46.

Figure 5-46
A gas laser can cut metals quickly and cleanly. Lasers are also used for drilling, welding, and a variety of other operations. (Rofin-Sinar, Inc.)

In a typical gas laser, a mixture of carbon dioxide and nitrogen is stored in a glass *lasing tube* at low pressure. This mixture is referred to as the *lasing medium.* When high voltage is transmitted from a power source to electrodes in the tube, the discharge excites the nitrogen molecules. These excited molecules, in turn, cause the single carbon atom in each CO_2 molecule to vibrate back and forth between the two oxygen atoms. The vibrating CO_2 molecule gives up energy as it changes from one energy vibration pattern to another. *Light* is then emitted in the form of photons. As each photon moves along to the next excited molecule, it stimulates that to give up a photon as it changes to a lower energy vibrational state. This causes a stimulated emission of still more photons.

Figure 5-47 illustrates what happens when mirrors are placed at both ends of the lasing tube. As light is reflected back and forth, the mirrors cause the photons to form into a highly *collimated* (focused) beam. One of the mirrors doesn't reflect all of the light, so some passes through (is *transmitted*) in the form of a laser beam. The beam is not continuous; it is produced in short, rapidly repeated pulses.

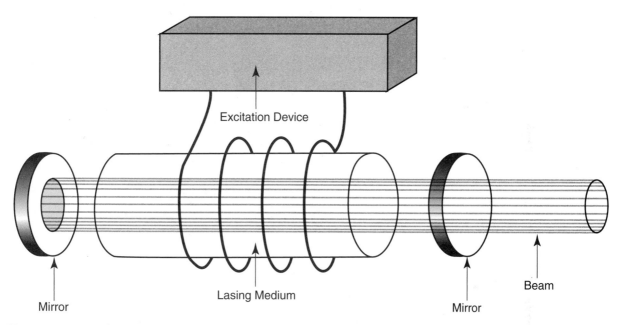

Figure 5-47 Basic laser setup. The mirror at the right side of the illustration is not fully reflective, so it allows the beam of laser light to pass through.

The laser beam that is transmitted for cutting is only 0.005 in. to 0.012 in. (0.127 mm to 0.30 mm) in width. This means that a minimal amount of heat will be generated in the area of the *kerf* (cut). This makes the laser a good means of cutting small holes, narrow slots, and closely spaced patterns.

The narrower the kerf, the faster that the laser can cut. Narrower kerfs require less energy from the laser and result in less heat on the workpiece. Figure 5-48 illustrates a typical laser cutting system.

One currently available CO_2 laser cutting system can be used to make a high-quality cut in steel up to 0.375 in. (9.5 mm) in thickness. A low-quality cut can be made in steel as thick as 0.5 in. (12.7 mm). Aluminum and brass up to 0.125 in. (3.17 mm) thick can be cut.

With laser cutting, there are no cutting tools to break or wear. Power capacities above 105 watts/in.^{2} are necessary for laser cutting of metal. The most common power range for commercial carbon dioxide lasers is from 400-1500 watts.

Figure 5-48
Parts of a typical laser cutting system. The lensholder is water-cooled to prevent heat buildup that would otherwise damage it.

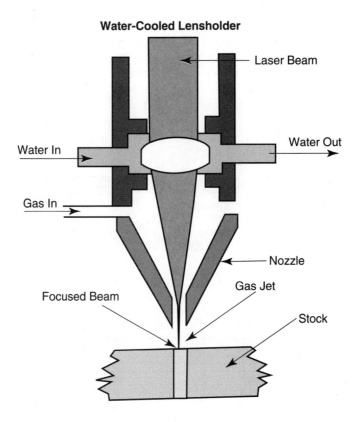

Lasers are often built as part of a turnkey (ready to operate) system that includes CNC and automated tooling. Like waterjet cutting systems, lasers can be dangerous if used carelessly. They should always be operated in a controlled environment.

Chemical Milling

Chemical milling is also known by a variety of other names: chemical machining, photofabrication, photoforming, photoetching, and photochemical machining. The terminology that is applied depends on the end use. When the process is used to etch or form metallic parts, it is usually called chemical machining, photoforming, or chemical milling. When it is used to prepare electronic substrates, such as printed circuit boards, it is referred to as photofabrication.

There are two different types of chemical milling, called chemical blanking and contour machining. *Chemical blanking* involves forming a part by etching metal away completely through the workpiece. *Contour machining* selectively etches a desired area to some specified depth.

Chemical machining can be used with virtually all metals and alloys, including stainless steel, beryllium-copper, aluminum copper-nickel, molybdenum, tin, titanium, and many other magnetic and special purpose alloys. The process is most effective when used with parts up to 0.020 in. (0.51 mm) in thickness. It is the most effective method for producing exacting detail in parts from foil under 0.001 in. (0.025 mm).

Chemical machining is basically a photographic process. The first step in the process is to thoroughly degrease and clean the workpiece. Next, a light-sensitive coating called a *photoresist* is applied to the surface.

There are two different types of photoresist, wet film and dry film. The wet film resist is a liquid applied by spraying, roller coating, or dipping the metal workpiece. Once the wet resist is applied and dried, a photographic master (negative or positive) is placed over the coated metal plate. The master and plate are then exposed to ultraviolet light. The light goes through the clear areas in the master and strikes the light-sensitive resist coated on the plate. The light hardens the resist in this area. The plate is then *etched* by running it through an acid spray bath. *Etchants* are normally sodium hydroxide (for aluminum), a hydrochloric/nitric acid blend (for steels), and iron chloride (for stainless steel). When the etching is completed, the part is removed and washed.

Since the wet method involves solvent-based materials, it is gradually being displaced by the dry method because of environmental and hazardous waste considerations. Despite their limitations, the wet films are preferred for microelectronic applications because they produce better resolution and image clarity.

Most companies involved in chemical milling now use the dry film method, since the water-based resists are easier to apply and more environment-friendly. The dry film is a light-sensitive polymer sheet that is laminated to the metal workpiece. As in the wet-film process, a photographic master is applied and the sandwich is exposed and etched. After etching, the polymer coating is chemically removed from the metal part.

Chemical machining can also be done by applying to the metal workpiece a *maskant* (another term for resist) that is *not* light-sensitive. The maskant can be a sprayed or brushed coating, a tape, or other type of material that will adhere to the surface. With this method, the maskant is scribed away in the area to be etched. Where the maskant remains, the etchant (acid) can't penetrate. Where it has been removed, the acid can attack the metal and eat it away.

Chemical milling is sometimes done to remove material or a section of the part. This might be done for ornamental purposes or primarily to reduce the weight of the part. Chemical milling can also be used for deburring (removing rough edges or burrs on the edges of the machined parts). When it is used in this way, the process is often referred to as chemical blanking.

Contouring is a method used to remove metal from surfaces of irregularly shaped parts. The part is selectively etched to a desired depth. Contouring can be done using a brushed-on mask. With this method, the steps of masking, scribing, and etching are the same. The difference is that the part is removed before it is completely etched.

Chemical machining has several advantages over producing parts by conventional stamping or punching methods. Since chemically machined parts do not undergo stresses and strains in forming, the parts are stronger and more consistent. There is no expensive tooling involved, and no retooling costs. Changes in design can be made quickly without the long lead times needed for tooling revisions.

The aviation industry is the largest user of chemically milled parts, particularly parts that are contour-machined. Other products include custom hardware, decorative panels, plaques, and instrument dials.

Ultrasonic Machining

Ultrasonic machining is a process that removes material by eroding it using vibrations generated by high-frequency sound waves. The sound waves are amplified in a funnel-shaped horn to create the desired vibrations. Attached directly to the horn is a cutting tool made in the shape of the hole or cavity that is to be cut in the workpiece. Ultrasonic machining has many similarities to ultrasonic welding, a process that is used extensively with plastics.

Ultrasonic machining can be used to replace conventional drilling, milling, turning, and surface grinding processes for very hard and brittle materials. The process isn't effective on soft materials such as aluminum and copper. It can be used most effectively in applications where conventional processes are not capable of handling materials such as germanium, ceramics, or glass. Ultrasonic machining is a slow and tedious process, but has the advantage of not generating any heat or distortion of the workpiece.

There are two methods of ultrasonic machining, both employing a vibrating metal tool. In the first method, sometimes called impact machining, the tool is covered with an abrasive slurry mixture. The tool itself never touches the workpiece. Cutting takes place through a constant vibratory action of the tool, which causes the abrasive particles to gradually erode away the workpiece, creating the shape pattern that is desired. The abrasive slurry is produced by mixing water with iron carbide, silicon carbide, or aluminum oxide powder.

The tool used for impact machining oscillates about 25,000 times each second. At the end of the tool, where it is held close to the workpiece, the actual movement due to vibration is about 0.003 in. (0.08 mm). The vibratory action of the tool produces gradual erosion in both circumferential and lateral directions.

The second method of ultrasonic machining utilizes a rotating, diamond-tipped vibratory cutting tool. The cutting action is produced by the tool itself; no abrasive slurry is required. There are several types of tools used in this method. Some resemble conventional grinding wheels, others look like twist drills or end mills. The cutting tool has the same shape as the hole or cavity that is to be produced in the workpiece.

An important consideration in ultrasonic machining is the specific frequency at which the tool must vibrate to be effective. This is called the tool's *resonance point.* Cutting takes place when the tool reaches the frequency that causes it to resonate. Experienced machine operators can tell when the tool has reached this point by listening to the sound of the machine.

Ultrasonic machining systems resemble ultrasonic welders. The workpiece is placed in a slurry pot, which is secured to the table. The vibrating tool is attached to an acoustic tool head that moves vertically. Other components of the ultrasonic machining system are the circulating pump and power supply.

Electrochemical Machining

Electrochemical machining (ECM) could really be called reverse electroplating. ECM can be used to shape a workpiece to a desired external contour, produce odd-shaped blind holes, or machine almost any shape in the cavity of a part to a specified depth. The process often provides advantages over conventional methods used to machine complicated shapes in hard and tough materials. Electrochemical machining is most cost-effective when used for high-volume production; it is not so practical for machining operations on limited numbers of parts.

Normally, ECM is selected only for those applications where conventional machining processes cannot be used. An example is the production run of aircraft struts is shown in Figure 5-49. The struts have pockets that must be machined on both sides of the arm. Due to an overhanging flange, however, the bottom pocket is obstructed, preventing the use of conventional milling. The solution was to machine the pockets electrochemically.

Figure 5-49
Aircraft struts machined with ECM. The pocket next to the left hand of the worker holding the part could not have been made with conventional machining methods. (Anocut, Inc.)

Electrochemical machining first gained popularity in the aircraft industry, where it was used to produce gas turbine engine components, Figure 5-50. The aircraft industry is still the major user of ECM, but the process is also being used in the making of dies and machine parts. The process is particularly attractive to these industries because the finished part is burr-free, eliminating the need for further processing. Parts produced by ECM retain their structural integrity, since no heat is involved in the machining process. A limitation of electrochemical machining is that the workpiece must conduct electricity. Cast iron, for example, does not work well with ECM.

Figure 5-50
This large turbine rotor was produced with the aid of electrochemical machining. ECM was first used in the aircraft industry. (Anocut, Inc.).

Here's how the ECM process works. The workpiece is the positive electrode, called the *anode*. The hollow cutting tool is the negative electrode, or *cathode*. Figure 5-51 shows the anode (an aircraft turbine rotor) and cathode in an ECM machine ready for forming.

Figure 5-51
An ECM machine ready to produce an aircraft turbine rotor. (Anocut, Inc.)

The machine is closed and a neutral salt electrolyte flows continuously between the anode and cathode. Low-voltage DC current is transmitted across the electrodes, removing electrons from the surface atoms of the workpiece. These *ions* (free electrons) are drawn toward the hollow cutting tool. The rapidly flowing electrolyte moves the ions out of the gap between the tool and the workpiece. This prevents *plating* (deposition of metal) from taking place on the cutting tool. The tool is moved automatically across the workpiece.

ECM equipment is available in both vertical and horizontal types. Figure 5-52 shows vertical model ECM machines that are used to produce small beryllium parts.

Figure 5-52
The vertical ECM units in the row at right are used to produce small parts from the metal beryllium. (Anocut, Inc.)

An important advantage of ECM is that there is no tool wear. Once the machine is set, the tool can produce thousands of parts. The simplest type of tool used is a tube made of titanium, copper, brass, or stainless steel. The outside of the tube is insulated with Teflon or epoxy. There is little standardization of cutting tools with this process.

While electrochemical machining has many advantages, it also carries with it a number of disadvantages. Setting up ECM systems for production is a time-consuming process, requiring a great deal of experimentation and testing. A more troublesome concern is the corrosive mist that ECM produces, particularly when sodium chloride is used in the electrolytic solution. Fumes must be carefully handled. However, systems for detecting leakage and handling exhaust vapors are standard on most machines. Approved safety procedures must be carefully followed for disposing of waste material associated with the process: sodium chloride-soaked cloth or paper is a fire hazard.

Electron Beam Machining

Electron beam machining (EBM) is a thermoelectric process that focuses a high-speed beam of electrons on the workpiece. The heat produced by EBM is sufficient to melt superalloys or any known material that can exist in a high vacuum. EBM is particularly useful for drilling small holes and for perforating or slotting materials that are difficult to machine, such as superalloys, ceramic oxides, carbides, and diamonds. The material removal rate is very slow.

Figure 5-53 shows the components of the electron beam machining system. The heart of the system is the electron beam gun. This generates *thermal energy* in the form of a high-velocity stream of electrons.

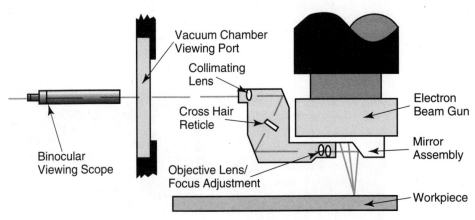

Figure 5-53 Components of an electron beam machining system.

Material is instantly removed by the melting and vaporizing action caused by concentrating the high-velocity electron beam on the workpiece. The electron beam strikes the workpiece with on-and-off pulsations several milliseconds in duration. Temperatures in excess of 12,000°F (6650°C) can be generated.

Electron beam machining is normally conducted in a vacuum. When electrons strike the workpiece, their kinetic energy is converted to heat energy. This causes vaporization. The heat is pulsed on and off, and it is this pulsing action that causes the cutting action. The equipment used for electron beam machining is the same as that used for electron beam welding.

Product applications include micromachining to manufacture computer memories and integrated circuits. Holes or slots as tiny as 0.005 in. (0.127 mm) in diameter can be drilled using EBM. Drilling time for a 0.005 in. hole would be instantaneous. Larger holes can be produced by rotating the workpiece.

One drawback of EBM is that work is limited to the size of the vacuum chamber. Another is that the process generates radiation, so X-ray shielding is necessary around the work area.

Electrodischarge Machining

Electrodischarge machining (EDM) is still another process that uses electrical energy to remove stock from metal workpieces. Electrodischarge machining bombards the workpiece with 20,000-30,000 electrical sparks per second, and thus is sometimes called *spark erosion machining*.

EDM can be used to cut hard metals and can form deep internal shapes or irregularly-shaped holes. It is an ideal process for removing material from hard-to-reach areas of parts. It is often used to produce intricate shapes on space-age materials that cannot be machined with other processes. However, EDM is a slow process, so it is not cost-effective for conventional machine applications.

Like electrochemical machining, electrodischarge machining can be used only with stock that is electrically conductive. Here's how it works:

The tool, or electrode, has the same shape as the hole that will be produced. A male electrode could be constructed in relief to resemble an intricately carved sculptured cameo. If the tool is moved into close proximity with the workpiece, the same intricate detail would be created, but in *reverse*, as a sunken-surface female form. The workpiece is placed in a dielectric fluid, such as mineral oil or a kerosene and distilled water mixture. The tool is brought close to the surface of the workpiece. The dielectric fluid flows between the tool and the workpiece. When DC current is transmitted to the electrode in short pulses, an electric arc is created between the tool and workpiece.

As the sparks strike the surface of the workpiece, they remove small bits of metal. The higher the energy in the pulse, the greater the amount of material that will be removed.

Sometimes the tool is moved downward into the workpiece; at other times, the workpiece is moved upward. This depends a great deal on the design of the electrodischarge machining equipment. Typical machines look like a vertical end mill, with a ram or quill replacing the cutter spindle. A tank is attached to the table to hold the dielectric fluid. Sometimes, vertical mills are converted to EDM machines.

One of the major uses for EDM is the machining of matched molds used for injection molding. EDM is also a popular process for shaping carbide tools, and for machining complex shapes in exotic metals.

The process can easily be adapted to computer numerical control (CNC) for automated operation. A number of CNC electrodischarge machines are commercially available. They can be programmed to automatically handle electrode changes, table movement, changes in power requirements, and dielectric flushing. The CNC operator can recall from memory a multitude of preset patterns and programs.

Travelling Wire EDM

The major difference between conventional EDM and travelling wire EDM is type of electrode used. In *travelling wire EDM*, cutting takes place using a round wire that travels through the workpiece. This type of EDM is also referred to as *electrical-discharge wire cutting*, and is sometimes compared to band saw cutting, because both processes can be used to cut intricate shapes from a metal workpiece. However, the two processes really have few similarities. The wire used in travelling wire EDM never touches the workpiece, and its movement is directed by computer numerical control. Band saw cutting is a process in which the saw blade remains in constant contact with the cutting surface, and the direction of the blade is controlled by an operator.

In travelling wire EDM, the wire moves down through the workpiece. As in conventional EDM, the sparks arcing to the workpiece from the electrode act like small teeth, each removing a small amount of metal. This results in a narrow tool path (kerf) being machined in the workpiece; the wire travels inside this kerf. The wire never actually touches the workpiece. Figure 5-54 shows a travelling wire EDM system cutting thick metal stock.

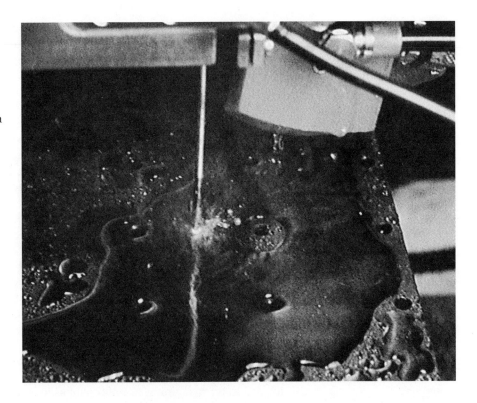

Figure 5-54
This travelling wire EDM machine is cutting a thick piece of stock. The EDM creates a narrow kerf, resulting in minimal stock loss during cutting. (Elox Corporation, an AGIE Group Company)

The position of the wire in a travelling wire EDM system is determined by computer numerical control. A typical CNC travelling wire system is shown in Figure 5-55. Some newer machines are equipped with an automatic wirefeed mechanism that will rethread the wire if a break occurs. The wire used is usually only 0.010 in. (0.254 mm) in diameter. For many years, copper was the only type of wire that was available. Now, other types of wire are gaining popularity including brass, molybdenum, tungsten, or copper coated with zinc. Previously, wire was only used once because its surface was worn away during cutting. Newer types of wire that can be reused are now being introduced.

Travelling wire EDM is a popular process for making blanking dies for sheet metal stamping. Any shape can be cut, in any direction. One of the advantages of the process is that the die metal can be hardened before it is cut to size with the travelling wire machine. This permits construction of both a punch and a die from the same die block.

One of the unique characteristics of travelling wire EDM is that thicker stock can be cut faster than thin stock. Stock 1 in. (25.4 mm) thick might be cut at a rate of about 10 linear inches per hour. This is expressed as 10 in.^{2}/hr. As a comparison, a 7 in. (17.8 cm) thick workpiece might be cut at a rate of 11.6 linear inches per hour (11.6 in.²/hr.). The thicker mass provides more area that can be attacked by sparks from the wire.

Plates more than a foot in thickness have been cut with travelling wire EDM. Travelling wire EDM is a popular process in tool and die work, or for gang cutting of stacked sheet metal parts. *Gang cutting* refers to the use of multiple heads which simultaneously cut single or multiple workpieces.

Figure 5-55
A travelling wire EDM machine operating under computer numerical control. (Elox Corporation, an AGIE Group Company)

Important Terms

abrasive jet machining
anode
arbor
arbor-type cutters
automatic feed
bed-type machine
blanking
blind hole
boring
broach
broaching
carriage
cathode
cemented carbide
centerdrilling

centerless grinding
cermets
chemical blanking
chemical milling
climb milling
collets
collimated
column-and-knee-type machine
compound rest
contouring
contour machining
cross slide
cylindrical grinding
dead center
depth of cut

drilling
electrochemical machining (ECM)
electrodischarge machining (EDM)
electron beam machining (EBM)
end effector
end mill
etchant
faceplate
facing
facing-type cutters
finishing cuts
fixture
flutes
gang cutting
gang drilling
grinding
gun drilling
headstock
heavy gage stock
horizontal milling machine
independent chuck
internal threads
ions
Jacobs chuck
kerf
knurling
lancing
laser
lasing medium
lasing tube
lathe
lathe dog
left-hand tools
light gage stock
live center
maskant
medium gage stock
milling
nesting
nibbling
notching
parting
perforating
peripheral turning
photoresist
planer
planing
plating
pot broaching

precision grinding
pull broach
punch-and-die shearing
punching
rake angle
reaming
resonance point
right-hand tools
ring-and-stick broach
ring broach
rotary shearing
rough grinding
roughing cuts
scrap
screw threads
shank-type cutters
shaper
shearing
shell broach
slab broach
slitting
slotter
spindle
stick broach
straight shearing
straight turning
surface grinding
swing
tailstock
tap
tapers
tapping
thermal energy
threading
tip angle
tool holder
tool post
travelling wire EDM
true
turning
turning processes
twist drill
ultrasonic machining
universal chuck
universal drilling machines
up milling
vertical milling machine
waste
waterjet machining
ways

Questions for Review and Discussion

1. Explain how electrodischarge machining would be used to remove material from the inside of a hollow part.
2. Discuss the relationship between the speed that an engine lathe is rotating, the depth of cut being taken, and the quality of the final finish.
3. When would up milling be preferred to down milling?
4. Why is tool wear not a consideration with electrochemical machining?
5. What type of grinding wheel would be best for use on cutoff saws that are used to cut metal? Why?

Workpieces up to 39.4 in. (1001mm) in length and 11.8 in. (300mm) in diameter can be mounted on this CNC cylindrical grinder. The computer control permits precise and repeatable operation, while monitoring tool and machine functions and recording information for later use. (Toyoda Machinery Co.)

6 Fabricating Metallic Materials

Key Concepts

△ Fabrication processes are used to join two or more materials, either permanently or semi-permanently.

△ Processes for permanently joining material typically use heat or pressure or both.

△ Permanent bonds may be made through adhesion (gluing) or cohesion (welding).

△ Mechanical joining is used when parts may have to be disassembled for adjustment or replacement.

△ Heat needed for cohesive bonding may be generated by burning a fuel or by various forms of friction.

Fabrication processes are used to join metals, either permanently or semi-permanently. Nearly every product has multiple parts that must be assembled or joined.

There are some fabrication processes which are used to permanently bond materials through adhesive bonding (glues). Others provide permanent bonding through cohesion (welding).

Many times, the assembly cannot be permanently joined. The product may need to be disassembled in order to remove and replace component parts once they become fatigued. Other products require disassembly for servicing. In cases such as these, mechanical assembly techniques may be necessary. There are hundreds of fabrication processes and techniques used in industry. The major techniques are presented below.

Mechanical Joining

Mechanical joining techniques include methods for physically attaching two or more materials to each other. Threaded fasteners, such as screws, nuts, and bolts, may be the best method of assembly on products that must eventually be disassembled. If the method of assembly requires a more permanent

167

linkage, rivets may be a more appropriate alternative. There are hundreds of unique mechanical fasteners that are used for specialized applications. There are texts on machine design that provide detailed information on which fasteners are best suited to a specific application. They are the best source of information for specific characteristics of fasteners, such as thread dimensions, pitch, and load capacity.

Since most mechanical fasteners must be inserted in a hole, a hole must be provided in each part to be fastened. Sometimes, the hole can be designed into the part and produced during such forming operations as casting or stamping. In other cases, the hole must be produced after the part is formed, so workpiece will have to be drilled. This is sometimes undesirable, because the hole can weaken the part and lead to early fatigue. If fatigue is a major consideration, a more permanent assembly may be necessary. When a permanent assembly is necessary, one of the welding processes is often chosen.

Welding Processes

Welding processes are used to join materials through the application of heat and/or pressure. In all welding processes, the edges or surfaces of the metallic parts to be joined are heated until they become molten, permitting them to be permanently fused together. The heat is created (in most welding processes) with a flame, an electrical current, or a chemical application. Some welding processes generate heat by a combination of means.

Welding can involve joining components so small that they must be viewed through a microscope. Welding is also used to fabricate structures that are too large to move and must be constructed in the field, where they will remain forever. See Figure 6-1.

Figure 6-1
High above the ground, welders fabricate the skeleton of a new office building. Welding is used extensively to join metals in all types of commercial and industrial building construction. (Lincoln Electric Co.)

In a welding application, the strength of the bond depends to a great extent on how well the surfaces have been prepared. In other words, the weld can be only as good as the surfaces that are being joined. Oxides and contaminants can ruin everything.

Two of the oldest and most important welding processes are *gas welding* and *arc welding*. They provide the foundation for many other welding processes that have evolved. Both of these processes are still extensively used today.

Gas Welding

Gas welding, often called ***oxyacetylene welding*** is accomplished by mixing two gases, oxygen and acetylene, in a welding torch. The heat that is generated is sufficient to melt and weld many soft metal alloys.

Gas welding can be used to join both similar and dissimilar metals. This includes steel, wrought and cast iron, brass, bronze, copper, and aluminum.

When two pieces of metal are melted together in the welding process, it is nearly impossible to fill all of the gaps at the *joint* (point of interface) with melted base metal alone. This is where fillers come in. **Filler metal,** such as brass or iron, is used to help fill any voids in the joint. Let's take a closer look at how gas welding actually works.

The heart of the system is the gas welding torch. The torch is held in one hand and the filler rod in the other. The pieces to be welded are placed together, and the torch is directed toward the joint. The flame melts some of the base metal.

The filler rod is then placed into the molten weld pool, and the filler melts into the pool. A bead of solid metal is created once the mixture cools. If a long bead is required, the torch and filler rod are moved along the interface joint.

Gas welding is popular for fabricating parts and structures in the field. It is particularly useful for repair and maintenance work. The only equipment that must be moved to the site are the oxygen and acetylene tanks (usually mounted on a wheeled cart for mobility), the torch, hoses to carry the gases to the torch, and regulators to control the gas flow. Gas welding is a relatively inexpensive process.

Torch cutting, using an oxyacetylene torch, is frequently done in industry. Oxygen and a fuel gas (usually acetylene or hydrogen) are mixed in the torch. Cuts can be made in materials as thick as 12 in. to 14 in. (30.5 cm to 35.6 cm), with a kerf as wide as 3/8 in. (9.5 mm). Torches mounted on cutting machines can automatically repeat cuts for production runs. The torches are normally guided along paths by programmable controllers or computer numerical control systems.

Arc Welding

Arc welding dates back to the mid-1800s. Today, this simple and versatile process remains highly popular.

There are many different types of arc welding. Some use a consumable electrode; others do not. The most common types of arc welding are shielded metal arc welding (SMAW), flux-cored arc welding (FCAW), gas metal arc welding (GMAW), gas tungsten arc welding (GTAW), submerged arc welding (SAW), and plasma arc welding (PAW). Of all of the these processes, **shielded metal arc welding (SMAW),** is most important, accounting for about half of all industrial welding work. Each of the other arc welding processes is a variation on the basic process of SMAW. Differences pertain to the type of electrode used, and the function of the flux in the welding process.

The arc is established by touching the tip of a coated electrode to the workpiece, then withdrawing it quickly to the distance required to maintain a suitable arc. The basic principle behind arc welding is the same as with most welding processes: heat from the arc melts the surface of the workpiece, and the electrode material introduces filler metal into the molten weld pool. The electrode coating, or *flux,* is vaporized by the heat of the arc to create the shielding gas. A *flux* is a material that is used to prevent the formation of oxides or other surface impurities in a weld, or to dissolve them when they form.

Shielded metal arc welding is often called "stick welding," since it uses a consumable stick-like electrode. Electrodes are approximately 0.025 in. (0.63 mm) in diameter. SMAW is fast and versatile, and requires less capital expenditure for equipment and materials than many other joining processes.

The arc that is produced in SMAW is created when electricity travels from the consumable electrode to the workpiece. The uncoated metal end of the electrode is clamped in the gripping jaws of an electrode holder.

The holder is attached to an alternating current (AC) or direct current (DC) power source. Anywhere from 50 to 300 amps of current are supplied to the holder. The workpiece is grounded, so an arc is created when the tip of the electrode strikes the workpiece. This melts both the coated tip of the electrode and the workpiece metal in the area of the arc.

The electrode consists of a filler metal alloy and a coating called flux. When the flux melts, it is changed into gas. Part of the gas combines with impurities to create a hard coating called *slag* on the surface of the weld. The gas and slag keep air out of the weldment to prevent the molten metal from oxidizing.

Careful monitoring of the amount of energy (current) supplied to the electrode holder is critical. Too much electricity will cause burning of the metal and poor fusion. Inadequate current will result in poor fusion.

The actual weld is created when an alloy mixture of base metal and filler metal from the electrode solidifies in the weld bead. Welds made are typically in the range of 9 in. to 18 in. (22.9 cm to 45.7 cm) long, and 1/4 in. (6.35 mm) wide. If additional layers are needed, the slag must be removed from the weld after each *pass* (new layer) to improve adhesion.

As in the case of all welding and cutting processes, eye protection with shaded lenses is required. Other protective clothing and equipment, such as welding gloves, should be worn as well.

SMAW is a fairly labor-intensive process, since welding is done by hand and slag must be removed after each application. In spite of this, the process is used extensively in maintenance, shipbuilding, petrochemical pipeline, and construction fields.

Some of the arc welding systems that are used in industry must be large enough to weld huge workpieces. The structural beam welder shown in Figure 6-2 is capable of handling structural steel beams weighing up to 12,000 pounds, producing massive structures for bridge construction, shipbuilding,

Figure 6-2
This large structural beam welder uses submerged arc welding to fabricate beams up to 6 tons in weight. (PHI CONRAC)

and tractor-trailer manufacturing. This machine utilizes two submerged twin-arc welders and automatically dispenses and removes flux. The system is able to weld a 1/8 in. (3.18 mm) fillet on 10 gage material at a speed of 100 in. (254 cm) per minute.

Gas Metal Arc Welding

Gas metal arc welding (GMAW) is also referred to in the welding field as "MIG" welding. **MIG** stands for *metal inert gas* welding. GMAW is similar in principle to SMAW, but uses a consumable wire electrode, rather than a stick. An inert shielding gas is supplied through a nozzle to protect the weld from contamination. This process was first developed in the 1950s, and has gained tremendous popularity. Since it is capable of productivity two to three times greater than SMAW, it has become one of the major welding processes used in industry today.

The terms MIG and TIG are sometimes confused. **TIG,** or *tungsten inert gas* welding is more correctly referred to as ***gas tungsten arc welding (GTAW).*** TIG and MIG (or GTAW and GMAW) function in basically the same way, but the tungsten electrode used in TIG is not consumed. A separate welding rod is sometimes used to provide filler metal. GMAW (MIG) is much more widely used in industry than GTAW (TIG).

There are several different shielding gases used with GMAW, but argon, helium, and carbon dioxide are most common. Welding methods vary somewhat, according to the type of gas that is used and the material that is to be welded. The shielding gas is generated to reduce contamination and oxidation of the weld area.

A consumable bare wire electrode is fed automatically through the welding nozzle, which also serves as the shielding gas dispenser. See Figure 6-3. The wire carries an electrical charge, while the workpiece is grounded. As the wire approaches the workpiece, an arc is created and the weld begins. The wire is fed by pressing a trigger on the electrode holder or nozzle. The operator moves the nozzle in much the same manner he or she would the electrode holder when using shielded metal arc welding.

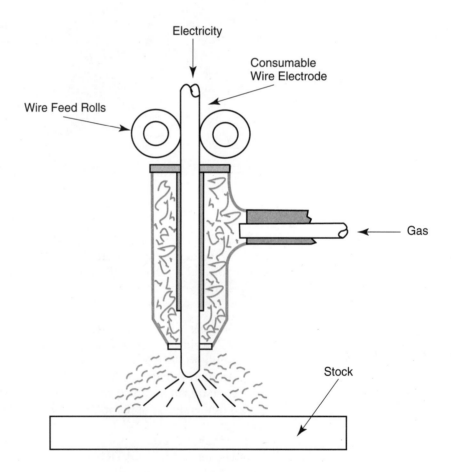

Figure 6-3
The GMAW nozzle serves a dual purpose: it feeds the consumable wire electrode to the weld pool while dispensing a shielding gas that prevents contamination of the weld.

One of the major advantages GMAW has over SMAW is that the gas effectively protects the area of the weld from contamination. The almost total absence of oxidation results in welds that have very little slag. Consequently, chipping and removal of slag between welds is no longer necessary.

Gas metal arc welding is a simple process to use, so little time is necessary to train operators. It is extensively used in the metal fabrication industry for joining both ferrous and nonferrous metals. GMAW is a process that readily lends itself to automated operation.

Resistance Welding

The simplest form of resistance welding is what is usually referred to as "spot" welding. This is a popular process for permanently joining thin-gage metal parts. See Figure 6-4. Spot welding is often used in place of riveting.

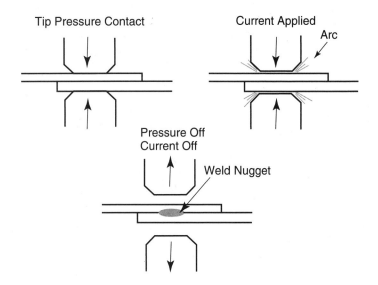

Figure 6-4
Spot welding is the most common form of resistance welding. The two pieces of metal to be welded are pressed tightly together by the electrodes and an electrical current applied. Heat from electrical resistance melts the metal and fuses it together to form the weld nugget or "spot weld."

Resistance welding is accomplished by placing the materials to be joined between two opposing electrodes. The electrodes are pressed firmly against the metal layers. When current is applied, an electrical resistance is created to the flow of that current between the electrodes. This resistance creates heat and causes the metal layers to melt and bond together. In spot welding, the result is a weld "nugget," typically 1/4 in. to 3/8 in. (6.4 mm to 9.5 mm) in diameter. Other types of resistance welding produce continuous or interrupted weld lines where the metal pieces are joined. The strength of the bond that is created depends on the temperature of the joint, the materials to be joined, and how well the materials were cleaned before welding.

One of the most extensive uses of resistance welding is *seam welding* of beverage cans. Such cans must be leakproof, and often must be able to withstand high pressures. The automatic seam welder shown in Figure 6-5

Figure 6-5
Automatic seam welding machines use wheel electrodes to resistance-weld a continuous longitudinal seam for cans. The printed cylinder form entering the machine from right is a tube that will be cut into individual can lengths after seaming. Note wheel electrode, upper center of photo. (Soudronic, Ltd.)

uses a pair of *wheel electrodes.* To make the longitudinal seam, one electrode rolls inside the can; the other, outside. Each wheel has a groove cut into its circumference. A thin copper wire can be seen just above and to the left of the center of the picture. The wire wraps around the electrode wheel. When the wheels rolls over the weld seam, the copper wire transmits current from the wheels to the work. The hot wire melts the surface of the can and then picks up molten tin from the weld seam. The purpose of the wire is to keep the wheels clean and prolong the life of the wheel.

The wheels on the welder shown in the photo are able to weld cans running through the machine at the rate of 197 feet per minute (fpm).

Butt Welding

Butt welding, also called *flash butt welding* is similar to resistance welding. Heat is generated from an arc that is produced when the ends of two parts to be welded make contact. It is an ideal process for joining rod end-to-end. A variation of this process, *stud welding,* is used to join threaded parts to plates.

Butt welding is really an arc welding process. Refer to Figure 6-6. When the parts to be joined reach the proper temperature, the materials are brought together. This upsetting action forces impurities from the joint and forge welds the two materials together. Linkage takes place through plastic deformation at the weldment. Some flash is created at the joint.

Flash butt welding machines are usually automated. Electrical currents required for steel range from 2000-5000 amps/in.^{2} The actual forging pressure can be as great at 25,000 psi (172 375 kPa).

Manufacturers commonly use butt welding to produce drills and machine tools. The process is particular useful for applications that require joining threads to shanks. Similar and dissimilar metals can be butt-welded.

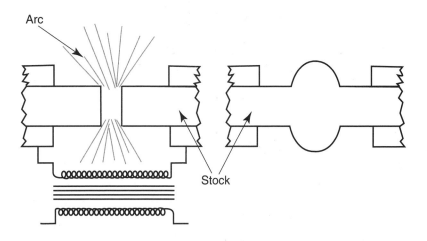

Figure 6-6
In butt welding, an electric arc formed when the two pieces are brought together provides the heat needed to melt and fuse the metal parts.

Inertia or Friction Welding

Friction welding was originally developed as a high-volume mass production process for joining similar alloys. The process has since been refined to become an effective means of joining many dissimilar metals and nonmetals. The principle of friction welding is simple: the elements to be welded are held in axial alignment while one element is rotated and the other is advanced into pressure contact. The friction developed at the point of contact creates heat. Welding occurs when the heat causes the materials to become plastic and fuse together.

The origins of friction welding dates to 1891, when a patent was issued to cover joining the ends of wire cable. Since that time, major patents covering friction welding processes have been issued in the United States, Germany, the United Kingdom, and other nations.

The process of friction welding has undergone several minor variations over the years to enhance its efficiency. This has created some confusion about terminology. The process may be called by any of a number of names, including inertia welding, stored energy friction welding, flywheel friction welding, continuous drive friction welding, or spin welding.

Two variations of the basic friction welding process are most frequently used today. One is based on the *stored energy* approach to friction welding pioneered by Caterpillar Tractor Company and AMF Corporation. This system relies on energy stored in a flywheel to rotate the workpiece and produce friction. The second variation uses direct power supplied from a continuous drive source. Both of these systems are used extensively in the automotive, aircraft, equipment manufacturing, and aerospace industries.

In inertia friction welding, one of the workpieces is clamped in a spindle chuck attached to a flywheel, and the other is held in a stationary holding device. The chuck is accelerated to a prescribed rotational speed. When the desired speed is reached, power is cut and the part held in the chuck is forced against the stationary piece. Friction between the parts causes the flywheel to decelerate, causing frictional heat. The parts are softened at the point of interface, but they are not melted. Just before the flywheel stops, the two parts bond. Force is still applied until the wheel stops. This forging force is useful in removing voids and refining grain structure.

In 1985, an Indiana company called Manufacturing Technology, Inc., acquired all patent rights to inertia welders and flywheel friction welders in the United States. The company manufactures inertia friction welders ranging in size from micro machines for workpieces smaller than 1 mm (0.039 in.) in diameter to giant models that apply a weld force of 750 tons to large workpieces. An example of the largest machines made by MTI is shown in Figure 6-7. This huge inertia welder is used to weld aircraft engine components.

Figure 6-7
Aircraft engine components are joined by inertia welding on this huge machine. For scale, note operator at left. (Manufacturing Technology, Inc.)

There are many advantages of inertia friction welding over conventional forming and forging processes. The power control drive shaft shown in Figure 6-8 was originally manufactured using upset forging, which required five straightening operations. When forging was replaced by inertia welding, the straightening steps could be eliminated, with resulting time and cost savings.

Inertia welding is a machine-controlled process that eliminates operator-caused defects. The process can easily be automated for high-volume production, or it can be adapted for use in flexible machining cells. There is no need to carefully prepare surfaces for welding. Joints can be machined, sawcut, cast, forged, or sheared. Most metals (cast iron is one of the few exceptions) and many nonmetallic materials perform well with friction welding.

Any solid forging in which a bar is upset at one end is a likely application for inertia friction welding. However, parts do not have to be solid. Figure 6-9 shows an application in which a tubular shaft has been welded to a stamped torque converter cover.

Many manufacturing firms are taking advantage of the benefits offered by inertia friction welding. Figure 6-10 illustrates the type of application that is perfect for inertia friction welding. In this operation, main injector ports for the engine of the space shuttle are welded. There are 600 welds per engine, and each shuttle has three engines.

Figure 6-8
The drive shaft blank shown at top was made by upset forging, and required straightening before it could be processed further. The two center views show the shaft and gear blank before and after being joined by inertia friction welding. Note the lack of distortion in the joined parts. At bottom is the finished shaft after machining. (Manufacturing Technology, Inc.)

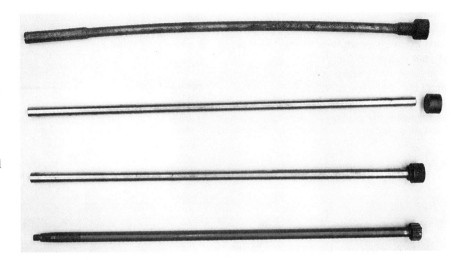

In another aerospace example, General Dynamics Corporation uses the process to weld tubular aluminum mandrels that adjust fuel-liquid oxygen ratios for the Atlas rocket engine. In an application such as this, a single surface spot of weld porosity, exposed by machining off flash, could crack a dielectric coating and impair control of fuel level. In this application, one part was spun at 4057 rpm before disengaging from the drive.

Figure 6-9
Inertia friction welding can be used with tubular parts, as well as solid components. This hollow shaft was joined to the stamped cover by the inertia friction welding process. (Manufacturing Technology, Inc.)

Figure 6-10
Installing the main injector ports in the space shuttle's engines is done efficiently with the inertia friction welding process. More than 600 welds must be made for each engine. (Manufacturing Technology, Inc.)

Friction welders can be fully automated. The CNC welder shown in Figure 6-11 receives input data on the outer and inner diameters of the workpiece, heating pressure, upset pressure, and amount of heating. All calculations and commands are then performed by the internal computer. This machine features a continuous drive (a motor, coupled to a clutch and brake) to rotate the moving piece.

Electron Beam Welding

High-quality deep and narrow welds can be produced on almost any type of metal with the *electron beam welding* process. The process generates heat by bombarding the weld site with a narrowly-concentrated beam of high-velocity electrons.

Electron beam welding can be used to weld almost any type and hardness of metal. Thicknesses range from thin foils of no more than 0.001 in. (0.025 mm) to plate as thick as 6 in. (15.2 cm). Welds can be produced at rates up to 500 inches (more than 40 feet) per minute.

Energy is produced by a gun that accelerates a stream of electrons, giving them tremendous penetrating power. Electron guns normally have a capacity ranging from 175 kilovolts (kV) to 1000 milliamperes (mA). The moving electrons produce kinetic energy, which is converted to heat as it strikes the workpiece. The concentrated beam of electrons is directed toward the workpiece in a vacuum. The greater the vacuum, the better the penetration of the beam.

Figure 6-11
This continuous drive friction welding machine operates under computer numerical control, permitting fully automated production of welded parts. (Rocon Equipment Company)

One of the major advantages of electron beam welding is that dissimilar metals can be joined with little distortion and shrinkage at the weld area. Electron beam welding also can be used to weld a wide range of metals from refractories to superalloys.

The process generates harmful rays, so careful monitoring, maintenance, and systems control is necessary. Electron beams can be projected for a distance of several meters, so the process should be isolated to ensure safe operating procedures.

A limitation of electron beam welding is its narrow field: it is not appropriate in applications where there is a wide gap between materials. Because of the narrow weld that electron beam welding produces, gaps should not exceed 0.005 in. (0.127 mm).

When first introduced, machines for electron beam welding were costly, often selling for more than $1 million. Smaller systems, at more affordable prices, are now commercially available.

Laser Welding

Lasers are gaining popularity for many welding applications, particularly where narrow and deep joints are required. Since it does not have to be done in a vacuum, **laser welding** has advantages over electron beam welding. However, in cases when *conventional* welding processes can be used, they are normally the most cost-effective choice. Laser welding is a process that should be saved for applications where the materials are difficult to weld, or where the parts are heat-sensitive.

In a sense, laser welding is really "poor cutting," and requires more power than is needed for laser cutting. Laser welding can be performed without filler material. All of the materials that are to be joined mix after they are melted by the beam.

Welding and heat treating use lasers with an output of 2000 watts or higher, because the beam that is created is not precise enough for good cutting. Both CO_2 lasers and Nd:YAG (normally called YAG) lasers are used for welding. Both produce pulsed beams. This makes them particularly useful for welding, because they can bring the metal to its melting temperature very quickly. This is an advantage when welding heat-sensitive parts, such as heart pacemakers and other electronic devices.

When lasers with higher power intensities (6 kW to 25 kW) are used, the efficiency of the system is greatly improved; deeply penetrating dense welds can be accomplished. The process is particularly suitable to producing deep and narrow welds.

For more than 20 years, the Chrysler Motors automatic transmission plant in Kokomo, Indiana, used traditional electron beam production systems to weld gear assembly components. In 1985, they replaced these systems with multi-kilowatt laser welding systems. Today, this Chrysler facility is one of the largest high-power laser welding installations in the world.

Laser and electron beam welding are similar, in that both are *non-contact fusion welding* processes. In laser welding, the heat for fusion is generated by directing a focused beam of photons (light) onto the workpiece. Laser welding does not require the use of a vacuum chamber, but it does depend on shielding the weld pool with inert gas to prevent contamination of the weld.

The laser systems typically used in the Chrysler facility feature a bottom-load/bottom-feed weld station, 6 kW CO_2 laser, static beam delivery system, pneumatic press, pick-and-place units for automatic parts transfer, PLC (programmable logic controller), and a 25-ton chiller and air compressor. The chiller cools the laser during processing.

These systems are fully automatic and do not require operator intervention or close monitoring. The two components to be welded are fed into the workstation and positioned for placement by a robotic pick-and-place system.

The part is transported to the weld station, where the welding takes place inside a safety enclosure. At no time is human exposure to the laser beam possible. Once the weld is completed, the assembly is placed on an output conveyor by the weld station pick-and-place robot.

Laser welding can be used for very precise work. For example, each of the narrow blades in the Gillette Sensor razor cartridge has 13 tiny welds. Eleven of these welds are visible as tiny dark spots on the blades. These precision welds are made by an Nd:YAG (neodymium-doped yttrium aluminum garnet) laser, operating at the rate of 3 million welds per hour.

Safety is of paramount importance when working with laser welding equipment. More eye injuries have occurred to set-up people than to operators. YAG lasers align their beam through laser optics. This requires the operators to use green-colored glasses. CO_2 lasers don't use glass lenses, so other methods must be used to direct the beam where desired. Sometimes helium-neon laser beams are used to locate the CO_2 laser beams. With these lasers, conventional plant safety glasses offer satisfactory eye protection. Few injuries have been reported from the deflection of the beam, or from actual contact with the beam.

Ultrasonic Welding

Ultrasonic welding works on the principle of changing sound energy to mechanical movement to generate heat for joining metals or plastics. The sound energy consists of frequencies well above the range of 20 Hz to 14,000 Hz (cycles per second) most humans can hear. A mechanism called a *transducer* is then needed to change the sound energy into mechanical vibrations.

For ultrasonic welding, a horn is attached to the transducer to carry these mechanical vibrations to the workpiece. The horn is tuned so it vibrates at 20,000 Hz. The horn and tip press together the workpieces to be welded, much like the electrodes in resistance welding. The vibratory energy first cleans away oxides on the surfaces of the metal. Welding occurs when the metals come into contact with each other. The welding tip is usually formed in the same shape as the final workpiece. Figure 6-12 shows a small to medium-duty ultrasonic metal welding system.

Machines are available for many different types of welding operations. The machine shown in Figure 6-13 is one of the most automatic wire splicing units available on the market. This machine has universal tooling that is automatically set up by inputs from an electronic controller.

Brazing and Soldering

Brazing and *soldering* are both old processes, dating to as early as 3000 BC. Today, both are frequently used for fabrication and joining applications in manufacturing. Both processes incorporate the use of filler metal to span the gap between materials. In soldering, the filler material is a lead-tin alloy called *solder,* while in brazing, it is normally *brazing rod*. The major difference in the two is the temperature required to melt the filler. The temperature point that differentiates soldering from brazing is generally considered to be 840°F (450°C). Solder melts below that point; brazing alloy, above it.

Figure 6-12
Ultrasonic welders, like the system shown here, convert sound energy into vibrations (mechanical energy) to heat and weld parts together. (Stapla Ultrasonics Corporation)

Figure 6-13
This automated ultrasonic welding unit is used for wire splicing. (Stapla Ultrasonics Corporation)

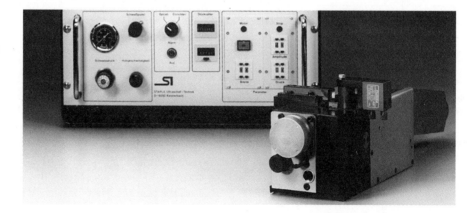

Brazing

Brazing is a joining method that requires the use of filler metal in the interface area. It joins the base metals by *adhesion* of the melted (and later cooled and solidified) filler metal—the base metals do not melt. Brazing rods are available in many different alloy compositions for use with different base metals. Brazing is a popular process for maintenance and repair of ferrous castings.

There are several different methods used for brazing. Parts can be dipped to apply filler material. However, the most common method involves heating the base metal with an oxyacetylene torch and letting the molten filler metal from the brazing rod flow into the gap between the materials to be joined.

Chapter 6   Fabricating Metallic Materials

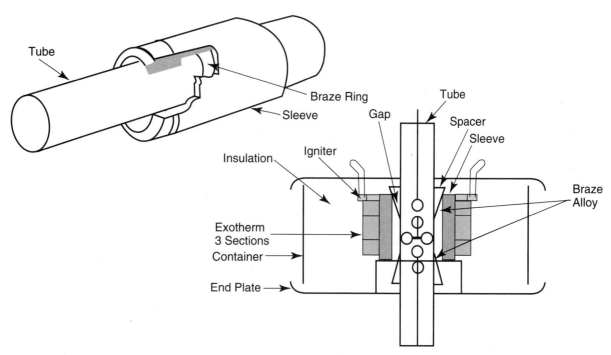

Figure 6-14   Exotherm brazing is a system in which electrical resistance to a high-frequency current creates the necessary heat for joining the parts.

Normally, filler metal is in the form of a wire or rod. It can also be purchased in sheets, in rings or other special shapes, or in powder form. *Flux* is used to prevent oxidation, remove oxides that have formed, and reduce fumes. When the filler metal is melted, it flows between the surfaces to be joined to form a bond. When the flame is removed and the joint cools, the materials are permanently joined.

When two tubes are to be joined, another type of brazing called ***exotherm brazing*** may be useful. See Figure 6-14. In this method, the heat is provided by an AC electrical coil situated close to the joint. High frequency current creates resistance and heats the joint area.

Soldering

As noted earlier, soldering and brazing are similar processes that have been in use for many centuries. Both are *adhesion processes* that form a bond between parts by filling joints with a material that melts at a temperature below the melting point of the base metal.

The filler metal known as *solder* is typically an alloy of tin and lead, although other metals are used in some cases. It is classified according to the percentage of the metals in the alloy: a 60-40 tin/lead solder would be 60 percent tin and 40 percent lead. Solder is available in several forms: solid wire, acid or rosin core wire, and bars. The alloy can be applied using a torch, induction, infrared, or dipping techniques. One dipping method often used to automatically solder electronic printed circuit boards is called wave or flow soldering.

The critical factor in *soldering* is to properly clean the metals to be joined. This is usually done with steel wool or emery cloth. Then, a flux is applied to clean and protect the work surface. The type of flux used depends on the metal being soldered. Zinc ammonium chloride is common.

Once the materials are cleaned, the metal to be joined is heated at the joint area, using a soldering gun, propane torch, or soldering copper. Solder is then applied to the work surface, at a point close to the heat source, so that it is drawn into the joint by capillary action. The heat is removed and the solder joint solidifies.

For industrial joining applications, specialized methods may be used to apply solder. Integrated circuit boards may be placed above a soldering bath, and the molten solder agitated to set up a wave that applies solder from below to form connections between components inserted in the boards. Soldering can even be done ultrasonically. A common method for coating wires on components or leads is dip soldering: the wire or component simply is immersed in a container of molten solder.

Sometimes, joints are prepared in such a way that the solder cannot show. This is called *sweat soldering*. This type of soldering is often done by applying a thin coating of solder to each of the surfaces to be joined. The materials are then clamped together and heat is applied to the joint. Heat is continued until all of the solder is melted and drawn into the joint.

Important Terms

adhesion processes
brazing
butt welding
electron beam welding
exotherm brazing
filler metal
flux
gas metal arc welding (GMAW)
gas tungsten arc welding (GTAW)
joint
laser welding
mechanical joining
MIG
oxyacetylene welding
pass
resistance welding
seam welding
shielded metal arc welding (SMAW)
slag
soldering
stud welding
sweat soldering
TIG
torch cutting
ultrasonic welding
welding processes
wheel electrodes

Questions for Review and Discussion

1. What are the disadvantages involved in designing a product that uses mechanical joining of parts, rather than adhesive or cohesive joining methods?
2. Describe the differences between butt welding and inertia welding. Which process would produce the stronger weld? Why?
3. What is the sequence of events that takes place when metal is joined with ultrasonic welding?
4. Under what circumstances would you select brazing, rather than soldering, to join parts? Why not choose welding, rather than brazing?

7
Conditioning and Finishing Metallic Materials

Key Concepts

△ Equilibrium or phase diagrams are important in selecting appropriate heat-treating processes.

△ Conditioning and finishing processes are used to full harden, partial harden, soften, and relieve stresses in metal workpieces.

△ Full hardening processes include heat/quench/temper and precipitation hardening or age hardening.

△ Case hardening processes include carburizing, nitriding, carbonitriding, flame hardening, and induction hardening.

△ Various processes are used to soften steel and relieve stresses caused by forming.

After metal parts have been fabricated, they often must be *conditioned* to adapt them for their intended use. Often parts must be *finished* for appearance or protection from corrosion. Conditioning processes alter the internal structure of a material to give it desired physical or mechanical properties. Finishing processes may convert (alter) the surface of the material by changing it chemically, or coating the surface with a protective layer.

The Crystalline Structure of Steel

To effectively use the various heat-treating processes, it is necessary to have at least a basic understanding of the phases, or changes in molecular structure, that take place when metal is heated, thereby changing it from a solid to a liquid.

Pure metals have a clearly defined temperature at which they melt, or change from a solid to a liquid. They also solidify (change from a liquid to a solid) at specific temperatures. *Alloys* solidify over a range of temperatures. When the temperature of a molten alloy drops below the **liquidus line**, it starts to change to a solid. At this point, the material becomes a mushy solid. If the temperature is lowered even more, the alloy will eventually harden as it reaches the **solidus line**. See Figure 7-1.

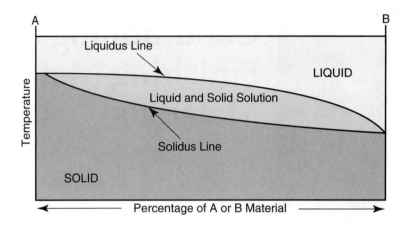

Figure 7-1
A typical graph representing phase changes in an alloy. Note that the liquidus and solidus lines reflect differing ranges of temperatures as the composition of the alloy changes.

With the many thousands of possible alloys and types of metals now available, it might seem next to impossible to accurately predict the effect that a heat-treating process would have on a particular metal. However, thanks to a tool referred to as a *phase diagram* or *equilibrium diagram,* it is easy to determine the structure of the material when it is exposed to a particular temperature. As iron or steel is heated or cooled, it goes through phase changes. The nature of the change is represented by terms that describe the structure of the material. The most common of these terms are shown in Figure 7-2.

Let's take a moment to think more carefully about each of these terms and what they mean to the process of heat treating. If you pull a piece of low-carbon (such as 0.4 percent carbon) steel from a storage rack at room temperature, it will be in solid form and will contain a small amount of carbon spread throughout the steel. At this point, the steel would be referred to as *ferrite.* Ferrite is the softest and most ductile form of low-carbon steel found on the equilibrium diagram. It is magnetic from room temperature up to about 1400°F (761°C). Ferritic iron may contain up to 0.8 percent carbon.

Figure 7-2
Phase changes take place as metal is heated or cooled. When steel is heated sufficiently, it changes to its austenitic form. Depending upon the speed of cooling and other factors, it may be transformed to one of the other structures shown.

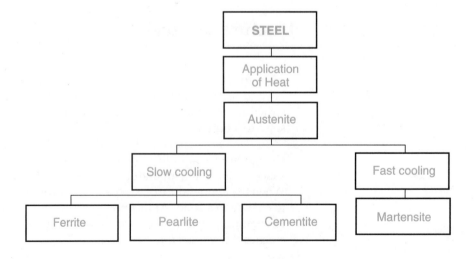

When steel reaches 0.8 percent carbon, it begins to change to another state, called *pearlite.* Pearlite is still soft and relatively ductile.

At the 1.0 percent carbon level, low-carbon steel has both pearlite and *cementite* structures. When it reaches 2.0 percent, it is truly cementite. The cementite phase is reached when steel is slowly cooled after being heated to above 1341°F (727°C), the point where it becomes *austenite.*

Austenite is the most dense form of low-carbon steel. One of the characteristics of austenite is that it has the ability not only to be deformed, but also to absorb carbon. Steel has less than 2.0 percent carbon, while cast iron has more than 2.0 percent carbon.

Another term to consider before applying what you have learned to phase diagrams is *martensite.* A martensitic structure is created when steel is rapidly *quenched* or cooled, usually by plunging it into a liquid bath.

Figure 7-3 shows a simplified form of the equilibrium diagram, which illustrates the relationship between iron and carbon. We can use this diagram to help us understand the phase changes of metal with different levels of carbon. Note that the illustration is divided into several parts. As noted earlier, metal with less that 2.0 percent carbon is considered steel; metal with more than 2.0 percent carbon is considered cast iron. The steel section of the diagram is further broken down into regions containing ferrite, pearlite, and some cementite. While the illustration shows definitive points where pearlite and cementite occur, it should be remembered that steel changes gradually from one state to another. There is overlap between ferrite/pearlite, and pearlite/cementite.

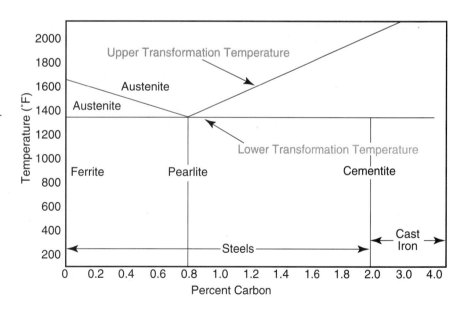

Figure 7-3
This simplified iron/carbon equilibrium diagram shows the phase changes of iron at different temperature and carbon-content levels. The lower transformation temperature and upper transformation temperature lines show where phase changes begin and are completed for iron with different carbon percentages.

Two factors are critical to phase changes in metal: the *percentage of carbon,* and the *temperature of the material.* We have discussed the percentage of carbon to some extent. Let's now think for a moment about the impact of temperature. Two additional terms are needed. The ***lower transformation temperature*** and *upper transformation temperature* represent limits for changes between phases, depending on temperature and percentage of carbon.

For example, when ferritic iron is heated it changes to austenite. The lower transformation temperature is the temperature where the material begins to change to austenite. The upper transition temperature is the temperature at which the change is complete: no ferrite exists any longer, and the iron is completely austenite. All of the area on the diagram above the upper transformation temperature line represents austenite.

The point where the two transformation temperature lines come together indicates that pearlite, with a 0.8 percent carbon content, has a *eutectoid* composition. The term "eutectic" is usually applied to an alloy that immediately changes from solid to liquid at a specific temperature. In the case of pearlite, however, the change that takes place immediately when the transformation temperature is reached is from solid pearlite to solid austenite.

Equilibrium diagrams can be obtained from steel manufacturing firms and metal suppliers. In many cases, these diagrams may seem quite complex; however, only three things really need to be known about an alloy in order to use the diagram. These are:

Δ   The sample's temperature.

Δ   The percent of carbon in the steel.

Δ   Information on any previous heat treatment that the sample has received.

Figure 7-4 shows a simplified version of an equilibrium diagram.

In the illustration, the horizontal scale represents the percentage of carbon in the sample. Temperature is shown on the vertical scale. Take a moment to locate the various regions shown on the diagram.

Figure 7-4
The relationship of temperature and percentage of carbon on material structure. The point where the upper and lower transformation temperature lines intersect represents the immediate transformation of solid pearlite to solid austenite, without any liquid phase.

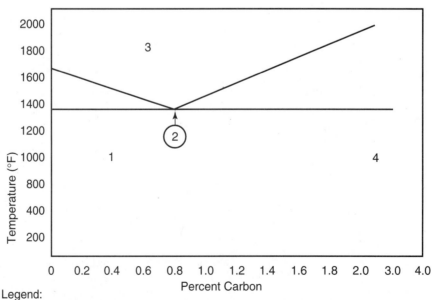

Legend:
1= ferrite,   2= pearlite to austenite transition,   3= austenite,   4= cementite

Processes Used for Conditioning

The conditioning processes used with metallic materials all involve heat, and are described generally as "heat treating." The Society of Manufacturing Engineers defines *heat treating* as "any process whereby metals are better adapted to desired conditions or properties in predictably varying degrees by means of controlled heating and cooling in their solid state without alteration of their chemical composition." Both mechanical and physical properties can be changed by heat treatment.

The controlled application or removal of heat causes the atoms of a metal workpiece to move, thereby changing the internal structure or conditioning the material. Another important aspect of heat treating is the rate and method used to cool the metal after it is heated. These also influence the internal structure of the material.

There are many different types of heat treatments, including case hardening, annealing, normalizing, and tempering. Each of these processes will be covered in this section. It is easy to get confused when thinking about heat-treating processes, because there are so many different types of alloys and methods for heat treating. One way to simplify all of this is to classify heat-treating processes by type. Generally, there are four major types of heat-treating processes:

- Full hardening of a workpiece.
- Partial hardening of a workpiece.
- Softening of a workpiece.
- Relieving of stresses in a workpiece.

Figure 7-5 lists specific processes that can be classified under each of these types of heat treating.

Heat treating is done to improve the toughness, wear resistance, machinability, tensile strength, ductility, bending quality, corrosion resistance, and magnetic properties of metal. Steel is the most frequently heat-treated metal, but iron, copper, aluminum, and many other metals can be conditioned by heat treating.

Heat treating of metals can be done using a conventional heat-treating furnace or any of a variety of other heat sources. Heat treatments can be conducted over a wide range of temperatures—the temperature required depends on the type of metal involved and the heat treatment desired. Some heat treatments are conducted at room temperature, but this is unusual. Heat-treating temperatures required for steel usually begin around 600°F (316°C).

Heat-treating furnaces are designed for either continuous or batch production. Parts are fed into continuous furnaces on a moving conveyor. *Continuous furnaces* are used for high volume production. *Batch furnaces* are used to heat-treat small quantities of parts carried on a rack or flat car that is loaded into the furnace. *Fluidized bed furnaces,* which suspend heated aluminum oxide particles in a chamber with hot gases, are also gaining popularity.

Heat treating can be accomplished with CO_2 lasers with power outputs above 500 watts. More powerful and larger diameter beams are better for heat treating of metals. Smaller beams are sharper and better used for such processes as cutting and drilling.

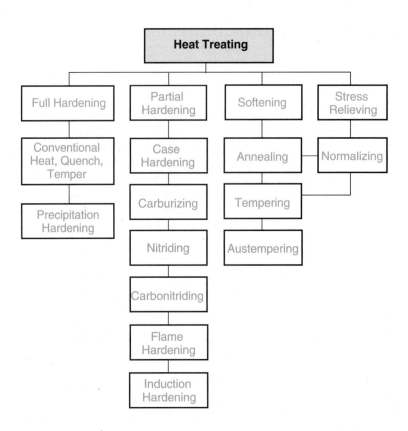

Figure 7-5
The major heat-treating processes can be classified into four types: full hardening, partial hardening, softening, and stress-relieving.

A focused CO_2 laser beam is approximately 0.002 in. (0.05 mm) in diameter. The diameter of the beam can be increased to approximately 1 in. (25.4 mm) by using parabolic mirrors.

In principle, the CO_2 laser works much like a fluorescent tube. The tube is filled with a gas that is heated as current passes through it between the anode and cathode. The gas is energized (excited) by the electrical current. Direct current (DC) is the most popular, but alternating current (AC) and radio-frequency-powered (RF) lasers are also available.

Full Hardening

Processes frequently used to full-harden metal workpieces are conventional heat/quench/temper (H/Q/T) and precipitation hardening, also called age-hardening or aging.

Heat/quench/temper hardening

The *heat/quench/temper process* is the most common way to fully harden steel. It is not practical to try to full-harden soft steel with less than 0.35 percent carbon (sometimes referred to as "35 points of carbon"). Look again at Figure 7-3. You will see that no matter how high the temperature, it is impossible to bring low-carbon steel out of the austenite region. Hardening really requires transformation to cementite.

Ferrite, or steel with less than 0.8 percent carbon, has very little ability to dissolve carbon. Carbon is what gives steel its ability to become strong and hard. In order to harden low-carbon steel, it is necessary to increase the amount of carbon on its surface by forming a thin outer layer or *case*. This outer case can then be hardened by heat treating the steel to a point slightly above its upper transformation temperature, then quenching it. This creates a part with a hard outer shell and a ductile center.

The process of hardening steel with the heat/quench/temper process begins by heating the workpiece above its upper transformation temperature and then rapidly cooling or *quenching* it in water, warm oil, or air. Other quenching media that are less frequently used are brine, molten salt, and sand.

Water is the most frequently used quenching medium, especially with low-carbon steels. Water provides a sudden quench that creates good hardness and strength. The problem with this type of quench is that the very rapid temperature change sometimes causes internal stresses and distortion in the part.

Oil is a less drastic quench than water, and is preferred for use on thin or delicate workpieces. Typical parts that are oil-quenched are thin and fragile cutting blades, such as razor blades. While oil is more gentle than water, it does not create the same degree of hardness that is possible with a water quench.

The slowest type of quench is achieved by letting the part slowly cool in air. The advantage of this quench is that it produces fewer stresses in the workpiece; the disadvantage is that it doesn't produce the hardness and strength possible with other quenches. High-alloy steels such as molybdenum or chromium have greater internal stress than milder alloys, and are more likely to crack or distort when quenched. At the same time, they do not require such a drastic quench to improve their hardness. Air is the preferred quenching medium for high-alloy materials.

After the part is hardened and quenched, it is usually tempered. In *tempering*, some of the hardness is sacrificed in order to improve tensile strength and toughness. A part is tempered by reheating it to just below the lower transformation point, then cooled (any rate of cooling can be used).

Precipitation hardening

To heat-treat nonferrous metals and some types of stainless steel, different methods must be used. These metals do not undergo phase transformations in the same way as iron and steel.

Some stainless steels and such *nonferrous metals* as aluminum alloy and copper alloy are hardened by adding *precipitates* (small particles of a different phase) into the original matrix. The process used to accomplish this is called *precipitation hardening*. Precipitation hardening is also called age-hardening, or aging.

There are three steps involved in precipitation hardening: solution treating, quenching, and aging. *Solution treating* involves heating the alloy, which exists as a two-phase solid solution at room temperature, to a temperature that causes it to become a single-phase solid solution. When an excessive number of particles from one phase are dispersed throughout the matrix of a different phase, many of these particles cannot be dissolved completely into the original matrix. This results in precipitates being formed.

The second step involves quenching this single-phase solution to create a supersaturated, but unstable, solid solution. Quenching is usually done in water. At this point, the material is softer than it might be in the annealed condition, and therefore, is in ideal condition for forming.

The last step, called *age-hardening* or *aging* is usually done after rapid cooling, either to room temperature, or to some range of elevated temperatures. Sometimes, the workpiece might even be *frozen* to prolong the time involved for age-hardening to occur. More will be said about such *cryogenic* conditioning later in this chapter.

Some aluminum alloys harden and become stronger when aged over a period of time at room temperature. This type of aging is called ***natural aging***. Usually, this is accomplished by first heating, quenching, and forming the part, then letting it age at room temperature for an extended length of time.

When the process is conducted at heat levels above room temperature, it is called ***artificial aging***. This method of age-hardening must be undertaken with care. Exposure to too much heat will result in ***overaging*** the part, making it weaker. One advantage of overaging, however, is that dimensional stability is improved.

Case Hardening

Often it is practical to use low-carbon or mild steel to make gears or other types of parts that must survive high-load and wear conditions. For a part such as a gear, it is undesirable to harden the entire workpiece. Such a fully hardened part would be excessively hard and brittle, and would lack the toughness necessary to withstand heavy loads. What is needed is a gear with hardened teeth and a softer, more ductile and tougher center. In order to accomplish this, carbon must be added to the outside surface of the part. This is referred to as ***case hardening***. Case hardening produces a hard shell or case on the extremity of the workpiece. The material becomes softer as one moves into the part from the outer shell. Figure 7-6 illustrates the effect of case hardening on various types of metal workpieces.

There are a number of different methods used for case hardening to produce various degrees of hardness. All involve heating the workpiece to a high temperature (typically about 1700°F or 927°C), exposing it to a particular gas, liquid, or powder, then quickly cooling (quenching) it. Some of the more popular methods include: carburizing, cyaniding, nitriding, carbonitriding, flame hardening, and induction hardening.

Figure 7-6
Case hardening results in a hard shell surrounding a core of softer material. The illustration shows the effect of case hardening on round and irregularly-shaped workpieces.

Of the various methods of case hardening, the oldest is called *carburizing*. This method involves covering the heated workpiece (by dipping or dusting) with a carbon-hardening powder. The powder will melt on the hot surface, forming a thin coating. If a *thick* coating is desired, the workpiece is packed in a container of carbon-hardening powder, called **Kasenit**, and kept in a furnace for an extended length of time. The longer the steel is kept hot, the thicker the coating that will be produced.

Carburizing treatments can also be accomplished with methane gas, liquids, or charcoal packs. With each method, the heated metal is placed in a carburizing environment for the required length of time.

At this point, case hardening is still not complete. The part now must be reheated to a bright red/orange color and held at this temperature for several minutes. Then, the part can be quickly quenched by immersing it in cold water, oil, or saltwater. The liquid used for the quench depends on the nature of the conditioning that is desired, and the type of steel.

Cyaniding

Cyaniding is a case-hardening technique used to create a high nitrogen-low carbon case. Here, an iron-based alloy is heated in contact with cyanide salt, so that the surface absorbs carbon and nitrogen. The process of cyaniding is followed by quenching and tempering to produce an outer shell that is tough and hard.

This process is also called the *liquid-salt method*, since the heated steel is quenched in a bath of liquid salt, heated to approximately 1600°F (871°C). After immersion, the part is then quenched in water. This is a popular case-hardening method when a thin case layer is desired. A typical layer of about 0.010 in. (0.254 mm) would be produced by leaving the part in the salt bath for about one hour before quenching it. This process is dangerous to use, however, because of the toxic fumes that are given off by the potassium cyanide bath. If the cyaniding method is being used, make certain that the area is properly ventilated, and that all personnel wear proper eye protection and protective clothing.

Nitriding

One of the hardest surface coatings that can be produced is accomplished using the process called **nitriding**. This method uses a gas, rather than the molten salt bath used in cyaniding.

The nitriding process begins by placing the workpiece in a sealed container in the heat-treating oven. The container is then heated to between 900°F and 1150°F (482°C and 621°C). High-temperature ammonia gas is then introduced into the container. The ammonia breaks down into nitrogen and hydrogen. The nitrogen then attacks the surface of the steel, combining with iron-forming nitrides. Nitrides are compounds that give the steel the desired hardness.

Nitriding produces the hardest shell of all of the case hardening methods. The major limitation associated with the process is that it is time-consuming. It would normally take about 50 hours to produce a case-hardened coating about 0.015 in. deep.

One of the advantages of this process is that it can be used to create a hard shell in excess of the 0.010 in. (0.254 mm) maximum thickness practical with cyaniding. Nitriding is particularly useful in situations where warpage, distortion, and cracking of the part may result from exposure to extreme temperatures. Nitriding can be used with very little heat; only 900°F to 1000°F (482°C to 538°C) is necessary. This is the only case-hardening process that permits the use of such low temperatures to produce a hardened case.

Carbonitriding

In the process of nitriding, a nitrogenous gas (usually ammonia) is united with the ferrous metal's surface to form iron nitride. With *carbonitriding*, low-carbon steel parts are hardened by exposing them to dry gases that are rich in carbon and nitrogen. Carbonitrided parts are harder and have greater wearability than carburized materials.

The process is conducted as follows. First, the parts are placed in an atmosphere of carburizing gas and ammonia at temperatures between 1400°F and 1700°F (760°C and 927°C). The ammonia gas creates nitrogen; both carbon and nitrogen are absorbed into the parts. Once the parts are saturated, they are quenched in oil to reduce internal stresses and improve ductility. Carbonitriding is a slow process. A coating of 0.010 in. (0.254 mm) would normally require about 90 minutes.

Flame hardening

Metallic workpieces are often *flame-hardened* with coatings up to 0.250 in. (6.35 mm) in thickness, using nothing more than oxyacetylene torches to generate heat. The surface of the iron-based alloy in the area to be hardened is heated to the upper transformation temperature. This changes its composition from ferrite to austenite. The metal is then quenched to create martensite. Both heating and quenching must be done quickly.

The advantage of flame hardening is that it can be used to create localized, or *zone* hardening. With this process, it is not necessary to heat the entire part. Heat needs to be applied only to the localized area that is to be hardened.

Induction hardening

This is another process than can be used to create localized case hardening. *Induction hardening* creates heat by passing a high frequency electrical current through a coil of wire. The part to be hardened is placed inside the magnetic field created by the coil. Eddy currents of from 3000 to 1,000,000 cycles per second (hertz) attempt to pass through the metal, generating a resistance that causes heat. The heated part is then quenched in water or oil.

One of the major advantages of induction hardening is its speed—hardening can be accomplished in seconds, instead of the minutes needed for other case-hardening processes. This makes it practical for use in many high-volume production situations. Another advantage is that it is well-suited to hardening such irregularly shaped parts such as gear teeth, crankshaft bearing surfaces, and roller bearings.

Softening Processes

Thus far, this chapter has covered processes that are used for full hardening and case hardening of ferrous and nonferrous metals. Now, we can examine the methods that are used to soften steel and nonferrous metals.

Annealing

The first process that usually comes to mind when thinking about softening metal is *annealing*. However, annealing is a process that accomplishes much more than this. Annealing is also used to refine grain structure, to restore ductility after hardening, to remove internal stresses caused by forming processes, to improve electrical and magnetic properties, and to improve machinability.

Annealing involves holding the material for an extended length of time at a temperature high enough to achieve an austenitic structure, then slowly cooling it. The temperature that is required varies depending on the material and type of annealing. There are various types of annealing such as full annealing, finish annealing, isothermal annealing, medium annealing, and soft annealing.

Full annealing is accomplished by heating to a temperature just above the upper transformation temperature. The workpiece is then cooled slowly in a furnace to a point just below the lower transformation temperature, followed by further cooling in still air to room temperature. Metal also can be annealed by submerging the red-hot workpiece in lime, ashes, or other noncombustible material until it is cooled.

A general rule for determining the rate of cooling is to allow approximately one hour per inch of material thickness at the largest section of the part.

Brass and copper can be annealed after they are cold-worked (formed) by heating them to approximately 1100°F (593°C) and then cooling the workpiece. The *rate* of cooling is not important.

Normalizing

Steels are normally annealed in still air to avoid excessive softening. This application of annealing is called *normalizing*. Normalizing requires heating the part to a temperature about 100°F (56°C) above the upper transformation temperature, holding it at this temperature until the steel is heated through, then removing it from the furnace to cool to room temperature in still air.

The purpose of normalizing is to produce a more uniform and fine-grained structure in the metal. It also provides greater strength and fewer stresses than does annealing. Normalizing arranges the grain structure into a uniform, unstressed condition with proper grain sizes to improve acceptance of further heat treating.

Stress Relieving

Complex parts that have been welded, cast, or machined often develop internal stresses that could cause distortion of their shape or affect their strength or serviceability. To relieve these stresses, parts made from carbon steel are heated to between 1000°F and 1200°F (538° C and 649°C), held at that temperature for at least an hour, then allowed to cool in air. Alloy steels usually must be heated to somewhat higher temperatures for effective stress relief.

Tempering

Tempering of steel is done to remove some of the hardness from metal so that it is less brittle. This makes it easier to form or machine, and also improves its *toughness* (resistance to breaking). ***Tempering*** is done by reheating the metal, after hardening, to a temperature slightly below the lower transformation temperature. The metal is then allowed to cool through various methods. When the steel is tempered, its microstructure changes from martensite to a softer ferrite or pearlite (a mixture of ferrite and cementite). Recall that martensite is formed by rapid cooling of a low-carbon steel part that has been heated, changing it from ferrite to austenite.

In manufacturing, it is sometimes necessary for tool-and-die makers, machinists, or technicians to fabricate parts for machine repair or maintenance. To determine the proper temperature for various heat-treating processes, they may use an interesting method that involves watching color change in the metal.

When heat is transferred to a polished part by a torch, the polished surface will change color. At 430°F (243°C), tool steel will turn yellow, at 530°F (277°C) it will become purple, and at 610°F (321°C), pale blue. When the desired temperature is achieved, the metal is quenched. Skilled tool-and-die makers know that a part heated to the purple color and quenched will be hard enough to withstand minor abuse.

They also know that if a part is to be bent or formed, it must be tempered. The ideal point for tempering is when it is heated to a cherry red, and then brought back down to a pale blue.

In actual production heat treating, more precise gauges and programmable logic controllers are used to monitor the temperature of parts held in batch or continuous furnaces. The parts are carried through the furnaces on elevators, trays, or other transporting devices for heat treating.

Another method of tempering is called ***austempering.*** This is accomplished by heat treating to a temperature below that required for case hardening. After heating, the workpiece is cooled in a salt bath kept at a temperature of about 800°F (427°C).

The advantage of austempering is that it increases ductility and toughness, together with hardness. It also reduces the possibility of distortion and cracks, because the method of quenching does not create such a shock on the material.

If the metal has already been hardened, and a martensitic structure has been created, then it is not possible to use austempering. In this case, the part can be tempered by heating it to a point just below the lower critical temperature. The actual temperature varies from 300°F to 1050°F (149°C to 566°C), depending on the type of steel involved.

Normally, if the application for the part requires it to be resistant to wear, it will be tempered below 400°F (204°C). If the part must be tougher, tempering will be done above 800°F (427°C).

Cooling, or quenching, of tempered metal is often done in a heavy oil bath. The oil is heated to between 500°F and 600°F (260°C and 316°C). The part is heated, immersed in oil, then reheated to the tempering point. Then it is dipped in a bath of caustic soda, followed by a quench in hot water.

Quenching is also accomplished in baths of molten lead or molten salt. With these quenches, the part is heated to the tempering point and is then immersed in the bath. It is left there until it is cooled to the bath temperature.

Cryogenic Conditioning

Conditioning at *cryogenic* (ultracold) temperatures as low as -300°F (-185°C) is purported to increase the performance, durability, and wear life of machine tools, drill bits, gears, cams, and end mills. However, there is still a great deal to learn about the new field referred to as *cryogenic processing*.

Metallurgists have studied cryogenic processing for approximately thirty years, but remain skeptical about what it accomplishes (mainly because it does not create any visible changes). It is known that ultracold temperatures are responsible for causing phase changes in metals. Metals that have been subjected to deep cryogenic cooling are also known to develop a more refined microstructure with greater density.

One major aircraft manufacturer claims to have doubled its output with carbide inserts for end mill cutting tools after giving the tools cryogenic treatment. In another test, the manufacturer claimed a 400 percent improvement when cutting stainless steel. Cryogenics are being used by other manufacturers to produce more durable parts for racing engines, gears and cams, progressive dies, and aircraft parts.

Some heat treaters are beginning to offer cryogenic manufacturing services. Most of these firms use liquid nitrogen; some use dry ice. However, liquid nitrogen is required for deep cryogenic treatment, below -300°F (-184°C), which appears to produce the best results.

An Illinois manufacturer of cryogenic equipment, 300°Below, Inc., is dry tempering gears by cooling them slowly to -317°F (-194°C) and holding them at this temperature for 20 to 60 hours. The parts are then heated to a temperature of 375°F (191°C) and allowed to cool to room temperature. Unlike many cryogenic processes, this system does not bathe the metal in liquid nitrogen. This reduces the chance of damage from thermal shock.

With cryogenics, everything must be handled with precision. Times are extremely critical, and temperature changes must be controlled with the utmost precision. Careful handling is vital: if a gear or tool is accidentally dropped in liquid nitrogen, it could shatter.

While cryogenic processing seems to offer many advantages in terms of the strength of the part and wear life, it does have one drawback: inconsistent results. Efforts are now underway to remove human factors, the primary source of variability in the process, through the use of automation.

Processes Used for Finishing

Before any surface finish can be applied it is necessary to prepare the surface for finishing. This must be done to remove unwanted scratches, burrs, and defects. Surface preparation is also necessary to clean the part so that the finishing material can properly adhere to the surface.

Surface Preparation

The conventional method for preparing metal surfaces is by mechanical means. Processes such as abrasive cleaning, tumbling, blasting, and brushing may be all that is necessary.

If the part has been machined, there may be rough edges that have to be removed before finishing can take place. This is normally done by deburring.

Deburring

Burrs are sharp edges that are produced when metal is deformed by shearing, trimming, stamping, or machining. Burrs would be created, for example, when a cutting tool on a mill ran off the edge of a workpiece. See Figure 7-7.

Typically, burrs would be no more than .001 in. to .005 in. (0.025 mm to 0.127 mm) high and .003 in. (0.08 mm) thick. In most cases, burrs must be removed before the part or product can be used.

Burrs are sharp and can cause injury to persons handling the product. They can interfere with proper assembly of parts, and can cause premature fatigue. In electrical environments, burrs can cause arcing and short circuits.

Deburring is the process of removing burrs through finishing processes. Burrs can be removed by filing, wire brushing, emery polishing, vibratory finishing, shot blasting, abrasive machining, ultrasonic machining, or waterjet machining.

In cases where there are only a few parts to be deburred, the most cost-effective method of burr removal is often filing, brushing, or emery polishing. In high-volume production, deburring is often done by *vibratory finishing.* This is accomplished in a closed container that holds the parts and an abrasive finishing media. Simple systems involve placing the parts and medium in a drum, then rotating the drum so that the parts and medium roll against each other long enough to remove burrs. Other production deburring machines are more complex, and some are fast enough to be used in line with

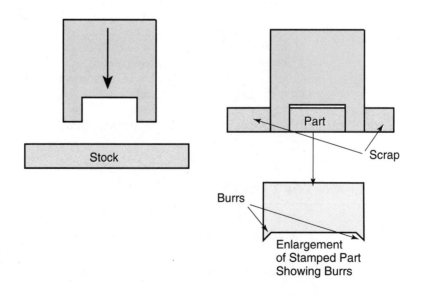

Figure 7-7
Burrs are produced when parts are stamped or a cutting tool slices across a flat workpiece.

a conveyor. Many incorporate automated feeding and unloading operations controlled by programmable logic controllers or computers. See Figure 7-8.

Sometimes deburring is not required, and other finishing methods may be used. With some products, chemical processes such as *pickling* (acid cleaning), ultrasonic cleaning, steam cleaning, or vapor degreasing may be desired. Normally, parts are conveyed to such chemical processes and cleaned automatically.

One interesting electrochemical finishing process is called *electropolishing*. With electropolishing, the workpiece is placed in a hot alkaline solution. The solution serves as the electrolyte, while the part acts as the anode. A cathode is added to complete the electrical circuit. Electropolishing is seldom used as a surface preparation process. This is because the work to be polished must be quite smooth to begin with. Electropolishing is not economical for removing more than 0.001 in. (0.025 mm) of material.

The process initially deplates the workpiece and then goes on to produce a highly polished, lustrous, mirrorlike surface. Here's how the process works.

When electricity passes between the anode and cathode, the current dissolves material from the part. Oxygen bubbles form on the surface of the

Figure 7-8
This centrifugal disk finishing machine uses treated walnut shells and corncob meal as the grinding media to deburr hard metals, such as cermet (ceramic/metallic) magnets. (Roto Finish Company, Inc.)

workpiece. When they burst, a scrubbing action is created that cleans dirt and scale from the part. In electropolishing, there is no mechanical contact between the polishing tool, which is the cathode, and the workpiece. This makes the process particularly useful when polishing irregular shapes. The electrolyte strikes the raised areas first and creates the polishing action.

Abrasives

Abrasive materials have been discussed in conjunction with grinding processes. However, a great deal of final finishing is accomplished using coated abrasives. Coated abrasives perform much like a grinding wheel, but the abrasive materials are attached to cloth or paper backing with a glue or resin binder. Coated abrasives are available in the form of sheets, discs, belts, rolls, and cones. Some abrasives are used for mechanical hand polishing; others are used with machines.

Emery paper and *crocus cloth* are two coated abrasive products that are used most frequently to polish and buff metals. Emery is made from aluminum oxide and iron oxide. Crocus is made from purple iron oxide. Both products are available with different types of backing, and with different°of coarseness.

Abrasives are both natural and synthetic. Natural abrasives include emery, sandstone, garnet, corundum, and diamond. The most common synthetic abrasives used in industry are silicon carbide, aluminum oxide, and bauxite.

Finishing Processes

Once the surface preparation and final finishing processes have been completed, the final finish can be applied. Finishing processes protect and/or beautify the surface of the workpiece. Finishing includes all of the surface-protecting and surface-decorating processes. Finishes can be applied through coating processes: brushing, rolling, spraying, or dipping. They can be applied by electrochemical means: plating or anodizing. Some of the finishing methods in general use for metals are described in the following sections.

Electroplating

Electroplating is a process that can be used to apply a metallic coating of one type of metal up to 0.002 in. (0.05 mm) thick on any other type of metal. Common coating materials used in industry are chromium, copper, nickel, zinc, and tin. Careful cleaning and surface preparation of the workpiece is critical to successful plating.

Parts to be plated are placed in racks or a barrel and are lowered into a water-based electrolytic plating bath. In electroplating, the workpiece serves as the cathode. The process is the opposite of electropolishing, in which the workpiece is the anode. In the *plating process,* electrical energy causes ions to be given off by the anode. These ions then combine with ions in the electrolyte and are attracted to the cathode (workpiece), where they are deposited as a metallic coating.

One of the major problems with electroplating is the difficulty of achieving a uniform plating (layer of metal) on workpieces that are irregularly shaped. Plating takes place on the raised areas of the surface before it is attracted to the cracks and depressions.

Electroless plating

Plating can also be accomplished without electricity. The process known as *electroless plating* is done entirely by chemical reactions. It is most often used to plate nickel, but it can be used on any metal. It can also be used with nonconductive materials, such as plastics and ceramics.

In electroless plating, the workpiece serve as a catalyst. When the process is used to plate nickel, sodium hypophosphite acts as a reducing agent, converting nickel salts to nickel metal which is suspended in a solid solution of phosphorous nickel. Since this is entirely a chemical process, the coating that is created is uniform in thickness.

Anodizing

Anodizing is an electrochemical process that turns the surface of the workpiece into a hard and porous oxide layer that is resistant to corrosion. The process is also called *anodic oxidation*. In addition to creating a hard coating, anodizing also produces an attractive finish for many product applications. Aluminum and magnesium are the most frequently anodized metals. The aircraft industry is a major user of anodized parts. An important reason for this is the ability to form and draw anodized coated parts without marring the coating or destroying its protective qualities.

Anodizing can be thought of as *reverse electroplating*. Instead of the coating being added to the surface, a reaction is created inward, resulting in a thin protective layer of aluminum oxide developing on the aluminum. The layer that is generated is normally from 0.0005 in. to 0.001 in. (0.0127 mm to 0.025 mm) thick. The workpiece is immersed in an acid bath and acts as the anode. Current flows from the cathode to the anode, causing a chemical reaction. Oxygen is absorbed from the bath, causing the reaction to create a thin layer of aluminum oxide on the surface of the metal. Sometimes, dyes are added to the bath to penetrate the pores of the metal and provide a colored finish. Anodizing may also be used to create a clear oxide layer that does not change the color of the metal.

Typical anodized products include aluminum utensils, furniture, automobile trim, and keys. An anodized surface also serves as a good foundation for paint application, which is useful when finishing hard-to-paint materials like aluminum.

Ion implantation

Ion implantation is a process that can be used to harden the surface of the metal, improving a part's ability to withstand wear, friction, and corrosion. The process, which is conducted in a vacuum, implants ions into the surface of the workpiece by accelerating them to the point where they are implanted into the substrate layer of the metal. Ion implantation is a rapidly growing process for producing semiconductor devices. When it is used in this way it is called *doping*, which means alloying with small amounts of different metals.

Doping is a unique process when used in manufacturing microelectronic devices. Microelectronic devices are normally made of substrates that must contain different types and thicknesses of materials. The addition or subtraction of material is accomplished through ion implantation.

The process can be compared, in principle, to exposing a photographic plate to a light source when preparing it for chemical milling. The difference is that, in doping, light is replaced by accelerated ions. The surface of the substrate can be masked to prevent exposure to ions where they are not desired.

Specialized equipment is required, often costing more than one million dollars. Here's how the process works.

Implanting ions is done by rapidly moving the ions through a high voltage field. A high-voltage electron beam is swept, through the use of deflection plates, across the silicon wafer that is the substrate for the semiconductor. The impact of the ions on the silicon wafer slows their movement. The dopants (ions) are embedded less than half a micron below the surface of the silicon wafer.

The substrate is then annealed at a temperature of between 750°F and 1500°F (399°C and 816°C). Annealing forces the molecules in the material to rearrange themselves and drives the implanted dopants to a depth of several microns below the surface.

Metallizing

Metallizing is the process of spraying or vacuum impregnating metallic coatings on various metals and nonmetals. Because the base material to be coated is not heated, metallizing can be used on paper, wood, and plastics, as well as on metals.

There are three major types of metallizing: plasma arc spraying, vacuum metallizing, and wire metallizing or flame spraying. Metallic coatings are also applied by using printing processes such as silkscreen, etching, or lithography. These processes are used primarily in the manufacture of integrated circuits and devices. However, they are not coating processes, and should not be confused with metallizing.

The ***plasma arc spraying*** process creates an arc between two electrodes to heat argon or another inert gas. The heated gas is accelerated to supersonic speed. A powdered metal coating material is fed into the stream of gas, where it is melted and blown onto the part.

Plasma arc spraying is often used to coat parts for use in high-temperature environments. This technique involves greater temperatures than other metallizing methods (in excess of 30,000°F or 16 650°C), so materials with higher melting points can be sprayed. Coatings of tungsten carbide, ceramics, nickel chromium alloys, and exotic alloys are commercially available. In addition to metals, materials such as cermets, oxides, and carbides can also be sprayed.

Vacuum metallizing is usually the preferred process when a coating of aluminum metal is desired. The coating material is placed in a vacuum chamber with the items to be coated. The chamber is then heated, producing vapors which coat the parts.

The vacuum metallizing process takes only moments to create the desired coating. Product applications include coating jewelry, automotive trim, and toys. The process is also popular for placing thin coatings on electronic substrates and devices.

Coatings applied by *flame spraying* (also called *wire metallizing*) are useful in corrosion protection. They are also used to rebuild worn wear surfaces on parts. Metallizing normally involves melting the coating material in a torch or gun. In wire metallizing, a wire of metal or ceramic coating material is fed into the spray gun, where it is melted by electric arc or flame. The molten metal is then atomized by a stream of compressed air, and is blown on the workpiece.

The spray gun used in this method is similar to one that would be used for paint spraying. The gun may be hand-held or mounted on the tool post of a lathe for automatic spraying. The stand-off distance between the workpiece and the spray gun is normally from 6 in. to 10 in. (15.2 cm to 25.4 cm).

Before the base material can be coated, it is necessary to roughen its surface. This produces a better bond. Coatings as thick as 1/2 in. (13 mm) can be applied with flame spraying, but most applications do not require more than a 1/16 in. or 1/8 in. (1.6 mm or 3.2 mm) coating.

Powder coating

Metal parts are often finished by coating them with powdered organic resins. The powdered coating is then fused with heat and cured. Common systems used for powder coating include fluidized bed coating, electrostatic fluidized bed coating, electrostatic powder spraying, and heat transfer coating. While the method of applying the powder varies from process to process, the principle is the same. Normally, the part is heated, the powder is applied, and additional heat is supplied to melt the powder, fusing it into a seamless coating on the surface of the material.

Important Terms

age-hardening
aging
annealing
anodizing
artificial aging
austempering
austenite
batch furnaces
burrs
carbonitriding
carburizing
case hardening
cementite
continuous furnaces
crocus cloth
cryogenic
cyaniding
deburring
doping
electroless plating
electroplating
electropolishing
emery paper
equilibrium diagram

eutectoid
ferrite
flame-hardened
flame spraying
fluidized bed furnaces
heat/quench/temper process
heat treating
induction hardening
ion implantation
Kasenit
liquidus line
lower transformation temperature
martensite
metallizing
natural aging
nitriding
nonferrous metals
normalizing
overaging
pearlite
phase diagram
pickling
plasma arc spraying
precipitates

precipitation hardening
quenching
solidus line
tempering
toughness

upper transformation temperature
vacuum metallizing
vibratory finishing
wire metallizing

Questions for Review and Discussion

1. What are the similarities and differences between anodizing and electropolishing?
2. When would water quenching be desired over oil quenching?
3. What is the difference between annealing and normalizing?
4. How is tempering accomplished?
5. What is the major process used to full harden low-carbon steel?

This microprocessor-controlled double-sided lapping system is used to produce finished parts with uniformly flat parallel surfaces. Diamond abrasives are used to rapidly and acurately smooth and polish the part surfaces. The machine can handle parts in thicknesses ranging from 0.004 in. to 2 in. (0.1 mm to 51 mm), and is equipped with automatic thickness control for optimum sizing. (Engis Corp. USA)

PLASTIC MATERIALS

Plastics are synthetic materials that are capable of being formed and molded to produce finished products. They exhibit many characteristics that provide design advantages over other industrial materials. These include light weight, corrosion resistance, strength, durability, thermal and electrical insulation value, and ease of manufacturing.

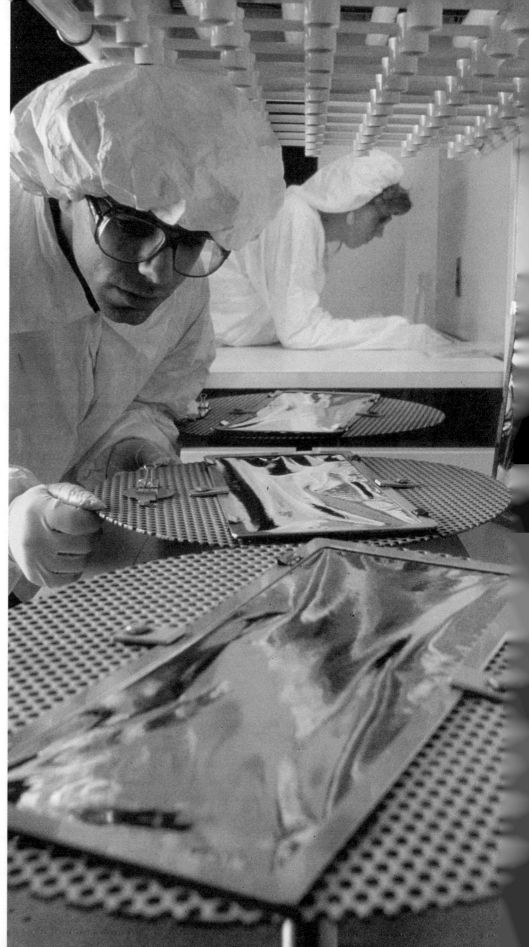

Introduction to Plastic Materials

Key Concepts

- △ Plastics are synthetic or engineered materials derived primarily from petrochemicals (oil and natural gas).
- △ Plastics offer cost advantages as a design material, replacing metal in many applications.
- △ Until a polymeric material is molded or formed, it is considered resin; after molding or forming, it is a plastic.
- △ Commodity resins are used for many common household products, engineering resins for products requiring special characteristics.
- △ Thermoplastics are formed by the process of addition polymerization, which allows them to be later heated and reformed.
- △ Thermosetting plastics are formed by condensation polymerization, involving a chemical change in which molecular chains become cross-linked.

Plastics are synthetic materials that are capable of being formed and molded to produce finished products. In fact, the term plastics is derived from the Greek word *plastikos,* meaning "to form." Although plastics can be derived from many types of organic and inorganic materials, they are most often made from petroleum (oil) or natural gas. For example, the ingredients used to make stock for the important polymer **polystyrene** are benzene and ethylene. Both of these chemicals are derived from petroleum.

Unique Characteristics of Plastics

Plastics exhibit many characteristics which provide design advantages over other industrial materials. Most plastics weigh about one-eighth as much as steel. This makes them ideal candidates for product applications where weight affects performance or for products that must be physically handled and moved over great distances. Plastics are highly resistant to corrosion, and for all practical purposes, do not deteriorate when exposed to the environment. See Figure 8-1. Plastics are also good insulators, making them ideal for cooking utensil handles or housings for portable electric tools.

Another design advantage of plastics is that they can be created with smooth surfaces for minimal friction. This makes plastics an ideal choice in applications where the product has moving parts, many of which may never be lubricated.

Figure 8-1
This mower deck is molded from an engineering resin that provides corrosion resistance, strength, molded-in color, and ease of forming. The plastic deck is only half as heavy as a comparable metal part. (GE Plastics)

The Development of Plastics

The first synthetic plastic, called Xylonite or Parkesine, was invented in 1862 by the English chemist, Alexander Parkes. The basic material of Parkes' plastic was nitrocellulose, softened with camphor and a vegetable oil to make it possible to form into the desired shape. But plastic, as a commercially successful material, had its real birth in the United States almost a decade later. In 1869, when the game of billiards was gaining great popularity, a New York manufacturer offered a $10,000 prize for a usable substitute for the ivory used to make billiard balls. A young printer, John Hyatt, experimented with many different chemical combinations. Finally, the search for a substitute material ended when he independently discovered the combination of camphor and nitrocellulose used by Parkes. Hyatt produced a plastic mass that he could form into any shape and allow to harden in a matter of minutes. On April 6, 1869, Hyatt patented his discovery, calling it "celluloid." The material was later used for such diverse products as dental plates, vehicle windshields, and motion picture film. Its greatest drawback was *flammability*. Celluloid has been replaced by plastics that do not burn so easily.

The really significant industrial application of plastics began with the development of Bakelite in 1909. The American chemist, Dr. Leo Baekeland, developed the thermosetting phenolic resin, a combination of the chemicals phenol and formaldehyde, while seeking a synthetic substitute for the natural coating material shellac. Bakelite, the first commercially successful thermoset plastic, was widely used for telephone housings, cooking pan handles, and similar applications. From this point onward, there were many innovations in the field of plastics. By 1940, many industrial and consumer products that had previously been made from metals, wood, glass, leather, paper, and vulcanized natural rubber were made of plastics. Today, plastics are an integral part of our society—it would be hard to imagine life without them.

Making Industrial Stock

The terms, **resin** and **plastic** are often confused with each other. A raw material may be processed a number of times before it reaches a form that is useful for a particular manufacturer. In the plastics industry, it is common practice to view all processed material, up to the point where industrial stock is created, as *resin*. The result of the forming or molding processes that change stock into a final product is *plastic*, Figure 8-2.

Figure 8-2
Plastic is the material that results when a resin is changed into a product or part by forming or molding. These returnable, reusable milk jugs are molded from a strong polycarbonate resin. They can be sterilized and refilled up to 60 times, then ground up and recycled. (GE Plastics)

Commodity resins

Today, most of the thermoplastic and thermoset resins produced are referred to as *commodity resins.* This includes standard-grade resins such as low-density polyethylene (LDPE), polypropylene homopolymer (PP), crystal polystyrene (PS), and rigid polyvinyl chloride (PVC). Most of the resin companies provide commodity resins, which are used for molding or fabricating such everyday items as household accessories and containers, refuse bags, toys, decorative items, or automotive accessories.

Engineering resins

Another category of resins is referred to as the *engineering resins.* Many firms process high-performance resins in various grades, such as advanced, intermediate, and commodity. Examples would be nylon, polycarbonate (PC), and polyphenylene sulphide (PPS). The advanced grade of resins would be those that are most resistant to chemical attack, extreme heat, and impact. Engineering resins might be used for such products as football helmets, scientific laboratory equipment, reheatable food containers, or industrial conveyor rollers. See Figure 8-3.

Figure 8-3
The design requirements for these industrial pallet flow rack wheels included impact strength and the ability to stand up to loads in both hot and cold temperature extremes. An engineering plastic was selected to precisely fit these requirements. The design shown uses steel bearings, but for use in corrosive environments, the wheels are available with bearings made from a different engineering plastic, rather than steel. (GE Plastics)

Resin is manufactured in pellet, granule, powder, or liquid form. Most makers of thermoplastic resins convert fine particles in suspension into BB-sized pellets. The pellets are then shipped in bags, drums, special containers, trucks, or train cars to the manufacturing plants that produce industrial stock in various forms—monofilament line, coatings on paper, sheets, or rods. See Figure 8-4. Most plastic industrial stock goes on to manufacturers who use it directly to make consumer products. Resins in liquid form may be combined with reinforcing material such as fiberglass mat, chopped fiberglass, cloth, roving (strips), or expandable foam.

Figure 8-4
Thermoplastic resins are shipped in many different types and capacities of containers. This artist's conception shows a bulk container designed for easy delivery and weatherproof storage of resin outside a manufacturing plant. (Polymerland, Inc.)

Structure of Plastics

The *Modern Plastics Encyclopedia* lists nearly 500 different types of compounds and resins used in the manufacture of plastics. There are many grades of each resin or compound. Consequently, there are many hundreds of different types of plastics, each formulated to meet the needs of a particular environment.

Plastics are Polymers

Despite the many variations in type, all plastics are based on a high-molecular-weight molecule called a ***polymer.*** The polymer is a long chain made up of thousands of smaller molecules linked together. These smaller, simple molecules are called ***monomers.*** See Figure 8-5.

Different polymers result from varying the combinations of monomers. For example, ***polyethylene,*** a polymer used for many types of containers, consists entirely of *ethylene* monomers. The polymer ***acrylonitrile-butadiene-styrene*** (usually referred to as simply "ABS") is widely used for plumbing pipe and fittings. It is made up of three different monomers: acrylonitrile, butadiene, and styrene. See Figure 8-6. When two or more kinds of

Figure 8-5
A monomer is a simple molecule of a substance. This ethylene monomer consists of two carbon atoms and four hydrogen atoms. If a number of ethylene monomers are joined together, end-to-end, the result is a long chain molecule (polymer) known as polyethylene.

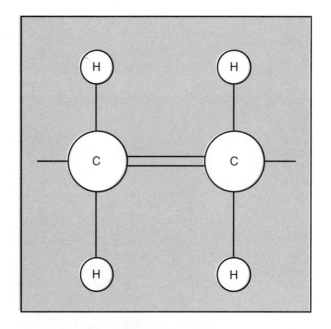

monomers are combined to form a long-chain molecule, such as ABS, the resulting molecule is normally called a *copolymer.* The individual repeating units in the polymer are often referred to as *mers*.

Natural and synthetic polymers

Polymers are found in both natural and *synthetic* (human-made) forms. Wood, rubber, and cotton, for example, are natural polymers. The group of commercial plastics called *cellulosics* are made from a natural polymer, cotton cellulose. However, most polymers used in industry are synthetic, and are based on chemicals derived from oil or natural gas.

Figure 8-6
A copolymer is a molecule made up of two or more kinds of monomers. If monomers of acrylonitrile, butadiene, and styrene join together, the result is the polymer ABS. The actual polymer chain, of course, would consist of literally thousands of repetitions of the A+B+S sequence shown here, all strung together in a chain.

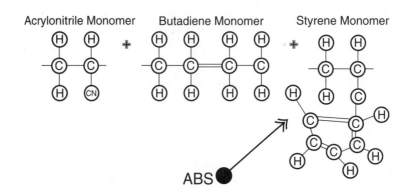

Another class of plastics is referred to as *elastomers*. These are highly resilient materials that are more like rubber than plastic. Under normal room temperature conditions, an elastomer can be elongated to at least twice its original length; upon release, it will return to its original dimensions. Elastomers can be processed with faster cycle times than other plastics in conventional injection molding machines. Neoprene, used for faucet washers, o-rings, and other types of sealing applications, is an example of an elastomer. Primary manufacturing related to the production of elastomers is considered to be part of the rubber industry, Figure 8-7.

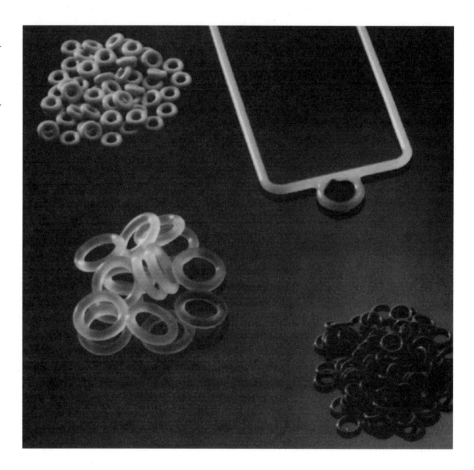

Figure 8-7
Elastomers are resilient materials that are often used for seals and gaskets. Different polymer formulations are used to meet specific application requirements. (Apple Rubber Products, Inc.)

The Polymerization Process

The chemical reaction by which polymers are formed is called *polymerization*. In order to create that chemical reaction, monomers must be subjected to heat, pressure, or a *catalyst* (agent that causes a chemical reaction). When polymerization takes place, the monomers blend chemically, linking their molecular chains. The polymerization process is critical; without it, there can be no plastic.

The *degree* of polymerization in a particular plastic influences the molecular weight of that material. In most cases, the greater the degree of polymerization, the greater the molecular weight. The weight of a particular sample of plastic is an important factor to consider, since it directly affects *viscosity*, the thickness or ability of liquid material to flow. This is especially important when products are being produced by casting or molding processes.

Types of polymerization

Essentially, there are two different types of polymerization: *addition* and *condensation*. In **addition polymerization**, the monomer molecules join together to form long chains. The chains may be *linear*, like beads in a necklace, or *branched*, Figure 8-8. Although they link with one another, the molecules do not change chemically—they neither lose nor gain atoms. There *is* chemical change involved, however, in **condensation polymerization.** When this type of polymerization takes place, chains become *cross-linked* to each other, Figure 8-9. In the process of cross-linking between chains, some atoms are lost by the monomer molecules. Those atoms combine to form water molecules or other compounds as a byproduct of the process. These new molecules do *not* become part of the polymer.

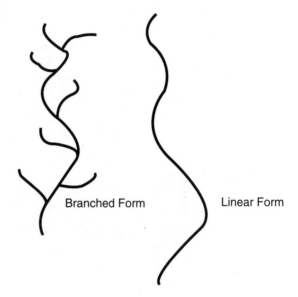

Figure 8-8
The polymers resulting from addition polymerization are long chains of either the linear or branched form. Polymers of the linear and branched forms are typically found in thermoplastics. These materials can be melted and recycled into new products.

The linear and branched polymers resulting from addition polymerization are typically thermoplastics like polyvinyl chloride or acrylic. Cross-linked polymers formed by condensation polymerization are usually thermosets, such as polyurethane or urea formaldehyde.

Figure 8-9
Cross-linking of polymers occurs during condensation polymerization. A chemical change in the monomer molecules takes place during the process. The thermosetting plastics that result, unlike thermoplastics, cannot be melted down and re-formed.

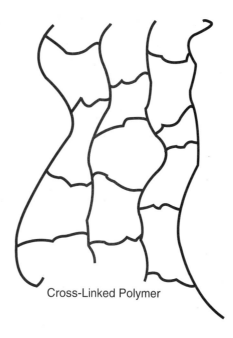
Cross-Linked Polymer

Major Classes of Plastics

There are two major classes of plastics: thermoplastics and thermosetting plastics (thermosets). Both types can be heated initially to make them fluid enough to be formed and molded. The thermoplastics can be reheated and re-formed again and again; thermosets cannot.

Thermoplastics

Thermoplastics consist of long, discrete chains of molecules that melt to a viscous liquid at the processing temperature, normally 500°F to 700°F (260°C to 371°C). After being formed and cooled, they become a crystalline, semicrystalline, or amorphous solid. The linear polymers allow molecular chains to pack more tightly together, resulting in plastics with a crystalline structure that is typically rigid, strong, and translucent or opaque. Branched polymers cannot pack so tightly together, so the resulting plastics have what is called an amorphous structure. They tend to be less rigid, have less strength, and be more transparent than crystalline plastics. Thermoplastics exhibit little attractive bonding between the molecular chains. Typical thermoplastics are polyvinyl chloride (PVC), polyamides or nylons, acrylonitrile-butadiene-styrene (ABS), polycarbonate (PC), cellulose acetate butyrate (CAB), and liquid-crystal polymer. See Figure 8-10.

Thermosets

Thermosets behave very differently from thermoplastics during polymerization. Thermosets are chemically reactive in their fluid state and harden through a further reaction called *curing*, in which cross-linking takes place. Cross-linking results in growing polymer molecules into a spatial network,

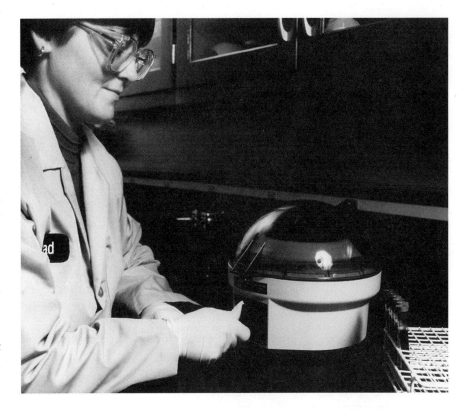

Figure 8-10
This centrifuge, designed for use in small medical laboratories, uses three different formulations of Lexan™ polycarbonate to meet different requirements. The basic housing used a foamable resin that provided structural strength, impact resistance, and sound-deadening. A retainer used to hold motor wiring safely away from spinning parts was injection-molded from a resin formulated to provide flame-retardant qualities. The resin from which the transparent cover was injection-molded was selected for high impact resistance. It was designed to protect workers from glass fragments and contents of test tubes if they shatter while being spun at high speed. (GE Plastics)

with strong bonding between the molecular chains. These three-dimensional bonds provide high dimensional stability, high temperature resistance, and excellent resistance to solvents. Heat is applied only after the partially cured polymer is formed or molded into the desired shape. After they have cross-linked, thermosets cannot be returned to their fluid state by heating. While heat may cause thermosets to soften slightly, they cannot be made fluid and then re-formed, as can the thermoplastics. Instead, they will typically char and decompose.

One of the first commercially important plastics, Bakelite, was a phenol-formaldehyde thermosetting resin created through a reaction of phenol and formaldehyde. Other thermosets include the unsaturated polyesters, melamines, epoxies, ureas, and silicones.

Thermoplastics are more resistant to cracking and impact damage than thermosets. However, they are usually inferior to thermosets in terms of high-temperature strength and chemical stability. An exception is one of the newer thermoplastics, *polyether ether ketone (PEEK),* which has a semi-crystalline microstructure with excellent high-temperature strength and chemical resistivity.

What should be remembered about all thermoplastics is that the softening and hardening cycles are reversible. The polymer can be formed into a desired shape—pellet, tube, bar, sheet, film, or rod—cooled, and then shipped as stock to the secondary manufacturing firm. There, the stock is used to make the final product by transforming it once again to the fluid state using heat, then forming and cooling. See Figure 8-11.

Figure 8-11

This is a taxonomy showing the major types of manufacturing activity in the plastics industry.

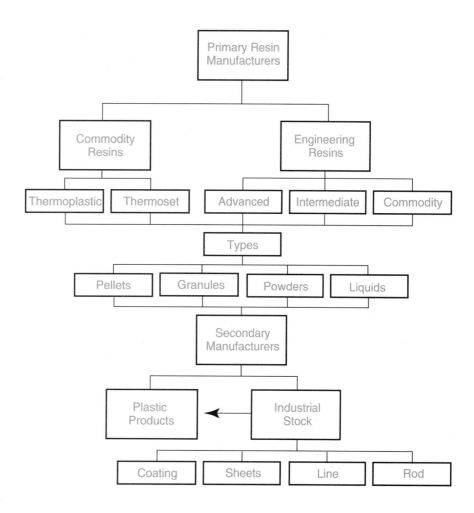

Plastic Memory

Imagine that you caught a favorite pair of sunglasses in your car door as it closed, severely deforming them. By taking advantage of the unique characteristic of thermoplastic materials called "memory," the glasses could be returned to their original shape by heating. *Memory* is the capability of a material to return to its original shape after it is bent or formed. Polyvinyl chloride (PVC) materials, for example, have very high plastic memory. If the material is distorted at slightly elevated temperatures and then cooled, it will maintain this shape until it is again heated to nearly the heat distortion temperature. Then it "remembers" what its shape was originally and returns to that shape. *Shape memory* materials are available in both plastics and metals. A spring can be formed from thermoplastic rod or straight Nitinol (nickel-titanium alloy) wire. Once that spring was installed in a hot environment, the heat would cause it to uncoil and return to its original straight shape. Shape memory materials can be used in applications where there is a need to cover joints, protect wires, or even apply tension to a window covering. Heat causes the material to behave like "shrink wrap," and become tightly compacted.

Creep

A major negative characteristic of thermoplastics is that they are subject to creep. *Creep* is the slow and continuous increase in length at the point of deformation, over a period of time, when a material is placed under a steady load and constant temperature. When creep takes place at room temperature, it is called "cold flow."

Let us consider how creep takes place. Imagine what would happen if you were to stretch a flexible strip of plastic horizontally above a work surface with support on each end. If you place a heavy weight on the center of the strip, the plastic will naturally sag to some degree. This is called elastic deformation. If the weight is allowed to remain in place for a period of time, the plastic will continue to sag at a slow rate. Displacement of material would eventually occur in the area of the bend. If you then remove the weight and examine the strip, you will find that it remains in the bent and deformed position. This is an example of *creep failure,* which occurs when the dimensions of the material change. With creep failure, the material will not return to its original size and shape once the load is removed. It is an important factor to consider when designing plastic parts that must withstand loads and stresses.

Creep failure cannot be accurately predicted by a short-term tensile or compression test. The creep stress limit that can be tolerated in a plastic material is often as little as 40 percent of the failure stress that might be shown in a short-term tensile test. This percentage would be even lower if the temperature of the plastic were increased above room temperature. Creep increases more rapidly at higher temperatures. Plastic materials can withstand high stresses for a short length of time. However, they must be used conservatively in applications where continued high loadings and temperatures are necessary.

Important Terms

- acrylonitrile-butadiene-styrene
- addition polymerization
- catalyst
- cellulosics
- commodity resins
- condensation polymerization
- copolymer
- creep
- creep failure
- curing
- elastomers
- engineering resins
- memory
- mers
- monomer
- plastics
- Polyethylene
- polymer
- polymerization
- polystyrene
- resin
- shape memory
- synthetic
- thermoplastics
- thermosets
- viscosity

Questions for Review and Discussion

1. What kind of effect would the application of heat from a molding process have on polymerization?
2. What are the major differences between thermoplastics and thermosets? When would you select one over the other?
3. List several typical thermoplastics and several typical thermosetting plastics.
4. What is the difference between a commodity resin and an engineering resin? List some uses for each.

Forming Plastic Materials

Key Concepts

△ There are two major forming processes used with plastics: casting and molding.

△ The basic closed molding processes are compression molding, injection molding, reaction injection molding, blow molding, and resin transfer molding.

△ Open molding processes are used to produce continuous lengths of tubing and solid stock.

△ Blow molding processes are extensively used for high-volume production of bottles and other containers.

△ Lamination and calendering are used to create polymeric "sandwiches" for thin-walled and sheet products.

One of the advantages of using an engineered material such as plastic is that stock can be purchased in just about any form desired—sheets, pellets, powder, granules, rods—and then, a manufacturing process can be selected that will produce a *near-net-shape part.* "Near-net-shape" means that a product can be made in the size desired, without significant shrinkage taking place during the manufacturing process.

With many materials, shrinkage is a condition that must be taken into consideration. Molds or dies often must be constructed oversize to compensate for this shrinkage during curing or drying. Sometimes, the estimate of shrinkage by the design engineer or manufacturing engineer is inaccurate, resulting in rejected parts.

Another important consideration with plastics is that there is normally no scrap (waste material) generated. When a part is injection molded, the *runners* (pathways that carry plastic to the mold cavity) must be trimmed off. In most cases, however, this material is ground up and reused. Trimmed material and rejected thermoplastic parts are melted and recycled. Rejected thermoset parts or scrap can also be ground and recycled for use as filler. This topic is covered in more detail in Chapter 10.

219

Forming Processes

While there are literally hundreds of processes that are used in industry to form plastic, most are really slight variations or hybrid combinations of two major forming processes. The major forming methods used by the plastics industry are typically casting or molding processes. The term *casting* generally is used to describe a process in which a liquid resin is poured into a mold and solidifies. In *molding*, the material is normally in a softened ("plastic") but not liquid state. It is formed into the desired shape using pressure and sometimes, heat. Most molding is done in a two-piece closed mold, but there are some open-mold processes. Extrusion, in which material is forced through an opening in a die to produce the desired cross-section, can also be considered a form of molding.

Closed Molding

Closed molding processes are those which compress stock, in a variety of different forms, between the halves of a two-piece mold. In the plastics industry, both mold halves are referred to as the die, or the tool.

When used with closed molding processes, the mold halves are opened to expedite filling and/or part removal. The closed mold halves are normally held in a press, which provides pressure on the part during cooling or curing (depending on the type of plastic). Once a thermoplastic part has cooled, or polymerization of a thermoset has occurred, the mold halves open and the part is ejected or removed.

With some closed molding processes, a mold with matched male and female halves is used. The female mold has a sunken surface (*die cavity*), while the male mold has a raised surface that functions as the core for forming the part. Other closed molding techniques incorporate the use of two-piece molds with halves that are identical. Each type of mold will be described in more detail in conjunction with the molding processes with which it is used.

There are five major closed molding processes:

- Δ Compression molding.
- Δ Injection molding.
- Δ Reaction injection molding.
- Δ Blow molding.
- Δ Resin transfer molding.

Each of these processes uses some type of press or support device (*fixture*) to compact the two mold halves tightly against the raw stock to produce the part. A variety of stocks may be used with closed molding processes, including:

- Δ Woven fiberglass mat, called *roving*.
- Δ Pellets or tablets compressed from powdered polymers. They are usually referred to as *preforms.*
- Δ Cores of partially cured resin with reinforcement, known as *prepregs.*

- △ Sheets of resin, usually referred to as *SMC (sheet molding compounds)*.
- △ Puttylike materials, called *bulk molding compounds (BMC)*.

It is important to understand a bit more about the different types of stock used with polymeric/plastic forming processes. Each type of stock has its own special characteristics, advantages, and weaknesses, and must be properly selected for use with a particular manufacturing process.

Fiberglass *roving* consists of twisted or woven glass fiber strands. Each of the individual strands is prepared by either slight twisting or straight drawing. The machine shown in Figure 9-1 is producing nearly 300 strands of fiberglass roving simultaneously. These strands may later be woven into roving fabric, made into a loose reinforcing mat, or fed to a chopper gun for spray application.

Figure 9-1
This machine weaves almost 300 strands of fiberglass roving at one time. Roving is used as reinforcement for many types of plastic products, including boat hulls, vehicle parts, containers, and sporting goods. (PPG Industries, Inc.)

Roving is different from the yarns used primarily for weaving cloth. Fiberglass roving is used in plastic fabrication processes because it has a higher yield than would be possible with woven yarns of other materials. The yield of a roving is normally from 197 to 980 yards per pound. Yield is related to the strength of the material, as well as the length of strand that can be produced from a particular weight of raw stock. Woven fiberglass roving is stronger than conventional yarns, and thus requires less weight than solid materials to provide the desired strength and support to a workpiece. This makes woven roving a good choice for many fabrication applications involving polymeric and composite materials.

Roving is frequently applied around the edges of large products to provide extra edge strength. Often roving is used in processes such as filament winding and pultrusion. It is also used in mat or woven cloth forms, with molding compounds, in the production of filament for winding, and in the production of rod stock.

Preforms are often used with the compression, injection, and extrusion molding processes. The polymer pellets are simple to store and flow easily though equipment.

A *coated preform* is considerably different in size and composition. It is a shape made by spraying chopped fiberglass and a binder material onto a form. Once the resin has set, the coated preform is rigid enough to be moved to a press for molding. Quite large parts, such as boat hulls or vehicle body panels, can be produced using coated preforms.

A *prepreg* consists of reinforcement material that has been impregnated with liquid thermosetting resin and cured to what is called the **B-stage.** At this stage, the preform is dry, but still slightly tacky. Since prepregs are only partially cured, they can easily be re-formed in a mold. The resin will behave like a thermoplastic until sufficient heat and pressure have been applied to cause cross-linking. The resulting product will be a thermoset.

Prepregs are sometimes made from continuous-strand woven fiberglass roving or mat. Other materials, such as paper, asbestos, aramid, and melamine are also used to produce prepregs. Integrated, fiberglass-reinforced composite prepregs can even be made from sheet molding compounds.

Sheet molding compounds (SMC) are sometimes called "flow mat" or "resin sheets." The compound is made by combining layers of resin sheets, reinforcement, fillers, fabric, and additives on automated, continuous-flow machines. SMC sheets are easily handled, since a film carrier (usually polyethylene sheeting) is left on the top and bottom surfaces until they are used.

Sheet molding compound is produced by chopping roving into strands up to two inches long and depositing them onto a paste coating of polyester resin spread on the lower film carrier. After the first layer of coating is applied, a second film carrier spread with more resin paste is applied, creating a laminated sandwich of fiberglass and resin. The semi-tacky SMC sandwich, with protective film top and bottom, is squeezed to the desired thickness between rolls on the winding station of the machine. The rolled SMC is stored on the film until it is needed.

Composite prepregs are also made using *bulk molding compounds.* BMC prepregs are produced in the form of a putty or dough consisting of resin, fillers, reinforcement, and other additives. The prepreg is normally produced in the shape of a slug or continuous rope, but sometimes in the form of H-beam-shaped slugs that can be automatically fed into forming dies. Bulk molding compounds are used with injection molding, compression molding, and extrusion processes.

Compression molding

Compression molding is one of the oldest and simplest of all of the closed molding processes. Compression molding can be accomplished using either thermosetting or thermoplastic polymers, but thermosets are popular.

There are two types of compression molding—hot and cold. However, this descriptive terminology may be deceptive. Both forms utilize heat to speed the rate of curing.

With *cold compression molding,* also referred to as *cold press molding*, the material to be formed is placed between the matching opened halves of unheated male and female molds. The material might be roving and catalyzed resin, sheet molding compound, or bulk molding compound. Figure 9-2 illustrates a typical setup for cold press molding. The mold halves are opened for

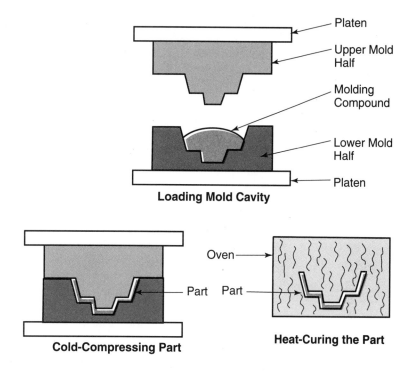

Figure 9-2
In cold compression molding, the unheated mold closes on molding compound or other material to form it into the desired shape. Curing is sometimes done by moving the still-closed mold into an oven.

loading, then closed for forming and curing. Sometimes, the mold is opened slightly during curing to permit gases to escape and reduce the formation of air bubbles in the part.

In cold press molding, the part remains in the closed mold and is transferred to an oven for curing. One advantage of cold press molding is that the part which results has two smooth, dimensionally accurate surfaces. If the color of the stock needs to be changed, pigment is mixed with the resin before molding.

Since heat speeds the action of the catalyst, many manufacturers prefer to use *hot press compression molding.* In this process, a plastic mixture of resin, reinforcement, filler, and additives is placed between die halves that are heated to between 225°F and 325°F (107°C and 163°C). The die is then closed by pressure that may range from 100 psi to 2000 psi (690 kPa to 13 790 kPa). Curing is accomplished while the mixture is held in the mold, and takes from a few seconds to five minutes, depending on the material. With hot press compression molding, the die halves are constructed with heating and cooling vents to first activate the catalyst and then increase the rate of cooling.

Compression molding is also done using preforms made from sheet molding compounds. This is a popular approach when complex parts with features such as ribs or bosses, or parts with intricate details, are desired. Pressures of 800 psi to 2000 psi (5516 kPa and 13 790 kPa) and temperatures of 275°F to 350°F (135°C to 177°C) are used as SMC material is squeezed between the die halves for several minutes. When compared to cold press molding, pressure and heat improve the flowability of SMC into small channels or voids in complex parts. Hot press compression molding is a process that can produce parts with good surface appearance and close dimensional control.

Compression molding is often used when parts are needed in medium- or high-volume quantities. Compression molding is used to make car body parts, appliance components, truck liner panels, office machine housings, tote boxes, building panels, dinnerware, and transformer cases.

Injection molding

Injection molding is a high-volume production process used primarily with thermoplastic materials, but it can also be used with thermosets. Injection molding is an important process for the plastics industry. If it were possible to retrieve a sample of each plastic product made today throughout the world and classify the products according to the process used to manufacture them, it is likely that injection molding would be listed as the top producer.

Injection molding is a process that forces a measured amount ("shot") of liquid plastic into a heated die cavity. While both compression molding and injection molding use a closed mold, the molds are quite different. When the process is used with metal materials, it is called die casting. Figure 9-3 shows how the injection molding process works.

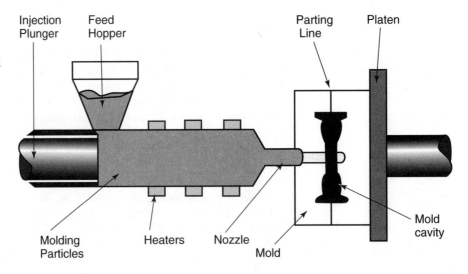

Figure 9-3
Thermoplastic materials that have been heated and softened are forced into the mold cavity by the injection molding plunger. After the part cools, the mold opens and it is ejected or removed. The molding cycle then repeats.

As shown in the illustration, pellets of thermoplastic stock flow from a funnel-shaped feed hopper into a heated compression cylinder. Once the pellets are melted to a liquid state, a plunger forces a controlled quantity of material through the injection molding nozzle into the closed mold. When the viscous liquid is injected into the mold cavity, it forms a part of the desired shape. Often, the mold will be cooled by water running through cooling channels. After the part cools (often a matter of only seconds), the mold is opened and the part is ejected or removed by robots or some form of simple automation.

Robots are gaining popularity for removing materials from injection molding presses. Robot unloaders are spinoffs of early devices called mold sweeps, which were used to push the runner out of 3-plate molds. *Sprue-picking robots* are also used today to remove sprues and runners (trimmed recyclable waste) from opened molds.

Both thermoplastics and thermosets can be injection-molded. The molding of thermoplastics, as described above, is most common. With thermosets, molding compound is squeezed from a low-temperature barrel into a closed mold. The matched halves of the mold are then heated to polymerize the thermoset material. After the part has cured, the mold is opened and the part ejected.

Some injection molding machines have a reciprocating screw-type barrel that transports pellets through heating stages before the material is injected into the mold. Other systems use a plunger that forces the stock around a heated mandrel called a torpedo. The reciprocating screw machine is the most popular.

Injection molding machines are rated according to the maximum number of ounces of heated polymer that can be shot, or injected, into the mold with one stroke of the injection ram on the press. Machines are also classified by the force that holds the molds closed. This is often several hundred tons. Figure 9-4 illustrates a plunger-type torpedo injection molding machine. Note the heating and cooling provisions.

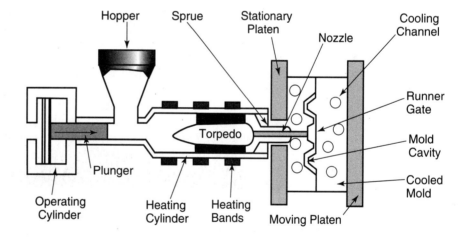

Figure 9-4
This injection molding machine forced material around a heated mandrel or "torpedo." Note the heating bands around the cylinder, and the cooling channels through the mold. Use of a cooled mold allows increased production rates.

Injection molding is a high-speed production process that can quickly produce large quantities of small parts. It is frequently used to manufacture electrical equipment housings, computer cases, appliance parts, automotive components, and microwaveable dishes.

In most cases, parts ejected from the mold require no finishing, other than trimming off sprues and runners (paths inside the mold carrying plastic to the molded part). This takes significantly less time than most types of finishing processes.

Sometimes two or more materials are injected into the mold to make an individual part. This is called *coinjection molding.* Coinjection molding is suited to manufacturing applications where a core must be made of one type of material and the outer shell from another.

Reaction injection molding

Reaction injection molding (RIM) is a high-production process used with thermoset plastics. The process involves combining two or more reactive liquids by aggressive mixing in a high-pressure mixing head just before they enter the mold. Once the liquids are in the mold, they quickly polymerize to form the completed part. The largest market for RIM parts is in the production of automotive and truck body panels, such as hoods and trunk lids. RIM is also used to produce thermal insulators for refrigerators and freezers, and stiffeners for structural components.

One of the important advantages of reaction injection molding is that fibers such as glass, graphite, or boron can be molded in to reinforce components. RIM parts have closed cells, with a continuous solid skin as thick as 0.080 in. (2.0 mm). This cellular structure makes it possible to produce a part that is stiffer than one made from solid plastics or metals.

Reaction injection molding works differently from conventional injection molding of thermosets. When injection molding a thermoset, polymerization sets the part's shape, with the heated walls of the mold activating the reaction. In reaction injection molding, the reaction is activated by impingement mixing in the mixing head. In RIM, polymerization takes place in the mold, during the process of cooldown. When polymerization is completed, the part is completely solid, and is cold to the touch.

Conventional thermoplastic and thermoset resins are not used for RIM. Most RIM production is with polyurethanes or ureaurethanes. Nylon RIM production began in 1983, and dicyclopentadiene and acrylamate materials were introduced in 1985. Research and development work is under way investigating the use of epoxies, unsaturated polyesters, and phenolic materials for RIM applications.

There are three major advantages in using RIM, compared to conventional injection molding:

∆   Lower pressures.

∆   Lower temperatures.

∆   Greater design flexibility.

These benefits often translate into lower part costs to the consumer.

In RIM, pressures in the mixing head are high, often exceeding 3000 psi (20 685 kPa). However, the actual pressure in the mold is only around 50 psi (345 kPa). Pressures required in the mold for conventional injection molding might be as high as 5000 psi (34 475 kPa). This makes the RIM process particularly useful in producing large parts. The actual clamping force in the mold of RIM vs. injection molding is also significantly less.

Temperatures for RIM usually range between 70°F and 120°F (21°C and 49°C). Normally, mold temperatures for injection-molded materials range between 150°F and 250°F (66°C and 121°C). This reduction in mold temperatures results in energy savings and translates into reduced part cost to the consumer.

One of the greatest advantages of the RIM process is the design flexibility that it permits. The use of low-viscosity liquids permits easy filling of molds, and makes it much simpler to produce complex parts. Many RIM parts are being produced with wall sections as thin as 0.010 in. (0.25 mm). The use of liquids makes it easier to change the mix to meet the demands of the product.

The pieces of equipment that make up a RIM system include:

Δ   Material conditioning tank.
Δ   High-pressure injection pistons.
Δ   Mixing head.
Δ   Mold carrier.

The reactive liquid is circulated in the conditioning tank, through the injection pistons into the mixing head, and back to the tank through heat exchangers. This keeps the temperature of the shot uniform and improves distribution of the liquid throughout the mold.

About 15 seconds before the mold closes, the pistons move backward, filling up with liquid. The pistons are then pressurized and the liquids fed into the mixing head for high-velocity impingement mixing. When the mixing is complete, the mixhead is opened and the mold is filled.

The mold carrier holds the mold, also called the "tool," in the proper position for molding. The carrier also provides clamping pressure, opens and closes the mold, and orients the mold as needed for various operations.

The range of materials and applications for RIM is diverse. Low-modulus elastomers are used in the automobile industry for fascia, bumper covers, trim parts, and window seals. High-modulus elastomers are used to produce large industrial parts. Reinforced elastomers use fillers such as glass fibers, flakes, or mineral fillers, to increase product strength. Integral skin foams are used in RIM to produce steering wheels and armrests. Polyurea RIM materials are used to make auto body panels. Polyacrylamate materials are being produced by Ashland Chemical Corporation, under the trade name Arimax®, for use with preforms and fiberglass mats.

Blow molding

Blow molding is a process used for producing thin-walled and medium-walled thermoplastic products. Most blow molding is used to manufacture packaging products, such as household bottles and containers for cosmetics, pharmaceuticals, chemicals, foods, and toiletries. It is also a popular process for making toys and other hollow objects.

Three major blow molding processes are now commercially used in the United States: extrusion blow molding, injection blow molding, and coextrusion blow molding. Extrusion blow molding is the most common of the three.

Each of the blow molding processes starts by placing a preformed *parison* (a hollow tube of heat-softened resin) inside a convex mold. Air is then blown into the parison to expand it against the walls of the mold. As shown in Figure

9-5, the parison is placed between the open mold halves, then the mold is closed, crimping both ends of the parison. Air is blown into the parison, expanding the tube against the walls of the mold. Finally, the part is cooled and the mold is opened.

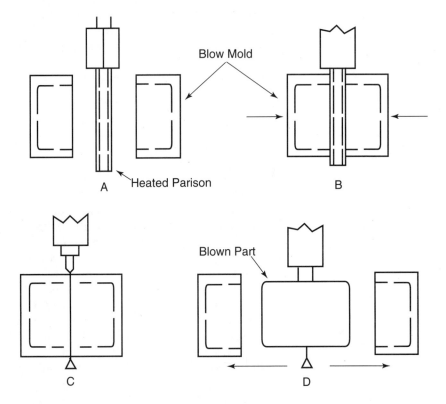

Figure 9-5
Operation of a typical blow molding machine. A—The heated parison is inserted between the halves of the opened mold. B—The mold halves close, pinching off the top and bottom of the parison. C—Air is blown into the parison, stretching its walls to fill the mold. D—After the part cools, the mold halves open to allow removal of the blown part.

Blow molding is a versatile process, used to make a variety of types of products. Figure 9-6 illustrates how blow molding can even be used to form two sheets inside a mold into an irregularly shaped hollow container. This process is called *indirect two-sheet blow molding*. With this process, an air tube is placed between two heated sheets in an open mold. The mold closes on the sheets, pinching off the ends. Air is blown into the mold through the tube, expanding the sheets against the walls of the mold. The same process can be used with plastic tubing instead of sheet stock. See Figure 9-7.

Another type of blow molding using heat is called *free blowing*. This process is popular for producing optically clear products, acrylic domed window panes, airline serving tray covers, and viewing panels. Free blowing doesn't use a mold. Instead, the heated plastic sheet is clamped between an upper forming ring and lower metal platen. Air is blown upward from a pressure box through a hole in the platen. The pressure forces the plastic upward into a bubble. See Figure 9-8.

Figure 9-6
The indirect two-sheet blow molding process is similar to normal blow molding, but uses two sheets of material, rather than the parison. As shown, the heated sheets and an air tube are clamped between the mold halves (Step 1), then air is injected to force the sheets to take the shape of the mold cavity (Step 2). After cooling, the mold will open and the part will be ejected.

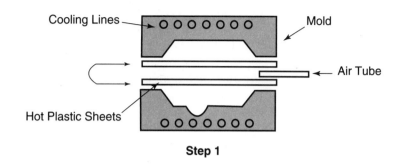

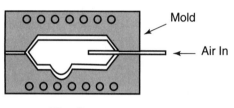

Figure 9-7
Indirect blow molding can also be used with plastic tubing, following the same sequence as the two-sheet method.

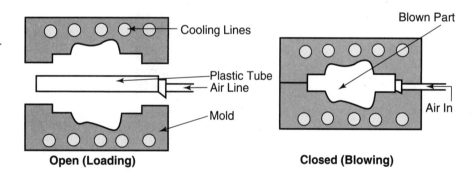

Figure 9-8
Free blowing is used to produce domed shapes of optically clear plastics. As shown, the heated sheet is clamped between the platen and a forming ring, then air is introduced from below to create the bubble or dome shape.

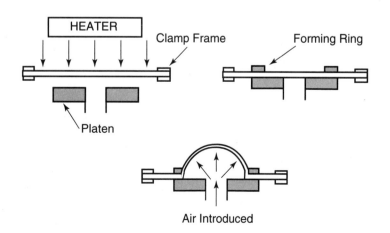

Blow molding is widely used by the plastics industry. In 1989, for example, 3.5 billion pounds of resin for blow molding was consumed in the United States alone. Blow molding machines range in price from $50,000 to $1 million.

Resin transfer molding

In *resin transfer molding (RTM),* catalyzed resin is transferred into a two-part matched mold. See Figure 9-9. In some RTM applications, a preform reinforcement or continuous strand mat is placed in the open mold. At other times, no preform is used. In either instance, the die is closed and preheated thermoset resin is then forced into the mold cavity from a separate chamber, called a pot. A plunger forces the resin into the mold cavity through a sprue, runners, and gates. The thermoset plastic is preheated, and is further heated in the mold to speed curing. Once polymerization has occurred, the mold is water-cooled to permit release of the part from the mold.

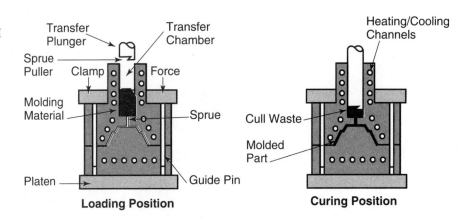

Figure 9-9
Thermoset resins used in RTM are heated to soften them for molding, then forced by a plunger into the cavity of a closed two-part mold. After curing, the part is removed from the mold.

Sometimes RTM resin is supplied to the mold by an external umbilical system, as shown in Figure 9-10. After the part cures (typically 10-20 minutes), the mold is opened and the part removed. The part that is produced will have two smooth surfaces, with attached sprues and gates. Pressures used in resin transfer molding range from 6000 psi to 12,000 psi (41 370 kPa to 82 740 kPa).

Resin transfer molding is preferred to compression molding for manufacturing parts that have significant variations in wall thickness. Without the preheating that accompanies RTM, such parts would be difficult to cure properly. RTM has become a popular process for molding low- and medium-volume reinforced plastic products.

Stamping

Stamping is a cold-forming process that uses matched molds in a stamping press to compact stock under pressure. Stamping is not considered to be a closed molding process, because no molding is involved. *Molding* involves shaping a substance by heating it until it is in a plastic (low-viscosity) state, then forcing it to conform to the desired shape in the mold. Stamping involves

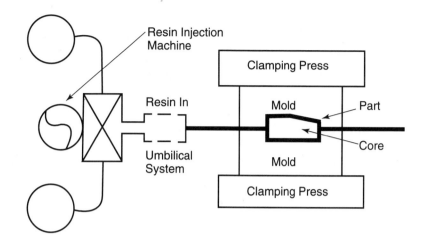

Figure 9-10
The use of an external umbilical system for RTM permits higher production rates, since a number of molds can be used. This is particularly important when using resins with curing times measured in minutes.

pressing solid or nearly solid sheet stock into the mold. It is different from compression molding in that it uses sheet stock rather than SMC, BMC, or preforms. Stamping is the fastest of all of the plastic processes used to compact stock in a mold. Here's how it works:

First, thermoplastic sheet material is heated until it is pliable. Then, the sheet is placed between matched mold die halves that are held in a compression or stamping press. The mold is rapidly closed and held in that position until the part cools. Stamped parts have good mechanical properties and high impact strength. Typical stamped parts are truck load floors, automotive ducts, fan guards, and fender liners.

Thermoforming

There are many different thermoforming processes. However, all involve heating thermoplastic sheet or film stock to its softening point, then forcing the material into or over the contours of a mold. See Figure 9-11. Some applications of thermoforming use vacuum; others do not. The best known of all of the thermoforming processes is called vacuum forming, which utilizes both heat and vacuum.

Vacuum forming

In vacuum forming, the plastic sheet is clamped and heated. A vacuum is then applied beneath the sheet, causing the atmospheric pressure to draw the sheet down against the cavity of the mold. See Figure 9-12. When the plastic touches the walls of the mold, it cools. In vacuum forming, a male (convex) or female (concave) mold can be used. See Figure 9-13.

Softening sheet stock for vacuum forming can be accomplished by heating the sheet in an external oven, then transferring it to the vacuum press. The material can also be heated on the vacuum press with infrared lamps.

The press used for vacuum forming consists of a box that holds the mold, with vacuum openings at the bottom. Clamps hold the sheet tightly against the top of the box. When the vacuum is applied the sheet is pulled

Figure 9-11
Thermoforming processes all involve forcing a heat-softened sheet of material into or over a mold, then allowing it to cool and assume the mold shape. The mold for this complex thermoformed part is cooled internally to speed production. The pipes around the edges of the mold base are tubes carrying coolant. (Aristech Chemical Corporation)

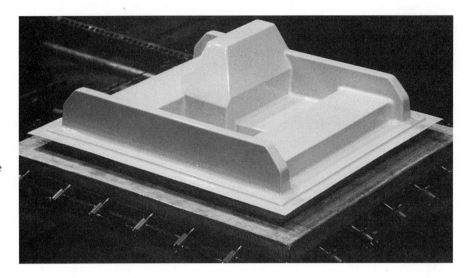

Figure 9-12
Vacuum forming is the most widely used of the thermoforming processes. As shown, a heated sheet of resin stock is clamped to the top of the mold, then a vacuum is drawn from below to permit the pressure of the atmosphere to force the material into the recesses of the mold.

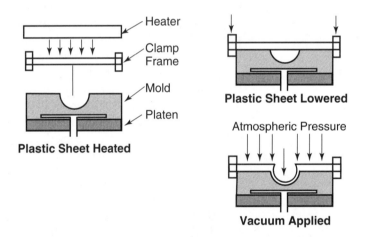

Figure 9-13
Vacuum forming can be used effectively even for fairly large products, such as this boat hull. (Aristech Chemical Corporation)

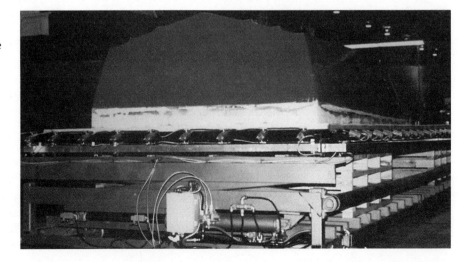

down over the mold, or the mold is pressed into the sheet. With vacuum forming, the side walls of the product are thicker than the area that touches the mold last.

One variation of the basic process of thermoforming, called *plug-assist thermoforming*, is shown in Figure 9-14. As is the case with most thermoforming processes, the plastic sheet stock is heated to permit it to drape down into a female mold. In the plug-assist process, vacuum is used in conjunction with a cylinder-activated plug to depress the heated sheet into the mold. The plug-assist process is useful for producing parts that have deep draws (a high depth-to-width ratio). Luggage, containers, and some auto parts are products that are made using the plug-assist thermoforming process.

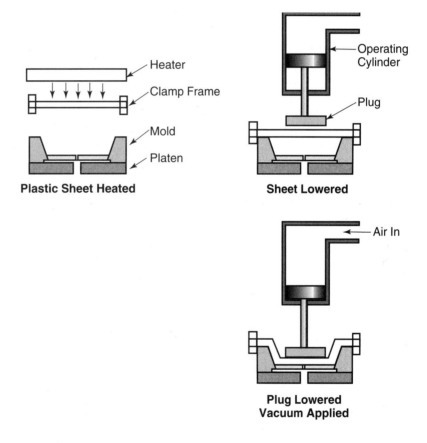

Figure 9-14
A variation of vacuum forming, plug-assist thermoforming uses an air-powered plug or plunger to reinforce the vacuum action when manufacturing a deep-drawn part.

Another variation of the thermoforming process is shown in Figure 9-15. This process is called *mechanical stretch forming*. It is very similar to plug-assist thermoforming, except that no vacuum is applied. The plug is used to depress and stretch the stock into the female mold.

Figure 9-15
In mechanical stretch forming, the air-powered plug is the primary force for shaping the part. No vacuum is used.

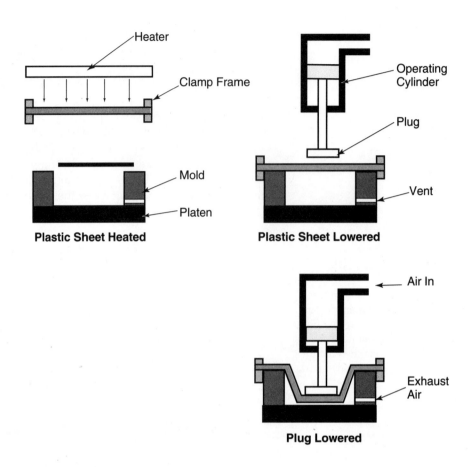

Figure 9-16 illustrates how the thermoforming process can be accomplished, without vacuum, by using a pressurized chamber. This process is referred to as pressure forming. Here, the sheet is softened with heat and placed in the pressure chamber. Air is blown in, providing the pressure to force the stock against the walls of the female mold. See Figure 9-17.

Figure 9-16
Pressure forming uses air pressure to force the softened sheet stock into the female mold. Vents in the mold allow air to be exhausted from beneath the sheet as pressure is applied from above.

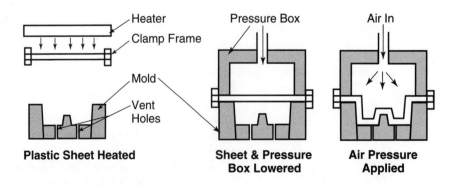

Chapter 9 Forming Plastic Materials 235

Figure 9-17
Pressure-formed products. A—The surfaces of these electrical panels have been formed with varied textures for appearance and mechanical reasons. B—Large parts, such as the dash and instrument console of this street sweeper, can be pressure-formed. (PMW Products, Inc.)

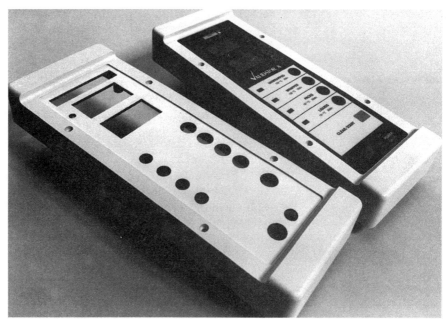
A

B

Thin-walled products, such as windshields, bowls, safety helmets, and luggage, are often produced with thermoforming processes. Many variations of the process enable the production of unique shapes by blowing or drawing softened sheets with air pressure.

Molds for thermoforming can be made of virtually any material. Popular materials are metal, phenolic paper laminate, plaster, wood, and ceramics.

Drape forming

Drape forming eliminates a problem that results from vacuum forming: parts with inconsistent wall thickness. With this process, the sheet is normally clamped, heated, then drawn down over the mold. Sometimes the tool or mold is forced into the softened sheet. When the sheet droops, a vacuum is applied.

The unique aspect of drape forming is that after the sheet is softened, the outer edges are draped, or pulled down, over the mold. The use of draping and vacuum results in more consistent wall thicknesses than other thermoforming processes. However, the points where the walls fold are still thinner than the other areas of the product. See Figure 9-18.

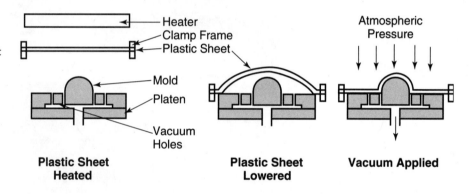

Figure 9-18
Drape forming is used to produce a thermoformed product that has a consistent wall thickness.

Pressure bubble plug-assist vacuum forming

Not all vacuum forming processes produce a vacuum from above the softened sheet. With pressure bubble-plug assist vacuum forming, air is introduced to the mold cavity beneath the sheet, causing the softened sheet to be blown upward in a bubble. This stretches the plastic evenly. Then a plug, shaped like the contour of the mold cavity, pushes the sheet downward into the mold as a vacuum is drawn on the part.

Matched mold forming

In *matched mold forming*, a two-part metal die is used to compact the softened thermoplastic sheet. Molding compound is pressed into the space left between the mating male and female dies. Sheet molding compounds, sheet, and preforms can be used with this process.

As in other thermoforming processes, the sheet is heated while being held tightly in a horizontal position between the two mold halves, Figure 9-19. Holes through the mold permit heat to circulate in the die halves. After forming is complete, these holes also provide channels for circulating water to hasten cooling.

Figure 9-19
Matched mold forming is the process of choice when fine surface detail and close product tolerances are involved.

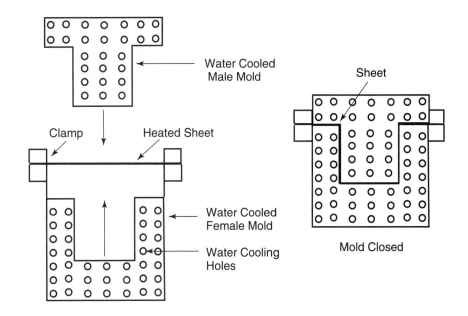

Matched mold forming is an excellent process when fine detail and close parts tolerance are desired. Grained surfaces and lettering can be produced with this process. Conventional metal stamping presses are often used for matched mold forming.

Open Molding

There are two types of open molding, hand layup and sprayup ("spray layup"). Basic materials used in open molding include: thermosetting resins (usually polyesters), glass fiber reinforcement (roving), and a catalyst (usually methyl ethyl ketone peroxide or benzol peroxide). Roving is used in woven or mat form for hand layup, or in strips several inches wide for use in chopper guns for spray layup.

With both methods of open molding, the glass fiber reinforcement roving is encapsulated in resin. It is the roving that provides structural strength to the product. When the catalyst and resin are applied to the reinforcement, cross-linking takes place and a structural laminate is formed.

Open molding is a process that is widely used for making prototypes, pools, tanks, boats, ducts, truck and bus components, housings, and corrugated and flat sheet stock. Figure 9-20 shows a typical hand layup of a boat hull, with resin being rolled into a layer of woven roving.

Virtually any size product can be open-molded. It is one of the simplest of the polymeric processes. A minimal amount of equipment is needed, and the molds can easily be fabricated of wood. The major drawback of open molding is that it is time-consuming, resulting in high per-unit labor costs.

Hand layup

With *hand layup,* the process essentially consists of applying a layer of roving to a mold, and then saturating the roving with resin. The process may be repeated with additional layers to provide the desired thickness and

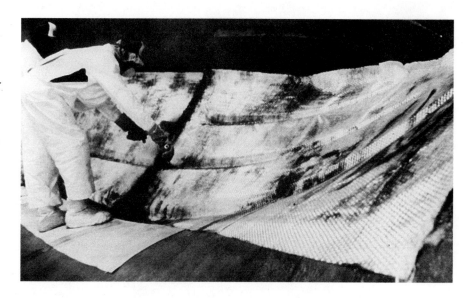

Figure 9-20
Hand layup is a widely used open molding method for such products as pleasure boat hulls. In this view, workers are impregnating fiberglass roving fabric with resin by rolling and brushing. The layers of resin and roving are applied over a thin gel-coat that will form the outer "skin" of the finished boat hull. (Grady White Boats)

strength. Molds used for hand layup are often made of wood or plastic. The tool may be complex, incorporating prepreg fabrics, foam plastic cores, or honeycomb cores.

To better understand how the tool works, imagine how a typical product such as a shower enclosure would be made. The final product has a smooth inside surface (the part you see) and a rough outside surface (the part that would face the wall of your house). Imagine that the male mold for making the enclosure is resting on supports on the floor before you, and that the mold extends upward to form the shape of the enclosure. The exposed surface of the mold is highly polished so that a smooth exposed interior surface will result when the part is completed.

Hand layup involves application of two distinct material surfaces—an aesthetic surface called gel-coat, and an interior structural laminate. The process of making a shower enclosure would begin with cleaning and coating the mold with a mold release compound. A thin film material, such as cellophane, might be applied to the mold to achieve a smoother finish. A specially pigmented resin, called *gel-coat*, could be mixed with a catalyst and sprayed on to form the smooth finished surface of the enclosure. The spraying of a gel-coat layer on a boat hull is shown in Figure 9-21.

The gel-coat layer would be applied to achieve a thickness of 15 to 20 mil (about the thickness of a piece of heavy paper). The gel-coat surface eliminates any need for any final surface treatment or finishing.

Next, fiberglass mat would be cut and placed in a single layer to cover the entire surface of the mold. Polyester resin and MEKP (methyl ethyl ketone peroxide) catalyst would be mixed together and applied to the mat. The resin mixture would be brushed or troweled on. In corners and other areas when saturation is difficult, a roller would be used.

The gel-coat, fiberglass reinforcement, and resin would cross-link and permanently bond, forming a structural laminate. After a short curing time, the finished product would be removed from the mold. The smooth-finished

Figure 9-21
To provide a smooth, attractive surface for the finished product, a layer of resin known as gel-coat is sprayed into the mold before hand layup of resin and reinforcement is done. The gel-coat is often pigmented to provide a permanent color finish. (Grady White Boats)

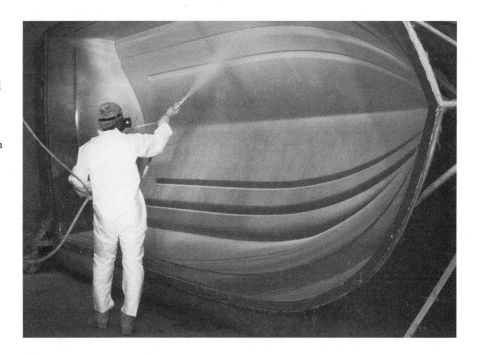

shower enclosure, an exact duplicate of the mold, would be ready for trimming and shipping.

Hand layup is preferred to sprayup by many manufacturers, because the resin is actually forced into the mat material by roller pressure. This results in a higher resin content, and a heavier and more durable product.

Sprayup

Like hand layup, sprayup is also achieved using an open mold. *Sprayup* is similar to hand layup, except that a chopper gun is used to spray strands of roving and catalyzed resin into the mold. Equipment required for sprayup is shown in Figure 9-22. The four basic components of the system are the resin pump, the catalyst dispensing unit, the spray gun, and the glass fiber cutter.

Figure 9-22
The sprayup method of open molding uses a gun that employs the principle illustrated here: streams of catalyst and resin are combined with short chopped lengths of roving and applied to the mold to build up a structural layer.

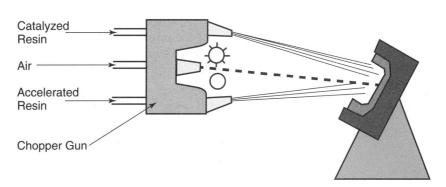

The equipment functions in the following manner: The resin pump feeds resin to the spray gun. The catalyst dispensing unit can either be a catalyst injector or a pump designed for use with MEKP. The resin pump and catalyst dispensing unit are connected so that both cycle at the same rate. There are several different types of spray guns: common types are airless, air-atomized, internal mixing, external mixing, air assist, or airless.

A glass fiber cutter is mounted on top of the spray gun. The cutter chops continuous strips of roving into desired lengths (typically 1-2 in.). Chopped strands are directed down into the resin/ catalyst stream, so that all three materials are deposited on the mold simultaneously.

Sprayup equipment is available with various tip sizes of tips and resin pump air delivery systems. Typical outputs of chopper gun systems range from 3 to 25 lb./min. The sprayup process is sometimes automated to maintain consistency of coverage. Sprayup is particularly popular for producing large tanks and pressure vessels.

Extrusion

Extrusion is an important process in the plastics industry. It is responsible for the production of more plastic products than any other process except injection molding. In principle, extrusion is a process that is similar to squeezing cake icing or toothpaste out of a tube.

In this process, a machine called an **extruder** converts thermoplastic powder, pellets, or granules into a continuous melt. The melt is then forced through a die opening to produce long shapes. Typical products manufactured with the extrusion process include auto trim, house siding, garden hoses, rods, soft drink straws, and profile molding. Plastic insulation is also applied to electrical wire and pipe with the extrusion process.

The *single-screw extruder* is the most common type of machine used for extrusion today. Extruders force plastic stock down a long barrel. Machines are sized by the barrel's inner diameter, and range from 1/2 in. to 12 in. (13 mm to 305 mm) or more inches in size.

The process begins with stock being fed by gravity into a hopper on the extruder. The granules, pellets, or powder flow down into a vertical screw which rotates feed stock through the heated barrel of the extruder and into the die. The barrel is heated and cooled by heating/cooling jackets surrounding its outer wall. Microprocessor controllers regulate the temperature throughout different heat zones in the barrel.

After the melt is forced through the die, it must be carefully handled and cooled to minimize distortion. The method used to cool the end product depends on the shape of the material. Sheet products are cooled on carefully polished liquid-cooled rollers. Figure 9-23 shows how pipe is formed by extruding molten plastic through a die, then pulling it through a cold water bath.

Film Blowing

The process for cooling blown film is more complicated. **Blown film** is extruded vertically in a tubular shape, in thicknesses from 0.015 in. to 0.025 in. (0.38 mm to 0.64 mm). After the tube is cooled by chilled air, it is collapsed by pinch rolls. Then air is forced in through the die mandrel to blow the tube into a bubble. This thins the walls to the desired final thickness. The tube is then slit to make thin sheets, and is wound up or made into bags.

Figure 9-23
Pipe is extruded by forming it over a mandrel as it exits the die. The mandrel is hollow, allowing air to be blown in to keep the soft plastic walls of the pipe from collapsing until it can be hardened by passing through a cold water bath.

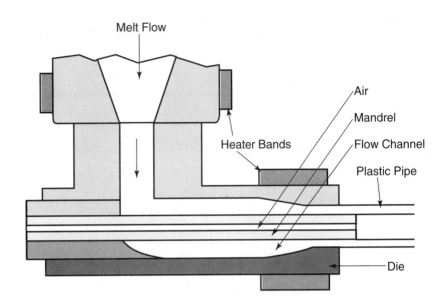

Extruded film can be differentiated from sheet stock in that anything less that 0.10 in. (2.5 mm) is classified as film. Film can be produced in either tubular (blown) form, as just described, or in cast form. Cast film is extruded through a linear slot die and cooled when it comes into contact with chilled metal rolls. Most film produced is blown (tubular) film, which is usually stronger than cast flat film.

Wire Coating

The production of coated wire is an extrusion process that uses a die with a tapered mandrel in the center to keep the inside of the product hollow. Wire is pulled horizontally through the mandrel and the liquid plastic surrounds it. After the coating is extruded, it is cooled and inspected, then the coated wire is wound with a coil-winding machine. Hollow tubing is made the same way, using a mandrel that extends through the die.

Pultrusion

Pultrusion consists of pulling continuous roving through three basic processing stages. The photograph in Figure 9-24 shows how raw material in the form of continuous roving is transformed into strands, or bundles, of pultruded material. The bundles are prepared by pulling the roving through a resin bath or impregnator. Next, they are pulled into preforming fixtures where they are partially shaped and excess resin and air are removed. Finally, the bundles are pulled through heated dies for forming, then into an oven for curing. Figure 9-25 shows pultruded bundles that have been processed into continuous lengths of plastic rod.

The manufacturing work cell used for pultrusion consists of a pultrusion machine, the forming dies, a pulling system, and a cut-off saw. The basic machine consists of a creel, resin tank, preforming fixture with heated die, and a puller. The creel is a shelf-like arrangement holding spools of roving. If mat reinforcement is used, it is dispensed above the machine.

Figure 9-24
A mat of continuous fiberglass roving, top, is being impregnated with resin and emerging (foreground) as thick bundles or strands. These strands will be further processed into reinforced plastic products. (Creative Pultrusions, Inc.)

Figure 9-25
A fluted rod of reinforced plastic is shown at center of this photo emerging from a pultruding machine. The large cutoff saw is used to trim the rod to uniform lengths. Inside the machine, bundles of resin-impregnated roving are pulled through heated dies to form the rod. After curing in an in-line oven, the finished rod emerges as shown, ready for trimming. (Creative Pultrusions, Inc.)

The process works like this: roving and mat enter the resin tank, where they are impregnated with polyester or epoxy resin. The resin bath, or "wet-out tank," is a trough containing rolls that force the reinforcement beneath the surface of the resin mixture. The impregnated resin leaves the wet-out tank through alignment slots in its end, and is pulled through a preform fixture to partially shape it. Preforming fixtures are often made of fluorocarbon or polyethylene because these materials are easy to form and to clean. Once the reinforcement leaves the preforming fixture it enters a heated matched metal die for final shaping. The shaped part can be cured in the die or in an oven. Then it is pulled into the pulling station. The puller consists of continuous caterpillar-type belts with pads to contact the pultrusions. After the stock leaves the puller, it is cut to length with a conventional cut-off saw equipped with an abrasive wheel.

Many products made with the pultrusion process go to the sporting goods market. Fishing rods, hockey sticks, bike flags, tent poles, CB antennas, skateboards, arrows, and golf club shafts are pultruded. However, the bulk of the market for pultrusions (since introduction of the process in the United States in the mid-1950s) has been for electrical applications. Pultruded products in this area include ladders, switch actuators, fuse tubes, transformer air duct spacers, and pole line hardware. One of the fastest growth areas for pultrusion is for corrosion-resistant products such as bridges and platforms, floor gratings, hand rails, and structural supports.

Extrusion Blow Molding

A process that combines features of extrusion and blow molding, called ***extrusion blow molding***, utilizes pellets instead of sheets or tubes. This process involves mixing and heating plastic pellets, powders, colorants, and additives in an extruder to produce a uniform melt. The plastic melt is forced through an extrusion die, which forms a tubular parison. The parison is then pinched together by the top and bottom of the closing mold. The parison is expanded by air pressure to force the plastic against the surface of the mold. The part is cooled and the mold is opened.

All extrusion blow-molded containers have flash. ***Flash*** is produced when excess material squeezes out of the seams of the mold. After the product is molded and cooled, the flash is trimmed and the product is completed. Automated flash removal systems are used with high-production blow molding machines. Bottles with necks are normally trimmed in the mold or machine.

Two different types of extrusion blow molding processes, continuous and intermittent, are popular. ***Continuous extrusion blow molding*** can be used with all blow molded resins, but it is most often used with polyvinyl chloride (PVC) and other heat-sensitive thermoplastic resins. The parison in continuous extrusion blow molding is continuously formed. This method is used to produce large containers up to one gallon in size.

With the continuous extrusion blow molding process, the parison clamp mechanism, which squeezes the parison shut at the top and bottom of the bottle, is moved back and forth between the blowing station and extruder. Shuttle and rotary clamping systems are widely used. With the *shuttle system*, once the parison reaches the desired length, the clamping mechanism shuttles from the blowing station to a position under the die head, where it surrounds the parison, trims it, and returns to the blowing station. The *rotary wheel system* provides as many as 20 clamping stations. This system enables a parison to be captured from the extruder while some parts are being removed and others are being cooled.

The *intermittent extrusion blow molding* process is used with polyolefins and other plastics that are not sensitive to heat. There are three basic types of intermittent extrusion blow molding machines: accumulator head, ram accumulator, and reciprocating screw. In this process, the parison is quickly formed by a clamp and part ejection system, immediately after the product is ejected from the mold.

The *reciprocating screw* machine pushes the melt backward and forward by hydraulic pressure. This causes the melt to be forced through into the die head for forming the parison. These types of machines are used to produce bottles smaller than three gallons in capacity. Usually, the "shot capability" in these machines is limited to less than 5 lbs. (2.2 kg).

The *ram accumulator* machine accumulates the melt in an auxiliary ram cylinder located beside the extruder. The disadvantage of this machine is that the melt that first enters the cylinder remains in the cylinder for the longest amount of time. Since the melt is heated for different lengths of time, there is lack of uniformity in the material. The ram accumulator system can produce any size parison, but normally it is used for parts from 5 lbs. to 50 lbs. (2.2 kg to 22.7 kg).

The *accumulator head* system is the most popular intermittent process for extrusion blow molding. The extrusion die head, called an accumulator head, is filled by the extruder. The accumulator head has a tubular plunger, which forces the melt out of the cylinder. With this process, the melt that enters first exits the cylinder first. In a typical operation, melt would fill the cylinder from the side, and would then flow around the inside of a mandrel, reuniting on the opposite side. Some accumulator head systems are capable of extruding shots of up to 300 lb. (136 kg).

Injection Blow Molding

Injection blow molding is another hybrid process, combining the best features of injection molding and blow molding. The process is used to make bottles from thermoplastic resins.

One of the major advantages of injection blow molding is that parts are produced with no flash. This is not the case with either injection molding or blow molding—both these processes generate parts with flash that must be trimmed. When thermoset materials are involved, this results in scrap which is often of little value (waste).

Injection blow molding is a two-step process. First, a tubular parison is formed by injecting resin into a metal die cavity. See Figure 9-26. The parison is kept hot until it is needed for blow molding. The blow molding phase of this process begins when the machine indexes (rotates) the preform from the parison station to the blow molding station. There it is enclosed by the mold, air is injected, and the desired shape is formed. The final step is cooling and the ejection of the part.

There are many different types of injection blow molding machines, which differ primarily in the number of stations that they utilize. Three- and four-station rotary designs are common. Costs for injection blow molding

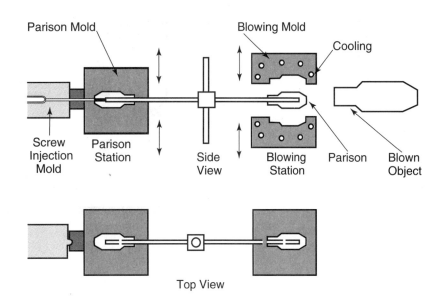

Figure 9-26
The injection blow molding process begins by using injection molding to form the parison, top left. The fixture holding the parison then rotates, and the halves of the blowing station mold close over it. Air is introduced to form the blown object. High production rates can be achieved with this process.

machines range from $100,000 to more than $400,000. Today 60 percent of all plastic bottles up to 16 oz. (0.47 liter) capacity are produced with injection blow molding.

Many producers are shifting from extrusion to injection blow molding because of its superior quality and productivity. Often smoother bottle surfaces can be produced from polished mold cavities. Parts do not have damage caused by removing flash, because there is no flash. Another advantage of this process is the uniformity of wall thickness that results.

The major disadvantages of injection blow molding are its size and design limitations. Normally, only bottles with capacities smaller than 16 oz. are produced. Also, the process is not practical for making bottles with handles. Another disadvantage is that the tooling needed is more expensive than that required for other blow molding processes.

Expandable Bead Foam Molding

Expandable bead foam products are popular for disposable food and cup containers, life support flotation systems, and packaging material. The plastic stock used for expandable bead foam molding is a lightweight, closed-cell polystyrene bead. Each bead contains a tiny amount of a compressed gas or volatile liquid. When the beads are heated, the gas or liquid expands, causing the plastic particles to "foam" to 50 times their original size.

The plastic stock is poured into a two-piece matching mold. Steam heat is normally applied to heat the mold and expand the beads to the shape of the mold. See Figure 9-27. There are several major producers of expandable polystyrene beads (EPS) used for this molding process.

Small, medium, and large beads are used with the process, depending on the minimum wall size that is desired in the product to be manufactured. Beads are expandable to as much as 50 times their original size, depending on the temperature and length of time they are heated.

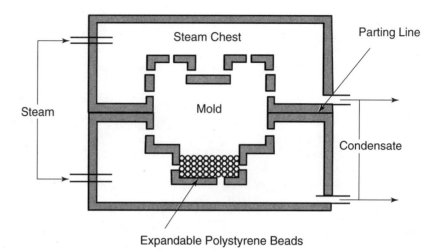

Figure 9-27
Expandable polystyrene beads increase in size by as much as 50 times when they are heated. This permits them to fill a mold quickly and completely. This process is used for products ranging from the familiar "foam" coffee cup to large blocks and sheets of rigid foam for building insulation.

Small beads are used for cup manufacturing. Medium-sized beads are usually used for molded shapes, such as packaging liners. Large beads are used to mold large solid blocks.

A light pastel-colored surface can be produced by adding dry color to the beads before they are expanded. If consistent color throughout the product is desired, colored beads must be purchased from a supplier.

It would be possible to immediately expand polystyrene beads in a mold. However, this is seldom the method actually used for producing EPS products. In most instances, beads are pre-expanded by exposing them to steam, outside the mold, until they are about the same density as required in the molded part. At this stage, the raw beads are unstable, and are easily deformed by shock. For this reason, they are then allowed to stabilize in a storage tank for a period of from 3 to 12 hours.

The EPS beads are then transferred by air conveyors to the molds in the molding machine. Steam is admitted from the steam chests through small holes in the molds. The heat expands the beads into the shape of the mold. The expanding beads close off the small holes, shutting off the steam. Finally, the mold is cooled and the part is ejected.

EPS bead molding is also popular in producing solid blocks that are later sliced to make sheets, slabs, or fabricated parts. The process is the same as that used to form shapes with EPS beads, except that the presses have large rectangular cavities instead of molds. Cavities in some of the machines are as large as 4 ft. x 3 ft. x 19 ft. (1.2 m x 0.9 m x 5.8 m). The cavities are filled with beads and steam is injected to expand them. After the product has cooled, it is cut into sheets with a band saw or hot wire. The hot wire method is most popular because of its speed and versatility.

Integral Foam Molding

Integral or **structural foams** have a surface layer or skin that has a density much like unfoamed plastic and a porous core with uniformly sized bubbles. The integral foam process is capable of producing parts strong enough for many structural applications. Structural foams can have glass fibers or carbon fibers incorporated to increase their strength.

Integral foam molding can be done using conventional injection molding equipment. It can also be accomplished by free foaming, cold molding, reaction injection molding, extrusion, or rotational molding. In these processes, a gas is added to the plasticized mix to create the foaming action. Either thermoplastic or thermosetting resins can be used.

Structural foams are created when a blowing agent gas expands, causing the polymeric melt to foam and fill the mold. The gas bubbles collapse when they contact the cold mold surface. This produces a solid skin on the formed part. Unlike the surface of other plastics, the thin skin of a structural foam is not cold to the touch.

Structural foams are different from laminated foams with sandwiched panels. Laminated foams incorporate the use of pre-fabricated sections in an assembly that is held together by foam. Structural foams consist of a single material with a density that gradually increases from the core to the surface skin.

The skin and core are integral components of any structural foam part. Integral foam is sometimes referred to as artificial wood, because of its similarity to the internal structure of wood. In fact, the first commercial application of integral foam was as a wood substitute in furniture. Integral foams have a higher stiffness-to-weight ratio than steel, with excellent impact resistance and dimensional stability. They are also being used as a substitute for concrete in some applications because of their significant weight-to-strength ratio. Integral foams are light in weight, while they exhibit greater strength than most other materials.

There are two basic types of structural foam molding: low-pressure processing and high-pressure processing. The low-pressure process is particularly useful when large and heavy parts with textured finishes are needed. There are two popular commercial processes for low-pressure structural foam molding, known as the Union Carbide process and the Beloit process.

With the Union Carbide process, thermoplastic resin and additives are fed from the injection molding machine hopper to the extruder barrel, where they are mixed and melted. Nitrogen gas is blown into the melt in the extruder barrel. The mix is then moved to an accumulator which maintains the proper temperature and pressure. When the melt is injected into the mold cavity, the gas expands and the foam fills the mold. As the foam strikes the cold surface of the mold, the bubbles collapse and form a skin around the porous core. When the part has cooled, it is ejected and the process starts over again.

The Beloit process is similar, except that nitrogen gas is not used as a blowing agent. Instead, chemicals that react to form the bubbles are blended into the thermoplastic resin and additives before they enter the extruder barrel of the molding machine.

High-pressure integral foam can be produced using conventional injection molding machines. What is different with this process is that the conventional injection molding pressure is enhanced by using rapid injection velocity to fill the mold. As a result, the mold is packed so full that its expansion capability is limited. This results in a denser skin on the part. Eventually, the gas on the inside of the part expands, causing the formation of a cellular core.

When the product is completed, it is trimmed to remove excess flash and runners. High-pressure parts normally require no surface finishing; low-pressure parts are often painted.

Typical product applications for integral foams include housings for computers, furniture, automotive bumpers, tennis rackets, wash basins, conveyor rollers, and window frames.

Free foam molding

The processes described in the preceding section used a closed-press mold. Structural foams can also be produced using a cold one-piece open press mold. This is called *free foam molding*. Granulated polymer, a chemical blowing agent, and other additives are fed into the cold female mold cavity. The mold is heated, usually by steam, and is then water-cooled. Foaming time normally takes 5 to 10 seconds; cooling time from 30 to 60 seconds.

This process is about ten times less expensive than conventional injection molding methods, but has some disadvantages: surface roughness, nonuniform thicknesses, and low productivity.

Rotational Molding

Rotational molding, sometimes called *rotomolding,* is used primarily to make seamless hollow products such as balls, containers, picnic coolers, floats, and toys. Other product applications are storage and feed tanks, agricultural sprayers, automotive dashboards, chemical storage tanks, hot tubs, trash containers, and fuel tanks. Products of almost any size can be made with this process. Historically, rotomolding has been used with thermoplastics, but the process is now being used with some thermosets, as well.

Rotational molding is accomplished by placing a polymer in a two-piece aluminum mold. The polymer may be a powder or in a liquid form known as a *plastisol*. The mold is then closed and heated, usually to between 500°F and 800°F (260°C and 427°C), while it is rotated simultaneously around two perpendicular axes. See Figure 9-28.

The powder or liquid coats the mold's inside surface, where it forms a thick skin or uniform layer. After the part is formed, the mold is indexed to a cooling station where air or water is used to cool it. The mold is then opened, the finished part removed, and another charge of powder or liquid is deposited for the next part.

Production equipment for rotational molding normally consists of a set of three arms. At any given time while the machine is operating, one arm will be in the load/unload cycle, another in the heating/rotation cycle, and the third in the cooling cycle. Some machines are capable of producing tanks as large as 500 gallons (1893 liters) in size. The latest shuttle-type machines have produced tanks with capacities exceeding 22,000 gallons (83 277 liters). With this type of machine, a large motorized shuttle cart travels on tracks from a heating oven to a cooling chamber.

One of the major advantages of rotational molding over injection molding and blow molding is that the molds are less costly. Aluminum is popular for molds, but other materials, including sheet metal, nickel, and fiber-reinforced epoxies are also used.

Figure 9-28
For hollow objects such as balls or storage tanks, rotational molding is ideal. The polymer, in liquid (plastisol) or dry powder form, is placed in the mold. As the mold is heated and rotated in two axes, the polymer forms a thick skin on the inside of the mold.

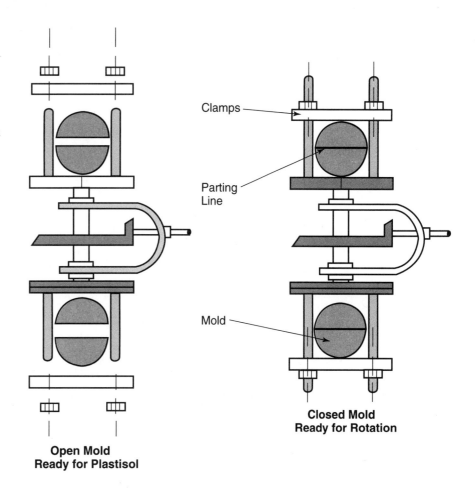

Vacuum Bagging

The *vacuum bagging* process is used as a method to improve the resin saturation of roving in open molding. Vacuum bagging is frequently used with sandwiched construction, such as honeycomb cores, foam plastic cores, and prepreg fabrics. It is also a process that can be used in the field for fabricating large structures, such as chemical tanks and boat hulls.

An end view of the tooling used for vacuum bagging is shown in Figure 9-29. Normally, cellophane or polyvinyl acetate is placed over the layup. The material to be bagged (resin-saturated roving or a composite) is placed against the cellophane. All joints are sealed with plastic. The air-tight vacuum bag is placed on top of the plastic, and the top edges are clamped against the supporting frame. A vacuum is drawn, eliminating voids and forcing out entrapped air and excess resin.

The use of vacuum results in less entrapped air and voids in joints and sandwiched areas. It also leads to higher *glass loading*. This means that the percentage of resin encapsulated in the fiberglass reinforcing mat is greater. This results in greater weight and adds strength to the final product.

The process has some disadvantages. The area next to the bag does not produce as good a finished surface as it would if it were next to the mold. There is more time and labor involved in vacuum molding, so cost is greater.

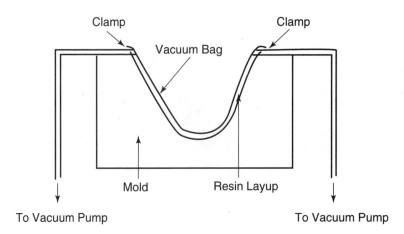

Figure 9-29
Vacuum bagging results in a denser, stronger sandwiched construction, since it eliminates trapped air in the resin/reinforcement layer.

Major products made with vacuum molding include domes, electronic components, aircraft parts, boat components, and engineering prototypes.

Casting

A number of thermoplastics (nylons and acrylics) and thermosetting plastics (epoxies, polyurethanes, phenolics, and polyesters) can be cast in rigid or flexible molds into sheets, rods, or tubes. Thermoplastics are more frequently used than thermosets for casting.

In order to cast thermoplastics, the monomer, catalyst, and necessary additives are heated and poured directly into the mold. After polymerization occurs, the part is ejected from the mold. One of the advantages of the casting process is that intricate shapes can be made using flexible molds. The molds are peeled off after polymerization takes place.

Sometimes, castings are made by pouring a partially polymerized syrup into the mold, and then heating the mold to complete the polymerization process. One of the disadvantages of this method is that normally *shrinkage* (a reduction in overall size) takes place during the cure. Shrinkage can be as great as 20 percent. One way to reduce the extent of shrinkage is to add up to 50 percent finely dissolved polymer to the syrup.

Sheet casting methods

The most common casting application for acrylics is the manufacture of sheet stock. Sheets are produced by either cell casting or continuous casting.

Sheets are produced in *job lots* (batches) through **continuous casting.** A highly viscous syrup is slowly cured between two continuously moving stainless steel belts. Width of the sheet is determined by the placement of flexible gaskets on the belts. Thickness of the sheet is dependent on the distance between the belts. Continuous casting is used for the production of thin sheets up to 0.375 in. (9.5 mm).

One interesting application of continuous casting is in the production of *films* (thin, usually flexible sheets of material). A film is cast by continuously drawing thermoplastic resin through a narrow opening in a die. Once the resin has left the die opening, it is cooled by passing it through quenching and chill rolls. Film thickness is controlled by varying the feed control rate and line speed. Film is produced in thicknesses up to 0.005 in. (0.13 mm). Thicknesses greater than this are classified as sheet stock.

There are two major pieces of equipment used in producing thermoplastic film: an extruder and a film-casting machine. Supporting process control systems, filtering, trimming, and material handling systems are also required. Since there is such a wide range of product applications for film, ranging from shrink wrap to food and medical packaging, there is often a need for specialized custom equipment to support specific applications.

Cell casting is accomplished using plate glass sheets slightly larger than the desired plastic sheet. The sheets are held together with spring clips that cause the glass to exert pressure on the plastic sheet, providing tension during cure. The thickness of the sheet is determined by a flexible gasket of PVC tubing that separates the glass sheets.

Cell casting begins with construction of the sandwich, or cell. The glass plates are clamped together against the gasket material, with one corner left open. The cell is tilted upward and filled with a weighed amount of catalyzed syrup. The syrup might consist of ingredients such as plasticizers, modifiers, release agents, colorants, ultraviolet absorbers, and flame retardants.

Thin cell-cast sheets are slowly cooled in the mold, using a forced-draft oven, typically for 12 to 16 hours. Thicker sheets are usually formed in a liquid bath, under pressure, in an autoclave. This keeps the monomer from boiling and reduces distortion.

The cell casting process is more versatile than continuous casting. Cell-cast sheets have better optical properties and smoother surfaces. Many different types of plastics can be cast.

Nylon casting

In applications requiring abrasion resistance and low weight, nylon is often used. Nylon casting is often more economical than injection molding or extrusion. Nylon cast parts can be created in almost any size and thickness. There are four steps in nylon casting: melting the nylon lactam monomer, adding the catalyst and activator, mixing, and pouring.

Melting and mixing is conducted in a temperature- and humidity-controlled environment, since the ingredients are extremely *hygroscopic* (water-absorbing). Water absorbed from the mixture would cause the catalyst to decompose.

The lactam flakes are melted under very precise temperature and environmental conditions. Any additives to be introduced into the melt must be carefully dried to eliminate moisture. The melt is then transferred by gravity or mechanical means to the mold. Parts that will require machining after casting are often annealed to reduce brittleness. Plastics are annealed by slowly cooling the parts in circulating air ovens or in mineral oil.

Potting and Encapsulating

Potting and encapsulating are casting processes that are often used in the production of electrical and electronic devices and components. Chopped fibers are combined with catalyzed resin and poured into molds to cast the desired product. Sometimes the mold is reusable. In other instances, the actual housing of the product serves as the mold. The plastic serves as a dielectric (nonconductor) and provides strength and protection to housings containing wires or electrical components. These processes are used when the wires, connections, or components do not need to be accessible at some later time —once potted or encapsulated, the components are sealed for life.

Potting is done by pouring polyester thermosetting resin into the housing of the product, therefore making the resin and case an integral part of the product. Most often, potting would be used to completely cover a component. A variation of this method is to cover only parts of the product with the polyester thermosetting resin.

There are several reasons for potting a product. One is to provide additional strength to the housing. Another is to have a reliable method for holding wires in position and provide additional protection in the case. Terminal blocks and electrical casings are among products that involve the use of potting.

Encapsulating is similar to potting, except that the mold is removed from the product in this method. The gel that is left serves as a permanent cover around the product. In this method, the entire component is covered with plastic. See Figure 9-30. Sometimes encapsulation is used to seal devices and connections, but it is normally used to embed coils, windings, transformers, chokes, resistors, transistors, diodes, and other electrical components.

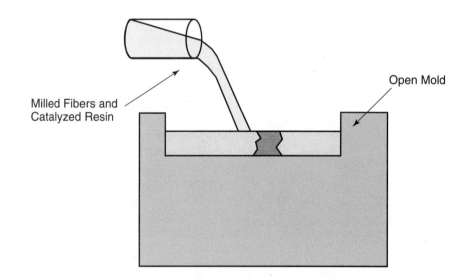

Figure 9-30
The processes of potting and encapsulation both involve pouring a resin (with or without reinforcement) over electrical components as a protective covering. Potting uses the component's housing as the mold. Encapsulation, shown here, uses an open, removable mold. The encapsulated component or assembly is surrounded by the cured plastic.

Laminating

Laminating is a process that involves sandwiching layers of plastics or composite materials into sheets or parts. Normally, laminating involves applying pressure to bond the layers together. Many thin-walled products are made by mixing and laminating plastic resins or films with reinforcing materials. There are two major categories of laminated products: high-pressure laminates (sheets, rods, and tubes), and low-pressure laminates (reinforced plastic moldings). Thermosetting resins, such the phenolics, silicones, and epoxies, are often used.

Low-pressure lamination is normally done using pressures of up to 1000 psi. Lamination occurs when layers of resin and fiberglass mat are sandwiched and the material is compressed under heat until polymerization occurs.

Low-pressure lamination is often used with thermoplastic film to provide a transparent protective covering for photographs and identification cards. The reinforcement can be paper, cotton, glass cloth, or any similar material. Even wood can be impregnated with plastic using low-pressure lamination. This is called the compreg process.

The ***compreg process*** is a unique variation of the basic process of lamination. Compreg is derived from the words, *compression* (com), and *impregnation* (preg).

The process begins by soaking a thin sheet of wood in liquid phenolic resin, and then letting it air dry. The impregnated veneer is then sandwiched with other pieces of impregnated veneer in a compression molding press. The heat and pressure causes the pieces to adhere to each other (laminate).

High-pressure lamination is used to adhere plastic coatings to particle board for making office furniture. Formica®, Micarta®, and Lamicoid® are common laminates. Continuous lamination is a high-pressure lamination process used to produce four-foot-wide construction panels at speeds up to 10 feet per minute. In continuous lamination, mat is pulled through a resin dip and between two cellophane sheets. The sandwich then passes through a heating zone, where it is cured. Finally, it is squeezed between two steel forming rollers. Resins used can be flame-resistant (acrylics) or weather-resistant (polyesters).

Calendering

Calendering is a process that was used more than 100 years ago to process natural rubber. Now, it is widely used in the plastics industry to manufacture ***polyvinyl chloride (PVC)*** film and sheeting. Most of the calendering throughout the world today is used to produce rigid and flexible PVC. Calendering squeezes pliable thermoplastic stock between a series of turning rollers to produce the desired film thickness. Calendering is normally conducted in three stages: mixing, feeding, calendering, and post-calendering treatment.

If PVC film is being produced, both solid and liquid raw materials may be required. Solids will be conveyed to a high-speed mixer from silos or bins. Liquids will be metered to the mixer. Mixers usually consist of planetary gear extruders. The mixing process is computer-controlled and fully automated.

Additives are introduced to the mixture, which is then heated to promote blending. The mixing of this dry blend is called ***fluxing***.

When the mixture has reached a consistency of soft clay, it is fed from the extruder through a metal detector or strainer to filter out contaminants from the stock. It is then conveyed onto the calender rolls. Calendering is normally accomplished on machines with from four to seven cast-iron rolls, depending on the design configuration of the machine. The rolls are heated to keep the sheet pliable during squeezing.

After calendering, the film that has been produced is conveyed by pick-off or stripper rolls in a vertical position to a unit called the embosser. The embosser consists of three rolls: the embossing roll, a contact roll, and a chilled rubber roll. It is the embosser that imparts texture or pattern to the film or sheet stock.

Next, the film travels through sets of rollers for tempering (hardening) and cooling. These rolls are carefully temperature-controlled. At this point, the product is nearly completed. The only operation left is entering the winding machine, where the completed film is trimmed, cut, and spooled. The winding machine has its own computer and programmable controllers that regulate the roll size according to the desired weight.

Centrifugal Casting

Pipe, tubing, and other round objects can be produced using the centrifugal casting process. Centrifugal casting is a process that is used to form ceramics, metals, plastics, and composites. The process is particularly useful with ceramics, and is discussed in greater detail in Chapter 21.

Fiberglass strands are saturated with thermoplastic polyester resin inside a hollow mandrel (tapered cylinder). Metal tubing is normally used as the mandrel. The mandrel is heated in an oven and rotated. Centrifugal forces throw the mix of resin and reinforcement against the walls of the mandrel. The mandrel continues to turn during the mixing and curing cycles. Curing is accelerated by pumping hot air through the oven. Products manufactured with the process include large tanks for agricultural chemicals, water softener tanks, pipe, tubing, and liquid storage tanks. Centrifugal casting can even be used to manufacture long cylinders with external threads.

The centrifugal casting process requires a minimum of labor and can be automated if high volumes are desired. It is a resourceful process with low tooling costs and little waste.

The major disadvantage of the process is that it is limited to the production of cylinders with uniform thicknesses. Tapered cylinders are difficult to produce.

Important Terms

blown film
B-stage
bulk molding compound (BMC)
calendering
casting
cell casting
closed molding
coated preform
coinjection molding
cold compression molding
compreg process
compression molding
continuous casting
continuous extrusion blow
  molding
die cavity

drape forming
encapsulating
expandable bead foam
extruder
extrusion
extrusion blow molding
films
fixture
flash
fluxing
free blowing
free foam molding
gel-coat
glass loading
hand layup
high-pressure lamination

hot press compression molding
hygroscopic
indirect two-sheet blow molding
injection blow molding
intermittent extrusion blow molding
job lots
laminating
low-pressure lamination
matched mold forming
mechanical stretch forming
molding
near-net-shape part
parison
plastisol
plug-assist thermoforming
polyvinyl chloride (PVC)
potting
preforms
prepregs
pultrusion
reaction injection molding (RIM)
resin transfer molding (RTM)
rotational molding
roving
runners
sheet molding compounds (SMC)
shrinkage
sprayup
stamping
structural foams
vacuum bagging

Questions for Review and Discussion

1. What are the similarities and differences between the resin transfer molding and reaction injection molding processes?
2. Select a plastic container, such as a parts bin. What steps would you perform if you desired to use this container as your mold and wanted to produce a part using hand layup? When and how would you add color, if desired?
3. What are the advantages and disadvantages of hand layup, compared to sprayup?
4. How is wire coating accomplished by using the extrusion process?
5. Sheets of thin film are produced from tubular stock. What keeps the walls of the tubing from sticking together?

Plastics can be easily formulated to provide desired characteristics and can be readily formed into simple or complex shapes by a variety of processes. As shown in some of these knobs and handles, a plastic can be combined with other materials, such as threaded metal rods. Plastics can also be finished to resemble metal or other materials. (Jergens, Inc.)

Separating Plastic Materials

Key Concepts

△ Proper selection of cutting speed and sawtooth shape is important for efficient material removal when cutting plastics.

△ Cutting speed is calculated in feet per minute, as measured along the circumference of the cutting tool.

△ Twist drill bits for plastic are ground to a sharper angle than those used for metal. Cutting speeds must be calculated carefully to avoid material breakage.

△ Plastic materials are cut using a variety of processes, including die cutting, extrusion cutting, laser cutting, waterjet cutting, and hot wire cutting.

Many of the processes that are used to cut metal and wood materials are also used with plastics. With separating processes such as circular sawing, the difference in approach between materials relates to the type of saw blade that is used.

Plastic sheet stock, rods, and tubes must be cut using fine-tooth blades, with approximately twelve teeth per inch. Specialty blades can also be purchased for cutting plastic. Band saw blades with four to six teeth per inch are ideal for use with plastics.

Most plastics can also be cut using traditional metal removal processes such as blanking, shearing, or die cutting. The removal process called routing, widely used with wood materials, can also be used with plastics. It is particularly useful in applications where the cutting tool does not go completely through the workpiece.

Plastics can be drilled with the same conventional twist drills used for metals and woods. However, since plastics are more brittle and easier to fracture than metals, the drills should be sharpened with a low lip relief angle, or even no relief. In practice, this means that the cutting edge of the drill should be almost perpendicular to the face of the material being drilled. The drill should be ground with a sharper point, Figure 10-1. This would keep the drill from gouging into the stock and cracking the workpiece.

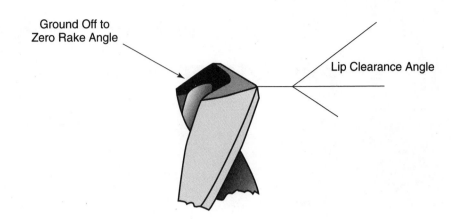

Figure 10-1
Twist drill bits for plastics should be ground so that the cutting edge has zero rake angle (is perpendicular to the material being drilled). Drills used for plastics have a sharper point than those used for metals. Typically, a drill used with plastics should be ground to a point angle of 60° to 90°. A drill for metals will have a 118° point angle.

Machining Plastics

Nearly all plastics can be machined — milled, cut, turned, planed, shaped, routed, stamped, drilled, and tapped. However, the approach taken with a particular manufacturing process must be adjusted to fit the unique characteristics of each material. Most often, these differences in approach relate to the types and speed of the cutting tool, and the use or absence of a lubricant or coolant.

Cutting Speeds

When the cutting tool is a saw blade, it is classified according to the number of teeth per inch. A fine-tooth cutter would have more teeth per inch than a coarse-tooth cutter. In most cases, a fine-tooth blade is used to cut plastics. The tooth *shape* is also important. When cutting with a bandsaw, the *skip-tooth* pattern is effective in clearing chips out of the kerf (saw cut). For some types of thermosets, the *hook tooth* blade is preferred. See Figure 10-2. Plastics have a machining speed of approximately 200 fpm (feet per minute). Fpm refers to the distance that the cutting edge of the tool travels in one minute (measured along the circumference of the cutting tool). Fpm can be calculated as follows:

$$fpm = \frac{\pi D \times rpm}{12}$$

$$rpm = \frac{fpm \times 12}{\pi D}$$

where:
  rpm = revolutions per minute,
  fpm = surface feet per minute,
  D = diameter of cutting tool in inches,
  π = 3.1416.

Figure 10-2
The skip tooth type of bandsaw blade is preferred for cutting plastics, since it more effectively clears chips from the saw cut, or kerf. The hook tooth design is used for some reinforced plastics that are harder or more dense. Compare the hook and skip tooth designs to the conventional tooth design.

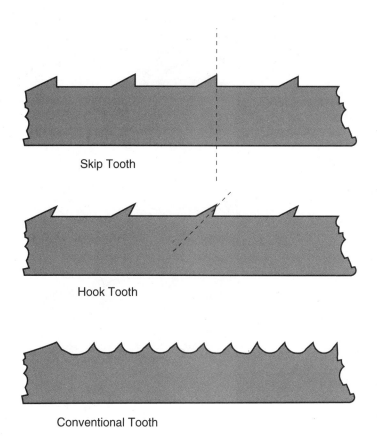

Skip Tooth

Hook Tooth

Conventional Tooth

To fully understand the importance of selecting the proper cutting speed for a particular process, consider the cutting speed (expressed in fpm) of plastics, against the speed used with other materials. Hard rubber has a cutting speed of 200 fpm, the same as plastics. Aluminum has a faster cutting speed, 250 fpm. The cutting speed of soft cast iron is 125 fpm. The ultra-hard metal titanium, has a cutting speed of only 15 fpm.

Not all plastics are cut in the same way. Normally, thermoset plastics are more abrasive to cut than are thermoplastics. Some high-pressure plastics even require diamond- or carbide-tipped tools. Cutting may also be done in some applications with waterjet or laser tools, Figure 10-3. Tools must cut cleanly without burning, since the coefficient of thermal expansion for plastics is nearly ten times greater than it is with metals. When plastic is heated, the workpiece exhibits three thermal effects:

- Δ   It absorbs heat.
- Δ   It expands.
- Δ   It transmits heat.

Expansion is described by the ***coefficient of thermal expansion*** α. In plastics, there is a close correlation between the coefficient α and the melting point. The higher the coefficient, the lower the melting point.

Figure 10-3
Plastics may be separated by many different means, including laser cutting done on a machine like this one. Computer numerical control allows the laser to rapidly perform cutting operations on simple or complex shapes. (Shibuya Kogyo Co.)

What should be remembered is that the harder the material, the slower the cutting speed. This is important to know when using a manufacturing process such as drilling. In drilling, the speed that the drill spindle turns is described in terms of revolutions per minute (rpm). However, this is not the same thing as *cutting speed*. The cutting speed of a drill is called its *peripheral speed*. This is the distance in feet per minute that the cutting edge of the drill actually travels. If soft materials, such as plastics, are drilled or cut using too great a cutting speed, breakage of the material is likely to occur.

Cutting Extruded Lengths

Cutting long lengths of extruded plastic is often done with an attachment to the extruder called an *extrusion cutter.* This is a machine that uses a high velocity rotating knife to cut the extrudate. Extrusion cutters are used to cut plastic hose, tubing, and various shapes with a diameter of less than 4 in. (102 mm). When cutting short lengths of material 1/8 in. (3.2 mm) or less in diameter, extrusion cutters can make several hundred cuts per minute. Machines are also designed to cut long lengths, often up to several hundred feet.

Die Cutting

Die cutting is a process that is used to cut plastic sheets, blister packaging, foam for upholstered furniture, and openings in molded parts. The process has long been used in the printing industry to cut packaging and containers using presses that force sharp-edged cutting rules through the paper or cardboard stock. See Figure 10-4. Plastics process applications range in complexity from simple hand-held dies to automated cutting equipment.

Figure 10-4
Continuous die cutting. A—Rotary dies are arranged around the circumference of a cylinder, allowing high-volume production. Two of the many sizes of cylinders are shown. (The Rotometrics Group) B—This full-head hydraulic cutting press can exert 80 tons of pressure to cut material using the dies shown on the open press bed. (Hudson Machinery Worldwide)

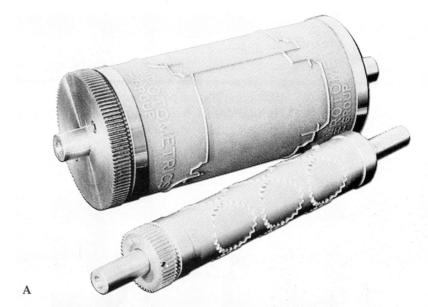

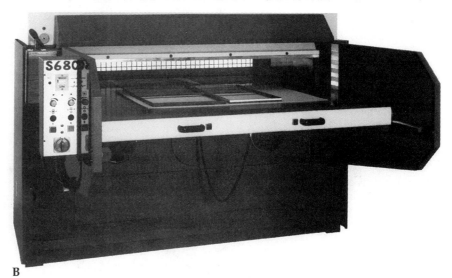

Hand-held dies consist of sharp-edged cutting rules bent to the desired configuration, much like a cookie cutter. The die is struck with a mallet to force the cutting rule through the material. Die cutting is a simple stamping process.

Die cutting machines are either single plane or rotary. With the ***single plane die cutting*** machine, a flat die set with multiple patterns is pressed into the stock. Normally the stock to be cut is placed on top of the press bed (a heavy shelf), and the die travels down on top of the stock. Some machines use a traversing roll to apply pressure.

Continuous die cutting can be accomplished with either a flatbed press or a rotary machine. In the rotary configuration, dies are arranged around the circumference of a cylinder. A back-up roll provides pressure for the cutting action; the material to be cut is fed between the die cylinder and the backup roll. Figure 10-5 shows a computer-integrated die cutting system capable of 120 cutting strokes per minute. The system used PLC microprocessor control. The machine shown in Figure 10-6 is a *traveling head press* designed to make multiple cuts in wide sheets or roll goods.

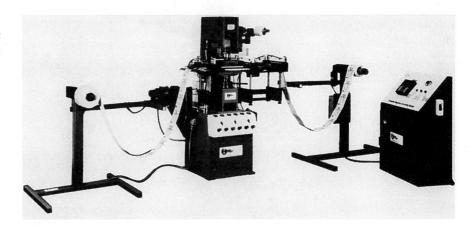

Figure 10-5
This computer-integrated flat bed converting system uses a programmable logic controller (PLC) to monitor and control high-speed die cutting operations. (Hudson Machinery Worldwide)

Figure 10-6
Travelling head presses like this one are designed to enable multiple-piece cutting from wide sheets or continuous rolls of material. (Hudson Machinery Worldwide)

Die cutting is used to cut parts such as gaskets and labels from various materials including plastics, rubber, foam, paper, and composites. Precise control of the die cutting process allows it to be used for such applications as *kiss cutting* of self-adhesive materials without penetrating the paper or film backing. See Figure 10-7.

Figure 10-7
Note how the material between the die cut shapes has been removed after cutting, leaving the backing intact. This mechanical platen press performs "kiss cutting," a precision operation in which the die cuts the material but does not penetrate the backing. It is often used for self-adhesive materials that will later be peeled off the backing and attached to another surface. (Hudson Machinery Worldwide)

Foam Cutting

The most common method for cutting plastic foam is a hot wire. **Hot wire cutting** is a simple process, which is accomplished with a specialty machine that is usually built for a particular application, often at low cost to the manufacturer.

This cutting method utilizes a thin wire that has high electrical resistance. When the wire is connected to an electrical power supply, it heats up, so it can be used to slice through foam plastic by melting the material. See Figure 10-8. Often, multiple wires are arranged side by side, so that several different parts or sections can be cut at the same time.

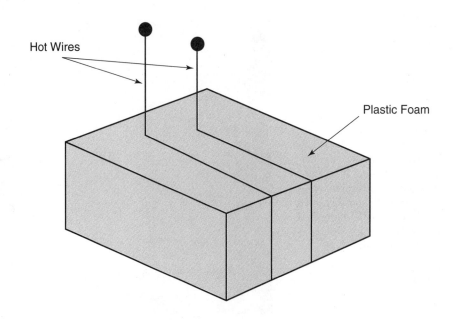

Figure 10-8
Plastic foam material can be cut at a high rate of speed by using a thin wire heated by electrical resistance. The wire cuts the foam by literally melting a narrow path through it. Some hot wire equipment will use a number of wires to perform simultaneous cuts.

Degating

To remove gate and runner systems from molded parts, specialized hand tools called degating cutters are used. These specially designed pliers are shaped so that the cutting surfaces are able to make a cut flush with the edge of the part. Cutters are available in various shapes to facilitate cutting in locations that normally would be difficult to reach. Degating tools are also used to cut structural foam molded parts.

Deflashing and Deburring

When a two-piece mold is used to produce the part, a thin seam or fin of material, called *flash*, often results at the point where the mold halves come together. Flash is produced with many casting processes, no matter what type of material is being molded. *Burrs* are sharp edges on parts that are produced by force, usually when a cutting tool removes material from the workpiece.

It is common practice to remove flash and burrs from parts. Sometimes this is accomplished by simple hand scraping. Another method for removing flash and burrs from the outside of a part is a process called tumbling.

Tumbling

Tumbling is a process much like the process of milling, which is used to mix ceramic powders and additives. When tumbling is used to deflash plastic parts, the parts are placed inside a canister or drum. As the drum turns, the parts slide against each other, gradually rubbing away the flash. No other grinding medium is used; the parts produce the abrasive action.

Sometimes flash is produced inside the part, around holes, slots, and grooves. It is more difficult to remove flash or burrs on internal holes or grooves. When this is the case, tumbling may not be an effective material removal process.

Vibratory deflashing

Another deflashing process, called *vibratory deflashing,* uses high frequency vibrations to improve the abrasive (rubbing) action between the parts. In this process, the cutting action is improved by adding an abrasive medium. The abrasive might consist of anything from powdered mineral material to metal pellets to wood chips or even corn cobs. Because of the abrasive cutting action, this process creates a dull matte finish on the parts.

Flash is also removed using a media gun. The media gun may be used for sandblasting, in which abrasive sand particles are carried in a stream of air from a nozzle to abrade away the flash. Many firms transport parts on a conveyor through a media stream. The deflashing media might consist of synthetic polymers, natural organic products, or minerals. Organic and polymeric materials are most frequently used as grinding media with plastics.

Grinding

In order to recycle plastics, it is usually necessary to break up rejected parts, runners, or other scrap material into small pieces or granular form. This is accomplished by grinding.

There are four different types of grinders that are used in the plastics industry to reclaim parts and scrap. These are generally referred to as the light-duty grinder, auger-fed grinder, medium-duty grinder, and heavy-duty grinder.

For light-duty grinding, the machine is normally placed in the manufacturing workcell, directly beside the equipment used to form the part. Small grinders, with motors of up to 40 horsepower (hp), are capable of handling scrap with a wall thickness of approximately 1/4 in. (6.3 mm). Such light-duty grinders are normally fed and emptied by hand, but can be attached to blowers and hopper loaders.

Another type of grinding machine is referred to as the auger-fed grinder or *granulator.* These machines are used when the manufacturing process is automated or semi-automated. They work in conjunction with other simple automation devices, such as sprue pickers (robot arms), and automatic load/unload systems. The material that is fed into these grinders typically consists of sprue and runner systems, as well as occasional reject parts.

Medium-duty grinders, with motors up to 100 hp in size, are larger and heavier than light-duty grinders. They have larger throats and cutters capable of grinding stock up to 1/2 in. (12.7 mm) thick. In most instances, the particles that these machines produce are loaded into drums, either by gravity feed or a pneumatic system.

Heavy-duty grinders, with motors of up to 400 hp, are used for large parts and tough grinding applications. See Figure 10-9. Because of the noise that they make when crushing and grinding parts, these machines usually are installed in a separate room, away from the work area. Heavy-duty machines are automatically loaded and unloaded.

Figure 10-9
This heavy duty granulator, driven by a 200 hp motor, is used to grind thick scrap pieces into small granules that can be melted and reprocessed. The cutting chamber has been opened to show one of the three knife blades mounted on the massive rotor. (Polymer Systems)

The finished product of each of these grinding operations is a material ranging from a coarse granule to a fine powder. The ground material can be recycled into new products at the injection molding station. Care must be taken not to mix unlike colors, or the ground material will have a lower value. Mixed colors are used to make stock that is used for filler and in applications where color of the product is not important.

There is really no end to the list of processes that can be used to separate plastics. Many additional separation processes are discussed throughout the text, in conjunction with separation of metals, woods, ceramics, and composite materials. These processes often can be used successfully with plastics. What is important to remember is that each material has its own unique behavioral characteristics. Adjustments in cutting speed, type of tools used, and type of process chosen may be necessary.

Important Terms

coefficient of thermal expansion (α)
continuous die cutting
cutting speed
die cutting
extrusion cutter
flash
granulator
hot wire cutting
kiss cutting
peripheral speed
single plane die cutting
travelling head press
tumbling
vibratory deflashing

Questions for Review and Discussion

1. Describe the type of blade that you would purchase if you were interested in cutting plastic sheet stock with a sabre saw or reciprocating saw. How would this blade be different from the blade needed for metal or wood?

2. When drilling plastic, should you use a cutting speed faster or slower than you would when drilling metal? What is the relationship between cutting speed and the coefficient of expansion of plastic?

3. If you were setting up your own plastics manufacturing firm, how important would you consider the purchase of a grinder or granulator? How would you use the recycled material? What would you do with the regrind from thermoset parts?

Resins used produce plastics are provided in a number of forms, including granules, flakes, and various kinds of pellets. Different resins may be combined to achieve desired properties or colors. These pellets are examples of color concentrate material that would be mixed with other resins. (Reed Spectrum)

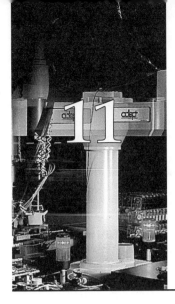

Fabricating Plastic Materials

Key Concepts

- △ Cohesive joining is used to permanently assemble plastic parts.
- △ Embedded wire, used with dielectric heating, is one means of welding plastics.
- △ High-frequency vibrations are used to generate frictional heat as a means of joining plastics.
- △ Molecular movement caused by radio-frequency waves creates the heat needed to seal products made of thermoplastic polyurethane or polyvinyl chloride.

There are many different methods used to *fabricate* (assemble, or make from component parts) products from plastic materials. One major method is mechanical fastening—there are literally thousands of different types of fasteners that are used to join materials. These range from rivets to Velcro® strips to quick release thongs to machine screws so small that they can be seen only under a magnifying glass. While the use of mechanical fasteners is an important aspect of fabrication, fasteners constitute a significant field in and of themselves, and are beyond the scope of this chapter. Instead, our focus in this chapter will be on other processes used to fabricate plastic products. Most of these involve cohesive joining that results in permanent assembly.

Cohesion Processes

Welding is a widely used fabricating process for thermoplastics. It should be distinguished from the process of sealing. While both are cohesion processes, *welding* is the joining of plastic components of relatively heavy thickness. *Sealing* is a process that is reserved for joining of films.

Hot Gas Welding

Most welding is done on polyvinyl chloride (PVC) and polyethylene (PE) materials, but all thermoplastics can be welded. Many of the techniques used with gas welding of metal materials also apply to plastics. There are some major differences, however.

Plastics welding uses a plastic filler rod, rather than metal, and the welding temperatures are much lower. Conventional gas welding flames have temperatures as high as 6000°F (3316°C). Such high temperatures would melt and burn plastics.

Plastics welding is accomplished using a welding gun that heats a gas (usually nitrogen or air), with an electric heating element or coil heated by a gas flame. Electric guns are preferred because of their light weight. The heated gas is applied to the material to be welded. A filler rod of the same material as that being welded is melted into the joint. The rod is held at an angle to the joint and heat is applied to the rod and the joint. Several passes are usually made, at a speed of about 1 in. to 2 1/2 in. (25 mm to 64 mm) per minute. Welding of a plastic container is shown in Figure 11-1.

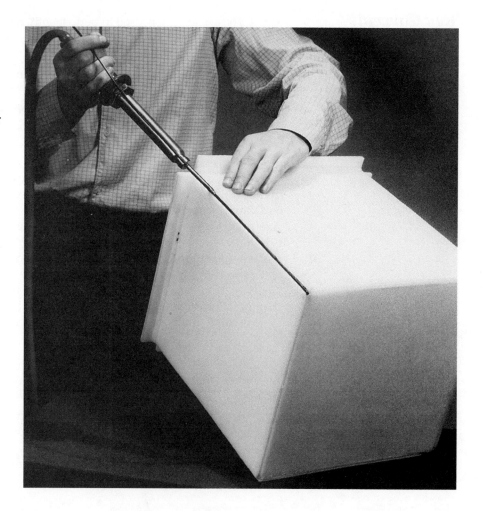

Figure 11-1
A welding gun that heats a stream of gas is used, along with a plastic filler rod, to weld plastics. This worker is welding a seam on a plastic container. Note how the filler rod is fed through a channel in the gun for melting at the tip. (Laramy Products Co., Inc.)

Hot Wire Welding

You may recall reading, in Chapter 10, about the separating process known as hot wire cutting. *Hot wire welding,* or dielectric heating, works on the same principle. Hot wire welding is a simple process in which a wire with high electrical resistance is placed between two thermoplastic surfaces that are to be joined. With this process, the wire is left in the joint after it is welded together.

The process can be used to join plastics, films, fabrics, or foams with a high dielectric loss characteristic (heat dissipation factor). Materials typically joined by this method include cellulose acetate, ABS, polyvinyl chloride, epoxy, polyester, polyamide, and polyurethane. Other plastics such as polystyrene, polyethylene, and fluoroplastics have very low heat dissipation factors and cannot be joined with this method.

Here's how the process works: the wire that is to be sandwiched between the two sheets of plastic is bent into a zigzag configuration or multiple loops to provide additional strength in the joint. Often, a slot or indentation is made in the plastic to hold the wire. A high-frequency (millions of cycles per second) voltage is then applied to the wire. Fusion occurs when the high frequency causes the molecules to rapidly realign themselves. This molecular movement causes frictional heat and the plastic is melted in the area surrounding the wire.

The Federal Communications Commission controls the transmission of high-frequency energy. Signals which are transmitted by hot wire welding are in the range of millions of cycles per second, similar to those produced by TV and radio transmitters. Conventional transmitters or generators normally operate at frequencies between 20 and 40 megahertz (millions of cycles per second).

Vibration Welding

Vibration welding is a quick method for joining most thermoplastics. Normally, parts can be vibration-welded in seconds. Vibration welding is practical for parts that are too large for ultrasonic welding. An important requirement for vibration welding is that the design of the joint permit clearance between the parts during vibratory action. The process is particularly effective in applications where leakproof pressure or vacuum joints are desired.

Vibration welding can be performed on machines that produce linear or angular vibratory action. Low-frequency vibratory machines are classified as those operating at from 120 Hz to 130 Hz. High-frequency machines operate at from 180 Hz to 260 Hz.

Electrical current is transmitted via wires to the joint area. The high frequency that is produced causes the plastic molecules to vibrate, thus creating sufficient frictional heat to melt the thermoplastic.

Induction Welding

Two major types of induction welding, or ***induction bonding***, are used in industry today: magnetic induction bonding, and Hellerbond induction bonding. Both processes can be used with either thermoplastics or thermoset materials. Heating cycles required for induction welding are short, normally only a few seconds.

As was the case with dielectric welding, a radio-frequency magnetic field is responsible for creating the heat at the joint. The difference is that no lead wires are required to transmit current to the weld area.

In magnetic induction bonding, an adhesive or bonding agent is placed in the joint between the two surfaces to be bonded. The powdered adhesive or bonding agent contains ferromagnetic particles dispersed throughout a polymer. The powder reacts (heats up) in response to a radio-frequency (RF) magnetic field generated by an induction coil.

The Hellerbond process is similar to magnetic induction bonding, except that the adhesive or bonding agent is replaced with a ceramic powder. The powder normally consists of finely ground magnetic iron oxide. As in magnetic induction bonding, the powder responds to an RF magnetic field. The advantage of this process is that the ceramic powder can be incorporated directly into parts, such as filaments, sealing gaskets, or liquid adhesives.

Many types of thermoset and thermoplastics—ABS, nylon, polyester, polyethylene, polypropylene, and the polycarbonates—can be induction-welded. Induction welding is also effective for joining materials such as glass, paperboard, textile, paper, and leather.

An important advantage of induction welding is that it can be used with heat-sensitive materials and with dissimilar materials. One drawback is that it can be used only with nonmetals.

The equipment required for induction welding is an RF power supply that can operate in the frequency range of 1.8 MHz to 4 MHz. The tooling that is required performs two functions. It holds the parts together, and also applies the RF field to the bonding agents through an induction coil embedded in the tooling.

There are many applications for induction welding, since products can be of virtually any size. Handles on kitchen knives and appliances are joined with this process. Thousands of induction-welded products are produced daily for the automotive, packaging, construction, and office equipment markets.

Spin Welding (Friction Welding)

Spin welding, or friction welding, is typically used to accomplish butt welding of cylindrical or tubular parts. One of the parts is held stationary in a nesting fixture, while the mating part is spun rapidly against it. Friction created at the point of contact generates heat and brings the two pieces to their melting point.

Immediately after the material reaches the melting point, the spinning stops and the parts are held together for about 1/2 second for cooling. After curing, the weld joint that is created is often much stronger than the substrate itself. Normally, the best weld results from welding times of no more than a second or two.

There are several points of caution that must be observed when using this process. When welding a solid bar, the heat generated is localized around the outer circumference of the stock, rather than the center of the bar. This leaves a relatively weak point of adhesion in the center. Also, the process develops internal stresses in the parts. Sometimes, these stresses must be removed through subsequent annealing.

Ultrasonic Welding

Ultrasonic vibrations can be used for welding and assembling metal and plastics parts, spot welding, transforming adhesives to a molten state, and stitching films and fabrics together without the use of needle and thread.

Ultrasonics can be used for many different types of thermoplastic sealing applications. Ultrasonic vibrations are also used for *staking,* a process that involves melting thermoplastic studs to mechanically lock dissimilar materials in place. Staking is illustrated in Figure 11-2.

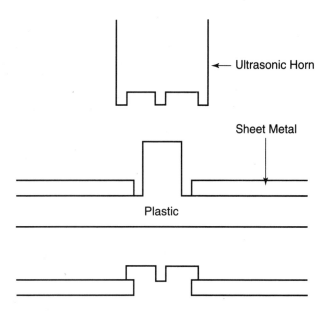

Figure 11-2
Staking is a variation of electronic welding. High-frequency vibrations cause plastic studs to heat up and soften so that they can be used to lock dissimilar materials together, as shown.

Another application of ultrasonics is *inserting,* embedding a metal component into a preformed hole in a plastic part. Ultrasonic welders are also used for spot welding and slitting.

Ultrasonic assembly is a fast and efficient method for assembling rigid thermoplastic parts or films and synthetic fabrics. Plastic can be joined to plastic, as well as to metal or other nonplastic materials. Ultrasonic welding eliminates the need for mechanical fasteners and adhesives.

Ultrasonic welding of thermoplastics such as ABS, acetal, polycarbonate, and polystyrene can be accomplished using high-frequency vibrations in the 20,000-40,000 cycles-per-second (kHz or kilohertz) range. These mechanical vibrations, applied to two mating parts that are pressed together, create rapid movement in the molecules at the joint interface (seam). This generates heat and results in fusion of the parts. The weld that is created often approaches the strength of the material in the original plastic part.

There are six major components in most ultrasonic welding systems. Power is generated through a *power supply,* which transmits 60 Hz energy to the converter. The *converter* changes the electrical energy to high-frequency (20-40 kHz) mechanical vibrations. A booster is attached to the converter. The purpose of the *booster* is to increase or decrease (fine-tune) the amplitude of the vibration to the point yielding the strongest weld with the least amount of flash. The horn is a key element. The *horn* is a metal wedge-shaped bar that travels down on a vertical spindle to contact the workpiece. The horn transmits the mechanical vibrations to the workpiece. The horn is made of aluminum, titanium, and hardened steel, and is sometimes highly polished and chrome-plated.

Attached to the base of the machine is a *fixture* or nest. The part to be welded normally consists of a two-piece shell that is snapped together and then ultrasonically joined. The fixture (often made of aluminum) is carefully machined to conform to one-half the shell that is to be welded.

The final component of the system is the *actuator.* The purpose of the actuator is to house the converter, booster, horn, and pneumatic controls. The actuator brings the horn into contact with the workpiece, applies the desired force for the correct length of time, and then retracts the horn after welding has been completed.

A microprocessor-controlled ultrasonic welder is shown in Figure 11-3. The power supply, converter, and booster are located in the control console at the top of the machine. The highly polished horn is visible at the bottom of the photograph.

A typical application for welding with the ultrasonic system would be a two-piece cover or shell housing electrical parts. The operator would begin by placing half of the shell into the nesting fixture. Normally, this half would include wiring and components. Then the other half of the housing would be snapped onto the first shell half. The operator would next place both hands on buttons on either side of the actuator and depress them, lowering the horn onto the two parts. The horn would hold the parts firmly together as ultrasonic vibrations are emitted, causing heat to be produced where the two parts fit together. After welding, the horn is retracted and the welded housing removed.

Joint design is critical in ultrasonic welding. The joint should have a small, uniform contact area, and include a method for aligning the parts properly. This is normally accomplished with alignment pins designed into the part.

Ultrasonic assembly techniques are used with a wide range of thermoplastic products. Products that are often sealed with ultrasonics include: head lamp assemblies, steering wheels, food and beverage containers, blister packaging, small appliances, electronic assemblies, toys, video cassettes, ballpoint pens, and medical components.

Radio-Frequency Heat Sealing

Radio-frequency (RF) heat sealing is often used to join thermoplastic parts that cannot be merged with processes such as ultrasonic welding, spin welding, and induction welding. Radio-frequency heat sealing is most frequently used to join rigid and flexible plastics, such as PVC and thermoplastic polyurethanes. Radio-frequency heat sealing has been used for many years, but is still one of the least-known methods for joining plastics.

RF sealing works much like ultrasonic welding: heat is generated in the joint, or area to be joined, by molecular movement. Radio-frequency heat sealing utilizes a frequency of approximately 27 MHz. In RF heat sealing, the high-frequency radio waves, combined with the energy and pressure exerted by pressing the parts together, cause the molecules to oscillate and create heat.

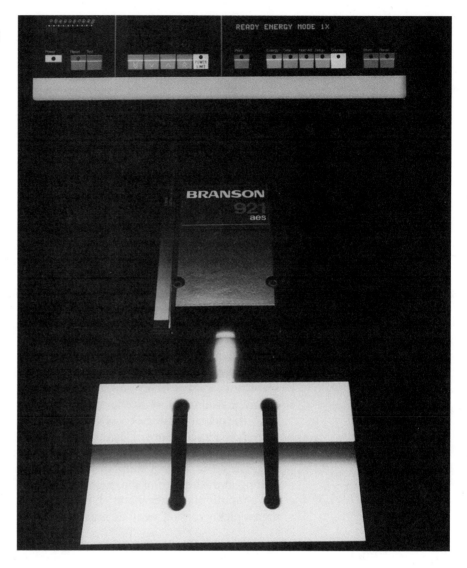

Figure 11-3
The microprocessor control of this ultrasonic welder allows precise temperature, time, and pressure regulation when welding joints in plastics. (Branson Ultrasonic Corporation)

Equipment required for RF heat sealing consists of a generating system, sealing press, and die or electrode. The generating system typically includes a generator (normally up to 25 kW in size), a controller, a power supply, an oscillator, and safety protection devices. The power supply converts AC to high voltage DC. The oscillator converts high voltage DC to high-frequency AC voltage. An arc detector device protects the system from overload, and will automatically shut down the equipment when preset limits have been reached.

The sealing press consists of a pneumatic ram that provides the pressure required during heating. The press utilizes a heated bottom platen with controls to regulate pressure during up and down strokes of the ram. The heated platen helps to keep the temperature constant during welding.

A brass die, or electrode, is constructed to provide the desired seal shape. The die is attached to an aluminum plate. A buffer material with high dielectric properties (such as electronic fishpaper or polyester film), is placed between the plate and the machine's bed. The buffer reduces heat loss from the workpiece to the bed.

This process is often used to seal blister packs and other packaging materials, such as medical bags, blood pressure cuffs, tarpaulins, pool liners, inflatable toys, life jackets, and eyeglass cases. It can also be used to bond together some nonplastic materials, such as cotton, paper, and woven fabrics. RF sealing can also be used to produce embossed signs, "tear seals," and decorative patterns.

Dielectric Sealing

Dielectric sealing, using RF vibrations, can be used to bond many thermoplastics, films, fabrics, and foams. Materials with high dielectric loss characteristics such as cellulose acetate, ABS, polyvinyl chloride, epoxy, polyether, polyester, and polyurethane are often joined with dielectric sealing. Other materials, such as polyethylene, polypropylene, polystyrene, and the polycarbonates do not have the proper electrical characteristics to be bonded by this process.

The thermoplastic to be sealed is placed as a dielectric between two flat electrodes or sealing bars. These electrodes normally consist of brass strips backed with steel or aluminum. The sealing bars are attached to a high-frequency power source, which is capable of emitting frequencies of between 20 and 40 megahertz (MHz). The FCC has assigned a frequency of 27.18 MHz for dielectric sealing. Most machines use this frequency.

The actual heating occurs when the molecules try to realign themselves with the RF oscillations. This movement of molecules causes frictional heat, causing the point of interface between the two materials to become molten.

Dielectric sealing supplies heat at the surface, causing more intensive heating at the interior of the thermoplastic. Thick material can be heat sealed more rapidly than thin material, because the heat created in thin material is more quickly carried away by the electrodes.

Thermal Heat Sealing

Heat sealing can also be accomplished *without* the use of high-frequency energy. **Thermal heat sealing** is done with a heated tool or die. The sealing bar or tool is held at a constant temperature, permitting the heat to flow through one of the layers of the film to the interface joint. This method is practical for thin film sealing. Some of the common types of seals that are used by the packaging industry are: lap, fold-over, step, tear, perforated, and peelable. The lap seal is used to join two wide sheets of film when high strength is required. Figure 11-4 illustrates some of the many applications for thermal heat sealing.

Thermal impulse heat sealing is a variation on this process in which the dies are heated intermittently. The temperature is increased during the heating cycle and is then decreased to permit the joint to cool under pressure. This intermittent heating produces improved appearance in the joint and is best-suited for film with a thickness of less than 0.010 in. (0.25 mm).

Heat sealing is used for many applications that require sealing plastic to other packaging materials, such as printed card stock. Figure 11-5 illustrates many of the different applications of thermal heat sealing for blister packaging.

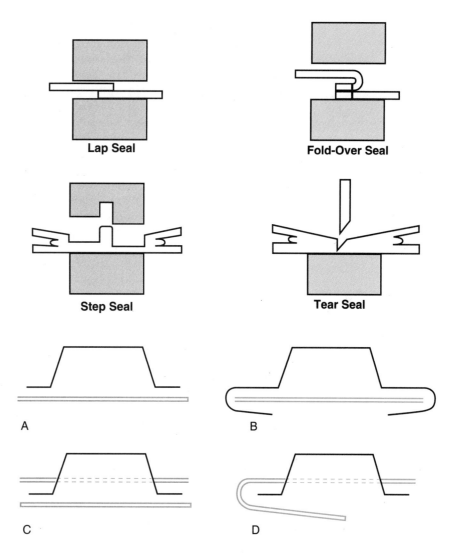

Figure 11-4
Some common types of seals that are made using thermal heat sealing.

Figure 11-5
Various applications of thermal heat sealing in blister packaging. This method allows plastics and printed card stock to be bonded together. A—Blister heat-sealed to card. B—Blister clipped and heat-sealed to card. C—Blister sealed between double cards. D—Blister sealed to fold-over card.

Cementing and Bonding

In addition to welding and sealing, cohesive bonding of plastics (softening and flowing together the two materials to be joined) can be achieved by the use of chemicals called *solvents* or solvent cements. In order to produce cohesive bonding in plastics, the surfaces must be heated to cause them to flow and allow the molecules to mix together. On thermoplastic polymers, this is normally accomplished by using solvents. The solvents create heat, soften the material, and permit a strong molecular bond.

With the acrylics (such as clear plexiglass), bonding is normally accomplished by soaking, capillary action, or dipping. With the *soaking* method, one of the pieces of acrylic to be joined is immersed in the solvent or cement. This "soaking" produces a softened cushion of acrylic material that will harden when the two pieces are pressed together and cured. *Capillary action* involves squeezing cement or solvent along the joint between the two acrylic surfaces. The cement is drawn into the joint by capillary action so that the softening and bonding can take place. The *dip method* is similar to the soaking method. It involves dipping one piece into cement and then pressing it against the piece to be joined until the two are bonded.

Thermosets and some thermoplastics do not soften easily when exposed to solvents, so they are typically joined by the process of ***adhesion*** or adhesive bonding. An ***adhesive*** is a bonding agent that forms a film between the materials to be joined. By "sticking" to both pieces, the adhesive holds them together. Some adhesives remain flexible; others form a rigid joint. Also, some types are best for bonding materials of the same composition, while others are best for joining dissimilar materials.

Choosing the correct bonding agent for adhesive bonding of plastics can be difficult. Adhesives must be selected which withstand the stresses of the environment where the product will be used. There are eight basic types of adhesives used for bonding plastics:

- Silicones.
- Cyanoacrylates.
- Reactive acrylate-based adhesives.
- Polysulfide sealants/adhesives.
- Epoxy resin-based adhesives.
- Polyurethanes and isocyanate-based adhesives.
- Rubber-based adhesives.
- Thermoplastic hot-melts.

Of all of the types of plastic adhesives listed, the cyanoacrylates have superior overall adhesive properties. They require no mixing, cure rapidly, can withstand tensile strength tests of 3000 psi (20 685 kPa), and are suitable for use in temperatures up to 180°F (82°C).

Hot Melts

Hot melts, also known as "reactive melts" or "reactives," are thermoplastics that become fluid when heated, then solidify upon cooling. Hot melts react with moisture in a substrate or in the air, and then cross-link. Polymerization forms an adhesive bond that resists heat, steam, chemicals, and harsh environmental conditions. When the reaction is completed, the melt will not resoften if heated.

The main polymers that are used in hot melts are polyesters, polyolefins, polyamides, and ethylene vinyl acetate copolymer. Hot melts are available in beads, granules, powders, ribbons, films, pellets, blocks, cards, pillows, slats, slabs, rods, cakes, and liquids.

Hot melts can replace screws, nut-and-bolt assemblies, nails, and other mechanical fasteners. They can also reduce product weight and improve appearance by eliminating exposed fasteners. Many large companies in the automotive, aerospace, and appliance industries are using hot melts.

One of the major advantages of hot melts is that they are nonvolatile. They don't pollute the air or water, and don't create problems with solvent emissions and hazardous waste disposal.

The future of hot melts seems very good. Many plastics products will use hot melts for assembly instead of mechanical fasteners. Hot melts can even be produced as foams, and used for applications such as foam-in-place gaskets and seals.

Hot melts seem to be the preferred method of assembly in the appliance industry, for applications ranging from sealing cabinets to bonding of plastics and metals. Today, the big five appliance manufacturers (Amana, Maytag, General Electric, WCI, and Whirlpool) all use robotic equipment to apply hot melts on their products.

Hot melts are applied with applicators that melt the adhesive and pump it to a handgun or spray head. In many applications on assembly lines, hot melts are applied automatically at high speeds as sensors signal the guns to squirt adhesive.

Typical applications for hot melts include positioning switches and wires in automotive products, assembling windshield washer bottles, securing refrigerator insulation, bonding aluminum jacketing to fiberglass pipe insulation, and foam-in-place gaskets for automotive air conditioner housings.

Filament Winding

Filament winding is a process that involves winding continuous fibers saturated with polyester or epoxy resins around a cylindrical mandrel. Filament-wound products have improved glass fiber strength, permitting thin walls and resulting in lighter weight. Many different types of fibers can be used for filament winding. However, the most popular are such high-modulus organic fibers as Kevlar®, which belongs to the family of fibers called aramids. Aramid fibers are different from conventional organic fibers. They are much stronger and have a higher modulus in terms of tension and compression.

The process lends itself to a broad range of product types. Simple cylinders and tanks, spherical tanks, even complex windmill blades and helicopter rotors are often fabricated using this process. Filament winding is also used to make cylindrical pressure vessels and tanks, missile bodies, pipe, pressure bottles, high-strength tubing, and shotgun barrels. The process is versatile. It is portable enough that on-site fabrication of large tanks is practical. Tanks with surface areas of up to 1500 square feet and cylinders 40 feet in diameter used as power plant stack liners are wrapped on single mandrels.

Let's take a closer look at how the process is conducted. The filament is drawn through a resin bath, then mechanically wound around a rotating mandrel. Mandrels can be removed after the resin cures or they can be left inside the windings as a permanent support liner. When a hollow vessel is desired, hard salt is sometimes used as the mandrel. The salt is then washed out with water. Low-melting-temperature materials also can be used to construct mandrels. Sometimes, balloons are inflated to serve as the mandrel, then collapsed after the resin and fiber combination cures. Wax, wood, plaster, rubber, and fiber-reinforced plastics (FRP) are also used to make mandrels.

Filament winding is effective for producing strong parts. Containers made using this process usually have a higher strength-to-weight ratio than those produced with other processes. Another advantage of this process is that it allows reinforcement materials to be placed in areas subjected to the greatest stress. Filament winding provides a high degree of control over uniformity and orientation of the fibers. Filament-wound products can be accurately machined.

Stereolithography

Stereolithography is an unusual process that was developed and patented in 1986 by 3D Systems, Inc., of Valencia, California. The process is being used to produce three-dimensional plastic design prototypes, models for testing, molds, dies, and patterns. See Figure 11-6. These objects are being produced directly from the computer keyboard, without machining. Stereolithography makes use of photopolymers, liquid plastic monomers that are changed to solid polymer when they are exposed to ultraviolet light. Here's how the process works.

Stereolithography builds an object, layer by layer, by exposing the liquid photopolymer to ultraviolet laser light. The photopolymer is cured and hardened when the light strikes it. Stereolithography relies on the integration of three major systems:

- Δ Computer-aided design (CAD) equipment and software.
- Δ A photopolymeric elevator.
- Δ A laser exposure and imaging system.

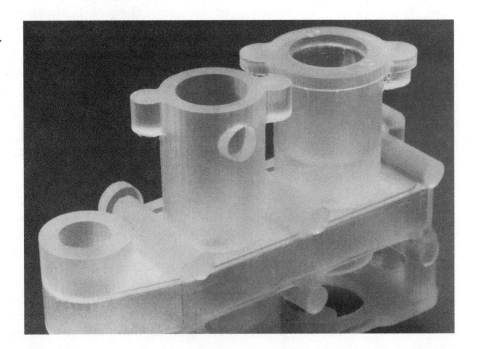

Figure 11-6
Prototype parts developed by the stereolithography process are highly detailed and dimensionally accurate. (3D Systems, Inc.)

A proprietary computer software program is used to slice the CAD drawing of the part into thin cross-sectional layers, typically from between 0.005 in. and 0.020 in. (0.127 mm and 0.508 mm) thick. The data is then sent from the computer to a laser, which draws each layer with ultraviolet light on the surface of the vat of liquid photopolymer. Where it is exposed to light from the laser, the polymer hardens.

After the thin layer is exposed and hardened, an elevator on the machine lowers the prototype slightly. This allows liquid polymer to flow over the part. The next layer is then exposed and hardened. The process continues until the part has been completed. The elevator platform raises the part out of the resin so that it can be removed from the vat and drained. Resin that is unexposed remains liquid, and is kept in the vat for later use. The part is exposed to ultraviolet light again to harden any liquid that has been trapped in crevices in the part. The final design prototype is finished by sanding, sandblasting, or painting.

The photopolymeric resin consists of two basic components. One is a *photoinitiator*, which absorbs laser energy and initiates polymerization. The other is an acrylic resin. The acrylic monomers and *oligomers* (polymers made up of two, three, or four monomer units) form solid polymers when exposed to light. The light, working in conjunction with the photoinitiator, causes the polymerizing reaction. This reaction stops abruptly with the removal of the light source.

Stereolithography is gaining popularity in the aerospace, computer, medical, consumer, and electronic component industries.

Important Terms

adhesion
adhesive
fabricate
filament winding
hot melts
hot wire welding
induction bonding
inserting
oligomers
photoinitiator
plastics welding
radio-frequency (RF) heat sealing
sealing
solvents
spin welding
staking
stereolithography
thermal heat sealing
thermal impulse heat sealing
ultrasonic
vibration welding

Questions for Review and Discussion

1. Why can thick plastics be heat-sealed with the dielectric method more rapidly than thin plastics?
2. Describe what happens when plastics are welded.
3. Why can't the dielectric sealing process be used with polyethylene and polystyrene?
4. Describe the differences between ultrasonic welding and radio-frequency heat sealing.
5. How do vibration welding and friction welding differ?

Various fabricating methods, including cohesive joining, snap fits, and mechanical fasteners, are used in the production of such consumer goods as these humidifiers. (Duracraft)

Conditioning Plastic Materials

Δ   Chemical blowing agents are used to produce lightweight cellular plastics.
Δ   To relieve stresses caused by processing activities, plastics are often annealed.
Δ   The rate of polymerization for certain plastics can be increased by exposing them to radiation.
Δ   Polyanisidine (PANIS) is a conductive polymer that can be used as a replacement for solder in some electronics applications.

The process of *conditioning* plastics involves changing the internal molecular structure of the material. This may involve heating to promote cross-linking of polymer chains or to relieve stress, exposure to ionizing or nonionizing radiation to improve such qualities as heat resistance, or the incorporation of reinforcements, fillers, or chemical additives to provide desired characteristics.

Since plastics are *engineered materials* that are designed and constructed by people, there are many ways to alter their internal characteristics. Many of the innovative processes related to conditioning plastics deal with use of the plastic as a support matrix carrying one or more different types of fillers or reinforcements. Some of the processes will be covered in this chapter, but most will be discussed in detail in the chapters on composite materials.

As a material, plastic has its own set of behavioral characteristics. Many of the surface-hardening processes used with metals, for example, have no impact on plastics.

Conditioning of thermosets and thermoplastics is also done through heat, which is usually generated or applied during the forming process.

It has been established that heat has a major impact on the internal molecular characteristics of polymeric materials. If the material is a thermoset, heat speeds the rate of polymerization. With thermoplastics, heat changes the form of the material from solid to liquid, allowing it to be formed.

It is sometimes necessary to alter the structure of a plastic material to form a part without breakage, or to relieve internal stresses. In such cases, a process that is commonly used with metals—annealing—can be useful with plastics as well.

Annealing

Like many other materials, plastics develop internal stresses when their external shape is changed. Some manufacturing processes bend, flex, or reshape a part in such a way that drastic realignment of the internal molecular structure takes place. At first glance, it might appear that only those processes which result in changing the external shape of the material would create internal stresses. This is *not* true: even some finishing processes and chemical treatments can create internal stress in a part. ***Annealing*** is a process that is commonly used with metals to relieve stresses created by forming or heat-treating processes. Annealing is also used with plastics to relieve the internal stress in the material (increasing its strength), and to improve its workability.

Plastic materials are annealed by subjecting them to prolonged heating at a temperature slightly above their molding temperature, then slowly cooling them under controlled conditions. The slow cooling lets the molecules gradually settle back into alignment. The length of time necessary to anneal a given plastic material varies according to the size of the part and the type of plastic involved. A typical heating/cooling cycle for 0.025 in. (0.635 mm) Plexiglas® would be heating for 2.5 hours at 230°F (110° C), then cooling for 1.5 hours to reach a removal temperature of 158° F (70° C). See Figure 12-1.

Radiation Processing

As you will recall from an earlier chapter, the process of polymerization typically involves the interaction of a resin and catalyst. Polymerization can also be created using radiation. This type of polymerization is accomplished with a process called ***radiation processing.*** Radiation processing produces internal change through exposing the stock to high-energy ionizing radiation.

There are two major types of radiation processing—ionizing and nonionizing. ***Ionizing radiation processes*** produce high-level radiation, either gamma rays generated from radioactive materials such as Cobalt 60, or high-energy electron rays created by electron accelerators.

Gamma rays are able to penetrate thick objects, but usually provide very slow processing speeds. High-energy electrons do not penetrate as deeply as gamma rays, but they are very powerful, often up to several hundred kilowatts (kW) of power. Industrial electron acceleration machines are presently available with voltages up to five million volts. Because of the versatility of these machines, electron acceleration has become the most popular radiation processing method. Electron beam acceleration is desired over other processes that expose the workpiece to radioactive material because it enables the operator to direct the source of radiation to the workpiece and switch it on and off as desired.

Today, some of the most popular uses of radiation processing are the cross-linking of insulation material for wire and cable, the cold sterilization of surgical supplies and medical disposables, and the grafting of polymers onto textile webs to improve soil resistance and the ability to be dyed.

Heating Times for Annealing of Plexiglas

Time in a Forced-Circulation Oven at the Indicated Temperature

Thickness (mm)	Plexiglas G, II, and 55					Plexiglas I-A				
	110 °C*	100 °C*	90 °C*	80 °C	70 °C**	90 °C*	80 °C*	70 °C*	60 °C	50 °C
1.5 to 3.8	2	3	5	10	24	2	3	5	10	24
4.8 to 9.5	2½	3½	5½	10½	24	2½	3½	5½	10½	24
12.7 to 19	3	4	6	11	24	3	4	6	11	24
22.2 to 28.5	3½	4½	6½	11½	24	3½	4½	6½	11½	24
31.8 to 38	4	5	7	12	24	4	5	7	12	24

Note: Times include period required to bring part up to annealing temperature, but not cooling time.
*Formed parts may show objectionable deformation when annealing at these temperatures.
**For Plexiglas G and Plexiglas II only. Minimum annealing temperature for Plexiglas 55 is 80 °C.

Cooling Times for Annealing of Plexiglas

Time to Cool from Annealing Temperature to Maximum Removal Temperature

Thickness (mm)	Rate (°C)/h	Plexiglas G, II, and 55				Plexiglas I-A			
		230 (110 °C)	212 (100 °C)	184 (90 °C)	170 (80 °C)	194 (90 °C)	176 (80 °C)	158 (70 °C)	140 (60 °C)
1.5 to 3.8	122 (50)	¾	½	½	¼	¾	½	½	¼
4.8 to 9.5	50 10	1½	1¼	¾	½	1½	1¼	¾	½
12.7 to 19	22 - 5	3¼	2¼	1½	¾	3	2¼	1½	¾
22.2 to 28.5	18 - 8	4¼	3	2	1	4	3	2¼	1
31.8 to 38	14 - 10	5¾	4½	3	1½	5¾	4½	3	1½

Figure 12-1 Charts used to determine proper heating and cooling times when annealing Plexiglas. Times are in hours. Note that annealing at higher temperatures on the chart may cause parts to show deformation to an objectionable degree. (Rohm and Haas Co.)

Induction and dielectric heating are common *nonionizing radiation processes.* Nonionizing radiation can also be generated by exposing the workpiece to microwave, infrared, or ultraviolet energy sources. Nonionizing processes are normally used when low-level radiation is needed to cure or dry adhesives or coatings.

Exposure to either ionizing and nonionizing forms of radiation, if carefully monitored and controlled, can improve the physical characteristics of a plastic material. Thermoplastics that have been irradiated are able to withstand exposure to higher temperature and caustic environments. Surface exposure to radiation can also be used to improve the material's resistance to static electricity.

Exposing formed parts to controlled irradiation can be used to achieve polymerization, cross-linking, and grafting. *Grafting* occurs when two or more elastomers attach to a molecular chain without the use of a catalyst.

If the radiation source is not carefully controlled, however, it can cause damage. Radiation damage can break molecular bonds, destroying the alignment of atoms. This can lower the molecular weight of the material. It can also cause cross-linking, polymerization, branching, or oxidation.

Some plastics are more sensitive to radiation than others. The polystyrenes, polycarbonates, and polyesters show good stability when exposed to radiation. The polypropylenes and acetals have poor stability. Plastics with poor stability can be damaged with far less radiation exposure than those with good stability. Epoxy has a particularly good resistance to radiation, requiring as much as 10,000 mrads (millirads) of exposure to damage the material. In contrast, acrylic plastic can be damaged with as little as 5 to 10 mrads.

Polymerization and cross-linking caused by radiation processing can be used to shrink wire and cable insulation, increase the bonding strength of adhesives, and improve the resistance of plastic to cracking. It is also used to cure plastic coatings and film layers.

Conductive Plastics

Plastics normally do not conduct electricity. However, experimentation with plastic/metallic composites has been ongoing for many years. Plastic is made conductive by adding metal to the plastic base material, resulting in a material referred to as *synthetic metal*. Metal fibers have been used as conductive elements in antistatic fabrics and yarns. Some of these applications have involved compounding stainless steel fibers with the polymer during molding. Other attempts to produce conductive plastics have involved the use of conductive webs.

In recent years, experimentation has concentrated on the actual mixing of materials to produce conductive plastics. This has led to new applications for plastics used as electronic devices, batteries, and chemically modified electrodes and sensors. A new field of research has emerged called **synthetic metal research.** The Department of Defense, the Office of Naval Research, the National Science Foundation, the Air Force Materials Laboratory, and such firms as Allied Chemical, IBM, Xerox, Bell Labs, Eastman Kodak, Exxon, General Electric, and Westinghouse are all investing heavily in synthetic metal research.

Polyaniline films and powders are particularly useful for these synthetic metal applications, not only because they are conductive, but also because they are relatively stable in air.

Dr. David MacInnes Jr. of Guilford College in North Carolina has been studying a derivative of polyaniline, called polyanisidine, or PANIS. This new type of conductive polymer has proven its ability for use as either a conductor or a nonconductor (resistor).

A conductive coating is produced when polyanisidine is blended with the nonconducting polymer, polyacrylonitrile. The blend provides a tough coating that is not susceptible to oxidation. PANIS can be produced in various forms, ranging from particles and sheets to a conductive gel that can be shaped much like putty. Research is even being conducted into using the conductive gel as a replacement for conventional solder in some industrial electronic applications. Wires can be inserted into the conductive gel and then easily removed as desired.

EMI/RFI Shielding

For a number of reasons, plastic has become the dominant choice for the construction of enclosures for computers, appliances, microphones, business machines, and other electrical/electronic products. See Figure 12-2. One important reason is the ability to add conductive materials to the polymer as a means of controlling the emission of EMI/RFI (electromagnetic interference/radio frequency interference). EMI/RFI interference is called "noise," and must be held within specified levels to meet requirements for electrical devices set by the Federal Communications Commission.

Figure 12-2
Conductive materials are added to the plastics used to fabricate enclosures for electrical devices to provide shielding against excessive EMI/RFI radiation. The enclosure shown houses a high-performance drive for an ac (alternating current) industrial motor. (MagneTek)

EMI-shielded products easily dissipate electrostatic charges and provide improved reception of electromagnetic signals through antennas and other receiving systems. They can be made using standard injection molding and blow molding equipment with conductive plastics created by adding metal modifiers. The stock that is used to modify the plastic is available in various forms: flakes, fibers, powders, or metal-coated substrates. Aluminum flakes, nickel-coated fibers, and stainless steel fibers are commonly used.

There are other methods which are also used to improve the EMI shielding of products. The most popular alternative to mixing conductive plastics is to use a paint that is heavily pigmented with a metal such as silver and nickel. Vacuum metallizing (described in the chapter on finishing) is also used to reduce EMI noise.

Plastic Additives

Various additives may be used to provide plastics with the characteristics desired for a finished product. Generally, an additive will make up no more than one-tenth of the finished product's weight; many additives represent a much smaller proportion of product weight. Commonly used additives include various types of coloring agents or ***colorants,*** ultraviolet light absorbers to retard damage from exposure to sunlight, flame retardants to slow or stop burning of the plastic, ***plasticizers*** to increase flexibility, lubricants to provide a slippery or nonstick finish, antistatic agents to dissipate static charges, preservatives to protect against attack by microbes, reinforcements to add strength, and fillers or foaming agents to change the density of the material.

Preservatives

Most plastics are relatively impervious to destruction by microorganisms, rodents, or insects. However, elastomers, polyvinyl chlorides, and other plastics that are heavily plasticized are susceptible to attack, especially when used in continuously warm and wet environments. Such products as shower curtains, boat covers, pool liners, and convertible tops are subject to attack by mildew, fungus, mold, and bacteria. Deterioration of the base material also may encourage attack by rodents and insects.

The best way to discourage such attacks is by adding chemical preservatives like mildewicides, fungicides, and rodenticides to the plastic. Use of these chemicals is carefully regulated by federal agencies to ensure safety of both production workers and end-users.

Franklin Fiber

Franklin fiber is a microscopic crystalline fiber, derived from gypsum, that can be used as a reinforcing filler in plastics. It is much more than just a filler, however. It also improves the tensile strength of plastic and, in most cases, improves its modulus of elasticity. This makes the material less brittle and increases its impact resistance.

Franklin fiber is a crystalline form of calcium sulfate, with crystals so small that they must be observed using a microscope. The fibers are white in color and feel soft and silky. When they are compressed, they have a tendency to mat.

The material is produced by agitating gypsum in suspension for several minutes in a steam reactor. The thick slurry that results is then forced through a filter press to remove most of the water. The fibrous cake is then shredded, dried, and calcined at high temperature to remove the remaining water.

Franklin fiber has excellent temperature resistance (even better than asbestos), and provides increased strength for the plastic to which it is added. The fibers are nonabrasive and are easy to form using conventional plastic molding equipment. The fiber can be used with thermoplastics such as PVC, polypropylene, and nylon. It is also used with sheet molding compounds and other thermosets.

While it provides many advantages, the greatest value of using Franklin fiber with thermoplastic and thermosetting resins is in reducing costs and conserving petrochemical resources. The basic raw materials (petrochemicals) used for making plastics are much more expensive than gypsum, the raw material for Franklin fiber. Gypsum is also readily available in the United States. When the fiber filler is added to plastic, less plastic is needed. This reduces the cost of manufacturing, while making better use of limited petrochemical resources.

Microspheres

Hollow glass bubbles, called *microspheres* (or sometimes, "microballoons"), are mixed with many different types of plastics to reduce assembly costs, decrease density, reduce weight, and generally improve the physical properties of the part. Microspheres are chemically stable, water resistant, and nonporous. The spheres are normally made of glass, ureaformaldehyde, silica, or phenolics, and are embedded in a matrix made of polyester or epoxy.

These thin-walled, hollow glass spheres are typically around 75 microns (75 millionths of an inch) in diameter, but are available in sizes as small as 20 microns and as large as 200 microns. Wall thicknesses range from 0.5 microns to 2 microns.

The spheres can be used by mixing them with polyester sheet molding compounds. They can also be used with processes such as hand layup, sprayup, and casting. The addition of these spheres usually improves sanding and machining rates with products that are generated. Products made with microspheres also have improved dimensional stability at high temperatures, and improved water resistance.

Using synthetic hollow glass spheres as fillers also offers an additional advantage: the spheres are as much as eighteen times lighter than fillers like calcium carbonate. In addition to their light weight and high strength, they have the added advantage of being compatible with many types of resins. This makes it possible to design composite materials which are better able to survive extreme weather conditions.

Industry is using this technology for many applications, including marine flotation devices, automotive parts, compression molding compounds, high performance aerospace composites, adhesives, furniture parts, coatings, sporting goods materials, cultured marble, and carpet backing.

Glass Flakes

Glass flakes are used, in a manner similar to that described for microspheres, to create moisture barriers and corrosion-resistant coatings. Owens-Corning Fiberglas Corporation, one major manufacturer, produces type C (chemically resistant) and type E (electrical) glass flakes in sizes of 1/64 in., 1/32 in., and 1/8 in. (0.4 mm, 0.8 mm, and 3.2 mm).

The U.S. Air Force has been experimenting with glass flakes since the early 1960s to develop high-strength composites for rocket cases, fins, and missile radomes, as well as moisture barriers and corrosion-resistant coatings. It was found that glass flakes mixed with polymer increase the flexural modulus and reduce the coefficient of thermal expansion of the polymer. Glass flakes were used commercially to make the body panels of the Pontiac Fiero, one of the first passenger cars to use such plastic body components. See Figure 12-3.

Figure 12-3
Glass flakes were mixed with the resins used to produce body panels for the Pontiac Fiero, one of the first cars to use plastic panels.

Foam-Frothing

The *foam-frothing* process is also known as cellular, blown, or bubble expansion. Cellular plastics have the advantage of being light in weight, with a very high strength-to-weight ratio.

The basic process of foam frothing is a variation of the urethane foam pour-in-place method. What is unique about the frothing method is that a urethane chemical mixture is dispensed in a partially expanded state, rather than as a viscous liquid. Frothing is achieved by adding chemical blowing agents to the mix. When the blowing agents decompose at a particular temperature, gases are given off, creating a cellular structure in the polymer.

A special machine is required to create the frothing action. The ingredients are partially expanded, then mixed under a pressure of about 100 psi (690 kPa). The blowing agents create frothing, which facilitates filling of narrow-walled molds.

The foam-frothing process produces lower density foams with thin-walled sections. The major disadvantage of frothing is the cleanup problems presented by the process: a solvent such as methylene chloride must be used to flush the mixer at the end of the pour.

Frothing can also be achieved by forcing nitrogen gas into the resin or melt, just before it expands in the mold. This variation of the process is called *physical frothing*, or physical foaming. Frothing can also be accomplished by aggressively whipping air into the resin, and then rapidly cooling the polymer.

Important Terms

annealing
colorants
conditioning
engineered materials
foam-frothing
Franklin fiber
grafting
ionizing radiation processes
microspheres
nonionizing radiation processes
physical frothing
plasticizers
radiation processing
synthetic metal research

Questions for Review and Discussion

1. Describe the procedure used for annealing plastics.
2. How can conductive plastics be used to reduce EMI/RFI interference?
3. What is the purpose of Franklin fiber?
4. Both Franklin fiber and microballoons are used for fillers. What is unique about each of these materials?

A lightweight cellular plastic provides dimensional stability, firmness, and thermal insulation when used for liner panels in such products as refrigerators. This cutaway view shows the cellular material sandwiched between two layers of a styrene alloy specially formulated to resist blistering from the blowing agent used to form the cellular material. (Dow Plastics)

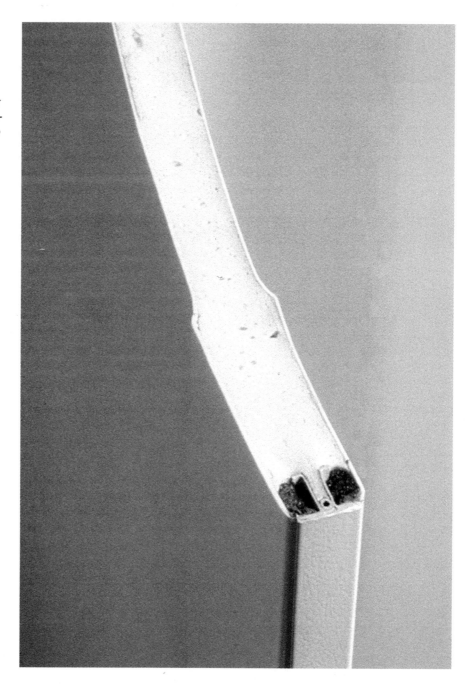

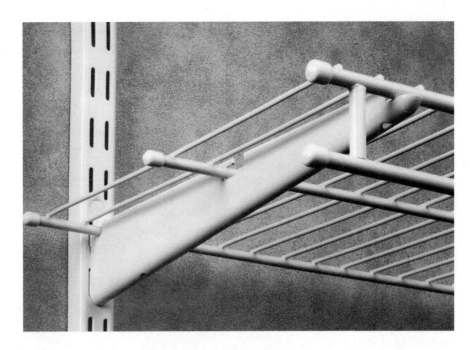

Electrostatic powder coating is an effective method for applying a uniform layer of plastic resin to products such as wire storage rack systems. The coating is used for its attractive appearance, ease of cleaning, and protection against corrosion. (Schulte Corporation)

13 Finishing Plastic Materials

Key Concepts

△ Phenolic resins can be site-applied by spraying to protect the interior of large chemical storage tanks.

△ Organosols and plastisols are widely used for dip coating and dip casting applications.

△ Extremely fine particles of thermoplastic resin can be suspended on air currents and act as a fluid for finish-coating of thermoset materials.

△ Metallic coatings can be applied by vacuum metallizing, electroplating, or electroless (chemical) plating.

Many of the processes used to finish metallic, ceramic, composite, and polymeric/wood materials—sanding, buffing, polishing, filing—are also used with plastics. Other processes, such as solvent-polishing or in-mold decorating with foils, are unique to plastics. Special surface treatments, such as electrostatic powder coating or electroplating, are used for plastics as well as metallic products. A number of the most important finishing processes are discussed in this chapter.

Material-Removal Processes

Many finishing processes involve the physical removal of material through abrasion to produce a smooth and lustrous surface. Other methods use heat or chemical action to soften and smooth the surface.

Tumbling

Tumbling, or barrel finishing, is one of the most economical methods for quickly finishing plastic molded parts. It is also used to remove flash, to smooth rough edges, to grind, and to polish. This process is often used with metallic and ceramic materials, as well.

The parts to be finished are placed in a drum with finishing materials known as **tumbling compounds.** These compounds may include abrasive particles, waxes, sawdust, wood plugs, and a variety of other materials.

293

Rotation of the tumbling drum causes the parts and finishing materials to rub against each other. This produces a polishing action. A small amount of material is removed, depending on the type of abrasive used, the length of the tumbling cycle, and the speed of the barrel.

Other methods used with tumbling to finish plastic parts include abrasive spraying and the use of dry ice. Dry ice is sometimes used in the barrel to improve removal of thin flash on the parts. The ice chills the flash and tumbling separates it from the product.

Smoothing and Polishing

The primary step in finishing rough plastic parts is often machine sanding using open grit sandpaper. The coarse, open grit of the paper will remove the maximum amount of stock without becoming clogged by plastic particles. Number 80 grit silicon-carbide abrasives should be used for rough sanding. When rough sanding is completed, the parts may need further finishing. Final finishing processes can be completed with fine 400 grit or 600 grit paper. Sanders normally run at speeds ranging from 1750 rpm for disk models to 3600 fpm for those using belts. Dry sanding must be done using light pressure to keep from overheating the plastic.

Machine processes such as polishing or buffing are sometimes used with a soft muslin or a chamois wheel. *Buffing* is done when small surface defects must be removed, or when a polished surface is needed. Normally, buffing is completed before polishing. Buffing is done using muslin disks sewn or collected together to provide either a firm-surfaced or flexible buffing wheel. Firm wheels are best suited to products with simple shapes. Flexible wheels are softer, and are used for buffing irregularly shaped parts. Both types of wheels are *charged*, or lightly touched, with a very fine abrasive material such as tripoli or red rouge. A buffing sequence often includes ashing, polishing, and wiping.

Ashing involves a wet abrasive, usually number 00 pumice mixed with water. The slurry is applied to a loose muslin wheel. The use of a wet abrasive reduces overheating of the plastic and improves the cutting action.

Polishing is sometimes referred to as "luster buffing." In order to polish plastics, a wax compound that includes a fine abrasive (such as whiting or levigated alumina), is applied to a clean chamois or loose flannel buffing wheel. The wax fills in small defects in the part and protects the highly polished surface. Polishing compound is applied to half the wheel, while the other half is left clean. The part is first held against the charged half, then against the uncharged part to wipe off the compound. Excessive wheel speeds and hard wheels should be avoided.

Wiping is accomplished by polishing the part with an uncharged wheel. Sometimes, cleaning compounds are applied to remove grease and wax.

When polishing or buffing, the part should be pulled toward the operator from beneath the wheel, using rapid even strokes. Care must be taken to avoid applying too much pressure, or the wheel will be distorted toward the edge of the part. This could cause the piece to be jerked from the operator's hands.

Flame Polishing and Solvent-dip Polishing

Flame polishing is sometimes used to smooth or reduce imperfections and to smooth rough surfaces on the edges of sheet plastic. Flame polishing is accomplished with a rapid back and forth motion of the flame from an oxygen-hydrogen welding torch. The flame is pointed downward to intersect the edge at an angle of approximately 45 degrees.

Cellulosic and acrylic parts can be polished by dipping or spraying them with solvents. This is often effective in removing the matte (dull) finish left from sanding, drilling, cutting, or other operations. A small amount of acrylic solvent, such as ethylene dichloride, can be applied to the finish by dipping or spraying. Contact time should be from 15 seconds to a minute, depending on the extent of the surface roughness. Solvents can also be used to polish edges or holes. The solvent is then removed and the surface is allowed to dry. The solvent-polished area will have a fine polished finish, but it will not have the clarity of a fine-sanded and highly polished (buffed) surface. Parts to be solvent-dipped should first be annealed to keep them from *crazing* (developing a network of fine cracks on or under the surface).

Coating Processes

Application of a coating is a common form of finishing with plastics, as well as many other materials. Coatings are applied to plastics for appearance, for protection, and for electrical conductivity. Some coatings are only thousandths of an inch in thickness; others may be up to one-half inch thick.

Dip Coating

Polyvinyl chloride powder can be dissolved in a large amount of plasticizer to create a viscous solution called a *plastisol*. An important use for plastisols is **dip coating** the gripping surfaces of metal products, such as tool handles. Other typical dip-coated items are laboratory ware, kitchen utensils, and dish draining racks. In addition to dip coating, plastisols can be used to manufacture products using slush casting or rotational molding processes.

Sometimes dip coatings are applied by dipping heated parts into plastisol. Materials are also available which do not require the part to be heated. Once the part is coated with plastisol, it must be cured. Depending on the type of plastisol used, this may be done by air-drying, or by heating the part to 330°F to 450°F (166°C to 232°C). After curing, the material becomes solid and rubber-like to the touch.

A related application to dip coating is the **dip molding** or **dip casting** process used to make products such as surgical gloves and spark plug covers. In this application, a heated metal tool, or *mandrel,* is immersed in a dip tank containing plastisol. The mandrel conforms to the internal dimensions of the part that is being molded. You will recall that a plastisol consists of PVC, or another resin, dissolved in a plasticizer. The resin usually is in powder form, and thus—even though a solution ("sol") has been created—fine particles of resin will remain in suspension. After the mandrel is coated with the thick plastisol, it is removed from the tank and placed in an oven. The heat of the oven cures the plastisol and fuses any remaining fine particles of resin together. When curing is complete, the mandrel is cooled and the part is stripped from it. Figure 13-1 shows typical PVC dip-molded parts.

Figure 13-1
These insulating covers for battery connectors were manufactured by the dip molding process. (Arbonite Corporation)

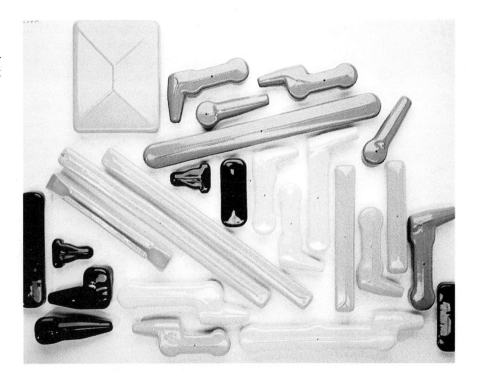

Organosols are also used for dip molding. Organosols are a suspension of polyvinyl chloride powder in a volatile solvent with a small amount of plasticizer. Since they are less viscous than plastisols, organosols provide better penetration into porous materials. They are commonly used to coat textiles, paper, and paperboard. Typical organosol-coated products are waterproof garments, gloves, and notebook covers.

Coating Storage Vessels in the Field

Sometimes, because of the size, weight, and location of large storage vessels and holding tanks, it is necessary to coat them in the field. Some manufacturers spray phenolic coatings onto large metal tanks in thicknesses up to 0.5 in. (13 mm) without removing them from the work site. Figures 13-2 and 13-3 show large chemical storage tanks about to be coated internally with a phenolic material. While this coating is not a true plastisol, it does behave in a similar way. Let's take a closer look at how this works: because the tanks are so large, preparation involves insulating their exterior surface, then baking the phenolic coating onto the inside surface using hot air generators. The hot air speeds polymerization and crosslinking. After the lining is inspected, the insulation is removed and the tank exterior is blasted and painted to suit the environment in which it is used.

Electrostatic Powder Coating

In *electrostatic powder coating,* particles of plastic powder are given a negative charge as they are sprayed on to a positively charged, grounded part. The charged powder is attracted to the part much like metal filings are

Figure 13-2
This reactor tower for use in chemical processing has been insulated to retain heat. A phenolic coating sprayed on the inside surfaces will be baked (cured) by the introduction of heated air. After the coating is cured, the insulation will be removed and the tower installed in a vertical position. (Arbonite Corporation)

Figure 13-3
Like the reactor tower shown in Figure 13-2, this large storage tank has been insulated prior to baking its protective interior coating of phenolic resin. The completed tank will be used to hold highly corrosive concentrated sulphuric acid. (Arbonite Corporation)

attracted to a magnet. The particles do not strike the part in a straight line, but follow the charged field. Consequently, the coating will be evenly distributed, even on the back side of a round part.

Once the part has been sprayed, it must be cured in an oven to fuse the plastic to the part. Without curing, the powdered plastic would simply fall off the part when it lost its electrical charge.

Many automotive and heavy equipment manufacturers use electrostatic spraying, since it provides excellent coverage with low material loss. Powder that falls off the part is captured and reused. Powder coatings are hard, tough, and attractive. They don't chip and peel like conventional paint coatings. Coated products can be bent or deformed and the coating will flex with the product.

An important advantage of electrostatic powder coating is that, unlike conventional paint spraying, there are no air pollution worries associated with the coating process. The powder contains no solvents, so no solvents are given off into the air during curing.

Fluidized Bed Coating

Fluidized bed coating is a process that is used for powder coating of thermoset plastics. Polyethylene is a popular polymer for use with this process. Fluidized bed coating changes powder to a fluidized state by mixing it with air. This holds the powder in suspension so that it behaves like a fluid. The fluidizer tank has a porous bottom through which air or inert gas is injected at low pressure. The powder is vibrated to make it mix more readily with the air. The process is illustrated in Figure 13-4.

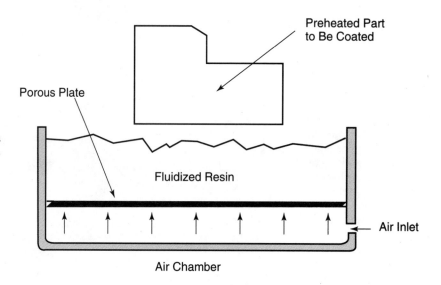

Figure 13-4
Rising jets of air in this fluidized bed system will suspend particles of resin, causing them to act like a fluid. An article to be coated with plastic is heated, then lowered into the layer of plastic particles suspended in the rising air. The heat causes the particles on the object's surface to fuse together in a smooth coating.

The part to be coated is heated to a point above the fusion temperature of the powder and is then dipped for a few seconds into the fluidized bed. Powder attaches itself to the part, then melts and bonds to the heated surface in a layer of uniform thickness. Sometimes, additional heating is used to improve smoothness and coverage.

The application of fluidized bed coatings usually takes place with a single dipping. Areas which are not to be coated are usually covered with a film of fluoroplastic material (such as polytetrafluoroethylene or polytetrafluoropropylene) to which the powder will not adhere. Coatings applied by the fluidized bed method are flexible and durable and are excellent electrical insulators. The coatings applied by this process are from 0.005 in. to 0.080 in. (0.127 mm to 2.032 mm) in thickness.

The disadvantages of the process are related to the difficulty in applying coats to thin workpieces. Thin materials are difficult to cover because they don't hold enough heat to build up a coat. Sometimes, it is difficult to produce a coating that is the same thickness across the entire surface of the part. Also, if the production environment is not temperature-controlled, the thickness of the coating will vary with seasonal changes in temperature.

Flame Spray Coating

Flame spray coating is a process that is often used to apply a plastic coating to tanks, vessels, or other products too large to cure in an oven. In flame spray coating, a fine plastic powder is dispersed through a specially designed spray gun burner nozzle. The powder is quickly melted as it passes through the flame. The flame also heats the part to be coated. Thus, the hot plastic is sprayed onto the heated substrate, to form a smooth and uniform layer. When the plastic cools, it forms a permanent coating.

Conductive Spray Coating

Zinc-arc spray metallizing is used to apply conductive shielding material for a variety of electrical applications. Conductive zinc coatings deposited by this method are dense, durable, and abrasion-resistant. The major limitations of this process are lack of uniformity in the layering, adhesion problems, and the creation of zinc dust.

The process is different from sputtercoating. In a zinc-arc spray gun, two zinc wires are fed through the gun and brought together so that they are almost touching. An electrical arc between the wires is created by applying DC current. This melts the wires, allowing air pressure from the gun to eject the molten zinc and deposit it on the plastic substrate.

The arc spray process is more efficient than flame spraying. It can produce a film 0.002 in. to 0.004 in. (0.05 mm to 0.10 mm) in thickness using only one-ninth the melting energy required for flame-spraying a film of the same thickness. Reduced heat results in less part warpage and distortion.

Vacuum Metallizing

Vacuum metallizing is a process used to apply very thin metal coatings to a plastic substrate for decorative or functional purposes. Coating of the plastic occurs by vaporizing the metal (usually aluminum, copper, or chromium) in a vacuum. The metallic film is deposited in thicknesses that are less than could be accomplished with electrolytic or electroless plating—only a few millionths of an inch thick. Coating adhesion to the substrate is not as good as that accomplished with electroplating. After the parts are metallized, a final protective coating of lacquer is normally applied.

Vacuum metallizing is used to frost and decorate glass, and to fabricate computer chips and web material for packaging. It is a popular process for coating alloys, costume jewelry, bottle caps, and automotive trim.

There are two basic type of vacuum metallizing, evaporative vacuum metallizing and sputtercoating. The hardware for these processes is quite different.

Evaporative vacuum method

In the *evaporative vacuum method*, the metal to be deposited on the plastic is evaporated from disk-shaped "targets" placed in a vacuum chamber. Usually, the targets will be made from chromium or copper. Evaporation is accomplished through resistance heating or electron beam bombardment. The vaporized metal molecules travel through the chamber and condense on the cooler plastic substrate to form the coating. The appearance of the metal coating that forms on the plastic is slightly different from that of the material that was first evaporated, because the original metal has vaporized and reformed on the substrate. Most applications require a basecoat and topcoats of lacquer to improve adhesion and protect the surface.

Sputtercoating

Sputtercoating is a process that is widely used to coat the surfaces of electronic components. Sputtercoating equipment that is magnetically enhanced creates a magnetic field in the vacuum chamber. The electromagnetic field directs chromium atoms from the disk targets to the substrate.

In this process, a high level of kinetic energy is produced in a magnetron sputtering gun. The gun has a water-cooled anode and a cathode which emits electrons. The electrons bombard the argon gas molecules and ionize them to create the sputtering action. When the argon ions are accelerated to the disk target, they become stronger, more dense, and exhibit stronger adhesive bonding. These ions liberate or dislodge ("sputter") atoms of the coating material, which are directed to the substrate. The sputtering process is used with high-temperature materials, or for more precise deposition of material on the substrate.

With both vacuum metallizing processes, coating of parts can be accomplished either in batches or continuously. In a batch coater, the parts are placed on racks in lots and are inserted into the processing chamber for coating. In the continuous coater, parts enter the processing chamber through an airlock. On the continuous machine, the processing chamber is always under vacuum.

Electroplating

The two processes generally used to plate plastics are electroplating and vacuum metallizing. Vacuum metallizing, as described in the preceding section, produces a very thin coating. This coating is not as durable as that produced by electroplating, but it is less costly.

You might think that, since plastics are not conductors of electricity, they can not be electroplated. This is not true—chromium, silver, nickel, gold, and other shiny metallic coatings can be applied to plastics. However, before electroplating can occur, a conductive coating must be applied to the plastic. This is accomplished by carefully cleaning the part, then chemically bonding a thin layer of copper to it. Once the copper coating is completed, electroplating can be carried out using the same methods employed for plating metals.

Electroplating is used to provide a protective and attractive metallic coating for a part. Since plastics are lighter than metals, plated plastic parts are desirable for many applications.

Approximately 50 percent of all of the plastics electroplated in the United States is used by the automotive industry, and an additional 25 percent by appliance manufacturers.

Electroless Plating

The term *electroless plating* is used to describe those processes that use chemical rather than electrolytic action to produce the desired metallic coating on a substrate. The fastest-growing areas for electroless plating are the construction of copper-plated printed circuit boards (photofabrication) and production of conductive plastics for electromagnetic interference shielding for the electronics industry.

Electroless plating is accomplished by autocatalytic reduction, using a solution of approximately 3 percent formaldehyde and 10 percent methanol stabilizer. The solution is further modified with caustic soda to adjust the pH (acid/alkaline balance). As the pH increases (the solution becomes more alkaline), the formaldehyde increasingly transforms the copper ions to copper metal for deposit on the substrate.

Electroless plating is generally a more expensive way to produce a copper coating than electrolytic deposition (electroplating). Because of the cost factor, thin coatings are normally produced by electroless plating.

Other Finishing Processes

There are numerous other types of finishing processes used in manufacturing. Many of these are used with materials outside of the major material families (metallics, plastics, ceramics, and composites). A number of these processes are used in the printing or textile manufacturing industries, and are thus outside the scope of this book. Two processes *are* worth mentioning here, since they are used with a variety of manufacturing materials.

In-mold Decorating with Foils

In-mold decorating can be done by applying foils to compression-molded and transfer-molded products made with melamine, urea, or ammonia-free phenolic thermoset and thermoplastic materials. A *foil* is a printed, melamine-impregnated paper that is introduced into the mold.

With thermoset materials, the molding cycle is interrupted at the earliest possible moment after curing, so that the printed foil can be placed on top of the part. The mold is then repressurized for several seconds at 2500 psi to 3000 psi (17 238 kPa to 20 685 kPa) and temperatures of 290°F to 350°F (143°C to 177°C). The pressure and temperature must be maintained long enough to cure the foil and produce the desired gloss.

Foils used may be opaque, or may be transparent with reverse lettering to achieve a color-on-color effect. They can be die-cut pieces or gaskets with multiple decorations for different parts of the product. Foils are capable of producing high quality pictures and intricate drawings. Any subject that can be commercially printed can be used as a foil.

With thermoplastics, insertion of the foil is accomplished at a different point in the process: the die-cut foil is placed in the mold just before it closes and the plastic is injected. The injected molten plastic bonds to the foil and suspends it inside the molded part.

Various methods are used to hold foils in correct alignment in the mold. They can be held snugly as a gasket in the mold. In some cases, they are held in place electrostatically or by the use of mechanical attachments on the mold.

Almost any molded product made with polystyrene, acrylic, ABS, polycarbonate, and polypropylene materials can be decorated using in-mold foils. Dinner plates molded from melamine resins are often decorated with this process. Highly reflective foils have even been made to produce unbreakable optical mirrors.

Coloring

While foils can be used to produce colored designs in many molded products, the most economical way to achieve color entirely through a product is to blend *pigments* directly into the resin base. Additives such as plasticizers and fillers, and the conditions of the molding process, can also affect the color of the final product.

Color is produced by mixing pigments in the form of dry powders, pastes, organic chemicals, or metallic flakes with the resin in a Banbury (batch-type) mixer or a continuous mixer. Mixing disperses the pigments throughout the resin. Water- and chemical-soluble dyes are used by some manufacturers. When such dyes are used, coloring is done by dipping the product into a dye bath and then letting it air-dry.

Important Terms

ashing	foil
buffing	in-mold decorating
charged	mandrel
crazing	organosol
dip casting	pigments
dip coating	plastisol
dip molding	polishing
electroless plating	sputtercoating
electroplating	tumbling compounds
electrostatic powder coating	vacuum metallizing
evaporative vacuum method	wiping
flame spray coating	zinc-arc spray metallizing
fluidized bed coating	

Questions for Review and Discussion

1. What is the difference between a plastisol and an organosol?
2. Describe the differences between electroplating and electroless plating.
3. How do dip coating and dip casting differ?
4. In electrostatic spraying, how is it possible for the coating material to be distributed completely around the object, even though the spraying is being done only from the front side?

Chapter 13 Finishing Plastic Materials   **303**

Plastisols are widely used in the hand tool industry to form dip coatings on handles of tools such as the pair of snips at top. The coating provides greater user comfort and helps to prevent slipping. Other tools use molded slip-on grips, like those at botton, for the same purpoes. (Scott Gauthier)

CERAMIC MATERIALS

Ceramic products are made of clay; the fusion of silica dioxide produces glass. Clay and glass are major divisions of the ceramics industry. When compared to metallic parts, ceramic parts exhibit many advantages. Their strong chemical bonding makes them hard and able to withstand high temperatures without melting, as well as surviving attacks by corrosive gases, molten metals, and acids. Ceramic materials can be made with densities so low that they will float on water, or can be mixed in such a way that they will be as dense and heavy as lead.

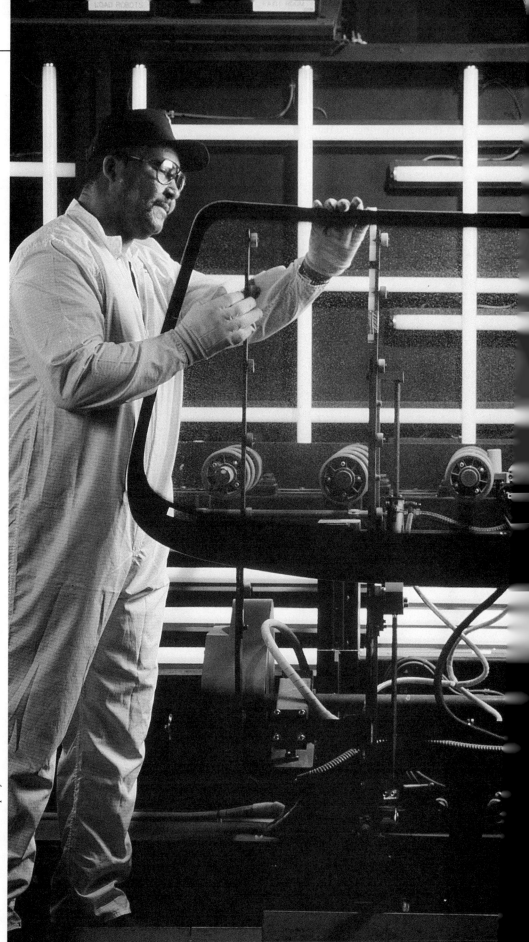

Ford Motor Company

Introduction to Ceramic Materials

Key Concepts

△ There are two major divisions of the ceramics industry, clay and glass.

△ Stock used by ceramic manufacturers includes synthetic materials such as carbide and alumina titania, and natural materials such as soapstone and clay feldspar.

△ Traditional ceramics and industrial ceramics are the two major sectors of the ceramics industry.

△ Bioceramic devices have been developed for use in the human body. They bond to bone, then are gradually assimilated into the bloodstream.

△ Although ceramics have been used by humans throughout history, major changes in both materials and processes were introduced in the 1960s and 1970s, leading to the birth of the industrial ceramics industry.

Ceramic products are made of *clay,* an inorganic, non-metallic solid material that is derived from naturally decomposed granite. Clay contains a white mineral, **kaolinite,** that is comprised of aluminum oxide, silica dioxide (quartz), and water. The fusion of silica dioxide produces glass. Clay and glass are major divisions of the ceramics industry. Both natural and synthetic minerals are used in the industry.

Unique Characteristics of Ceramics

Historically, ceramics have been viewed as materials that behave very differently from metals or organics. However, the scope of ceramic materials used in industry today is so comprehensive that it is difficult to describe ceramic parts in terms of their composition at all. It is easier to talk about ceramics in terms of how they behave as materials — that they become hard and resistant to their environment when exposed to a sufficiently high level of heat.

Ceramic products, such as brick, pottery, and artware have been made from sand and clay for thousands of years. It is important to realize that the raw material used for making such products bears little resemblance to the stock that is used to produce ceramics in industry today. Most ceramic products are now made from very pure, microscopically fine powders that are pressed in molds at high temperatures to yield dense, durable compacts.

The clay and glass industries manufacture a diverse offering of products—from glassware and tiles to concrete, to high-temperature refractories and oxide ceramics, to electronic glass and laser crystals produced from a melt.

When compared to metallic parts, ceramic parts exhibit better wear resistance and greater strength at high temperatures. They also are more able to withstand contamination in harsh chemical environments. Ceramic materials are also *versatile:* they can be made with densities so low that they will float on water, or can be mixed in such a way that they will be as dense and heavy as lead. Their strong chemical bonding makes them hard and able to withstand high temperatures without melting, as well as surviving attacks by corrosive gases, molten metals, and acids.

Ceramics in Space

Ceramics have been instrumental in the American space program. The most widely known application of ceramic materials in this program is the tiles used to provide a protective covering on space shuttles. The tiles have had to withstand temperatures produced during reentry into the earth's atmosphere of approximately 2500° F (1400° C).

The structural skin on each shuttle is made of aluminum, which is light in weight and easy to form. Without the protective ceramic coating, however, the aluminum skin would be unable to survive reentry, let alone contact with flying debris in space. NASA's design solution involved bonding 31,000 carbon composite ceramic tiles, each with a unique shape, directly to the aluminum skin. See Figures 14-1 and 14-2. Tests verified that the tiles would be able to withstand temperatures as high as 4500° F (2500° C).

In the electronic field, ceramic components can be created that have lower electrical and thermal conductivity than parts made from most other materials. Capacitors for storing electricity can be made of ceramics. Electrical insulators that are virtually unaffected by excessive voltages are also being made from ceramics. Ceramic materials can even be used for making electrical conductors.

Materials for Ceramics

One of the most attractive features of ceramics is that they are made from readily available materials. Historically, the United States has been dependent on other nations for supplies of critical minerals, especially such metals as chromium, manganese, cobalt, and platinum. These metals are important ingredients in making many steel alloys and superalloys.

Ceramics are made largely from such plentiful and widely available natural materials as clay and kaolin, sand, feldspar, carbonate, and soapstone. Some of the most plentiful minerals are the silicates and aluminum silicates. These minerals, together with oxygen, constitute the majority of naturally occurring ceramic raw materials.

Most ceramic products are made from natural silicon (sand), but synthetic materials are also used to produce special purpose glass or porcelain ceramics. Many industrial ceramic products also use new raw materials such as carbide, fused cordierite, and pure alumina titania.

Chapter 14 Introduction to Ceramic Materials 307

Figure 14-1
Each of the approximately 31,000 carbon composite ceramic tiles applied to the aluminum skin of the space shuttle vehicles is shaped to fit in a specific spot. Note the identification numbers on the tiles already in place. (NASA)

Figure 14-2
A worker cleans adhesive residue from a tile after placing it in position on the wing of a space shuttle vehicle. The tiles protect the spacecraft and its occupants from the heat of reentry to the Earth's atmosphere. (NASA)

Although ceramics have many advantages over other materials, they also have some distinct disadvantages. Ceramics are typically less tough than metals and many composites, exhibiting a tendency to *brittleness.* This is a fault which may cause them to break without warning, after stress causes cracks that start as microscopic flaws in the material.

It might appear that the characteristic of brittleness would severely hamper the usefulness of most ceramic products. However, this tendency can be engineered out of the material. New technology, such as the development of *ceramic-matrix composites,* offer methods for toughening the material. These composites incorporate reinforcing fibers or whiskers (high-strength single crystals with a length at least 10 times their diameter). Another popular technique to provide strength is to apply ceramic coatings to metal substrates.

In other cases, *fatigue* (the tendency of a material to crack and fail due to repeated stress) can be reduced by using processes that compact powder under pressure to improve the density of a part. Most ceramic forming processes involve compacting finely ground powders, then firing them at high temperatures in a furnace called a *kiln.*

Not all ceramics require heat to create a molecular bond, however. Some ceramic materials develop their strength at room temperature, usually through chemical reactions occurring with the addition of water. Cement paste and concrete materials behave in this manner.

Structure of the Ceramics Industry

There are many manufacturing firms producing different types of ceramic products. In most cases, they can be classified in terms of the type of product they manufacture; however, because of the tremendous diversity of ceramic products, it is difficult to develop a simplified structural model of the industry.

The best way to describe the industry is by analyzing it in terms of two major sectors, traditional ceramics and industrial ceramics. All ceramic products can be classified under one of these major headings.

Traditional Ceramics

Manufacturing firms in the *traditional ceramics* group are generally concerned with the production of clay and glass products. While this includes potteries and glassmaking establishments, it also extends into other product lines, such as concrete, sandpaper, and "E" (electronic) glass. Figure 14-3 shows the different types of products manufactured in the traditional ceramics sector. The major processes used to manufacture each of these products will be discussed throughout this section.

Industrial Ceramics

The *industrial ceramics* sector is sometimes called "structural ceramics," but this title is not really accurate. What industrial ceramic firms have in common is that they manufacture products from engineered (human-made) materials. The ceramic stock used in industry is designed to meet the needs of unique and highly specialized environments.

Figure 14-3
The structure of the traditional ceramics sector of the ceramics industry. Raw materials used in this industry segment, such as clay and sand, are naturally occurring and plentiful.

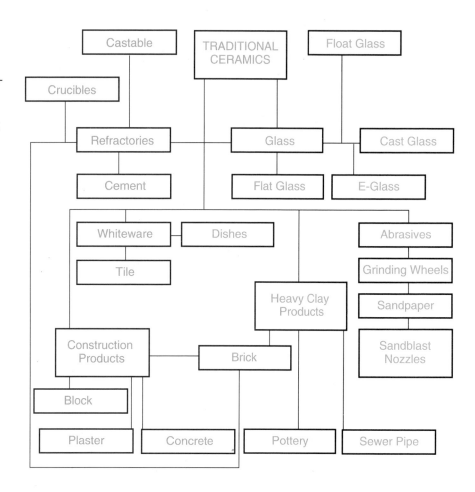

For example, there are many different types of ceramic components and devices made for use in electronics. However, the range for industrial ceramics reaches far beyond electronic products into a variety of other fields, as shown in Figure 14-4.

Figure 14-4
The structure of the industrial ceramics sector of the ceramics industry. The more sophisticated nature of the products in this segment require the use of industrial stock that is engineered (human-made) to meet varying requirements.

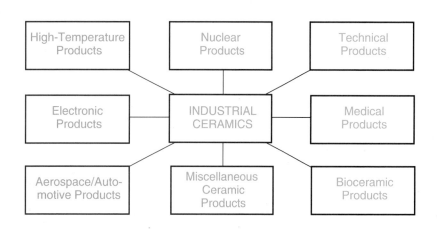

Take a moment to think more carefully about the types of products produced by those manufacturing firms involved in industrial ceramics. This is where there is the most significant growth in the field of ceramics. The U.S. Office of Technology Assessment, a forecasting arm of Congress, estimates that the U.S. market for advanced structural ceramics (industrial ceramics) is between $1 and $5 billion annually.

Two of the fastest-growing product applications for ceramics are heat engines and medical devices. The use of alumina, silicon nitride, and silicon carbide for *wear parts* (moving parts exposed to extensive friction) is gaining popularity in the automotive and aerospace industries. See Figure 14-5.

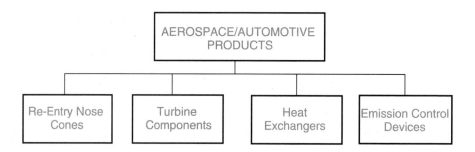

Figure 14-5
One of the fastest-growing segments of the industrial ceramics industry is aerospace/automotive products.

Ceramics in the automotive industry

On the horizon is the development of the *ceramic engine.* Research on this innovation indicates that such a development is probable, and that it will be able to perform at higher operating temperatures with better efficiency than conventional engines. The ceramic engine may be refined sooner than one might think. Today, all industrialized countries are manufacturing engine components that are either entirely ceramic or ceramic-coated metal. Countries which appear to have the lead on this technology are the United States, Japan, Germany, and Russia. The Toyota automobile provides evidence of the many applications which ceramics have in the automotive market. Ceramic components in the typical Toyota engine include the oxygen sensor, knock sensor, backup sensor, electric buzzer, water temperature sensor, exhaust gas sensor, blower resistor, fuel level switch, spark plug insulators, hybrid integrated circuits, and light-emitting diodes.

New materials are being rapidly developed for use in manufacturing products such as cutting tools, coatings, and bearings. See Figure 14-6. Manufacturers combine metals and ceramics to make *cermet* (ceramic/metal composite) cutting tool inserts from aluminum oxide and titanium carbide. The ceramic insert acts as the cutting tool. The insert is attached to the tool holder or inserted into the cutting tool. The advantage of ceramic inserts is that they can be replaced when they are worn or chipped. See Figure 14-7.

Cermets are also used for high-temperature applications in jet engines and aerospace systems. The U.S. Office of Technology Assessment anticipates that cermets alone will be a $160 million industry by the year 2000. The market for miscellaneous industrial ceramic products, as a category, is expected to exceed $2 billion.

Figure 14-6
Cutting tools and cermets (ceramic/metal composites used as cutting inserts in machine tools) represent one area of technical advancement in ceramics. They are among the miscellaneous products produced by the industrial ceramics industry.

Figure 14-7
Cermet and carbide cutting tool inserts are extensively used in aerospace and other metalworking industries. The inserts, shown in front of the tools in which they are used, improve productivity by eliminating the need for tool sharpening and permitting rapid replacement in the event of tool breakage. (Sandvik Coromant)

Bioceramics

In the medical field, ceramics materials (referred to as ***bioceramics***) offer tremendous potential. See Figure 14-8. There are three major types of bioceramic applications in medicine. These are referred to as *nearly inert, surface-active,* and *resorbable.*

Nearly inert ceramic devices can be implanted in the body without causing toxic reactions. These materials include silicon nitride-based ceramics, zirconia, and alumina.

Figure 14-8
Bioceramic products, produced by industrial ceramics firms, represent a huge growth market. Bioceramics are used in both medical and dental applications.

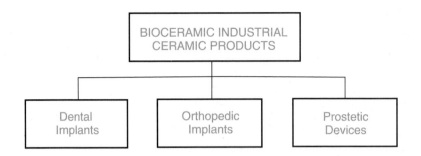

Surface-active ceramics are those that form a chemical bond with the surrounding tissue and encourage growth. They allow the implant to be held in place and help prevent rejection due to dislocation. Surface-active glass-ceramics and surface-active glass bond directly to the bone.

Resorbable ceramics materials have the ability to dissolve and be assimilated into the blood stream. Here, ceramic materials act much like bone tissue in the body.

The worldwide market for biocompatible materials is expected to reach nearly $10 billion by the year 2000. In the next several decades, bioceramics will account for a significant percentage of the world ceramic market.

Other product segments within the industrial ceramics sector are shown in Figures 14-9 through 14-12.

Figure 14-9
Specialty glassware for scientific and technical fields makes up the technical products segment of the industrial ceramics industry.

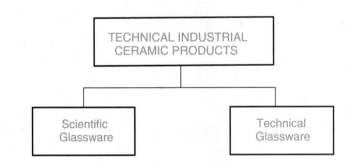

Figure 14-10
The ability of ceramic materials to withstand high temperatures in furnaces and similar applications has lead to growth in the high-temperature structural ceramics area.

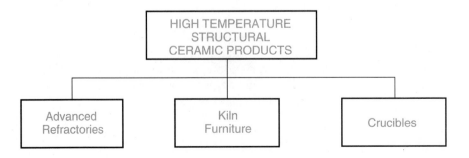

Figure 14-11
The use of ceramic materials for fuel rods and control devices has grown along with the shift toward nuclear-powered electrical generation and the increased production of nuclear isotopes for medical use.

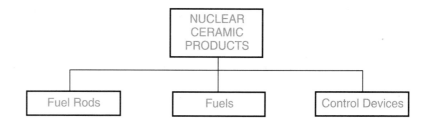

Figure 14-12
Ceramic products are used for many applications in the field of electronics.

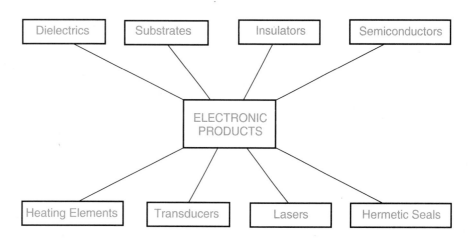

The Nature of Industrial Stock

To understand the nature of stock used to make ceramic products, it is important to briefly trace the historical uses of ceramics in manufacturing. We know that long before the building of the Great Pyramids in Egypt, craftsmen made ceramic containers or "pottery" to carry and store food and water. They combined local clay and water, shaped the plastic material by hand, and then permanently removed the water with fire, or in heated furnaces.

For centuries, there were few changes in the materials or manufacturing processes used to make ceramic products. The major change was from shaping the product by building it up with coils of clay to "throwing" and shaping the clay mass on a rotating potter's wheel.

Major changes in both materials and processes were introduced in the 1960s and 1970s, when new discoveries in science and technology led to the birth of the industrial ceramics industry. Manufacturing processes were developed to produce the great diversity of products demanded by expanding markets. New processes, such as thin film tape casting on belts of Mylar® (used to make electrical capacitors) and isostatic hot pressing of fine powders under a vacuum, were introduced. Other processes, such as blow molding and injection molding, were refined and modified.

As the demands for new product applications increased, new types of ceramic raw materials had to be made available for use with the more sophisticated manufacturing processes. In many cases, the wet clay that lent itself so beautifully to the processes used in the ceramic industry of 1940 or 1950 did not work well with the automated processes and improved technology of the new era.

Products with a high water content were very elastic and easily formed, but they were difficult to move without destroying their shape. Once the pieces were placed in an oven (kiln) for firing, the most difficult task was still unfinished: the water had to be gradually removed from the workpiece. If the product was too wet when fired, it would crack or explode in the kiln. Shrinkage occurring during firing was excessive, and achieving a perfect product was often a combination of skill and luck.

Today, ceramic stock is usually purchased by the manufacturer in a dry, or semi-dry, plastic state. The type of stock that is selected depends on the type of manufacturing process that will be used, the end use for the product, and the manufacturer's preference for the basic raw material.

If the basic material used is powdered clay, it is possible to compact the dry powder to create a pressed part. This is a process that is also used to press powdered metal parts. When metal powder is used, the manufacturing process is called powder metallurgy, and the pressed part is referred to as a green compact. In the field of ceramics, processes used to compact the powders into parts are known by a variety of names, but can be generally referred to as *dry pressing.*

One of the advantages of pressing ceramic powders to form parts is that such compacted parts have unusual strength and toughness. The finer the size of the individual grains in the powdered mix, the stronger and tougher the finished product.

Not all manufacturers use powdered stock with their manufacturing process. Many firms use liquid or semi-liquid clay. Normally, they start with powder and then add water, binding agents, and other additives. The more water that is added, the thinner the clay mixture.

There are other types of materials that are used in making ceramic products. In place of conventional clay, many companies work with materials such as silicon carbide, alumina, uranium dioxide, and silicon nitride. The ceramic manufacturing industry also includes many firms that manufacture glass products.

Glass manufacturers do not start with dry, semi-liquid, or liquid ceramic stock. Glass is a solid that can be turned into a liquid when it is heated. Glass as a raw material is created by fusion. It has been heated and then cooled too rapidly to permit crystallization. Glass has no known melting or freezing point, and consequently behaves much like plastics. Glass manufacturing companies usually purchase stock in the form of beads or sheets as their basic raw material. See Figure 14-13.

Figure 14-13
The float glass process is used worldwide to manufacture glass for residential and commercial window glazing, automotive use, and other applications. Molten glass is literally floated on a bed of molten tin. As the continuous ribbon of glass moves along, it gradually cools and solidifies. This inspector at a plant in France is checking the glass ribbon as it moves along a conveyor in a cooling area. (PPG Industries)

Important Terms

bioceramics
brittleness
ceramic
ceramic engine
ceramic-matrix composites
cermet
dry pressing
fatigue
industrial ceramics
kaolinite
kiln
traditional ceramics
wear parts

Questions for Review and Discussion

1. What is the major problem related to the use of ceramics as a design material? How can this be engineered out of the material?
2. What is the most common form of stock used in industrial ceramics?
3. Why must water be removed from products made with highly liquid clay before they are fired?
4. What is the melting point of glass?

A variety of oxide ceramic cutting inserts used for dry (no coolant) machining of cast iron. The ceramic materials are durable and easily replaced when they wear or break. The development of such tools has made possible decreased production costs and improved productivity in metalworking operations. (ITW Woodworth)

Forming Ceramic Materials

Key Concepts

- △ There are three types of ceramic forming processes: those that compact dry powder using heat and pressure, those that melt solids, and those that produce a liquid slurry from powder.
- △ Most industrial ceramic parts are formed by the dry pressing method.
- △ Material preparation, forming, densification, and finishing are the four process actions involved in ceramics.
- △ Deflocculants are used to reduce the percentage of water in slip and to maintain formability of the material.

There are many different types of forming processes used with ceramics. One group of forming processes uses heat and pressure to compact dry powders mixed with binders. *Binders* are chemicals that are added to reduce the *viscosity* (thickness of the mixture), and hold the mixture together.

Another group of processes utilizes powdered clay suspended in water. The liquid clay is mixed with binders and other chemicals, called *lubricants.* Lubricants are added to improve the shaping characteristics of the material and to aid in removal of the product from the mold. Once this is accomplished, the liquefied ceramic material is referred to as *slurry.* When binders are added to the slurry, the amount of water that must be used to produce a viscous mixture can be reduced. Binders provide body. Many of the ceramic processes in this category are used to produce cast products by pouring slurry directly into a mold.

Clay mixtures are engineered so that they can be used with a particular process, as well as conform to the design requirements of the product. Many of the processes that use either compacted dry powders or binders and lubricants in a slurry employ heat to improve the binding characteristics of the ingredients. Other processes involve cold-pressing.

A third category of processes involves no pressing. Here the product is created by melting stock in solid or particle form until it becomes a viscous (thick) liquid. Then, the liquid is shaped as it cools and solidifies. Glass products are made this way.

There are many imaginative uses for ceramics as a design material. For example, Russian ceramics engineers have been experimenting with using ceramic compounds to line the inside of metallic tubes. In this application, the ceramic material is heated to a molten state, and then hurled against the walls of the rapidly spinning tube, using a process called centrifugal casting. The process appears to have applications in producing ceramic gun barrels or ceramic liners for metal barrels.

Producers of metal or plastic parts normally begin by purchasing **stock** (raw material), in the form of powders, pellets, sheets, or bars. This stock is then transformed into the desired shape, using a manufacturing process such as drawing, forging, or bending. In the field of ceramics, the stock normally has to be *further modified* before it can be formed or shaped.

By now, you should be developing an appreciation for the tremendous diversity of product applications in ceramics. There are hundreds of unique products and process applications with glass products, traditional clay products, electronic and high-temperature ceramics, and ceramic glasses. In spite of this diversity, ceramic processes do have some commonalities.

Most ceramic processes involve four basic processing actions:
- Δ Material preparation.
- Δ Forming.
- Δ Densification.
- Δ Finishing.

It is important to understand why ceramic products must incorporate these basic processing actions.

Products can be manufactured using appropriate processes and techniques, and the end result can still *fail* quality assurance tests. Often, this is due to careless treatment of one or more of the basic processing actions. The problem is further complicated by the many stages of processing required before stock can be used with a particular process. Processing actions will be discussed in a later chapter.

Preparing Ceramic Materials for Processing

Each particle of finely ground clay powder that is used in a dry mixture contains many individual ceramic crystals. When these crystals are crushed or deformed under pressure, they will retain their new shape when the pressure is removed. At the same time, the crushing process must be conducted with care. Unlike metals, ceramic crystals are brittle and will fracture, rather than flow, under applied pressure.

When clay must be reduced to a semi-liquid plastic state, a process called *doping* is often used. This process is also used by industrial ceramicists to create a plastic condition in non-clay materials. Doping is done by adding organic binders and lubricants to the mix. Binders provide wet strength during forming and dry strength after drying. Lubricants decrease friction during forming, thus enabling forming to be accomplished with lower pressures.

Adding water to doped ingredients improves the plasticizing action of the clay mass. Binders that are used with water are polyvinyl alcohol, various starches, methylcellulose, and protein. Waxes and water-soluble oils are lubricants used with water.

For injection molding and other processes that use ceramic stock in a liquid state, synthetic thermoplastic resins can be used with organic solvents or plasticizers. Here, the resin acts as a lubricant as well as a binder.

Most ceramic products are formed and air-dried. Then, they are placed in an oven called a kiln for firing. Firing heats the product to a point where it is completely dry; it also binds the clay molecules together to create a usable product. Without firing, clay pieces would break easily, and thus would be unusable. Before they are fired, parts are referred to as *greenware*. After being fired, they are said to be in the *leather-hard* state.

Before being fired, ceramic parts can be machined using conventional metalcutting tools. Because of the fragile nature of these parts, however, extreme care must be taken when securing the workpiece and using the tool. Often, parts are fired before they are machined. However, when this is done, the machining is limited mainly to grinding, lapping, and honing processes using specialty tools or abrasives. Diamond-impregnated wheels are sometimes used.

Many ceramic pieces, particularly those made from slurry or viscous plastic material, contain a high percentage of water after forming. In these cases, it is particularly important to remove most of the water from the product before it is fired. This is normally done through air-drying or controlled-heat drying of the part.

Mixing the clay powder with the correct amount of water, binders, and lubricant is important. Clay that is mixed with too much water, binder, or lubricant retains a high percentage of moisture after the initial drying cycle. If that part is then placed in the kiln and fired, heat will cause it to quickly release water. This would result in drastic shrinkage, and could cause the part to crack or explode in the kiln.

The proper degree of dryness of ceramic parts is also important. If the part is too dry, then it will be excessively fragile, and will tend to crumble when it is transported to the kiln. Excessively dry parts also require greater forming pressures. Now that you have a basic understanding of ceramic forming, you can take a closer look at some of the processes.

Dry Forming Processes

As noted at the beginning of this chapter, one group of forming processes for ceramics uses pressure to compact dry powders that are then fired (heated) to produce the final product. There are a number of variations of the dry forming process.

Dry Pressing

Most industrial ceramic parts are formed by ***dry pressing***. Automated dry pressing is normally accomplished using ceramic powders with two percent or less moisture content. Stock used for dry pressing consists of the dry powder, binders, and lubricants.

The dry pressing process was first used in Germany in the late 1800s. The first use in the United States was in the 1930s when this process was employed to make electrical insulators. Dry pressing has a number of major advantages over other ceramic processes. Dry pressed parts have less flash, and can be produced to closer tolerances. The most significant advantage, however, is that dry pressing is a *near-net-shape (NNS) process:* parts undergo little shrinkage during firing. See Figure 15-1. "Green" (unfired) pressed parts are nearly the same size as they will be after removal from the kiln. This eliminates uncertainty and approximation in the manufacturing process.

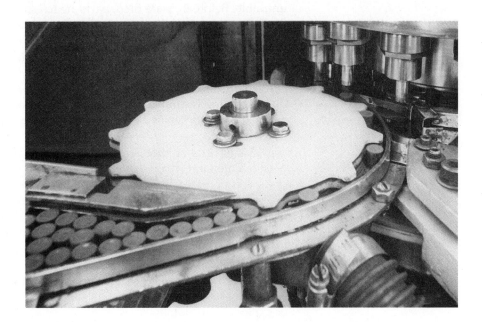

Figure 15-1
The ability to produce near-net-shape parts is one of the advantages of dry pressing. This rotary press is turning out nuclear fuel pellets. (General Electric)

Dry pressing, and other processes which use pressing to create NNS parts, are particularly attractive in applications with stringent electrical requirements. The process is used extensively for producing computer memory cores, refractories, tiles, spark plug insulators, nuclear fuel pellets, and many electronic components. Capacitor dielectrics as thin as 0.020 in. (0.508 mm), and thick- and thin-walled insulator boxes for protecting miniature electronic circuits, are also produced with this process. A variation of dry pressing, called *tablet pressing*, is also popular.

Some of the first electrical products made by dry pressing used a type of ceramic material called *steatite*. This natural crystalline form of magnesium silicate or talc was mixed with clay and barium carbonate flux. Steatite electrical products are usually very tiny. In composition, they contain a small amount of clay, mixed with artificial plasticizers like wax or polyvinyl alcohol. When they are fired at around 2460°F (1350°C), their normal firing temperature, a glassy bond is formed that holds all of the particles together. Fired steatite insulators are as strong as porcelain.

Dry pressing is one of the simplest high production processes used today for producing dense parts. Production speeds often reach 5,000 ppm (parts per minute). The dies are compacted in a press using pressure from 3,000 psi to 30,000 psi (20 685 kPa to 206 850 kPa).

Preparation of the powder for dry pressing takes place by spray drying slip to produce hollow solid particles, or by screening the slip through a fine mesh. Once the fine powders are separated and mixed with binders and lubricants, they are ready to be compacted in hardened steel or tungsten carbide dies.

The selection of binders is a critical factor in the dry pressing process. The ideal binder improves flowability of the powder, reduces abrasion, improves bonding strength, increases internal lubrication, and aids in removal of the part from the die. The binder burns off during the initial stages of the firing cycle, without disturbing the structural density of the compacted part. The compacting process causes the powdered particles to shift and deform, reducing the porosity of the material and thus improving the density of the part. After the granulated powder is pressed, elastic compression of the granules continues through compaction. This stored elastic energy produces an increase in the dimensions of the compacted part on ejection. This is referred to as *springback.*

In addition to this deviation in the size of the part, there are other limitations to the dry pressing process. A major concern is a variation in the amount of shrinkage throughout the part during firing and a loss of original tolerances. That is, not all areas of the part undergo the same amount of shrinkage. Despite these limitations, dimensional tolerances of + or - one percent are common in most applications.

Steps in dry pressing

There are four steps involved in the dry pressing production cycle:
1. Filling the die.
2. Compacting the powder.
3. Ejecting the part.
4. Firing or sintering.

When the die set opens, the matched dies are then filled with free-flowing granules of fine powder. Powder flows from a hopper to a feed shoe, which rests on top of the lower die. The shoe slides across the top of the die body and distributes powder into the lower die cavity. The die punches then close to compact the part.

Once the part is compacted, the lower punch is raised and the part is ejected from the die. The process continues when the die shoe, filled with a new load of powder, pushes the part out of the way, and on to a conveyor for transportation to the kiln for firing. The firing of dense ceramic compacts is referred to as sintering.

Normally, firing causes ceramic products to shrink to about two-thirds of their green (before firing) volume. This shrinkage makes it difficult to produce parts which will be the desired size (net shape). One method for obtaining near-net-shape products is to machine the part prior to sintering.

A variation of dry pressing is called hot forging. ***Hot forging*** involves heating to high temperature a part that had already been dry pressed, then pressing it in a cold steel die. The formed part is next returned to the furnace for annealing, a process that relieves the strain created by the additional pressing. This hot-forging technique provides improved density and strength to the part, when compared to conventional cold pressing processes.

Refractory materials

Hot forging is often used to produce *refractories*, ceramics that can withstand continued exposure to high temperatures. Refractories can withstand temperatures that often approach 5000°F (2760°C), and are also resistant to chemicals, thermal shock, and physical impact. Because of their ability to survive in high temperature environments, refractory materials are particularly useful for lining the inside of kilns and ovens, and for crucibles to carry molten metal.

There are many different types of ceramic materials used for making refractories. The most common are fire clay and silica, but kaolin, magnesite, bauxite, alumina, and limestone are also used.

Refractory products are usually made in brick or tile form. The manufacturing process for refractories begins with stock preparation: separating, crushing, grinding, and screening. The powder is then mixed and compacted in the same manner as it is in other dry press applications.

Refractory coatings are also produced for use on metals and ceramics. In this type of application, the coating is sprayed directly onto the refractory brick and then fired to facilitate curing. Ceramic coatings are being used effectively in high-temperature applications, to provide energy savings and extended service life of components. Ceramic refractory coatings prevent oxidation and fluxing (burnout) of refractories at elevated temperatures. In Figure 15-2, a black-body ceramic coating is being sprayed on the refractory lining of a reheat furnace at a steel mill.

Figure 15-2
A ceramic coating being sprayed onto the refractory lining of a reheat furnace. Such applications are used in the steel industry and others where high-temperature furnaces are involved. (Ceramic-Refractory Corp.)

Another interesting variation of dry pressing is in the production of *nuclear fuel rods.* In this process, finely crushed radioactive ceramic powders are combined with binders and an oil lubricant. The powders are highly abrasive, so oil is added to protect the punches and dies from excessive wear in the pressing process.

Dust Pressing

Dust pressing is very similar to dry pressing. The major differences are the amount of moisture that is added to the powder, and the types of products that are generally produced.

You will recall that in dry pressing, approximately two percent water was added to the mix. The moisture content in *dust pressing*, however, is normally five to ten percent. Like dry pressing, dust pressing is accomplished by compacting the powdered mixture under high pressure in steel dies. The process is illustrated in Figure 15-3.

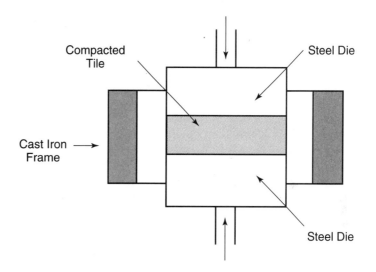

Figure 15-3
A major application of dust pressing is the production of ceramic wall and floor tile. The powdered material is compressed between steel dies before being fired.

Dust pressing is the preferred process for manufacturing ceramic wall and floor tile. For many years, the process was used primarily for the manufacture of tile. It was not until the late 1970s that dust pressing was applied to the manufacture of more complex shapes of flat tableware.

From the illustration of the dust pressing process (Figure 15-3), it appears that the process is so simple that it must be trouble-free. While it is true that the process eliminates many of the variables associated with handling plastic ceramic bodies, it *does* have some limitations. Dust pressing requires the addition of expensive spray driers and presses. Like dry pressing, this process has difficulty in producing parts with uniform wall thicknesses, and there is some inconsistency of shrinkage throughout the part during firing.

Experimental programs are presently underway at the U.S. Army Materials Technology Laboratory and the Ballistic Missile Research Laboratory to investigate the feasibility of combining the dust pressing process with explosive forming to produce even more dense tiles. Results indicate that powder can be pressed into a green compact with approximately 60 percent density. After pressing, the compact is then placed in a forming chamber, similar to that used for dust pressing. An electrically-detonated charge forces punches against the compact and compresses it against the walls of the metal die. The gas produced during compaction escapes through vertical grooves on the internal surface of the die.

As noted, in both dry pressing and dust pressing, water and binders are added to the mix. Some dry pressing processes require only pressure; other variations of the process require both heat and pressure.

Cold Isostatic Pressing

Isostatic pressing is also known as *uniaxial pressing, hydrostatic pressing,* or *hydrostatic molding*. Higher pressures are required for isostatic pressing than for conventional cold or hot pressing. Isostatic pressing is used for parts that, because of intricate configuration or extreme density requirements, cannot be produced with conventional hot or cold pressing methods.

Isostatic pressing can be done either cold or hot. **Cold isostatic pressing** at room temperature, using pressures ranging from 5000 psi to 20,000 psi (34 475 kPa to 237 900 kPa), is the process most often used in industry. Molds are normally filled with a free-flowing spray-dried powder. Vibration is used to eliminate voids (holes) in the powder mass inside the mold.

In isostatic pressing, pressure is applied to the powder from three directions, rather than the single direction used in linear pressing. Isostatic pressing is based on Pascal's Law. This law states that when pressure on a liquid in a closed container is increased or decreased, the resulting change in pressure occurs uniformly throughout the liquid. As applied to powder compaction, this means that powdered ceramic particles in a flexible, air-tight container, when placed in a closed vessel filled with pressurized liquid, will receive an applied force uniformly from all sides. The result is that material compacted by isostatic pressing will be uniformly pressed in every direction.

Dry bag isopressing

There are two basic types of cold isostatic pressing: compacting powders in a two-piece matched mold, or using fluid and a rubber liner to create pressure. The first method was discussed in conjunction with other pressing methods. The second method, called **dry bag isopressing,** requires additional discussion. See Figure 15-4.

In dry bag isopressing, a male die punch is used with an open-cavity lower die block. A rubber bag is inserted into the lower die block, which has no female impression, only an open cavity containing fluid. The fluid used is usually hydraulic oil, water, or glycerin. The bag is sealed after it is filled with powder and the forming punch is inserted. After the bag is submerged in the liquid, pressure is applied. The ceramic powder is compacted as the liquid in the mold transmits the pressure, pushing the rubber bag into the powder and against the forming punch.

Hot Isostatic Pressing

Hot isostatic pressing can be accomplished by pressing ceramic powder at 15,000 psi (103 425 kPa), while heating the compact to approximately 3000°F (1649°C). Hot isostatic pressing is also called *gas pressure bonding*. Heating the pressed compact improves the quality of the part that is produced.

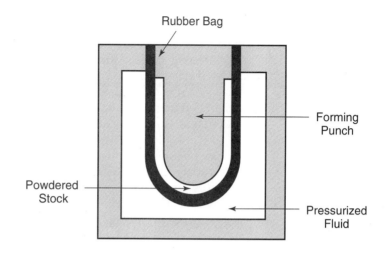

Figure 15-4
The ceramic powder and a forming punch are sealed inside a bag in the dry bag isopressing method. Pressure is exerted on the bag from the outside by fluid to compress the powder against the forming punch.

The heating action, called *sintering*, causes the loosely bonded powder to be transformed into a dense ceramic body. During sintering, moisture and organic materials are removed by burning them out of the greenware body. A process called *diffusion* then takes place, causing material to move from the particles into the void spaces between them. This improves the density of the part and causes less shrinkage during firing.

In hot isostatic pressing, the container is placed in a pressure vessel inside a high-temperature furnace. The pressure vessel is purged with gas and then evacuated to remove air and moisture. Helium is injected into the vessel at the desired pressure, and the furnace temperature is raised and held at the final bonding level until the part is properly sintered.

Explosive Forming

Explosive forming is used more frequently to shape refractory metals and carbides than ceramics, but it can be used with ceramics, as well. When ceramics are to be formed, the material used is in the form of granulated powder.

The powder is poured around a forming mandrel, which is encased in a pliable envelope. Often, this envelope is made of a thin-gauge metal. Explosives are placed around the envelope, and in some instances, the entire assembly is placed under water. The powder is compacted against the mandrel when the explosive is detonated.

A new field of study, called *dynamic compaction*, has evolved from the process of explosive forming. The Lawrence Livermore National Laboratory, Lewis Research Center (NASA), Los Alamos National Laboratory, and Defense Advanced Research Projects Agency are all conducting research on this technology.

Dynamic compaction is a process that can be conducted either at room temperature or with heat. When heat is used with dynamic compaction, a much lower temperature is needed than is required for hot pressing or sintering of ceramics. Firing at lower temperatures results in a part with fewer contaminants. Dynamic compaction can be used in conjunction with cold and hot pressing, reaction bonding, hot isostatic pressing, and injection molding.

Wet Forming Processes

A second large group of forming processes used for ceramics utilizes powdered clay suspended in water. The liquid clay is mixed with binders and lubricants to improve shaping characteristics and aid in removal from the mold. Depending upon the water content, the liquefied ceramic material may be referred to as slurry or slip.

Wet Pressing

Dry pressing and dust pressing both involve using relatively dry mixes. In comparison to these processes, **wet pressing** exhibits both similarities and differences. One would infer from the process name that the clay mixture must be viscous, or at least plastic, in nature. It is true that the mixture for wet pressing is fluid, but it is not necessarily wet.

There are two variations of wet pressing. The first method uses powdered stock, with a moisture content of approximately 15 percent. The water creates plastic deformation during compacting, when the mixture is forced against the contour of the die cavity. Compacted parts will have *flash*, the thin ridges of material that seep out at the parting line of the two dies when they are pressed together.

Wet pressing is not very suitable to automation, because parts are easily damaged and are difficult to transport after pressing. However, the process does have one major advantage over dry pressing: because of the higher moisture content, less pressure is involved.

The most significant disadvantage of the process is the dimensional accuracy of wet-pressed products. Wet processing is not as good a near-net-shape process as dry pressing, since dimensional tolerances can vary by as much as two percent.

A second form of wet pressing, called **ram pressing**, was developed and patented by Ram, Inc., of Springfield, Ohio. Ram pressing requires the use of a plastic (pliable) mass, rather than powder. The material is mixed and extruded into a storage canister until it is needed at the machine.

The material to be formed is pressed in a steel or plaster die. As pressure is applied, the plastic mass flows into the various contours of a two-piece mold. Air entrapped in the clay mass is removed by applying a vacuum through holes in the mold. As in the previously discussed method of wet pressing, flash is also created at the parting line of the mold. Ram pressing is often used for making oval-shaped dishes and contoured parts.

Hot Pressing

The major difference between cold pressing and hot pressing is temperature: in pressing, powders are pressed at high temperatures in heated graphite dies. Hot-pressed parts exhibit strength and density superior to products made by cold dry pressing.

Hot-pressing dies are usually heated by an induction or resistance coil, Figure 15-5. The heating of the dies makes it possible to use lower pressures (only a few thousand psi) to compact the part. This results in cost savings for the press. However, several additional factors also must be considered.

First, the dies are normally heated by electricity. This costs money. However, the main expense of the process is the graphite dies, which are soft and must be discarded after a single pressing.

Figure 15-5
The water-cooled induction coil of this 100-ton hot press heats the graphite die to a temperature of approximately 3600°F (2000°C). (Ceradyne, Inc.)

Hot pressing can also be accomplished using a more plastic ceramic body. In this method, a heated die lowers to sear the surface of the ceramic material, causing a cushion of steam to form between the ceramic body and the die surface. This cushion acts as a lubricant to keep the part from sticking to the mold. The heated pressing ram is made of metal, but the molds themselves are normally made of plaster.

One interesting application of hot pressing is in the field of electro-optic ceramics, where solid-state (no moving parts) switches are being made. A unique type of ferroelectric material, lead zirconate-titanate doped with lanthanum, is used. This material, called *PLZT*, has the ability to shift its polarity under the influence of an electric field, permitting the solid ceramic part to act as a switch.

The manufacturing process for producing the switch begins by placing a prepressed slug of PLZT in an alumina-lined, silicon carbide mold inside a pressurized chamber. The mold rests on an alumina plate and is surrounded with a thin layer of refractory material. Oxygen is backfilled into the pressurized chamber, which is heated to approximately 1300°F (700°C). The temperature and pressure are maintained for 18 hours while a pushrod presses the part and holds it in alignment in the mold.

After the hot-pressed slug is cooled and extracted from the mold, it is cleaned of refractory grain, annealed, and polished for optical testing and evaluation. This hot pressing application has allowed a very significant reduction in the size of switches with superior capabilities. Figure 15-6 illustrates the hot pressing of PLZT.

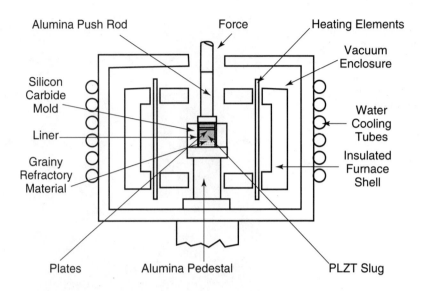

Figure 15-6
Solid-state switches can be dry pressed from the ferroelectric ceramic material PLZT by applying pressure and heat for a lengthy period. A typical pressing setup for PLZT is shown.

Throwing

Any discussion of contemporary manufacturing processes must include mention of one of the foundational processes in ceramics, called **throwing**. While throwing dates back to antiquity, it is still used today by firms producing round objects such as large electrical insulators, production prototypes, and dinnerware and pottery. In addition, many of today's applications still depend on variations of this basic process.

Throwing involves hurling onto a revolving potter's wheel a body, or mass, of clay that is in a plastic state. Once the clay is attached to the wheel, alignment (centering) of the mass is done to assure a symmetrical product.

Shaping of the product is accomplished by hand. The body is formed by pulling the clay upward and outward with moistened fingers, thumbs, and palms. This ancient process is still used today to make large electrical insulators that are later joined together in sections.

Jiggering

Jiggering is a ceramic production process that is used to make flatware, such as plates, saucers, and oval dishes. The process begins with a slice of clay being placed on a revolving convex mold that will form the inside of the plate. The mold typically is rotating at a speed of about 400 revolutions per minute (rpm). A template or forming tool then forces the clay over the mold. The process can be accomplished with or without automation.

In manual jiggering, the operator wets the plaster mold with a sponge, places the clay, then uses a forming template to press the plastic mass down over the contour of the rotating mold.

When the process is automated, the forming template is normally a dome-shaped profile tool that automatically contacts the surface to compress the clay and form the second side of the product. See Figure 15-7.

Jollying

The difference between jiggering and jollying is that in *jollying*, the plaster mold forms the outside surface of the product, and a forming template shapes the inside. A simple jollying operation would involve placing the plas-

Figure 15-7
Jiggering machines like this one are used to form greenware. (Lenox China)

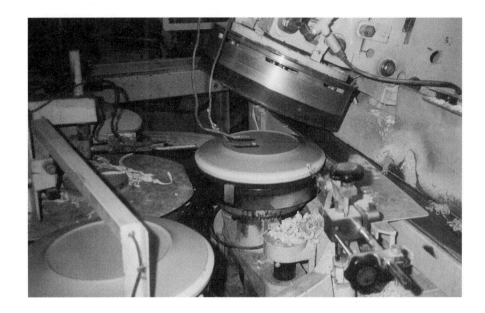

tic clay, in the form of a ball, onto the spinning mold. The sides would then be drawn upward, as in throwing and shaping a cup. The template is pushed down into the center of the ball of clay to shape the inside of the product. China and porcelain cups are made this way.

Today, many jiggering and jollying operations are performed by a roller machine. The major difference here is that instead of using a forming template, a heated die, called a "bomb," is used to perform the shaping operations. Figure 15-8 illustrates how a roller machine works.

Extrusion

Extrusion is a process used with metals, plastics, and composites to form continuous lengths of irregularly shaped stock. Examples of extruded products would be moldings and door casings. As you may recall from earlier chapters of this book, the stock is reduced to a semi-plastic liquid form in the barrel of a machine called an *extruder,* and is then forced with pressure through an orifice in a rigid metal die.

When extruding ceramics, the process is considerably simpler: since ceramic stock can easily be transformed into a pliable plastic with water and additives, no heating is necessary to produce stock that can be forced through the die. Figure 15-9 illustrates a simple piston extruder.

Ceramic products that are formed by extrusion range in size from hollow furnace tubes and small tubular electronic and magnetic ceramics to graphite electrodes weighing more than one ton. Extrusion of ceramics is used to fabricate brick, tile, rods, and elongated items that have a cross-sectional shape running the length of the product. Extrusion is also used as an intermediate step to other plastic forming processes.

Sometimes extrusion is used to *de-air* (remove the air from) highly viscous ceramic stock, and produce slugs, called **blanks** or **pugs,** that will be used

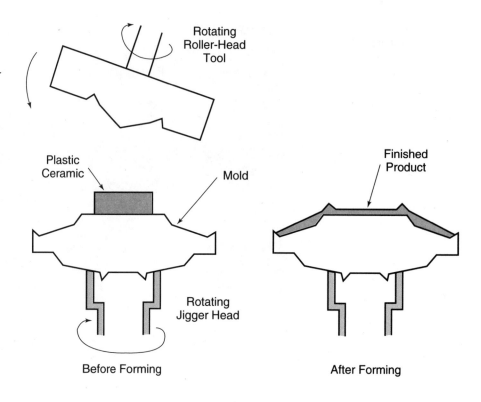

Figure 15-8
This roller machine is used to form a ceramic product, using the process called "jiggering." The ceramic material, in the plastic clay state, is pressed into shape between two rotating surfaces, left. After forming, right, the product can be removed from the jigger head.

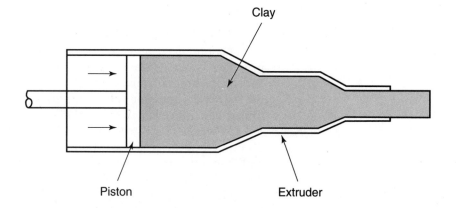

Figure 15-9
Clay, in a pliable plastic state, can be forced through a die to form an extruded shape. This simplified drawing shows a piston extruder.

with pressing and jiggering processes. Figure 15-10 shows extruded body pugs that have been stacked prior to being sliced into individual blanks. When high-temperature refractories are produced, extruded bricks are made and then re-pressed in dies before drying.

Figure 15-10
These body pugs are being stacked on a pallet by automated equipment prior to being sliced into disk-shaped blanks. (Lenox China)

Extrusion is a useful process for making the control rods and ceramic fuel for use in the nuclear fuel industry. See Figure 15-11. Before the advent of ceramic nuclear fuel, the pure isotope Uranium 235 had to be extracted from uranium metal. Earlier reactors used this type of fuel.

Figure 15-11
A technician monitors production of ceramic pellets that will be used to assemble nuclear fuel rods. (General Electric)

Today, ceramic fuels have become popular because of their stability, high heat-resistance, and their superior ability to ward off corrosion. Ceramic forms of uranium are made from oxides or carbides. ***Uranium dioxide (UO_2)*** is popular. UO_2 has a melting temperature of 5070°F (2800°C), making it one of the most refractory (heat-resistant) materials used today. Uranium dioxide must be enclosed in a container to keep it from further oxidizing.

Uranium carbide is more frequently used as a uranium substitute. It has a thermal conductivity higher than uranium dioxide, and has no pores. This makes it a very useful material for use as a fuel rod in nuclear reactors, Figure 15-12. Any increase in the temperature of the rod is consistent throughout the rod.

Figure 15-12
Ceramic fuel pellets are assembled into bundles like these for installation in nuclear reactors. Uranium dioxide and uranium carbide are both used for such applications. (General Electric)

Other interesting uses for ceramic extrusion to produce nuclear fuel is in the manufacture of cermets, uranium oxide, and boron carbide powders that are suspended in a stainless steel matrix. In another application, mounds of extruded strands of morillonite clay are being used to absorb radioactive waste. After the waste is drawn into the clay, the mass is fired to permanently lock in the radioactive atoms.

Injection Molding

Injection molding is a process used extensively in the plastics industry. The same process is used by the ceramics industry for making spark plug insulators and electronic parts. There are several differences in the way that the process is conducted.

Injection molding of ceramics requires mixing of special plasticizers and resins with the dry ceramic powder. The clay mix is heated and formed into pellets by extrusion. The pellet mix is loaded into a hopper on the heater which feeds the heated injection molding cylinder.

Hydraulic pressure then forces a cylinder forward on the heated plastic mass, and the material is squeezed through an opening in the cylinder into the cooled metal mold. With ceramics, the dies must be made of harder, more wear-resistant metal alloys than the dies used for injection molding of plastics.

Often, thermoplastic or thermosetting polymers are mixed with the ceramic powder to act as a filler. The polymers fill the pores of the ceramic body. When the resins are heated, they soften and are injected into a cold die. The plastic additives are then removed from the product by careful thermal treatment before firing. Sometimes, the removal of the resin takes several days. Kiln firing is normally the final stage of the process.

Band/Tape Casting

Band casting, or *tape casting,* is a continuous production process used to produce thin strips of ceramics for use as electrical substrates and heat-exchanger devices. Some of the most frequent uses of this process are for production of multilayer titanate capacitors, piezoelectric devices, thin film insulators, ferrite memories, and multilayer electronic packaging. Electrical substrates, called fabrication boards, are normally green in color. Resistors and metal circuit patterns are usually printed on the substrate and co-fired.

In tape casting, controlled thickness film is produced by flowing a ceramic slurry down an inclined surface, or passing it through forming rolls. Figure 15-13 shows two different types of tape forming processes. In the top illustration, thin film is cast on low-ash paper by pulling the paper through ceramic slurry. Other types of carriers can also be used. The film and carrier is then dried and collected on a take-up reel.

The lower illustration shows thin film casting using rollers to regulate the thickness of the film. With this variation, the ceramic powder is mixed with organic binders.

Another type of tape casting, called the ***doctor blade casting process,*** involves the use of a doctor blade, or metal squeegee, to regulate the amount of ceramic slurry that is allowed to flow on to a metal belt. In doctor blade casting, organic binders are added to the powder. A solvent makes the mixture fluid so that it can be cast on a continuous steel ribbon.

Figure 15-13
Two of the several types of tape casting processes. At top, a casting process using a paper substrate is shown. In the process shown at bottom, the material is formed to final thickness by being squeezed between rollers.

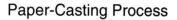
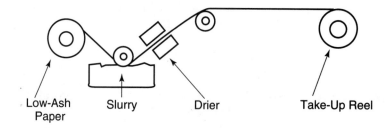
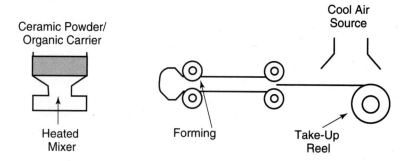

The basic difference between doctor blade and tape casting is *motion:* When the blade is held stationary and the surface moves, the process is called tape casting. When the doctor blade moves across a stationary surface holding the slurry, it is referred to as doctor blade casting.

With either type of casting, a flexible, smooth-surfaced, white leathery film is produced in thicknesses of approximately 0.010 in. to 0.060 in. (0.254mm to 1.524mm). The process is continuous and automated.

The casting process

Let's take a closer look at how the process works. First, the casting slurry is prepared by mixing powdered ceramic stock with binders, plasticizers, and dispersants. Mixing is usually accomplished by *milling,* a process similar to the barrel tumbling used to finish metallic materials.

After milling, the slurry mixture is heated, filtered, and de-aired to remove air bubbles. The de-airing process is accomplished by transferring the slip from the ball mill to a vessel where the slip is mechanically de-aired under vacuum for several minutes.

Slurry is then cast on a very clean Mylar, Teflon, or acetate film. The film is continuously supplied from a spool at one end of the machine. The ceramic tape is applied to the top surface of the film, where its thickness is regulated by the size of the opening between the bottom of the doctor blade and the film carrier. The film dries slowly as it is being transported through the machine on the film carrier.

The tape that is produced is flexible, and does not permanently bond to the carrier belt. When the tape reaches the end of the machine, it can be cut or stored in a reel with the film carrier.

When it is time for further processing, the tape can be removed from the carrier and stamped, slit, or scored to the desired size and configuration. Parts are produced approximately 16 percent oversize to compensate for shrinkage during firing.

Calendering

Calendering is a process that can be used in place of extrusion or tape casting to prepare thin plates. In the calendering process, powder mixed with a polymer is dissolved with a solvent in a paddle mixer. Stirring continues until the solvent is removed and the clay is transferred to mixing rollers. Next, the clay mix is squeezed between the two mixing rollers and a corrugated calendering roller. The rollers function as a cutter, turning in opposite directions and producing a shearing action on the material being pinched between them.

The rollers squeeze the clay powder and polymer until the film has the desired thickness. Then it is collected on a paper or polymer reel and stored for further processing.

One of the major advantages of calendering is that the final product is solvent-free. In the band or tape casting process, the evaporation of solvent creates pores in the final product.

Slip Casting

You will recall, from its description earlier in this chapter, that *slip* is a very liquid clay and water mixture. Slip often involves adding as much as 25 percent water to dry or nearly dry powder. The water content is kept as low as possible to reduce the extent of shrinkage that occurs during drying.

Slip casting, also called drain casting, is a process that has been used to make figurines and intricately designed products since the early 1700s. See Figure 15-14.

The process begins by pouring slip into a hollow two-piece plaster mold. When the slip has been drawn by capillary action into the porous walls of the mold, the mold is inverted and the undried slip is poured out. This results in solid layers of clay being cast, or layered, on top of other layers to form the walls of the product. The walls become thicker with each pour.

Theoretically, the walls could keep getting thicker and thicker until a solid cast product was developed. Actually, however, as the walls become thicker, it takes longer for the liquid slip to be absorbed into the clay. Trying to cast a solid clay body with this process would require further modification of the slip material. The desired thickness produced with conventional slip normally would not exceed 0.050 in. (1.27mm). After the product has dried, the mold is separated and the product is further dried before it is fired.

The formation of slip is often a complicated process. Generally, it is not practical to produce slip by just adding water to clay. Such an approach would require excessive proportions of water, perhaps as great as 50 or 60 percent. This would result in slip that would quickly saturate the pores of dry plaster molds. The wet molds would then have to be dried before more castings could be made. Time would be wasted and mold life would be shortened.

An even more critical factor, however, would be the high percentage of shrinkage: excessive amounts of water result in distortion and cracking during drying and firing.

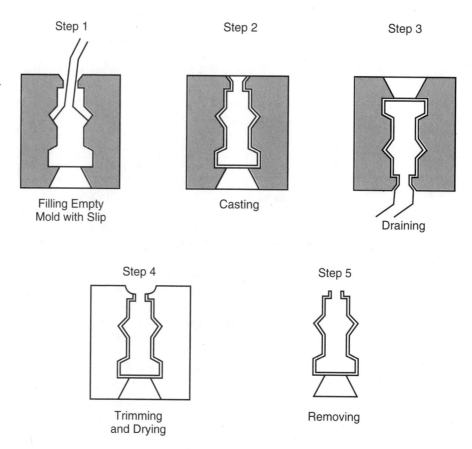

Figure 15-14
Slip casting is a process that has been used for hundreds of years to produce hollow ceramic products. Wall thickness can be built up in several layers.

For these reasons, it is necessary to use a *deflocculant,* a chemical that helps to make slip fluid. Deflocculants can reduce the water content of slip to 25 percent or less. Sodium salts of oxalic and tannic acid are common deflocculants.

The effect of a deflocculant is often dramatic. If clay is mixed with enough water to create a stiff paste and a tiny amount of deflocculant is added, then a highly liquid slip will be immediately produced upon stirring. The reverse action also occurs; if too much deflocculant is added, the result is a thickening of the slip.

Sometimes, a slip consists of coarse-grained materials that cannot be dispersed effectively with water and additives. In this case, a process called *freeze casting* is sometimes helpful. In freeze casting, thick slip is poured into a smooth-walled, rigid rubber mold. The mold is then frozen to force the clay to expand and press against the walls of the mold. Drying of the frozen part is accomplished through the use of absorbent sand.

Another unusual variation of the slip-casting process is called ultrasonic casting. With this method, low-amplitude audio frequency vibrations of 15-20 kilohertz (kHz) are transmitted to the plaster mold during the casting process. Some of the heat generated in the plaster mold by the ultrasonic vibrations is transferred to the workpiece. A number of firms are using this process to slip-cast cermet and alumina plates. Cermets are composite materials consisting of ceramic and metal particles held in suspension in a metal matrix. Heat causes the cermet to be sintered, bonding the particles together.

Casting

Casting is a relatively simple process that is often used to create solid ceramic objects. It is a variation of the process of slip casting. A liquid suspension of clay (slip) is poured at room temperature into a porous plaster-of-paris mold. The dry mold absorbs moisture into its pores, and a hard layer is built on its walls. The process is continued, building layer by layer, until the interior of the mold is completely filled. In this case, it is referred to as solid casting. For solid casting, slip with a high percentage of deflocculants is used. The addition of deflocculants creates a fluid suspension with as little as 20 percent liquid. If the slip was produced with a high percentage of water, excessive shrinkage would occur during drying, and the part would be distorted and cracked.

A less-common method of casting involves using molten ceramics to manufacture high density refractories, grinding wheels, and other forms of abrasives. Molten ceramic casting is done by pouring the melt into cooled metal molds to accomplish rapid cooling, or quenching. The quenching process causes stresses that result in the formation of fine-grained crystals and a very tough product.

Glass Forming Processes

This third category of processes does not usually involve pressing. The product is created by melting stock until it becomes a viscous (thick) liquid, then shaping it as it cools and solidifies. Various forms of glass are made this way.

Drawing

Drawing is a process that was used many years to produce sheet glass. The first unsuccessful attempts to draw sheet glass directly from the furnace date back to the 1850s; it was not until 1905 that a Belgian named Fourcault perfected the process so that it was practical for manufacturing. The Fourcault process, first used in industry in 1914, involved pulling molten glass from an oven and upward through rollers to a horizontal conveyor. After being transported through an *annealing lehr* (oven), it is cooled and cut.

Since the early days of glassmaking, manufacturers have developed their own unique processes. Popular at various times were the Colburn process and the Pennvernon process. Today, most sheet glass is made using the *float glass process*. Drawing is more frequently used for producing glass fibers and tubes.

Float Glass Process

The ***float glass process*** produces sheet glass with the qualities of thicker plate glass. Today it is the most popular process for making safety glass, mirrors, and transparent glazing for commercial buildings. A schematic illustration of the basic process is provided in Figure 15-15.

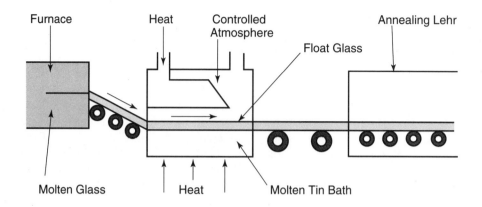

Figure 15-15. Much of the glass produced today is made by the float glass method, in which the sheet of melted ceramic material gradually cools and solidifies while floating on a bath of molten tin.

In this process, molten glass is fed into a bath of molten tin. The glass floats on top of the tin. When the glass first contacts the molten tin, its temperature is about 1800°F (1000°C). After floating along on top of the molten tin, it reaches its exit point at around 1100°F (600°C). Tin is used to float the glass because it is the only metal that is liquid at 1100°F (600°C) and does not vaporize and burn at 1800°F (1000°C).

To reduce oxidation of the tin, the bath is enclosed in a protective gas. When the glass leaves the float tank, rollers take it to the annealing lehr, where it is slowly cooled. When it leaves the lehr, its temperature is about 390°F (200°C). See Figure 15-16.

With the float glass process, the entire operation is almost totally automated. Today, throughout the world, almost all traditional plate glass manufacturing has been eliminated and float glass manufacturing has taken its place. See Figure 15-17.

Figure 15-16. This quality control technician is performing a visual inspection of a wide ribbon of float glass as it is conveyed through a cooling section of the processing operation. The glass ribbon runs back through the annealing lehr (shed-like structure in background) to the molten tin bath where molten glass first assumes the flat ribbon form. (PPG Industries)

Figure 15-17
These workers, with the aid of an overhead crane, are moving a bundle of float glass sheets weighing four tons. The sheets will be fabricated into architectural glass panels. (PPG Industries)

Fiber Drawing

Glass fibers are made by drawing heat-softened glass marbles through small orifices, called bushings, in the bottom of a tank on the remelting furnace. Drawing fibers is an ancient process that is actually older than glassblowing. Today, the process has been automated and upgraded by firms such as Corning Incorporated. See Figure 15-18.

Here's how the process of fiber drawing works. First, molten glass is pulled by suction through the holes in a melting tank. As soon as the fibers leave the furnace, they are immediately cooled. Then, the threads are coated with sizing by a roller to protect them from becoming scratched and damaged. Once the fibers are sized, they are gathered together on a spinning drum. After collecting, they are further processed into yarns and roving material, using conventional textile manufacturing processes.

Figure 15-18
This droplet of molten glass will be drawn into a fine optical fiber. (Corning Incorporated)

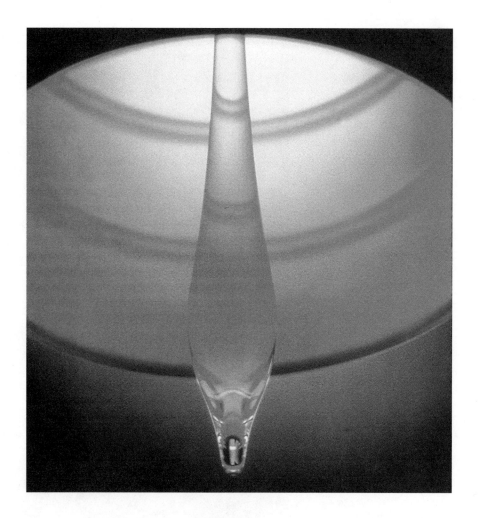

Two basic types of fibers are *staple fibers* (6 in.-15 in. in length), and *continuous fibers.* The fibers are sometimes sprayed with binders to hold them together in a continuous mat.

Tube Drawing

Glass tubes are made using the *Danner process*, illustrated in Figure 15-19. With this process, molten glass flows down over a hollow mandrel. The mandrel is positioned at an angle to the glass, so the molten material runs down its length and off the lower end. Air is blown in through the center of the mandrel to keep the walls of the tubing from collapsing. The tubing is carried by rollers into an annealing lehr. The diameter of the tubing depends on the temperature of the glass, the amount of pressure blown through the tubing, and the speed at which the tubing is pulled off of the mandrel.

Centrifugal Casting

You will recall, from an earlier chapter, that the centrifugal casting process was used with metallic materials. The same process is also used to form ceramic pieces, particularly cone-shaped glassware. Figure 15-20 illustrates how the process works with ceramics.

Figure 15-19
In the Danner process, glass is formed into a tube by flowing over a hollow mandrel. Air is injected through the mandrel to keep the tube from collapsing.

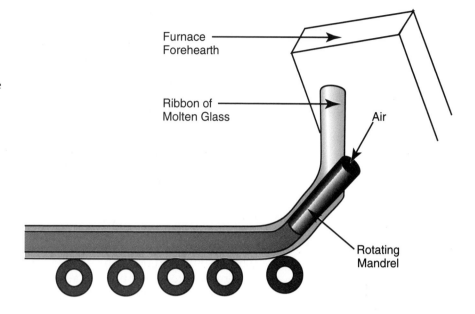

Figure 15-20
Centrifugal force causes molten glass to flow outward and upward along the walls of a rotating mold, forming a cone shape. As the glass cools, it is first trimmed, then removed from the mold.

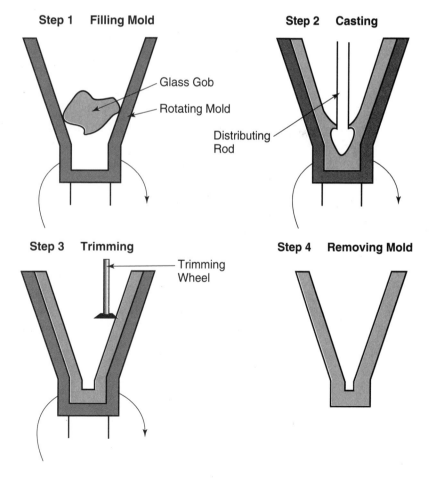

The process begins by filling the rotating, funnel-shaped mold with molten glass. Centrifugal forces created by the rapidly turning mold throw the molten glass against the walls of the mold, and cause it to creep upward as the mold continues to turn. A distribution rod is then inserted into the mass of molten glass, helping to force the glass against the walls of the mold. Once the walls are coated to the desired thickness, the glass funnel is trimmed to height with the trimming wheel. When the product is cooled, the finished part is ejected from the mold.

Glassblowing

You are probably familiar in a general way with the ancient process of glassblowing. In this process, a glob of molten glass, called a *gather*, is formed into the desired shape using air pressure. Much scientific glassware is still produced in this way.

Most glassblowing in industry today, however, is done on automated high-production glassmaking machines. The first automated blowing machine was invented by Michael Owens in 1903.

There are three major types of automated glassblowing machines in use today: the press and blow machine, the blow and blow machine, and the ribbon machine. Some machines are used primarily for the manufacture of bottles, jars, or containers. Others are used to manufacture such products as light bulbs and Christmas tree ornaments.

Press and blow machine

The *press and blow machine* is used primarily for container production. The process is illustrated in Figure 15-21. Container production is accomplished in two major operations. The neck of the container is made by pressing or blowing, then the rest of the container is formed by blowing.

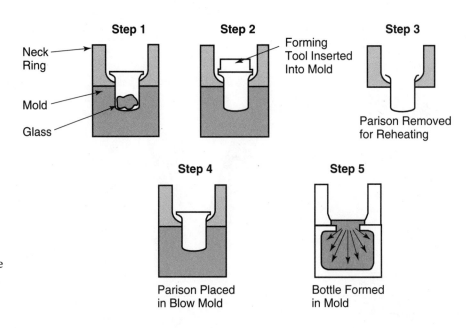

Figure 15-21
The press and blow process. Step 1—Molten glass is placed in a mold. Step 2—A forming tool is inserted in the mold. This presses the glass up and against the walls of the mold. Step 3—The tool is then removed and the preform, or parison, is transported by the neck ring to the blow mold. Step 4—The parison is placed in the blow mold. Step 5—Air is blown in, forcing the molten glass against the walls of the mold to form the bottle.

After the part is blown, the mold is opened and the product is transported to the annealing lehr. The annealing process permits the glass product to gradually cool down, thus relieving stresses and strains created during forming.

The press and blow process is ideal for producing containers with thin walls, such as disposable jars and bottles, Figure 15-22. The process is not used for bottles with very narrow openings, such as those for medicines and pharmaceutical preparations, because it does not produce a surface quality and wall thickness suitable for many of these types of applications.

Figure 15-22
Glass bottles that will eventually hold salad dressing are shown emerging from their molds in this blow and press operation. Note the rapid cooling of the glass, as indicated by the color difference between the bottles in foreground and those just removed from the molds. (Ball Corporation)

Blow and blow machine

The *blow and blow process* is shown in Figure 15-23. First, molten glass is introduced, step 1. When the blow head is closed, a blast of air blows the molten glass down into the mold, step 2. This forms the neck of the bottle.

At this point, an air blast is introduced from the bottom, blowing the glass upward and against the walls of the mold. The basic shape of the preform has now been created. Next, the parison is removed from the first mold, inverted, and placed in mold 2 (step 3). Here it is heated, then expanded by blowing a blast of air back through the mouth of the container (step 4).

Ribbon blow molding machine

The ribbon blow molding machine is used by Corning Incorporated to manufacture incandescent lightbulbs. The ribbon machine is one of the fastest machines in the industry, capable of producing enclosures for bulbs at speeds up to 2000 parts per minute.

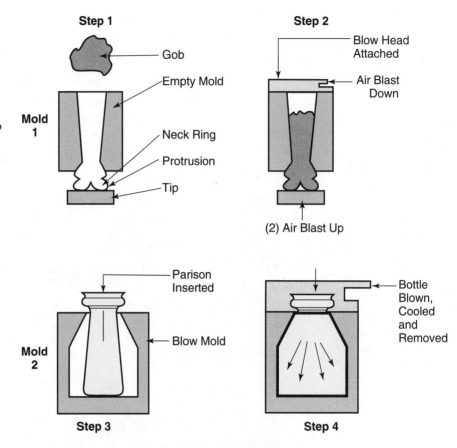

Figure 15-23
The blow and blow molding process for glass containers involves the use of two molds. The first mold is used to blow the neck, then the parison is transferred to a second mold to be blown into final form.

In the ribbon blow molding process, a thin (3 in. wide) ribbon of molten glass flows continuously from an overhead melting tank through two water-cooled forming rollers. As the ribbon travels along a steel track, air heads from above contact the ribbon and blow the glass to shape the parison. At the same time, split molds come up to surround the molten glass. Air blows the parison against the walls of the mold to shape the final product. After the mold halves separate, the completed bulbs are cracked off the ribbon and fall onto a transporting conveyor.

Glass rolling

If flat glass does not have to be totally transparent, it can be poured and rolled. **Rolled glass** is translucent, with a light transmission capability of 50 to 80 percent. Rolled glass is often used in skylights, bathrooms, and for interior lights.

The molten glass flows out of the furnace over a refractory barrier called a *weir*, onto the machine slab, and on through two water-cooled rollers. It is these rollers that establish the thickness of the plate. A refractory gate permits glass to flow onto the slab. If wire-reinforced glass is being made, the wire is introduced into the hot glass using a locating roller. If the glass is to be decorated with a design, shaping rollers are used to emboss the glass, creating the desired surface relief.

Once the glass is shaped, it is carried by rollers to the annealing lehr. When it enters the lehr, the cast glass is reheated to about 1470°F (800°C). When the glass comes out of the lehr, it is cool enough to handle. Finally, it is cut to size and packed for shipment. The rolling process is used to manufacture wire reinforced safety glass, colored plate glass, greenhouse glass, and opaque glass.

Glass pressing

Automatic presses are also used to press molten glass into molds. Once the molds are loaded with glass, they are transported down conveyors to the pressing station. When the mold containing a molten gob of glass moves beneath a plunger, it is pressed into the desired shape. While this is happening, another molten gob of glass is being placed in the next mold. After each piece is pressed, it is air-cooled, ejected, and transported by a conveyor to the next work area.

Sagging

Sagging is a technique for shaping glass that has been used commercially for many years. Sagging is accomplished by placing sheet glass over a mold, and then applying heat until the glass softens and sags downward. The sagged glass conforms to the shape of the mold and takes on the texture of the mold surface.

The process of sagging is simple, but not all shapes can be effectively sagged. The mold must have a gradual slope and not be too deep. Molds for sagging glass can be made of metal or soft fire brick. Often *grog mix,* (clay mixed with fine particles of previously fired clay), is used to make the mold.

Important Terms

annealing lehr
band casting
binders
blanks
blow and blow process
cold isostatic pressing
continuous fibers
Danner process
de-air
deflocculant
diffusion
doctor blade casting process
doping
dry bag isopressing
dry pressing
dust pressing
dynamic compaction
extrusion
float glass process
greenware
grog mix
hot forging
hot isostatic pressing

injection molding
jiggering
jollying
leather-hard
lubricants
near-net-shape (NNS) process
nuclear fuel rods
PLZT
press and blow machine
pugs
ram pressing
refractories
rolled glass
sagging
sintering
slip
slip casting
slurry
springback
staple fibers
steatite
stock
tablet pressing

tape casting
throwing
uranium carbide

uranium dioxide
viscosity
weir
wet pressing

Questions for Review and Discussion

1. What causes shrinkage in a clay product? Describe several methods that can be used to reduce the amount of shrinkage.
2. Distinguish between the processes of dust pressing and hot forging.
3. What is the advantage of hot pressing in terms of consistency of material?
4. Why does calendering produce a better quality product than band/tape casting?

16

Separating Ceramic Materials

Key Concepts

△ Glass, ceramic, or steel media are used in the various types of mills that are employed to grind clay into extremely fine particles for mixing as a slurry.

△ Filter pressing, spray drying, and other methods are used to remove excess water from bulk ceramic material.

△ Stock for dry pressing is ground and milled with a variety of processes.

△ Removal of excess or unwanted material from sintered workpieces is typically done by grinding.

Raw clay must be ground and refined many times before it is useful in manufacturing. The first phases of grinding typically occur after mineral deposits are mined from the earth by primary manufacturing firms. The raw materials are often *beneficiated*, which means that, to reduce transportation costs, the raw material is processed or refined by plants located near the mine. When the material is first mined, pieces can be as large as a yard in diameter or as fine as a granule of powder.

You should recall the structural model described in Chapter 1, which differentiated between the functions of primary and secondary processing. As shown in that model, *refining* is a process usually conducted by primary manufacturing firms. Since we are interested in the processes used by *secondary* manufacturers, our study of grinding will begin at the point where a ceramic plant receives the raw material.

Grinding

When raw material is purchased from the processing plant, it usually must undergo additional blending and crushing before use. Often, a pulverizing machine called a *mix muller* is used. The muller consists of a large circular pan in which two large steel-tired wheels revolve. Some models use rollers suspended in the middle of the pan from a revolving pivot arm, rather than wheels.

Clay placed in the muller pan may be in chunks up to several inches in size, or as small as a grain of sand. The wheels or rollers grind and crush the material until it is reduced to a size where it can drop through openings in the bottom of the pan. After the clay particles are reduced to the desired fineness, they are passed through a series of screens to eliminate those that are oversize or undersize.

In many cases, a manufacturer purchases stock that has already been crushed, screened, and refined or purified. One of the preferred methods is to receive dry powdered stock in railroad tank cars or enclosed tractor trailers that can be unloaded by vacuum. Sometimes, the clay is mixed with water, with the resulting slurry shipped by tank truck.

If the stock is received in powdered form, the first step in making it useful for manufacturing is usually tempering. *Tempering* is the process of mixing and kneading liquids into a dry material to produce stock that is pliable enough for forming. In most cases, tempering consists of mixing water with the dry clay stock while continuously cutting and kneading the mixture. The clay mixture that is produced can vary from almost dry (very little water) to a paste suitable for plastic forming.

When the clay emerges from initial mixing, it has the desired water content, but is not compact. At this point, the clay mixture still contains air bubbles. When the clay is to be used for extrusion, these bubbles must be removed to increase the density and strength of the part. This is done in a machine called a *de-airing pug mill.* The pug mill consists of knives on a rotating shaft. The knives cut and fold the stock in a shallow mixing chamber. This kneading action traps air bubbles and compresses and compacts the clay. Extruded clay leaves the pug mill in a continuous column.

Milling

Ceramic stock is usually ground in order to produce fine particles for use in making slip or with pressing processes. In this case, the grinding process is usually called *milling.* When hard clay in lump form is used, it is first passed through roller crushers to break it down into smaller particles. The crushing operation would most likely be followed by milling to further reduce the size of the particles.

There are many similarities in milling the basic ingredients for the different types of clay and product applications. Clays exhibit varying degrees of plasticity, and all contain nonplastic elements such as flint or quartz, and mineral fluxes such as feldspar. All of these ingredients must be thoroughly milled and mixed to ensure that the body will have uniform composition throughout.

Composition will differ, depending on the type of ceramic product to be made. For example, if bone china tableware is being made, the raw material consists of calcined cattle bones mixed with china clay and feldspathic stone. About 50 percent of the mix is bone, 25 percent is clay, and the rest is feldspathic stone. The china clay adds plasticity to the mix, while the feldspathic stone serves as a flux that makes the clay body flow when it is heated. The bones provide a refractory element that helps to reduce the impact of the heat on the clay in the furnace. Nonplastic materials such as bone or stone must be subjected to the process of *calcination* to mix properly with the clay. In this process, the material is first broken down by heating it to a red heat. Then, it is milled to produce particles that are sufficiently fine to be mixed with the clay.

Milling is normally accomplished in a tumbling, or rotating **ball mill**. Industrial ball mills normally consist of a steel cylinder, up to ten feet in diameter, that is partially filled with spheres of heavy steel or dense ceramic **grinding media**. Grinding media consists of small pieces of siliceous rock or rubber, usually less than an inch in diameter. Porcelain balls, flint pebbles, or alumina balls are also used.

In a laboratory or research environment, a smaller-scale mill referred to as a "jar mill" is used. These small mills are made of porcelain, with a removable top. They are made in capacities ranging from one pint to several gallons (0.5 liter to 8 liters or more).

Conventional milling

The process of milling can be done wet or dry, but it is easier to pulverize the mixture when the clay is suspended in a slurry.

In conventional milling, the process involves introducing flint, stone, feldspar, or other materials with water as the mill turns. The amount of powder that is added is usually around 25 percent of the total mill volume for dry milling, and about 40 percent for wet milling. The balls or grinding media lift, turn, and crush the mixture as the cylinder rotates. The turning, mixing, and crushing action may take many hours to reduce the particles to the desired size.

Often the milling process is more complex than just adding the ingredients and grinding them in a ball mill. For example, if color is critical in the finished product, metallic iron contaminants that would create dark spots during firing must be removed using powerful magnets. As noted earlier, nonplastic materials may have to be calcined before milling so that they will properly mix with the clay.

A variation of the ball mill that has gained popularity in industry is called the *Hardinge conical mill*. The major advantage of the Hardinge mill is continuous operation—it never has to be stopped to be filled. The mill turns around a horizontal axis, while the mix is continuously fed in at one end and automatically discharged at the other. Because of the conical shape, the material slides down the walls of the mill, and is ground in a graduated fashion. The largest and heaviest particles are pushed to the bottom of the mill, where they are lifted and dropped with the greatest impact at the widest diameter of the cone. The smaller particles move automatically to the narrow discharge end.

Other wet milling processes

Continuous grinding and dispersion of slurry, consisting of solids suspended in water, is also done using a John mill, a Molinex mill, or an attrition mill. All three of these mills accomplish grinding action through the use of glass, ceramic, or steel media. In the *John mill*, short agitator pegs of tungsten carbide are set into the inner and outer walls of a grinding chamber. See Figure 16-1. As the grinding cylinder is turned at high speed, agitation is introduced by the protruding pegs, producing a uniformly ground product. To dissipate the heat generated by friction, cooling water is often circulated in the large-diameter agitator shaft and in the chamber's cooling jacket. The John mill is a high-energy mill, requiring only one pass for most grinding operations involving solids and high viscosity slurries. Materials typically ground in a John mill include ferrites, toner for copying machines, printing inks, conductive coatings, and metal oxides.

Figure 16-1
The John mill, as it would be seen from above. The agitator pegs on the walls help move the grinding media through the slurry to accomplish fine grinding.

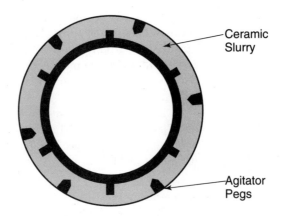

In the *Molinex mill*, eccentric grinding disks (rather than the agitator pegs of the John mill) are used to obtain a superfine grind. The disks are mounted on a rotating shaft and are staggered to function as an auger. See Figure 16-2. This auger arrangement helps move the grinding medium against the flow of material to be ground. The disks are sectioned so that the grinding media can be agitated from both inside and outside. The Molinex mill is used in industry for grinding a wide array of material ranging from chocolate to microorganisms to zirconium oxide to clay.

Figure 16-2
In the Molinex mill, openings in the eccentrically mounted disks permit the grinding media to circulate and reduce the particles of clay to small size.

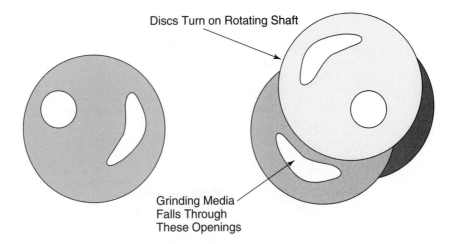

Attrition milling is a batch operation that involves pumping the materials to be ground into a stationary chamber, along with a grinding medium. A rotating internal shaft with extended arms, Figure 16-3, is used to agitate the grinding medium and material to be ground. The beating action, and the rubbing of the material against the grinding media produces a uniform particle size. Attrition mills are extensively used to grind food products, cosmetics, graphite dispersions, many oxides, and ferrites.

Figure 16-3
The attrition mill uses a rotating shaft with arms to continually agitate the mixture of grinding media and clay. A high-pressure pump helps to keep the mixture moving.

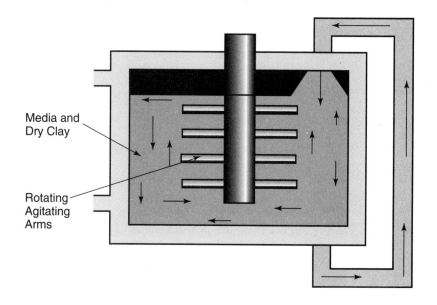

Dry milling

Dry milling is normally accomplished with a different type of milling process, such as the fluid energy mill. During the past several years the **Trost fluid energy mill,** Figure 16-4, has gained popularity in the industry. Commonly known as the *jet mill*, it has the advantage of producing fine powders of uniform particle size while maintaining purity.

Figure 16-4
In the Trost fluid energy mill, opposing jets of air create a turbulence that causes material particles to collide and fracture. The classification chamber separates materials by particle size.

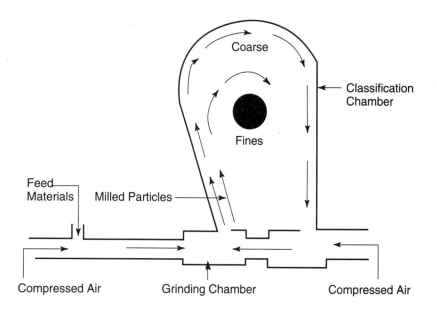

In the Trost mill, a vibratory hopper feeds solid clay into a stream of compressed air that is made turbulent by the action of two opposed jets. This aggressive action accelerates the particles to sonic velocities. The particles collide with each other, fracturing as a result of the impact. When the particles are reduced, they travel through an upstack channel at reduced velocities and are carried into a flat classifying chamber. The mill has essentially no moving parts. Jets of superheated steam, nitrogen, carbon dioxide, or water can also be used. Both hard and soft materials can be milled by this process, producing particles of micron to sub-micron size. The output capacity of Trost mills ranges from grams per hour to thousands of pounds per hour.

The major advantage of the jet mill is its ability to break dry material into the ultrafine particles that permit dry pressing products with improved strength, hardness, and density. See Figure 16-5. Gears and bearing inserts made by powder metallurgy, cermets, and heat-resistant devices are typical products that must be tough to continue functioning in hostile environments. To ensure high-quality products, finely ground powders must be used. See Figure 16-6.

Figure 16-5
The exposed dies shown in this photo are used to apply 40 tons of pressure and form a dense compact. (Gebruder Netzsch Maschinenfabrik GmbH)

Figure 16-6
An assortment of dry-pressed workpieces formed from finely ground ceramic and metal powders. (Gebruder Netzsch Maschinenfabrik GmbH)

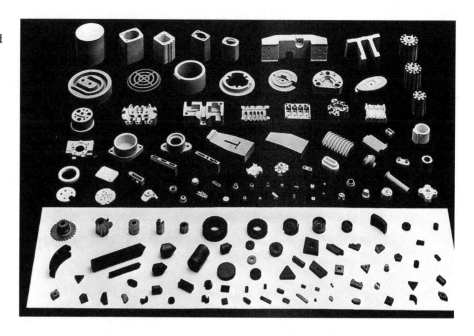

Filter Pressing

We have discussed grinding of ceramic materials with many different viscosities. The desired viscosity of the material naturally depends upon the process that will be used to shape the product.

Potteries often require the ceramic body to be in the plastic state. A high percentage of water is added to improve mixing. Sometimes, the water content of the slip may be 40 to 50 percent; before it can be used for forming, the prepared slip has to be dewatered. This is accomplished by forcing it through a filter press. If a product is to be made from slurry and shaped plastically, the slip is dewatered only to the point where it can be shaped, but still has a plastic consistency. Tableware is made from this type of material.

Filter presses used to dewater slip consist of an iron frame with nylon filters, and are closed with hydraulic pressure. The filters remove part of the water to produce squares or thin slabs of material. What is left after pressing is called a *filter cake.*

When the filter cake is first produced, it is not useful for processing. It is wetter (and thus more plastic) at its center than on its surface or outside walls. The cake also contains air, which would cause holes in the product that was to be pressed or formed. After curing and inspection, the cake is normally further refined using a process called pugging.

Pugging

Filter cake may be stored for some time before it is used. Products such as porcelain or bone china, or other products to be extruded, require careful treatment of the raw material that is used. Consequently, a process called *pugging* is often used to remove air bubbles from the cake and further refine the clay body. The equipment used in this process is referred to as a *de-airing pug mill*. See Figure 16-7.

Figure 16-7
The de-airing pug mill tempers a clay mixture by mixing and extruding it, then removes entrapped air by applying a vacuum to the mixture.

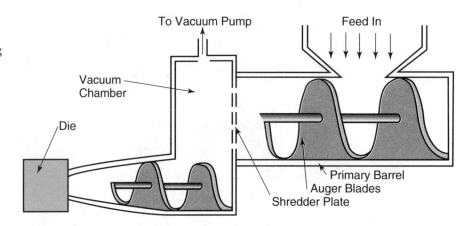

The pugging process is used to temper the mixture (give the clay uniform consistency), remove air bubbles, and improve the workability of the mass for use in plastic-state processes such as jiggering or jollying. Often an extrusion press is attached to the delivery end of the pug mill.

When cake is fed into the pug mill, it is taken into the barrel of the machine and shredded by rotating auger blades, which then carry the clay to a shredder plate. This plate contains small holes through which the plastic clay is extruded. The "worms" that are produced are cut into short lengths by a rotating blade in the vacuum chamber where de-airing takes place.

After the clay is de-aired and extruded, it is then cured and inspected and is ready for further processing. With many processes, material in the form of a damp powder is desired. This can be accomplished by using one of a variety of drying processes.

Spray Drying

Spray drying is another method for reducing the amount of water in slip or semi-dry plastic body. Spray drying is used to produce granular particles with a moisture content of seven percent or less. This is a dewatering method that is particularly useful when dry forming processes are to be used.

The advantage of spray drying is the forming of powders from ceramic slip with a very uniform particle size. This precision in terms of uniformity of powders cannot be obtained with any other drying method. The rate of concentration of products that are dried by this method is approximately 60 to 70 percent. The solid granules do not contain any air bubbles or dust. Particles can be produced that are either solid or hollow, depending on the pressure and type of spray nozzle used. Here's how the process works:

There are two basic methods of producing spray-dried particles. The first of these involves pumping slip onto rotating disks located at the top of the spray drier. The disks then throw the collected droplets of slip outward through heated air and on to the walls of the drier. When the slip hits the walls, it falls down through a cloud of hot air. The second type of system uses a spray nozzle at the top of the drier, rather than rotating disks. In both cases, the droplets are completely dry by the time that they reach the bottom of the chamber. The particles that are produced are small, free-flowing granules that are perfect for dry pressing. The size of the granules will vary depending on such factors as the type of drier, temperature, binder used, and type of clay. Figure 16-8 shows the complexity of air-handling systems in a large commercial spray-drying installation.

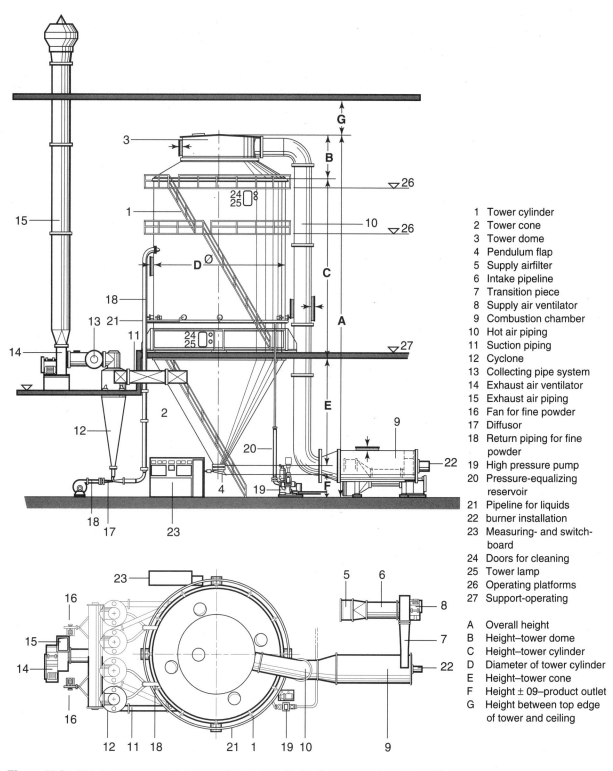

Figure 16-8 This large commercial spray-drying installation has a complex air-handling system. (Gebruder Netzsch Maschinenfabrik GmbH)

Green Machining

Green machining is the process of machining a dry-pressed ceramic part while it is still "green" (before it has been sintered or fired). Machining is done on the loosely compacted powdered part. Since the part has not been sintered, it is quite soft and can easily be machined using conventional metal-cutting equipment. Diamond tooling is not required.

Machining greenware requires skill and careful handling. The formed part, or *compact*, is fragile; care must be taken to avoid collapsing the material when securing the part in the chuck or workholder on the machine tool. In addition, machining must proceed slowly to prevent cracking the part and producing unnecessary stresses. Securing the ceramic workpiece is often accomplished by fastening the part to fixturing using beeswax.

Grinding for Material Removal

It is also possible to grind ceramic parts that have been fired. You will recall that grinding was a basic material removal process used with metals and other materials. Ceramic parts can also be ground to remove undesirable material. Many different types of grinding wheels are used. Hard materials such as alumina may require diamond grinding wheels. Softer materials may be ground using silicon carbide or alumina.

An example of a grinding application is a part with an unusual shape that was supported in the kiln with struts, called **kiln furniture.** Often, there will be imperfections in the surface finish due to furniture interfering with the final coating (glaze). It may be necessary to grind or polish the part after it is fired to remove the imperfections.

Heat generated by grinding can be a problem: temperatures as high as 1800°F (1000°C) can be produced at the grinding surface. This often results in some microscopic cracking, which can extend into the body of the ceramic part. The use of cutting fluids generally improves the cutting efficiency of the wheel, while helping to prevent overheating.

Nonabrasive methods are also used to remove ceramic material or to polish a ceramic part. These include ion beam, laser beam, flame, and chemical polishing techniques. Most machining of ceramics, however, is done using abrasive grinding processes.

Etching Processes

Material removal from a glass surface, resulting in a frosted or opaque appearance, is done by the use of etching processes. This can be done through the use of chemicals that attack the glass surface, or by mechanical abrasion.

Chemical Etching

Glass can be frosted, or made opaque, by **chemical etching.** An entire pane of glass can be frosted with an acid (**etchant,** or etching agent), or portions of the glass can be frosted by covering any area that is not to be etched with a protective coating of wax. When designs are desired in the glass, one method that can be used is to coat the entire glass with wax, then scratch the design through the hardened wax. A weak etching solution made with hydrofluoric acid (or a stronger commercial etching solution) is then applied. The acid eats into the glass wherever the wax was removed. When the etching reaches the desired depth, the acid is drained off and the glass is washed in hot water. This stops the etching action of the acid and melts the wax coating.

Frosted products are desired in applications where obscured vision is desired to ensure privacy, or to produce artistic designs directly on the glass product. Frosted incandescent lamps are made with this method.

Photographic resists, similar to those used with the process of chemical milling and photo fabrication on metals, can also be used. With these resists, masks are used to block exposure to light. The exposed resist is hardened by light and is unaffected by the acid. The unprotected areas of the glass are frosted when immersed in the acid bath.

Mechanical Etching

Etching or frosting of glass can also be accomplished by *mechanical etching,* by sandblasting, or by bombarding its surface with an abrasive and lead shot mixture. With the *abrasive shot method,* a mixture of abrasives and lead shot is placed inside of an agitation frame, along with the sheet of glass. The abrasive particles cling to the lead shot. When the mix is agitated, these coated particles of shot act like individual cutting tools. The abrasive and shot combination can be used over and over again.

The mechanical etching process called *sandblasting* can also be used with glass. Because of the danger of inhaling poisonous abrasive dust, sand is seldom used today; instead, alumina oxide abrasive is normally used. Unique designs can be produced on sheet glass with frosting methods. The opaque frosted surface can even be shaded with sprayed colors or oil colors. The frosted surface holds the color well and adds a beautiful sheen to the glasswork.

Important Terms

abrasive shot method
attrition milling
ball mill
beneficiated
calcination
chemical etching
compact
de-airing pug mill
etchant
filter cake
filter presses
green machining
grinding media

Hardinge conical mill
John mill
kiln furniture
mechanical etching
milling
mix muller
Molinex mill
pugging
sandblasting
spray drying
tempering
Trost fluid energy mill

Questions for Review and Discussion

1. Think for a moment about creating a hollow granule using the process of spray drying. Why does the pressure and size of the nozzle used affect the solidity of the particle?
2. How is it possible to produce ultra-fine particles from dry stock, using the fluid energy mill? Why do you think this is called a fluid energy mill, without any fluid being present?
3. List three processes presented in this chapter that can be used to remove air in clay.

Grinding is a commonly used method for removing ceramic material after the part has been fired. Cut-off wheels like these are used to trim off unwanted material and cut parts to length.
(Norton Company)

17 Fabricating Ceramic Materials

Key Concepts

Δ Ceramic materials often must be joined to other materials with joints that are not only mechanically strong but air-tight or water-tight, or both.

Δ Metallizing techniques are used to provide electrically conductive pathways on ceramic substrates.

Δ Glass-to-glass and metal-to-glass bonds usually involve heat application to form fusion seals.

Δ Metal coatings are applied to ceramics for decorative, electrical, and sealing purposes.

Δ Refractory ceramic materials, in forms ranging from bricks to spray coatings, are used to line industrial furnaces and for other high-temperature applications.

The fabrication of ceramic materials frequently involves joining the ceramic part to a part made from another material, especially metal. Often the joint must be not only mechanically strong, but air- and water-tight. Ceramic materials are also used as refractories, either through being formed into bricks for furnace-lining, or sprayed on as a coating.

Metallizing

In addition to creating a decorative effect, such as gold or platinum designs on dinnerware, metal coatings are often applied to the surface of ceramic products to perform special technical functions. For example, *metallizing* is often used to provide a transmission path for electrical current on electrical devices and components. In electronic applications, the metals silver, molybdenum, and palladium are often preferred because of their superior conductivity.

Glass surfaces can also be metallized to form a base for sealing unlike materials together. Joining is done by coating the ceramic substrate with metal and then soldering or brazing the desired metal to the metallized surface. This is a process that is often used to construct specialized glassware for medical and technical applications.

When pieces of glass are to be joined (*sealed*) to each other, one of two techniques is commonly used. Sometimes, glass seals are created by melting and fusing the two pieces of glass together. In other applications, powdered glass solder is used to form the joint. The use of glass solder is described in more detail later in this chapter.

Spray Metallizing

In addition to metallizing through the use of solder, ceramic parts can also be coated by using such methods as plasma arc spraying or oxyacetylene spraying. Metallizing can be done on cold or hot ceramic/glass surfaces.

When spraying is the desired mode of application, the metal that is to be applied is fed, in wire form, through a special torch or spray gun. The metal wire has a low melting point. As it is fed through the oxyacetylene flame, it is melted by the flame and atomized by compressed air. The fine droplets of molten metal are then blown out the nozzle of the spray gun onto the substrate.

One of the most sophisticated methods, ***plasma arc spraying***, uses temperatures as high as 30,000°F (16 650°C). This permits the spraying of materials with high melting points.

In addition to metals, other types of materials (such as cermets, carbides, ceramics, and plastic-based composite powders) are sprayed onto various substrates. Ceramics, for example, can be sprayed at high temperatures, using high-velocity oxyacetylene flames. This process is sometimes used to apply porous protective ceramic coatings to metals.

Other Processes

Coatings are also applied through other processes, such as dipping or powder coating. When the application involves dipping or coating, the material is normally applied to a heated substrate. For permanence, it is then fired on the glass or ceramic part at high temperatures. Figure 17-1 shows a technician applying a coating that will be used to produce a ***hermetic seal.*** Such seals prevent movement of gases or liquids into or out of a product. Many medical products require hermetic seals.

Metallizing can also be done by mixing metal powders, organic solvents, and binders, then spraying or painting the material onto the ceramic part. Some manufacturers then fire the coating in a kiln containing a ***cracked ammonia atmosphere*** (hydrogen plus nitrogen). This improves the quality of the coating.

The process of metallizing is used on a variety of different products. Electrical meters are sprayed with brass. The brass is then soldered to a metal bracket to produce a moisture-proof seal. Medical encapsulating tubes, glass-to-metal seals, and instrument windows are also produced with fired-on metallized coatings.

The techniques for producing ***glass-to-ceramic seals*** requires heating the glass and ceramic together until they are hot enough to melt the glass solder applied to the area to be sealed. Glass-to-ceramic seals are used to bond unlike materials in products such as spark plugs and electronic components.

Chapter 17   Fabricating Ceramic Materials

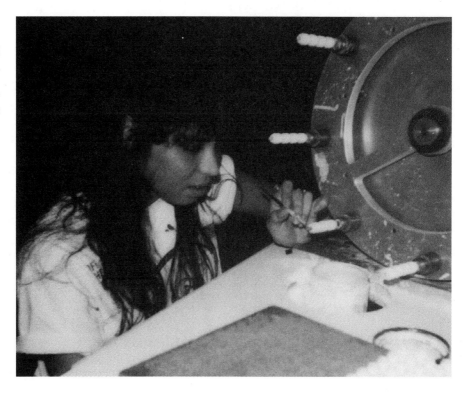

Figure 17-1
This technician is brushing on a ceramic coating that will be fired to hermetically seal a part. Hermetic sealing prevents infiltration of liquids or gases. (Pekay Industries, Inc.)

Fusion Sealing

Many electronic products require pressure-tight seals between metal and ceramic materials. The bonding of metal to some crystalline ceramics involves the use of a molybdenum-manganese layer. When fired under partially oxidizing conditions, the molybdenum-manganese layer creates an oxide that reacts with the ceramic material to produce an adhesive bonding layer. When metals are to be soldered to the sealed area, the part is plated with nickel, silver, or gold before soldering.

When a metallic layer is applied to a ceramic material, there are two different levels of thermal expansion operating against each other. Normally, metals have a higher *coefficient of thermal expansion* than do ceramics. This means that, to avoid problems caused by differential expansion, thin applications of metal with small surface areas are desirable. Otherwise, excessive stresses could be created in the ceramic material.

The *fusion sealing method* uses special glass compositions to make direct glass-to-metal seals in an oxidizing environment. The molten glass combines with a previously formed oxide on the metal to produce a fairly strong bond. The critical sealing temperature required must take into consideration the expansion and contraction requirements of both of the materials. If this is not done, stresses may break the seal.

An alternative to this process is the *compression seal,* where parts are pressed into their position in the completed assembly. This type of seal is used in applications such as a product where small leads must join one point on the part to another. The problem with this seal is that such leads are sensitive to vibration.

Encapsulation

In an earlier chapter, the process of encapsulation as used with polymeric/plastic materials was discussed. *Encapsulation* is also an important process in the manufacture of ceramic capacitors. Such capacitors deteriorate when they are exposed to moisture, so encapsulation provides a functional method for sealing them. Small capacitors that form a part of a hybrid circuit can be sealed into ceramic containers, along with other components of the circuit assembly. The leads are typically coated with a polymer. Encapsulation is done by dipping in liquid polymer, by injection molding, or by immersing the assembly in a fluidized bed of molten polymer. Epoxy and thermosetting resins are used for dipping; polypropylene with injection molding.

Laminating

Laminating is a process that is often used to sandwich polymeric and composite materials together. A typical product of this type would be a laminated glass automotive windshield. First, two pieces of curved glass for the windshield are formed by heating two sheets of glass in a stainless steel mold. After the glass melts and droops into the mold, it is annealed to relieve stresses and then cooled. A polyvinyl butryl plastic *interlayer* is placed between the two pieces, and the sandwich is pressed together in an autoclave furnace to bond under heat and pressure. The vinyl interlayer serves two essential functions: it holds the glass together, and it keeps the glass from shattering upon impact. See Figure 17-2.

Theoretically, there is no limit to the number of layers of glass that can be fused together. In practice, however, laminating normally involves two sheets of glass. Laminated glass is permanently bonded.

Laminating is also used to join two or more layers of glass with metals or other types of materials. In Figure 17-3, a worker at an architectural glass assembly plant inspects sealant being applied to spacers that will be sandwiched between two pieces of glass. The sandwich will form an energy efficient window for a commercial building.

Glass Soldering

In some applications, it is desirable to seal glass together in such a way that the parts can be later separated and repaired. One such application is in the manufacture of a television picture tube. Owens-Illinois Glass Company has developed a *glass solder* to join the face plate of the tube to the rear funnel section. The solder enables the installation of new parts without destroying the tube. Several different types of glass solder—vitreous, devitreous, and conductive—are available. *Vitreous solder* is a low-melting-point glass that can be applied to glass with a higher melting point to create a seal. *Devitreous*

Chapter 17   Fabricating Ceramic Materials 363

Figure 17-2
A "sandwich" of two layers of glass with a "filling" of polyvinyl butryl plastic is being made by these workers in a plant that manufactures automobile windshields. Heat and pressure will be used to permanently bond the layers together. (PPG Industries)

Figure 17-3
An energy efficient window for a commercial building will result from sandwiching these aluminum spacers between sheets of glass. The worker shown is inspecting sealant application on one of the spacers. (PPG Industries)

glass solder is much like a thermosetting plastic. It can be reheated without becoming highly liquid and flowing freely. ***Conductive glass solder*** is used to provide both a seal and an electrical path.

Refractory Materials

Refractory ceramics are nonmetallic ceramic-like materials that remain solid and intact when they are subjected to high temperatures. Refractory products also are capable of withstanding corrosive conditions in high-temperature environments. Typical applications for refractories are industrial furnace liners, crucibles, and similar high-temperature areas.

Refractories are made from many different types of ceramic materials, including high alumina fire clays, alumina, magnesia, and silica. There are many other materials which are used in the manufacture of refractories. These include chrome and chrome-magnesite, carbon, and pure oxide refractories. A mixture of sawdust, coke, and silica sand has even been heated in an electric furnace to produce refractory materials. The silica becomes silicon and combines with the carbon to produce ***silicon carbide***. The sawdust burns out, leaving pores in the material for the escape of gases. Because of its hardness, silicon carbide is a popular abrasive used for manufacturing grinding wheels.

The most common material for making refractory products is clay. Pure fired ***kaolin clay***, which is about 46 percent alumina and 54 percent silica, has a melting point of over 3000°F (1700°C).

Other materials, such as high-melting-point alumina, are used in applications requiring even higher temperatures. Alumina has a melting point of 3722°F (2050°C). For many years, the basic refractory materials used by the steel industry have been magnesium oxide (magnesia), chromium oxide, and lime.

Refractory products are installed as linings for industrial furnaces or in similar high-temperature environments. The metal casting industry is the largest user of refractories. Refractories are also used for many special applications, such as rocket launching pads and nuclear reactors, and in the technologies involved in processing fertilizers, refining oil, and manufacturing petrochemicals.

Most furnaces use refractory bricks that are produced by dry pressing. ***Fire clay*** brick is the most commonly used, comprising about 75 percent of all refractories produced in the United States.

Refractory Spraying

Refractory ceramic materials, in the form of a liquid slurry, can be sprayed onto the inside of the open-hearth furnace or the firebox of a boiler. ***Refractory slurry*** can also be used as mortar in joints between bricks in a furnace lining.

Refractory ceramic coatings can be applied to almost any base material by using the process of flame spraying, or metallizing. As noted earlier, coatings are applied by dispersing them through a flame and blowing the atomized material at high speed onto the receiving surface. Ceramic coatings are

Chapter 17 Fabricating Ceramic Materials

popular for application to metal components that must withstand extreme environmental conditions, such as acids, corrosion, abrasion, erosion, and oxidation.

The most common processes used to manufacture refractory products are dry pressing and extrusion. Other processes, such as extruding and then shredding ceramic fibers, are also used to produce refractory mats and heat-resistant blankets.

Important Terms

compression seal
conductive glass solder
cracked ammonia atmosphere
devitreous glass solder
fire clay
fusion sealing method
glass solder
glass-to-ceramic seals
hermetic seal
interlayer
kaolin clay
laminating
metallizing
plasma arc spraying
refractory
refractory slurry
silicon carbide
vitreous solder

Questions for Review and Discussion

1. Explain why and how sawdust is used in the manufacture of refractory materials.
2. Describe several methods used to apply metallic coatings to ceramic workpieces.
3. What is the difference between fusion sealing and the conventional lamination process used to seal vehicle windshield glass?
4. Describe the process that would be used to make heat-resistant thread for electric heating blankets from refractory materials.

A scientist at Sandia National Laboratories Active Ceramic Materials Division displays a sample PLZT ceramic that will be fabricated into windows for combat aircraft. The material can be switched from transparent to opaque in 50 microseconds to protect crews from flashblindness and burns. (Sandia National Laboratories)

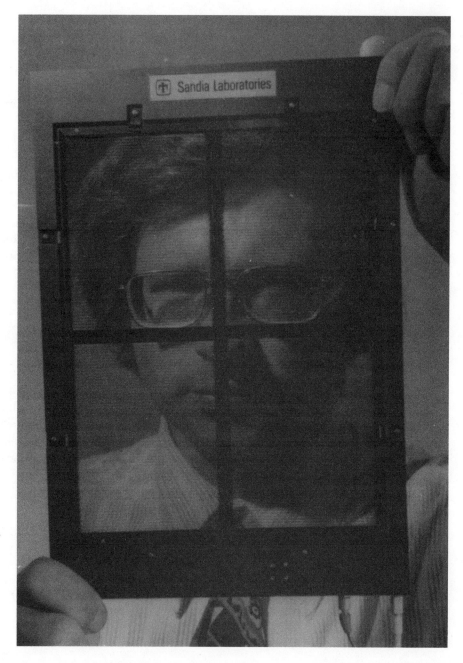

Conditioning Ceramic Materials

Key Concepts

△ Sintering is the most common method of increasing the density of a ceramic workpiece.

△ Calculation of the bulk density of a ceramic workpiece is important in identifying its porosity and potential for absorbing moisture.

△ Most ceramic products must be dehydrated (have moisture removed) before they can be fired.

△ There are four major types of sintering used in producing ceramic products.

△ Reaction sintering, used to produce ultra-dense workpieces, results in a near-net-shape product.

△ Pyrometric cones, developed in the 18th century, are still widely used to identify the proper firing range for ceramics.

The behavior of any solid, liquid, or gas material is often described in terms of ***density***, or *weight of the material per unit volume.* The behavior of a material depends on such factors as grain size, distribution of the atoms, and surface area. These factors are used to describe how powder will pack, flow, react, and yield to pressure.

Unit volume is normally described in terms of pounds per cubic foot (lb/ft^3) or grams per cubic centimeter (gm/cc). Density is one of the most frequently recorded measures of the morphological characteristics of ceramics.

Densification

The process of increasing the density of a ceramic part is called ***densification***, and it is most often accomplished by *sintering*. Sintering was previously discussed in Chapter 15. As you will recall from that chapter, sintering transforms the loosely bonded green compact formed by dry pressing into a dense ceramic product.

Sintering is simply diffusion on an atomic scale. Moisture and organic materials are burned out of the green body during the initial stages of firing. Then, the temperature of the furnace is raised to the point where diffusion

367

occurs. At this point, matter is channeled from the particles into the openings between the particles. This results in densification and shrinkage of the workpiece.

The type of density that is often measured on ceramic parts is referred to as bulk density. **Bulk density (P_b)** is based on the bulk volume of the part. It can be calculated as follows:

$$P_b = \frac{W_D}{V_b} = \frac{W_D \times P_L}{W_S - W_{SS}}$$

Where:
W_D = the unsaturated dry weight of the powder;
V_b = the bulk volume;
W_S = the saturated weight of the powder;
W_{SS} = the weight of the saturated sample when it is submerged in the liquid, and
P_L = the density of the saturating and submerging liquid.

It is important to be able to calculate the bulk density of a product. Density relates directly to the porosity of the ceramic workpiece. The greater the porosity of the part, the more likely it will be to absorb liquids and vapors into the material. During curing of a freshly molded part, the individual grains are drawn together until they contact each other, resulting in shrinkage. Usually, absorption of moisture into the porous compact will result in excessive shrinkage and structural damage to the material during drying and firing. The higher the moisture content, the greater the shrinkage that will result.

Depending on the product's end use, pores may or may not be desirable. The density of the grains and the number and sizes of the pores in the ceramic product must be carefully controlled. For example, unglazed tableware would be undesirable, because it would become permanently stained and would be unsanitary. On the other hand, for applications such as residential construction, a product such as a highly porous, lightweight brick may be desirable.

The major factor influencing the conversion of powder particles into a solid body is the reduction of surface energy. Thus, the alignment of ceramic particles in the green compact is critical to the thermodynamics of densification. This is not always a simple matter to achieve.

One of the most difficult problems encountered in preparing fine ceramic powder is achieving homogeneity in the powder. The individual grains (particles) are so small that even after thorough mixing, inconsistencies arise. Several approaches are being used to obtain powder consistency and improve the capacity for densification of the final product.

Chemical processes are sometimes used with materials such as uranium, zirconium, thorium, and several other metal oxides. All of the desired metal ions are prepared in solution, then precipitated as hydroxides or oxalates. Chemical precipitation of materials is performed during intensive mixing. One of the problems with this process is the low concentration of solids. Often, a mixture of 10 percent solids and 90 percent water will form a gel that is difficult to dewater by filter pressing. Normally, the solid must be dried, then compacted or pelletized before it can be further used.

The fluid energy milling process discussed in a previous chapter is also used to produce the finely ground particles needed for densification. The mill utilizes a superheated stream of air—1500°F to 1700°F (800°C to 900°C)—to grind the particles. This enables the grinding operation to be performed at temperatures that are above the decomposition temperatures of many nitrates, sulfates, hydroxides, and oxylates. The high temperature helps break down the particles much faster and produces finer grains in the mixture that is milled.

Materials that are ground in the fluid energy mill can be mixed slurries, salt solutions, molten salts, or even solid compounds. The oxide particles that are produced are so fine that it is difficult to separate them from the heated gas in which they are suspended. The particles that are generated from fluid energy milling are like thin-walled bubbles or hollow spheres, many of which are so fragile that they will be destroyed before they can be collected. To make these tiny crystals useful for ceramic processing, a process called calcination is used to grow the particles to the desired size.

Calcination

Calcination is a process that is used to achieve uniformity in particle sizes, and to prepare ceramic powders for later processing. Calcining improves the interaction between the constituents in a ceramic body, prior to sintering, and results in improved densification. A calcined part is more dimensionally stable, and is more desirable for many electrical applications.

Calcining is essentially high-temperature heat treating of ceramics. The temperatures used and approaches taken depend on the material characteristics of the powder being calcined. Calcining can be used to decompose a salt or a hydrate to an oxide.

Calcining can also be used to **dehydrate** (remove moisture from) materials, as is done when producing plaster from gypsum. Heat treatment dehydrates the gypsum and results in plaster-of-paris. If water is added, the material once again forms gypsum. If too high a temperature is used to dehydrate the gypsum, all of the water will be removed, and an anhydrous powder will be created. Such a powder cannot be easily rehydrated.

In other instances, calcining is used to **coarsen** the powder, increasing the size of the ceramic particles so that the powder can be more easily compressed. Coarsening is usually accomplished by loosening particles to create aggregates that function like larger particles.

Freeze Drying

The conventional method of drying ceramic materials before firing is to use either ambient air, or heated air from a kiln. Drying is sometimes done in a vertical oven that moves ceramic pieces up and down on a conveyor through heated air.

A less conventional method is *freeze drying*, also known as *cryochemical drying*. The process rapidly freezes the water in the clay particles, encasing drops of salt solution that hold the desired metal ions. The cryogenic process produces spheres (frozen droplets) that are 0.004 in. to 0.20 in. (0.101 mm to 5.08 mm) in diameter.

To obtain usable oxides from these frozen spheres, they must be carefully dried by heating them in a vacuum. The moisture is removed through *sublimation* (a process in which a substance passes directly from the solid state to the gaseous state, without becoming liquid); the sphere remains intact. The resulting particles are anhydrous sulfate or another salt. The salt spheres are then calcined to remove the sulfate and grow the oxide particles to the desired size.

Firing

Ceramic products that have been dried and finished are called *greenware*. At this point, the products are fragile and vulnerable to cracking and damage through handling. *Firing* consists of heating the part to an elevated temperature in a carefully controlled environment. It is normally initiated in a kiln or furnace to fuse the materials together, providing strength and permanence to the product.

Firing is necessary for all types of ceramics and glasses. It improves the densification of the materials. The strength or hardness which is developed in the ceramic product is due to the glassy bond between the oxide particles in the ceramic material. When heat removes moisture from the part, the pores close, resulting in shrinkage as much as 15 to 20 percent with some processes.

It is common to fire some ceramic parts *twice*. The first firing is done without a glaze (coating) applied to the product, which is then is referred to as *bisque ware*. This firing, called the *bisque firing*, solidifies and fuses the body so it can be handled without problems in glazing. The second firing is referred to as the *glost firing*. In this firing, the clay body is sintered and the glaze develops.

Generally, there are three stages in firing:
- Δ Presintering (bisque firing).
- Δ Sintering (glost firing).
- Δ Cooling.

Presintering

The actual process of sintering usually does not occur until the temperature in the product exceeds one-half to two-thirds the melting temperature of the material. The *presintering* heating process causes drying, the decomposition of organic binders, vaporization of water from the surface of particles, changes in the states of some ions, and the decomposition of additives. During presintering, stresses from the pressure of gas produced or from thermal expansion must not cause cracks in the fragile greenware.

Sintering

Sintering can be defined as the process of densifying a ceramic compact with the application of heat. Sintering eliminates the pores between the particles of powder, and therefore results in shrinkage of the part. The first melting phase in sintering normally takes place between 700°F and 750°F (1200°C and 1300°C). There are four major types of sintering:

- Δ Vapor-phase.
- Δ Solid-phase.
- Δ Liquid-phase.
- Δ Reactive-phase.

In *vapor-phase sintering,* vapor pressure causes material to be shifted from the surface of the particles to the point where the particle contacts another. Since the particles have a high radius of curvature, considerable force is created to increase the speed of the transportation of material from the surface to the contact region.

Vapor-phase sintering reduces both the size of the particles and the pore space between them. Sintering is accomplished by evaporation and condensation or by vaporization pressure. The strength of the product is increased by reducing the pore space.

Solid-phase sintering involves movement of surface material by diffusion. Diffusion can occur as movement of surface atoms or may take place throughout the material. This process may or may not result in shrinkage, depending on the sintering time and temperature. Solid-phase sintering results in material being fused where the particles are in contact with each other. See Figure 18-1.

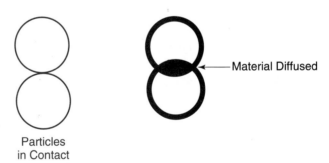

Figure 18-1
In solid-phase sintering, material diffuses from one particle to another through physical contact.

Liquid-phase sintering involves wetting the ceramic particles with a viscous liquid before heating the part to the sintering temperature. Liquid held between the grains of the material creates capillary pressure. The capillary pressures that can be generated by silicate liquids can exceed 1000 psi. This enhances densification by reorienting the particles to improve compacting. It also increases the contact between the particles, resulting in plastic deformation.

A variation of liquid-phase sintering is called **reactive-phase sintering.** It is also known as *transient sintering.* As in liquid-phase sintering, a liquid is present during sintering. However, in reactive-phase sintering, the liquid breaks

Reaction Bonding

down or disappears. Because of the effect of the liquid on the sintering process, parts processed by this method sometimes can even be used at temperatures exceeding the initial sintering temperature.

Reaction bonding, or reaction sintering, is a process that yields a part with less shrinkage than conventional sintering processes. Reaction-sintered silicon nitride can be produced from silicon powder. Pressing, injection molding, or slip casting is used to produce the green compact. Once the compact is formed, it is placed in an oven with a nitrogen, nitrogen/hydrogen, or nitrogen/helium atmosphere. Initially it is heated to a temperature of approximately 700°F (1200°C).

The heat forces the nitrogen into the pores of the compact, and the reaction begins. As the temperature is increased, the reaction rate increases. Reaction-sintered *silicon nitride* is a material with tremendous potential for manufacturing. Silicon nitride has a relatively low elastic modulus and coefficient of thermal expansion, and relatively high thermal conductivity. This makes it a material with good *thermal shock resistance*. Possible product applications include turbine engine and space station components.

Silicon carbide is another material that has been produced using reaction sintering. A mixture of powdered silicon carbide and a thermosetting phenolic resin is formed using plastic processes such as casting, pressing, extrusion, or injection molding. After the molding process is completed, the plasticizer is removed by charring. This produces the carbon that is used later in the process.

The preform is then exposed at high temperatures to molten silicon. The silicon reacts with the carbon and bonds the original silicon carbide particles together. The material that is produced is a nonporous silicon/silicon carbide composite with a high modulus of elasticity. This makes it particularly useful for forming complex shapes. The process is good for near-net shapes, as well —final products normally have less than one percent shrinkage during densification.

Vitrification

In some product applications, such as brickmaking, the temperature used to sinter the green product is particularly important. When brick is fired at a temperature of approximately 1800°F (1000°C), a glassy bond begins to form and pores begin to close. If the temperature is increased, more pores will close. Consequently, the brick is made denser and stronger.

The process is known as *vitrification.* Finding the proper balance between density and strength is important. If the firing temperature is too low, the bricks will be too porous and fragile; if too high a temperature is used, they will be too dense. The span of temperatures between the point of sufficient vitrification and too much vitrification is called the *firing range.*

The proper firing range depends on the composition of the material being fired, and cannot be left to chance. To determine the proper firing range, a system using what are referred to as *pyrometric cones* was developed in the 18th century. The system is still used today.

Pyrometric cones are wedge-shaped pieces of clay, each with a specific melting point. Cones are purchased in a set or series. The series is arranged so that a cone will slump over at a temperature about 20 degrees below that of the next one in the series. Thus, the higher the number of a cone in the series, the greater the temperature that is required to get it to slump over. See Figure 18-2.

Figure 18-2
Pyrometric cones are purchased in a set or series. The series is arranged so that each cone will slump over at a temperature about 20 degrees below that of the next one. The kiln operator uses the successive slumping of cones as a visual indicator that the product being fired is approaching the melting point.

For use, the cones are seated in a base plaque of clay and placed in the kiln at a point where they can be observed through a viewing port in the kiln door. The cones are selected by the operator so that the highest number in the series will slump at a temperature close to the melting point of the material being fired. The successive slumping of cones as the temperature rises is a visual indicator that the operator uses to determine when the product is approaching the melting point.

Transformation Toughening

Transformation toughening is a process that is used to increase the strength and toughness of ceramics in products which must exhibit high wear resistance. Today, this process is used most frequently to toughen zirconia, the principal ceramic material in zirconium oxide. Transformation toughening is accomplished by carefully controlling the heat treatment cycle, composition of the powders, and particle size. Here's what it accomplishes.

Zirconia goes through a phase transformation from its tetragonal crystal form to a monoclinic crystal form while cooling through a **temperature of approximately 2100°F (1150°C)**. The transformation results in an increase in the volume of the material of about three percent. By controlling the conditions of compacting and processing, zirconia can be densified at higher temperatures and cooled with the tetragonal phase being carried all the way down to room temperature. What this means is that if a fault appears in the part, the crack will start to propagate. Immediately, the high stresses in the tip of the crack will halt the transformation of the adjacent tetragonal crystalline grains into monoclinic form. This will cause them to expand by approximately three percent, compressing the crack tip and preventing the crack from expanding. Transformation toughening produces a ceramic that is tough and strong.

Vapor Deposition Coating Processes

Various vapor deposition processes can be used to deposit nonporous ceramic coatings on a substrate. Chemical vapor deposition (CVD) and sputtering are two of the most widely used chemical deposition processes.

Chemical Vapor Deposition

Chemical vapor deposition is a ***thermochemical*** process that is typically accomplished by heating the part to be coated in a vacuum chamber while flowing a controlled gas mixture over it. See Figure 18-3. In chemical vapor deposition, the gas reacts when it contacts the heated part. This results in a very fine-grained coating that is harder than those achieved with most other ceramic fabrication processes. Ceramic coating is applied at a rate of about 0.010 in. (0.254 mm) per hour. Care must be taken to obtain a uniform coating.

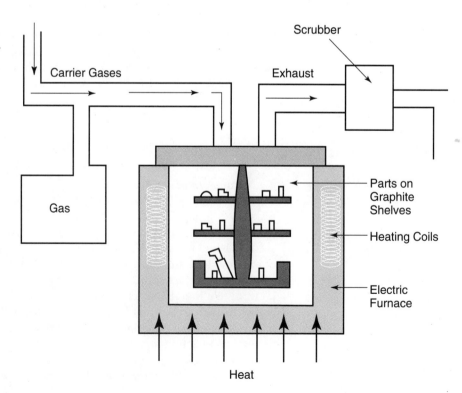

Figure 18-3
In chemical vapor deposition, gas reacts when it contacts the heated parts inside the vacuum chamber, producing a hard, very fine-grained coating.

Sputtering

Sputtering is done by placing the part to be coated next to a flat plate of coating material inside of an evacuated chamber. The plate, called a ***target***, is bombarded by a beam of electrons. The electrons break atoms off the target so that they can be deposited on the part. Only the area of the part facing the target receives the coating. Thin coatings are applied consistently, but the coating rate is very slow.

In the sputtering process, the part itself is not heated during coating. This makes it particularly useful for coating electrical substrates where insulation and protection are desired.

Ion Implantation

Ionic bonds occur when an atom of one material releases an electron that bonds to an atom of a different material. Ion implantation is a process that introduces ions into the surface of the part to improve hardness, wearability, and corrosion resistance.

In the semiconductor industry, crystals are "grown" using a form of ion implantation called doping. Doping involves adding ***dopants*** (alloying elements) to liquid metal to produce unique properties.

Doping Methods

Two major processes are popular for growing crystals. The first of these, called ***crystal pulling***, is shown in Figure 18-4. With the crystal pulling method, a small, single-crystal seed is brought into contact with molten semiconductor material. The seed crystal, attached to a rotating pulling bar, is then slowly pulled away. Where the semiconductor material touches the seed, it influences the growth of the seed's crystalline structure. Single crystals of materials such as titanium and zirconium are made with this process.

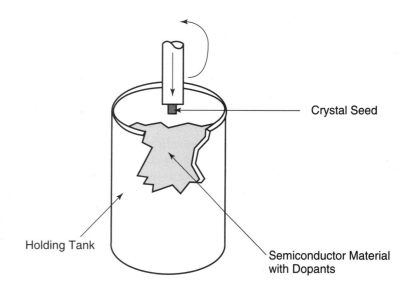

Figure 18-4
In the crystal pulling method, a single-crystal seed is brought into contact with molten semiconductor material, then slowly pulled away. Where the semiconductor material touches the seed, it influences the growth of the seed's crystalline structure.

Another process that is used to produce single crystals is called the *floating zone method.* This process is used with materials that have high melting points, such as silicon. See Figure 18-5.

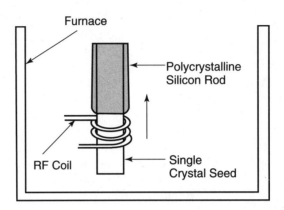

Figure 18-5
In the floating zone method, energy from a radio frequency (RF) coil is used to create a molten zone around the interface between a single crystal seed and a rod of polycrystalline silicon. The coil (and thus the molten zone) is slowly moved upward along the rod. The material that solidifies below the coil has the same crystalline structure as the seed.

In this process, a rod of polycrystalline silicon is brought into contact with a single crystal seed. Both are suspended in a furnace. Energy from a radio frequency (RF) coil is used to create a molten zone around the interface between the seed and rod. By slowly moving the RF coil upward, the molten area moves along the originally polycrystalline rod. The material that solidifies below the coil will have the same crystalline structure as the seed. The long crystal rod is then sliced with a diamond saw into wafers about 0.020 in. (0.508 mm) thick. The wafers can then be cleaned and further cut into tiny "chips" for electronic applications.

Another crystal-growing process used in industry is the *Czochralski method.* With this process, the seed is dipped into a crucible containing melted silicon, then withdrawn. The seed and crucible are then rotated in opposite directions. The molten material freezes to the crystal, as described earlier.

Annealing

Annealing is a process that relieves the strains in ceramic or glass products, just as it does in metals, through reheating and gradual cooling. Annealing in ceramics could be likened to the glost firing.

Annealing is a much more *critical* process with glass. After shaping, all commercial glass products must be reheated to a temperature at which frozen-in stresses can be relaxed by internal flow of the glass. The temperature for annealing is below that at which the products would seriously deform under their own weight. Annealing is accomplished in an annealing lehr, and is followed by slow cooling.

Tempering

Tempering is a heat-treating process used with glass products. It gives these products three to five times their annealed strength.

Glass is tempered after forming by heating it to a temperature close to the softening point. Then it is removed from the heat and quickly chilled from about 1200°F (650°C) by rapid air-jet cooling. This places the outside surface under high compression stresses. For example, tempered borosilicate glass products are produced with working stresses of 4000 psi (27 580 kPa). This is about the same strength as ordinary gray cast iron.

Glass breaks only when it is subjected to tension, not compression. When tempered glass shatters, it breaks into small harmless pieces that hold together. Glass cannot be cut or drilled after tempering, since the cutting or drilling action would cause it to shatter.

Important Terms

annealing
bisque firing
bisque ware
bulk density
calcination
chemical vapor deposition
coarsen
cryochemical drying
crystal pulling
Czochralski method
dehydrate
densification
density
dopants
firing
firing range
floating zone method
freeze drying
glost firing
greenware
liquid-phase sintering
presintering
pyrometric cones
reaction bonding
reactive-phase sintering
silicon carbide
silicon nitride
sintering
solid-phase sintering
sputtering
sublimation
target
tempering
thermal shock resistance
thermochemical
transformation toughening
vapor-phase sintering
vitrification
zirconia

Questions for Review and Discussion

1. What is the difference between reaction bonding and transformation toughening?
2. Which process is easier to accomplish, chemical vapor deposition or sputtering?
3. Select one of the methods used in industry for growing crystals. Explain how it works.
4. What is the purpose of glost firing?

Glass is a versatile ceramic material, since it can be formed and conditioned for many different applications. In these compact fluorescent lamps, glass has been formed into a double-curved hollow tube, and the interior surface coated with a phosphor material. The tubes mount in an adaptor that allows them to be used in lighting fixtures with standard sockets. (GE Lighting)

Finishing Ceramic Materials

Key Concepts

Δ There are two major types of finishing processes used with ceramics.

Δ If surface flaws in a ceramic piece are minor, polishing and lapping are adequate; grinding must be used for larger defects.

Δ Glazes can be applied by spraying, dipping or brushing; spraying is most common.

Δ Decorations can be applied to ceramics before glazing (underglaze decorations) or after glazing (overglaze decorations).

In many cases, ceramic products are ready for the consumer when they leave the kiln, and no final finishing is necessary. Most of the time, however, some additional processes or postfiring operations are necessary.

There are essentially two different types of finishing that apply to ceramic products. The first of these includes processes that address the final surface preparation of the workpiece after firing. It includes such processes as scribing, etching, grinding, and the application of decals and decorative stencils.

A second type of finishing includes processes that are used to place a decorative or protective coating or design onto the bisque ware. This includes processes such as silk screen printing, glazing, and the application of decorative frit to the clay body.

Grinding

The process of grinding has already been presented as a basic material removal process for use with ceramic products. Grinding as a finishing process is normally required when the design of the ceramic product requires support by refractory kiln furniture to keep it from slumping in the kiln.

In some cases, solid masses of clay called *setters*— clay of the same composition as the part—are placed under certain areas of the workpiece to reduce warpage. Some types of ceramic products, such as columns, rods, and tubes, are suspended from the ceiling of the kiln by special fixtures.

In cases such as these, where the part cannot be produced without damage to the surface finish because of contact with the support device, then grinding is to be expected. In other cases, where the extent of surface marring is minimal, processes such as polishing or lapping may be adequate.

Flame Polishing

Flame polishing can be used to reduce the size and quantity of surface defects in small-diameter rods and filaments, especially in such products as single crystals of sapphire or ruby. Flame polishing is done by rotating the rod or filament while passing it through a helium/oxygen flame.

Flame polishing melts the thin surface layer of the product. It can also be used to polish two finished edges on sheet window glass in a forming machine. Today, there is very little polishing necessary, since glass is normally made by the float glass process. In this process, flat glass is formed by flowing the material onto a bed of molten tin. See Figure 19-1.

Figure 19-1
This float glass machine produces a continuous ribbon of flat glass by floating the molten material onto a bed of molten tin. The process virtually eliminates the need for grinding and polishing. (PPG Industries)

Flame and Plasma Spraying

As you will recall from an earlier chapter, ceramic powder can be changed to molten droplets by passing it through a high-temperature plasma, such as an oxyacetylene flame. The molten particles flow out of the gun through the nozzle and strike the substrate to be coated, at a velocity of approximately 150 feet per second (fps). Most carbides, borides, oxides, nitrides, and silicides that do not decompose can be applied by molten particle techniques.

The first widely used molten particle process was the oxyacetylene powder gun, called the *flame spray gun.* Another approach is the *oxyacetylene rod gun.* With this gun, ceramic powder is not used. Instead, a sintered rod of coating material is fed into the oxyacetylene flame. Air bursts blow the molten ceramic at the tip of the rod onto the substrate at a rate of approximately 600 fps.

A newer version of the molten particle process is accomplished with the *plasma arc gun.* A high-intensity, direct-current arc is produced in a chamber. Helium or argon gas is passed through the chamber, heated by the arc, and forced through a water-cooled nozzle in the form of a high-temperature plasma. Ceramic powders are ejected into the plasma, where they are melted and sprayed onto the workpiece. Velocities as great as 1500 fps have been reported.

The major disadvantage of this process is the high temperature of the gas. This means that metallic parts have to be cooled to keep them from melting, while ceramic parts must be preheated to reduce thermal shock and keep them from cracking.

The process has many present-day applications, ranging from spraying chromium on ship propellers to providing thermal coatings on superalloys. Molten spray techniques are particularly advantageous in applying refractory linings and coatings that are chemical-resistant or wear-resistant.

Laser Processing

Because ceramic substrates readily lend themselves to laser processing, there are many processes that can be performed on ceramic parts using laser technology. The most common general-purpose laser for such applications is the CO_2 laser.

The most popular of the processes that use laser technology are scribing, drilling, and cutting. These will be discussed here as final finishing processes, since laser processing is normally done after firing. This enables final tolerances to be obtained in a single step.

Lasers are used to perform such processes as *scribing* on a continuous basis, without concern for material thickness. An automated laser scribing system for use with ceramic workpieces is shown in Figure 19-2.

To successfully perform scribing on ceramic materials, it is necessary to minimize the heat transmitted to the workpiece to avoid thermal shock. This is accomplished by drilling a series of tiny holes into the substrate, at a depth of 20 to 30 percent of the material thickness, to form the scribe line. Laser manufacturers such as Coherent General produce equipment capable of varied rates and scribing depths. In a typical application, a line with a width of 0.006 in. to 0.008 in. (0.152 mm to 0.203 mm) can be scribed in 0.025 in. (0.635 mm) alumina at a rate of 10 in./sec. (25.4 cm/sec.).

Lasers are also used to scribe glass, a material that calls for even more careful precautions against thermal shock. This is accomplished by using very short, rapid pulses. Quartz glass tubes are often cut and sealed in a single operation using lasers.

Figure 19-2
This laser scribing system for ceramic workpieces is precisely operated by a CNC (computer numerical control) system housed in the console at left. (Coherent General)

Glazing

A *glaze* is a specially formulated glass that melts on the surface of the ceramic workpiece, and adheres to the body after cooling. Glazes are often applied to provide an impermeable surface on what would normally be a porous product. This results in the creation of a surface coating that can be easily cleaned.

Glazes are also applied to improve the appearance of the product by giving it a glossy surface. A glaze can also serve as a base for a decoration, or as an overglaze to protect a decoration. Firing melts the glaze and helps it to flow evenly. See Figure 19-3.

Glazes can be likened to porcelain enamels, which are applied with dipping, brushing or spraying directly on the surface of the ceramic piece to be finished. Both glazes and porcelain enamels contain silica glass, flux, colorants, and opacifiers.

Before 1920, ceramic floor and wall tiles were glazed by hand dipping or on a roller machine. Today, most glazing of tile is done automatically on a conveyor or by spraying.

Application of Glaze

When glaze is sprayed, the method of application is referred to as the *waterfall method*, Figure 19-4. In this method, a typical product, (such as a floor tile) would move through a curtain ("waterfall") of continuously-flowing glaze. Glaze can be applied to either bisque ware or greenware. It can also be applied to completely vitrified ceramic parts. The most common approach is to apply the glaze to pieces that have been bisque-fired.

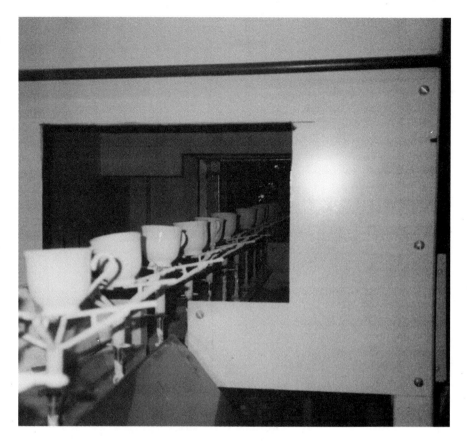

Figure 19-3
These teacups have had glaze applied by the spraying method. They are shown exiting from a glaze dryer prior to firing. (Lenox China)

The waterfall method has gained considerable popularity in recent years because it reduces the amount of airborne contaminants produced. After the finish flows down onto the product, any excess falls into a trough of water; the water is then collected in drums for disposal. This method contrasts sharply with conventional spraying, which generates a mist of finishing material in the air. Such mists can result in explosions or fires, as well as posing health hazards to workers.

Once the glaze is applied, pieces are normally loaded into *saggers*. These are refractory boxes that protect the workpieces from contamination in the firing operation. The pieces are then pre-sintered, fired, and cooled in the kiln. See Figure 19-5.

Glaze Characteristics

Glaze is classified according to the percentage (by weight) of the oxide it contains. Glaze compositions vary a great deal, but the major constituent in most is *silica*. Exact recipes depend on the thermal expansion required of the glaze, whether it is to be colored or not, whether it is to be transparent or opaque, and the temperature that is needed to cause it to adhere to the clay body. The composition of glaze is carefully formulated so that it will expand at a slightly lower temperature than the ware on which it will be applied.

Figure 19-4
The waterfall system, shown here being used to apply a glaze to ceramic tile, eliminates harmful and potentially explosive spray mist. A stream of water carries away excess material for proper disposal.

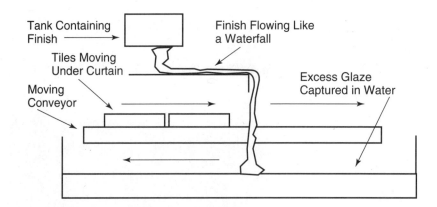

Figure 19-5
This worker is stacking small plates on saggers—fixtures used to prevent the plates from sticking together when they are fired. (Lenox China)

Glaze additives

Glaze is prepared with additives to color it or make it opaque.

One of the ingredients frequently added to the glaze mixture is called frit. **Frit** consists of particles of glass that have been milled to the desired size. The addition of frit lowers the necessary firing temperature of the glaze.

Some ceramic products must be decorated using a coloring compound. This is usually a metal oxide, which is mixed together with the glaze.

Colors can be applied *before* glazing (an underglaze decoration), or after glazing (an overglaze decoration). Machine application of decorative *gilt* (gold) lines to fine china after glazing is shown in Figure 19-6.

Figure 19-6
Gold decoration (gilt) is applied to the rim of a plate by this automated system. (Lenox China)

Underglazes are protected from wear by the overlying glaze, but do not provide the extensive range of colors available with overglazes. Colorants for underglazes are fine powders that can be easily mixed with feldspar or china clay.

Overglazes consist of powders ground with frits containing soft fluxes such as soda and potash. Overglazes must melt at temperatures lower than the glaze over which they are applied. Overglaze decorations can be applied by hand painting, by silk screen printing, or by applying decals or paper transfers. The most popular method is to apply printed designs with ***decals*** (paper transfers), Figure 19-7.

Figure 19-7
Paper transfers are applied by hand to fine china to provide the desired overglaze decoration. (Lenox China)

The printed image for paper transfers consists of varnish applied to a paper backing, usually by lithography. Dry ceramic powdered colors are applied and adhere to the wet varnish. After the varnish sets, the decal can be stored until needed. To apply the decal to a product, it is wetted to release it from the paper backing. The image is slid off the backing and onto the workpiece. When the piece is fired, the colors are permanently affixed.

Important Terms

decals
flame polishing
flame spray gun
frit
gilt
glaze
overglazes
oxyacetylene rod gun
plasma arc gun
saggers
scribing
setters
underglazes
waterfall method

Questions for Review and Discussion

1. Why are glazes necessary?
2. What types of products might not require a glaze?
3. How are decals applied? Are they overglaze decorations or underglaze decorations?
4. What are "saggers?" Why are they used?

Chapter 19 Finishing Ceramic Materials 387

Ceramic dinnerware typically has decoration applied either before glazing (underglaze decoration) or after glazing (overglaze decoration). Finishing processes for ceramic products often provide both surface protection and decorative appearance. (Caleca USA Corp.)

WOOD MATERIALS

Wood is one of the few natural materials that humans have used throughout history without finding it necessary to drastically change its properties. Wood requires little modification to make it useful for most industrial applications, because it can easily be formed, shaped, and smoothed with a multitude of manufacturing processes. In recent years, wood has even been used in product applications that previously required plastic or metal. The open pores of wood can be impregnated with synthetic polymers to improve stiffness, water repellency, strength, and stability.

Georgia-Pacific Corporation

20 Introduction to Wood Materials

Key Concepts

△ Hardwoods are not necessarily hard, nor are softwoods always soft.

△ Cellulose and lignin are the principal components of wood, with cellulose molecules accounting for about 70 percent of the volume.

△ The moisture content of wood affects its strength and its dimensions (swelling and shrinkage).

△ Wood is classified into various groups, depending upon its end use. It is further divided into quality grades.

△ There are seven major industries making up the broad category of the woods industry.

There are many different types of polymers, but the two most important for making hard good consumer products are plastic and wood. Plastics are synthetic (human-made) polymers, while woods are one of our oldest natural polymers.

Unique Characteristics of Wood

Wood is one of the few natural materials that humans have used throughout history without finding it necessary to drastically change its properties. Wood requires little modification to make it useful for most industrial applications, while its warmth and beauty are unmatched by other materials when used for residential and furniture applications. Thanks to careful harvesting and reforestation programs, wood has become a bountiful renewable resource, Figure 20-1.

Wood as a Manufacturing Material

Wood is a particularly useful material in manufacturing because it can easily be formed, shaped, and smoothed with a multitude of manufacturing processes. At first glance, one might think that wood has utility as a manufacturing material only for the furniture, cabinetmaking, and construction industries. While it is true that these industries depend on wood materials,

Figure 20-1
When carefully managed, wood is a renewable resource. Reforestation programs are an important activity for most logging companies, helping to ensure lumber supplies for the future. (Weyerhaeuser Co.)

nearly all manufacturing firms capitalize on the unique processing characteristics of woods to produce design prototypes, molds, jigs and fixtures, or other needed items.

During the past few years, wood has even been used in product applications which previously required plastic or metal. The cellular structure of wood makes it practical to impregnate the open pores of the material with synthetic polymers, thus improving stiffness, water repellency, strength, and stability.

There are hundreds of different species of wood used commercially throughout the world. They exhibit a wide variety of grain patterns, tones, and surface textures. Some types of wood are strong and durable, others are fragile and have little structural strength.

Classification of Wood

Wood is classified, according to its cellular structure, as either a **hardwood** or a **softwood**. However, this classification is sometimes misleading—just because a wood is classified as a hardwood, it isn't necessarily hard in physical terms. For example, balsa wood is very soft and lightweight, but is still classified as a hardwood.

Hardwoods are produced from ***deciduous*** trees, broad-leafed species that typically shed their leaves each fall. Oak, walnut, maple, birch, and ash are well-known hardwoods.

Softwoods come from ***conifers,*** the cone-bearing trees that have needles which remain green all year long. Pine, fir, hemlock, cypress, redwood, and red cedar are common softwoods.

Depending on their cellular structure, woods differ in terms of hardness, porosity, density, moisture content, and strength. It is the structure, variety, and arrangement of wood cells that makes one type of wood different from another. Wood fibers are different from synthetic polymers (plastics). Wood fibers are hollow, while synthetic polymers are solid. The cells are arranged in a bundle of hollow tubes, much like soda straws bound together with glue. Cellular structure affects the appearance of the wood, forming its *grain* (appearance of the annual rings and fibers viewed longitudinally).

Wood Structure

There are two major ingredients in wood, cellulose and lignin. About 70 percent of the volume of the wood is ***cellulose.*** The long-chain cellulose molecules are arranged in a nearly-parallel orientation, into units called ***crystallites.*** Crystallites are linked together in bundles called microfibrils. About 100 microfibrils join in the cell wall to form ***fibrils,*** or lamella. A major factor in determining the strength of the cell is the angle of the fibrils against the long direction of the cell. The smaller the angle, the stronger the cell.

Adjacent layers of cells are bonded together with a natural adhesive called ***lignin.*** Lignin constitutes about 25 percent of the total volume of the wood. When a tree trunk or limb is viewed in cross-section, the layers of cells appear as concentric circles. Each year's growth results in a new ***annular ring.*** Cells that are formed in the spring and early summer have a thinner wall and are lighter in color. Growth is slower in late summer and fall, so the cells formed then are thicker and darker in color. The darkness of these cells and the pattern of the annular rings creates the attractive grain in wood.

In addition to cellulose and lignin, minerals and extractives in the tree also influence the composition of the wood. ***Extractives*** consist of starches, oils, tannins, coloring agents, fats, and waxes.

Cellular structure of hardwoods

Wood is made of long thin cells with tapered ends. Figure 20-2 is a drawing that shows what a block of yellow poplar about 1/32 in. (0.8 mm) thick would look like under intense magnification. Yellow poplar is a hardwood. The horizontal plane labeled TT on the drawing corresponds to a tiny area of the top surface of a stump or end of a log. This is often referred to as end grain.

The vertical plane labeled RR, at lower left, corresponds to a cut made close to the surface and parallel to the radius. The vertical plane TG, at lower right, shows what the cells look like when they are exposed tangentially within the log. The annual growth rings, the amount that the tree grows each year, are labeled AR.

Hardwoods have cellular structures called *vessels* for carrying sap vertically. The vessels (labeled SC on the drawing) are made up of large cells with open ends positioned one above the other and continuing as open passages or tubes for relatively long distances. The area labeled K is where the sap drips from one vessel to another. When exposed on the end of a log, these vessels appear as holes or ***pores.*** This is why hardwoods are often referred to as *porous* woods. The pores vary considerably in size. They are readily visible in some species of wood, but cannot be seen without a magnifying glass in others.

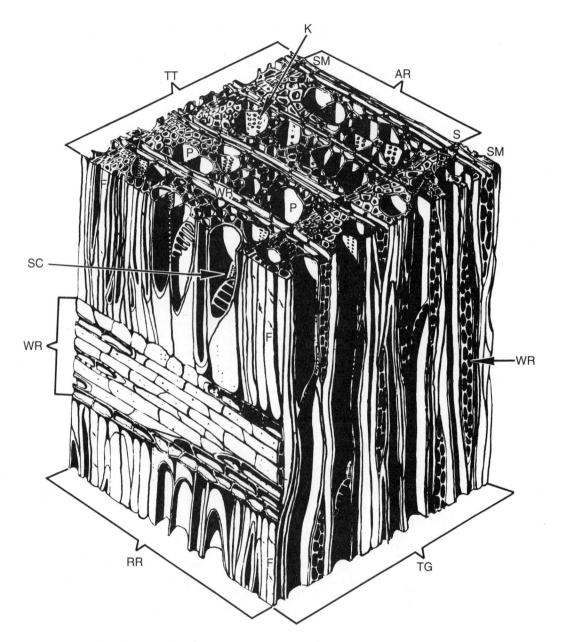

Cell structure of a hardwood

Figure 20-2   Cell structure of a hardwood. Note the varied sizes of cells, especially the large vessels (labeled SC) that permit vertical movement of sap. In open-grained woods like oak, these are visible as large pores. (Forest Products Laboratory, USDA Forest Service)

Hardwoods with large pores (ash, mahogany, oak) are classified as *open-grain woods*. Small-pore hardwoods (birch, cherry, maple) are classified as *closed-grain woods*. Normally, open-grain wood pores must be closed with *wood filler* to achieve a smooth surface before finishing materials are applied.

Cellular structure of softwoods

The structure of softwood is quite different from that of hardwood. The drawing in Figure 20-3 shows a block of white pine approximately 1/32 in. (0.8 mm) in thickness. The top of the block, labelled TT, shows the area parallel to the end surface of a log. In softwoods, sap is transferred vertically through cells called *tracheids*, labeled TR in the drawing. In addition to transporting sap, the tracheids provide strength to the wood.

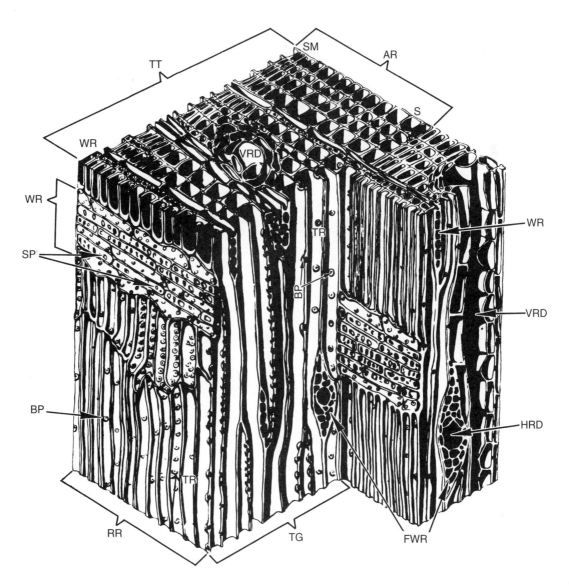

Cell structure of a softwood

Figure 20-3 Cell structure of a softwood. Note that the cells are much more uniform in size; sap movement is through heavy-walled cells called tracheids (labeled TR) that also provide strength to the wood.
(Forest Products Laboratory, USDA Forest Service)

The walls of the tracheids are thicker than the walls of cells in hardwoods. In effect, they form the bulk of the wood substance in softwoods. As you can see in the illustrations, cells are arranged in more orderly rows in softwoods than they are in hardwoods. The thin *intercellular layer* (adhesive material between the cell units) in softwoods can easily be dissolved using certain chemicals. This is what is done when making paper, Figure 20-4.

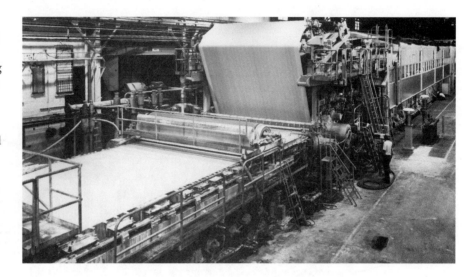

Figure 20-4
In papermaking, the fibers of softwoods are separated by dissolving the "glue" holding them together. They can then be rearranged and formed into a thin sheet on huge papermaking machines like the one shown. (Hammermill Paper Group)

Keep in mind that it is not the type or physical hardness of the wood, but the *cellular structure* which determines whether a given wood can be classified as a hardwood or softwood.

Effects of moisture

We know that hard and softwoods differ in terms of their porosity. Hardwoods have pores, while softwoods do not. Another factor, which is also a result of cellular structure, is the ***moisture content*** of the wood.

Because of its porosity, hardwood is particularly susceptible to the absorption of water, in both vapor and liquid form. Softwood doesn't have pores, but is also influenced by water: any wood that has not been filled, treated, or finished will absorb and lose water with changes in temperature and humidity.

It is sometimes necessary to know the moisture content of the stock being used. Moisture content is calculated as follows:

$$MC = \frac{\text{weight of water in sample}}{\text{weight of kiln-dried sample}} \times 100$$

Knowing the moisture content of a particular piece is important information, because (as was the case with ceramics) wood *shrinks* as it loses moisture. Wood also expands or swells with the absorption of moisture.

Most shrinkage in wood occurs *tangentially*, or around the tree, in the direction of the annual rings. There is little shrinkage *radially* (across the rings) or *longitudinally* (lengthwise, or in the direction of the grain). See Figure 20-5. The drawing in Figure 20-6 shows how tangential shrinkage and distortion of flat, square, and round stock occurs in relation to the annual rings.

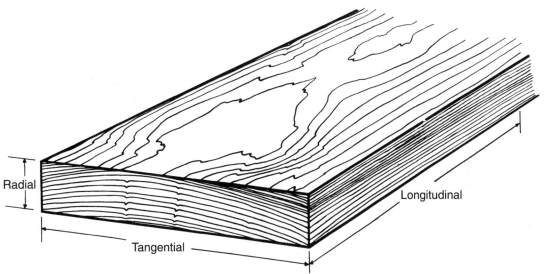

Figure 20-5 Directions used when discussing wood shrinkage. The greatest shrinkage is tangential (around the tree, in the direction of the rings). Radial (across the rings) or longitudinal (lengthwise) shrinkage is much less extensive. (Forest Products Laboratory, USDA Forest Service)

Figure 20-6
Tangential shrinking of wood is more or less severe, depending upon the way the annual rings are related to the board's dimensions. Note the severe cupping effect of the flat-sawn board at top, and the even shrinkage and minimal distortion of the flat-sawn board at center left. (Forest Products Laboratory, USDA Forest Service)

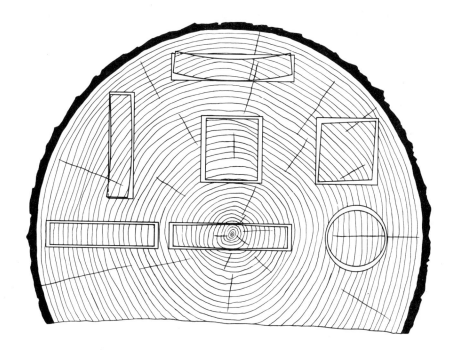

The *strength* of wood normally increases as it dries. Stock that has just been cut from a log is called **green wood,** and contains a great deal of moisture, or sap. It has only about half the strength (in endwise compression) of wood that has been dried to 10 percent moisture. However, reducing moisture content does not improve *all* characteristics related to strength. For example, shock resistance or toughness decreases as the wood dries. Dry wood cannot be bent as far as green wood without breakage.

Drying methods

As noted, when wood is first cut into lumber it is classified as green wood. If the wood is used at this stage, it will shrink and undergo drastic shifts and distortions as it gradually dries through exposure to air. For this reason, wood must be dried or seasoned before use. Sometimes, however, the drying process isn't enough too keep the stock from further drying and warping. Stock should always be checked for warpage prior to purchase and use.

Wood is usually air-dried (AD) or kiln-dried (KD). ***Air drying*** reduces the moisture content to around 15 percent. Normally, air drying is done for wood that is to be used outdoors. See Figure 20-7.

Figure 20-7
Wood that is to be used outdoors, such as the siding on this attractive contemporary house, is normally air-dried. Often, products like siding will be treated with preservatives during manufacture to provide longer service life when exposed to weather. (Benjamin Moore)

Kiln drying is accomplished in temperature- and humidity-controlled ovens. Kiln-dried wood has only about 7 percent moisture content. In addition to removing a larger amount of moisture from the wood, kiln drying relieves much more stress in the material than would be possible with air drying. Kiln-dried wood is intended for indoor use.

One of the newer processes gaining popularity for drying hardwood is called *radio frequency (RF) dielectric drying.* RF drying results in improved wood quality. The RF process also improves the shear strength, impact resistance, checking resistance, and surface finish of the material.

One of the major advantages of radio frequency dielectric drying is that it is much faster than kiln drying. A typical sample of wood might take 4 or 5 days to be kiln-dried. RF drying can dry lumber with a moisture content of 20 percent down to 6 percent in 24 hours.

The radio frequency vibrations generate frictional heat within the wood to accomplish the drying. The heat drives off moisture and results in uniform shrinkage of the stock with no distortion.

Structure of the Woods Industry

There are more than 2 million people employed in the woods industry. Wood materials are one of our most abundant renewable natural resources. More than 700 million acres, or nearly one-third of the continental United States, is either in forests or is well-suited for growing trees.

The woods industry can be subdivided into seven major industries: forestry, lumbering, millwork, furniture making, construction, wood processing, and distribution of wood products. This includes firms that manufacture veneer and plywood, wood composition board, paper, wooden containers, and many others. The paper industry is almost totally dependent on wood. About 70 percent of our paper is made from small logs, 30 percent from recycled material.

The *forestry* industry manages our forests so our supply of timber will continue. The Forest Service, a part of the United States Department of Agriculture, is the governmental agency concerned with the overall management of our timber reserves. Foresters and forester assistants are responsible for planting and conservation programs.

The lumber industry is responsible for selecting and cutting trees, and transporting them to mills. Typical occupations in this industry include lumberjacks, fellers, limbers, and buckers.

When the timber has been transported to the sawmill, it is cut into boards and graded. Then it is air- or kiln-dried. The sizes of the cuttings that are made depend on the purpose of the mill. Some mills produce *veneer* (thin sheets of hardwood for laminated surfaces) and ship it to furniture plants for final processing. Other mills specialize in hardwoods or plywood.

There are more than 5000 industrial woodworking firms in the United States and Canada. Some of these are furniture-making companies. Others make panels, doors, windows, cabinets, sporting goods equipment, and other consumer products. There are more than 100 companies that manufacture machinery for use by the woods industry.

Construction firms use woods to build structures on site for commercial, residential, and industrial purposes. About two-thirds of all people in the construction industry are carpenters. This is the largest of all the skilled construction occupations.

Wood is also processed for use by other industries. The paper industry is one of the largest users, consuming more than 15 million cords of pulpwood each year to make paper and boxes.

Trees are also used as a source of material to make finishes and other chemical products. Gum and resin harvested from living trees are used to make varnish, paint, printing ink, insecticides, pharmaceuticals, chewing gum, and wax. Chemical processes are used to change wood cellulose to lacquer, synthetic fibers, photographic film, floor tile, and solid fuel for rockets.

The wood distribution industry is concerned with the distribution and sale of lumber and wood products. There are more than 30,000 retail building supply centers, Figure 20-8, that are concerned specifically with the distribution of wood materials.

Figure 20-8
Large building supply centers stock many kinds of dimension and structural lumber. Often, they prefabricate building components, such as these roof trusses, to meet customer demand. (Hoge Lumber Company)

Nature of Industrial Stock

Wood *stock* is normally purchased by a manufacturing firm as sheets requiring no or little finishing, as rough-sawn boards, or as yard lumber. Yard lumber is the material normally purchased at a lumber yard. Yard lumber consists mainly of softwoods, particularly white pine. There are three major classifications of yard lumber: dimension lumber, factory and shop lumber, and structural lumber. There are several quality grades in each of these classifications.

Dimension lumber is normally purchased in standard-length boards up to 1 inch thick and 12 inches wide. Dimension lumber is ready to use, without any additional sanding or surface preparation.

Factory and shop lumber is used primarily for remanufacturing purposes in mills that produce fabricated doors, windows, cabinets, moldings, and trim items. The poorest grade of softwood, referred to as number 4, is used for this purpose.

Structural lumber includes light framing material, such as 2 x 4s, and widths up to six inches and wider for joists, rafters, and framing uses. See Figure 20-9. Structural lumber is purchased by the construction industry for use as beams, stringers, posts, and timbers for heavy structural applications, and for factory and shop applications.

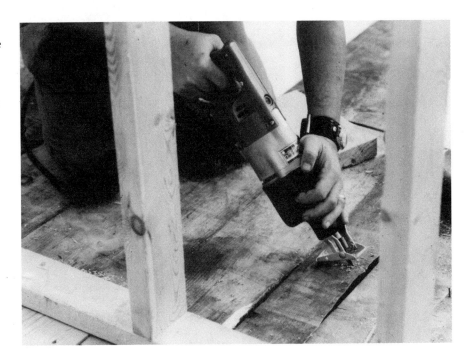

Figure 20-9
Structural lumber, such as the 2 x 4s used for wall framing in this photo, is the basic stock of the construction industry. (Milwaukee Electric Tool Corp.)

Both hardwood and softwood stock used for manufacturing fine quality wood products are normally purchased from the mill in rough-sawn form. *Rough-sawn lumber* requires subsequent planing or surfacing to smooth the board prior to use. The dimension across the board or face is smoothed by running it through a planer. Normally, the rough-sawn edges are smoothed by another process, called jointing. These processes will be discussed in a later chapter.

Wood stock is also produced in chip form for making particle board, wafer board, and other sheet material. Wood fibers are used to make hardboard. Panels are laminated together to produce plywood and veneered stock. See Figure 20-10. Wood is also used as a raw material to manufacture a variety of products, including explosives, synthetic fibers, turpentine, and waxes.

Figure 20-10
Plywood is a versatile material created by laminating together several thin sheets of wood. It is easily cut and shaped with either hand or power tools. (American Plywood Association)

Wood Grades

Wood is classified according to its quality. The standards for assessing the quality or grade of lumber is set by the woods industry. The **grade** of lumber refers to the appearance, strength, and lack of defects in the stock. With softwoods, standards are regulated by the *American Softwood Lumber Standard PS 20-70*, produced by the U.S. Department of Commerce. Trade groups such as the California Redwood Association, Western Wood Products Association, and Southern Forest Products Association publish rules for grading various species of lumber in their specialty areas.

For softwoods, there are two subclassifications within the basic classifications (dimension, factory and shop, and structural). These subclassifications are *select* and *common*.

Select lumber is graded from A to D, with A presenting the best quality surface appearance. Structural ***common lumber*** is graded by number, ranging from 1 to 4. Number 1 common is the best grade, with no knots or knotholes. Each lower grade (2, 3, 4) exhibits an increase in the number of defects. Defects would include knots, stains, split areas, holes, and warp.

Hardwoods standards are established by the National Hardwood Association. The best grade of hardwood stock is **FAS**, meaning "firsts and seconds." Thus, the highest quality grade is called *firsts*, and the next best grade, *seconds*.

In order to meet the exacting standards for FAS grading, a board must be as least 8 feet long and 6 inches wide. It must also have at least 83.33 percent clear cuttings. No knots or defects are acceptable in FAS-grade lumber.

The next hardwood grade is "select." ***Select hardwood*** boards must be at least 6 feet long and 4 inches wide. One side, called the *face,* must be FAS quality; the other can have some defects and blemishes.

The next lower grade of hardwood is called Number 1 Common, which must have 66.66 percent clear cuttings. Number 2 Common must have at least 50 percent clear cuttings; Number 3 Common, 33.33 percent clear cuttings.

When wood is purchased from the mill, or in quantity from a supplier, it is sized according to the number of *board feet* in the stack. A ***board foot*** is the basic unit of measurement, and is 1 inch thick by 12 inches by 12 inches. To determine the number of board feet in a piece of lumber, multiply the thickness in inches by the width in inches by the length in feet, then divide by 12. When stock is less than 1 inch thick, it is figured as 1 inch. The formula for computing board feet is:

$$\frac{\text{Thickness} \times \text{Width} \times \text{Length}}{12} = \text{board feet}$$

Important Terms

- air drying
- annular ring
- board foot
- cellulose
- closed-grain woods
- common lumber
- conifers
- crystallites
- deciduous
- dimension lumber
- extractives
- factory and shop lumber
- FAS
- fibrils
- forestry
- grade
- grain
- green wood
- hardwood
- intercellular layer
- kiln drying
- lignin
- longitudinally
- moisture content
- open-grain woods
- pores
- radially
- radio frequency (RF) dielectric drying
- rough-sawn lumber
- select hardwood
- select lumber
- softwood
- stock
- structural lumber
- tangentially
- trachieds
- veneer
- vessels
- wood filler

Questions for Review and Discussion

1. What is the major difference between a hardwood and a softwood?
2. What type of surface preparation must be done before finishing hardwoods? What will happen if this is not done?
3. What advantages does RF drying offer, when compared to kiln drying?

Wood is used as a material in a wide variety of sizes and shapes, from massive glue-laminated beams and arches in building construction to fine veneeer inlays on furniture or picture frames. This vocational school student is cutting molding for application to shop-fabricated kitchen cabinets.
(Howard Bud Smith)

21 Forming Wood Materials

Key Concepts

- △ There are two basic woodforming processes: bonding and bending.
- △ Hardboard, insulation board, and particleboard all can be grouped as composition boards.
- △ Particleboards of various types all are formed from pieces of wood and a synthetic resin binder.
- △ Reconstituted wood is one of the newest forms of engineered wood materials.
- △ Steam or soaking is often used to plasticize (make pliable) wood that is to be bent.

The processes used to *form* wood materials, as distinguished from such *separating* processes as sawing, can be divided into two categories: bonding and bending. Bonding is used primarily to form sheet-type materials that are used in construction or as industrial stock. Bending uses various techniques for forcing solid or laminated wood materials into desired shapes.

Bonding

Bonding is a forming process used in the production of wood composition board. It involves the use of heat and pressure to compact particles or chips into sheet stock. *Composition board* is made from wood that has been broken down into particles or fibers, then reconstituted to form products such as insulation board, hardboard, waferboard, particle board, or oriented strandboard.

Commercial development of composition board began as a means of using waste byproducts from the manufacture of paper. The manufacture of wood composition board takes advantage of many parts of the tree which would have been considered unusable for making other types of wood products. Sawdust, planer shavings, and other wood residues are also used.

Composition Boards

There are three major types of composition board—hardboard, insulation board, and particleboard. Each of these composition boards is made through bonding processes. Both hardboard and insulation board are made

403

by compacting fibers. The fibers in hardboard are so fine and are so tightly compacted that it is difficult to see them when looking at a piece of stock that has been sawn to make the interior composition visible. Insulation board is more loosely compacted, and thus the fibers are easy to see. Particleboard is not made from fibers. It is made using larger particles of wood and synthetic resins. Other types of stock are also made from particles and pieces of wood pressed together. The most popular types are waferboard and oriented strandboard.

Hardboard

Hardboard was first produced in 1924 by William H. Mason, who later formed the Masonite Corporation. Mason discovered that small wood chips could be blown apart in a digester with high-pressure steam. The fibers that were produced were perfect for producing a continuous mat of pulp that could be pressed into a sheet of strong, hard material. The basic process developed by Mason is still used today. Figure 21-1 shows how the process works.

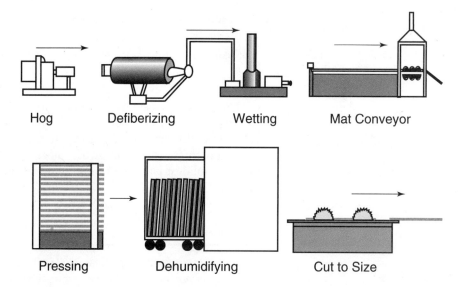

Figure 21-1
The processing stages used in manufacturing hardboard. Wood is broken down into fibers, which are mixed with water to form a pulp that is pressed into sheet form. After further dehumidifying, the sheets are cut to size.

First, the wood material is broken up in a gigantic chipper called a *hog*. The next step is to **defiberize**, or dry out, the particles so that most of the chemicals are removed. Water is then added to the dry particles so that they can be mixed further and form a pulp that is easier to handle. Once all of the pulp material is in dispersion, it is run off onto a conveyor. Here, some of the water is removed through absorption. The next step, pressing, removes most of the water. Finally, the material is placed in a dehumidifying oven to dry it further. The final stage in the process is cutting the hardboard to size in 4 ft. x 8 ft. panels.

The U.S. Department of Commerce, in commercial standard CS 251-63, defines hardboard as a panel "manufactured from interfelted lignocellulosic fibers consolidated under heat and pressure in a hot press to a density of at least 31 pounds per cubic foot." Hardboard has high tensile strength, high density, and low water absorption. Often, the board has greater hardness and weight per cubic inch than the wood from which it was made.

Hardboard can be purchased in either tempered or standard form. *Tempered hardboard* has been impregnated with a resin/oil blend, which is stabilized during drying. Tempered hardboard has better strength, stiffness, and resistance to water and abrasion than standard hardboard. Tempered hardboard has a density of 60 to 75 pounds per cubic foot. Hardboard is used for floor underlayment, door panels, and in a multitude of product applications, ranging from cabinet shelves and furniture backs to core materials overlaid with veneers or films. See Figure 21-2.

Figure 21-2
This door is covered with a thin molded facing of hardboard that can be painted or can be stained for a natural wood look. As shown in the cutaway view, the facing is applied over a solid core of particleboard. Solid wood edging is used for additional rigidity. (Masonite)

Insulation board

Another type of composition board is called *insulation board*. This material has been used since 1914, when it was developed by Carl Muench, a refrigerator manufacturer who was working to find a use for residue from a friend's pulp mill. The machine designed by Muench was an adaptation of the papermaking machine, and was capable of producing 3,000 square feet of insulation board per day. His process went through many refinements. The first plant producing Muench's Celotex insulation board was built in 1920 near Marrero, Louisiana. It is still the largest producer of insulation board in the U.S. Today, insulation board plants have achieved daily capacities of more than 3 million feet.

Insulation board has high resistance to heat transfer, and is also low in density. Small air pockets in the board slow down the passage of hot or cold air which passes through the product. Insulation board is also good for reducing noise. As an acoustical barrier, it can absorb more than 70 percent of the noise that strikes it. Billions of feet of insulation board are produced for use by the housing industry each year. See Figure 21-3.

Figure 21-3
The many small air spaces between the fibers of insulation board make it an efficient sound-absorbing material. Ceiling tiles and suspended ceiling panels like those in this photo are forms of insulation board. The exposed surfaces are often painted or laminated with a plastic material for a finished appearance. (USG Interiors)

Insulation board is manufactured using pulpwood and *bagasse*, a cellulosic byproduct produced during the pressing of sugar cane. The process of making insulation board begins with feeding logs into a chipper, where they are first reduced to chips, then to fibers. Water and chemicals are added to turn the fibers into a continuous mat. The mats are fed into a drier, then cut and trimmed to size, in sheets from 16 to 26 feet long.

Particleboard

Particleboard is also popular in the construction and furniture-making industries. *Particleboard* was developed in Germany in 1943; it was first sold in the U.S. for construction use in the mid-1950s. As a manufacturing material, particleboard met with only limited success until the 1960s, when users began to learn how to work with this new material. Today, more than one billion square feet of particleboard is produced annually.

About 40 percent of the particleboard used today is purchased by the construction industry for flooring *underlayment*. Most of the remaining 60 percent is used as core material by the furniture industry, Figure 21-4.

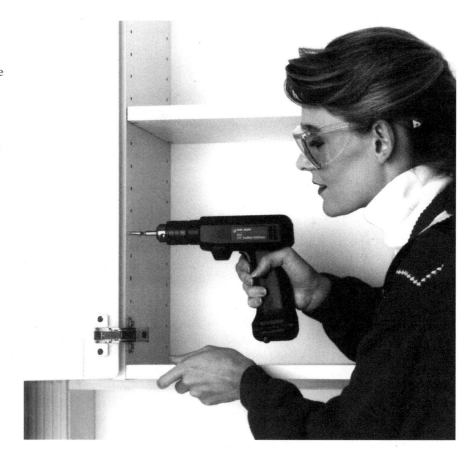

Figure 21-4
The largest use of particleboard is for core material in furniture manufacturing. The kitchen cabinets being installed in this photo are constructed of particleboard with plastic laminate facings on all visible surfaces. (Black & Decker)

The U.S. Department of Commerce, under the commercial standard CS 236-66, defines particleboard as "a panel material composed of small discrete pieces of wood bonded together in the presence of heat and pressure by an extraneous binder." Particleboard can also be defined in terms of the method by which it is pressed. When the pressure is applied in a direction perpendicular to the faces, as in a conventional multi-platen hot press, the board is defined as *flat-platen-pressed*. Most of the particleboard produced today is made using flat-platen pressing.

When the applied pressure is parallel to the faces, the board is defined as *extruded*. Extruded board is weaker across the length of the board, and exhibits more lengthwise swelling during pressing. Most extruded hardboard is used as core material supporting veneered surfaces.

Particleboard can be treated with chemicals to resist attack by insects or fungi. It can even be embossed or stamped with special surface treatments. At the same time, it should be remembered that this engineered product has the ability to absorb water readily, and thus can swell and warp. Without surface treatment, it is not very useful in high-humidity environments.

Manufacture of particleboard

About 90 percent of the materials used to manufacture particleboard come from wood processing mills. The ingredients consist of planer shavings, sawmill chips, veneer wastes, and other residues from preparing wood stock. Sometimes, logs that are too small to be useful for other purposes are also used to produce wood flakes. After the wood flakes are reduced to the desired size and shape, they are mixed in a blender with synthetic *binders* (normally phenolic resins) and other chemicals. The resin-coated chips are then fed into a forming machine, which deposits them directly on a metal belt to form a mat. Mats are made in layers, using different-sized particles. Large particles are positioned in the middle to provide strength and the small particles are layered on the outside to produce smoothness. The forming machine produces a dry mat. The mat is then conveyed to a hydraulic press, where heat and pressure are applied. This is normally about 400°F at 1,000 psi (205°C at 6895 kPa).

Waferboard and oriented strandboard

Waferboard and *oriented strandboard* are also types of particleboard. The process used to make these materials is the same but the materials that are used are slightly different. See Figure 21-5. Waferboard, or *waferwood*, is made using high quality wood flakes about 1.5 in. (38 mm) square. The flakes are bonded together under heat and pressure with phenolic resin that serves as a binder and waterproof adhesive. Both sides of waferboard have the same type of textured surface appearance. Waferboard is produced in 4 ft. x 8 ft. sheets, mainly for use in the construction and furniture-making industries. Waferboard has a density of about 40 lbs. per sq. ft.

Oriented strandboard is made from wood *fibers*, rather than particles or flakes. The fibers are large and irregularly shaped. Consequently, they are easy to see after the mat material is pressed into a sheet. As is the case with all particleboards, the materials are bonded together with resins and glues. Each successive layer of fibers is arranged at right angles to the preceding one. As is the case with other types of particleboard, hardboard, and waferboard, oriented strandboard can be cut using regular woodworking machines and tools.

Figure 21-5
Three of the common forms of composition board. From top to bottom: waferboard, oriented strandboard, particleboard. All are widely used in both building construction and in manufacturing. (American Plywood Association)

Reconstituted wood

One of the newest developments in the field of wood-based engineered material is *reconstituted wood.* This type of material can be purchased as a veneer, or in solid blocks. Veneer comes in very thin sheets cut from expensive wood. These sheets are used as the exposed (top, bottom, or side) layer that is glued over less expensive wood. Veneer is used when a fault-free, decorative wood with a particular color or grain is desired. It is less expensive than using solid hardwood.

Reconstituted wood is made from plywood that is cut apart in pieces and reassembled. The strips that are glued together are often smaller than 0.030 in. (0.762 mm). Strips this size are so small that they can scarcely be seen with the naked eye. When the wood is glued together, it is hard to tell that it is not "real" wood.

Lamination

One of the most important processes to the woods industry is lamination. It is the same process that was used with plastic materials, but in this case, involves sandwiching sheets of *wood* together to make larger pieces of wood. Lamination is used extensively for the construction of plywood.

Plywood

Plywood is made by gluing a number of layers, called ***plies,*** together at right angles to each other. The plies on the surface (top and bottom) are arranged with the grain approximately parallel. An odd number of plies (3, 5, 7) is used so that the material will be *balanced,* with the same number of plies on either side of the center core. This makes it possible to run the grain on both sides in the same direction. See Figure 21-6.

Figure 21-6
Plywood is laminated from 3 to 7 layers of veneer for strength and dimensional stability. The top sheet is plywood with a brushed surface to emphasize the wood grain. The bottom sheet is a variation of plywood called "composite board." It has facings of veneer applied to a core of particleboard, and can be used for many of the same applications as traditional veneer plywood. (American Plywood Association)

Plywood is manufactured in accordance with the U.S. Product Standard PS 1-74/ANSI A199.1. This standard, established by the American National Standards Institute, designates the species, strength, type of adhesive, and appearance of plywood.

Lamination is also used for processes other than making plywood. It is used, for example, to produce bent or irregularly shaped parts in the furniture industry. Typical applications would be in making bent forms for tables and chairs.

Advantages of lamination

When shapes such as gunstocks or huge beams or arches are desired, there are really only two ways that they can be constructed: by sawing them to the proper shape, or by laminating them (building up the desired shape in layers). Lamination is preferred, since cutting is usually more wasteful of material and the grain does not run parallel with the flow of the part's surface. Another advantage of laminating is that it produces a much stronger product than cutting to shape.

There are other reasons why lamination may be the ideal process for forming many wood products. With laminated parts, the highest quality materials can be used for the face surface and material of lower quality or appearance for the inside. This results in a better-looking product for less money.

Another advantage of lamination is that the process enables manufacture of more-complex and more-intricately shaped larger pieces. There is little waste with lamination, since smaller pieces can be glued together to make wood molds. The process is often used in the boat manufacturing industry for making original wood molds. Thousands of individual wood pieces are glued together in contact with each other to form the mold. Since the mold is often large (60 ft. or more in size), a great deal of precision work is necessary to fit all of the pieces together in perfect alignment.

The sequence of operations normally followed for laminating irregularly shaped parts, such as large architectural beams, might be something like this:

First, the stock (boards) would be cut to length and planed to remove the rough surface from both faces. The ends of each board then would be bevelled to facilitate joining when they are glued together end-for-end. *Beveling* is the process of changing the sharp (90-degree) angle where a vertical and horizontal surface join into a (usually) 45-degree angle.

After beveling, each piece of stock would be run through a glue-spreading machine, where an adhesive would be applied to both sides. The glued pieces would be joined end-for-end and placed in plies in a jig that conforms to the desired shape. The jig with glued plies would then be placed in a forming press and heat and pressure applied. After the adhesive cures, the piece would be removed from the press. Finally, it would be machined on a specialty-type planer designed to handle the desired shape.

Thick pieces of wood tend to crack and split over time. Lamination is often a useful method for eliminating these types of worries—products can be made in any size, and they will retain their structural integrity without cracking.

Bending

Bending is an effective process for producing curved shapes, whether solid or laminated materials are used. Hardwoods such as white oak, elm, hickory, ash, birch, maple, or walnut are particularly suited to the bending process. Softwoods are difficult to bend, so they are seldom used.

Wood can be bent either *across* the grain or *along* the grain. Bending of solid wood can be done either when the wood is dry or when it is wet. Laminated plies are bent when they are wet.

Some woods are more prone to fracture (breaking) than others, so they are bent when they are wet. Moisture and heat are applied to the dry wood. The workpiece is then bent and clamped over a form until it cools and dries. It will then retain its new form.

Wet Bending

There are two basic methods of **wet bending:** steaming or soaking, and laminating in a two-piece mold. Steaming or soaking will soften (plasticize) the wood so that it can more easily formed. Laminating is accomplished using adhesives in a forming jig. Plywood, solid, or laminated wood can be bent using either method. It is more difficult to bend solid pieces of wood than laminated stock. See Figure 21-7.

Steaming or soaking is the better way to bend wood. Here's how the process works:

First, the stock is subjected to steam, or is soaked in boiling water, until it reaches a moisture content of about 20 percent. Dry wood normally has to be soaked or steamed about one hour for each inch of thickness. After it is soaked or steamed, the stock is placed in a two-piece form or jig that conforms to the desired shape. The part is kept in the jig until it cools and dries. Once it has dried, the part will hold its desired shape.

Sometimes, sheet metal is clamped to both faces of the stock during bending to provide additional support and help in holding the desired shape. The inside of the wooden mass can be compressed as much as 25 percent, but the exterior surface can't be stretched more than one or two percent. For this

Figure 21-7
The curved seat and back of this chair were laminated in a forming jig. (Herman Miller, Inc.)

reason, bending must be done slowly, or the outside layers will split. After the stock is bent, it should be allowed to dry on the form for at least 24 hours before it is removed.

Important Terms

bagasse
bending
beveling
binders
bonding
composition board
defiberize
flat-platen-pressed
hardboard
insulation board
oriented strandboard
particleboard
plies
plywood
reconstituted wood
tempered hardboard
underlayment
waferboard
wet bending

Questions for Review and Discussion

1. What are the major steps used to plasticize wood prior to bending?
2. What are the differences between hardboard and hardwood?
3. Which of the three major types of composition boards would you specify for use in a very humid environment?
4. How would you be able to recognize a sheet of reconstituted wood?

Separating Wood Materials

Key Concepts

- △ Edged tools are the primary means used to separate wood materials.
- △ Planers and surfacers are used to provide parallel, smooth-planed surfaces on wood.
- △ For material removal using a lathe, the rotating workpiece may be attached to a faceplate or held between centers.
- △ Sawing is a chip-producing separating process, with each succeeding tooth of the saw cutting away a tiny chip from the wood.
- △ Drills, boring machines, and mortising machines are used to produce holes in wood that are round or irregular in shape.

Separating activities involving wood materials involve a variety of edged tools, used in both hand- and power-operated equipment. The edged tools— knives, sawblades, drills, and chisels—are used to cut pieces of wood apart, make holes or shaped openings in the wood, smooth the wood's surface, or reduce its dimensions.

Planing

Machine planing, or surfacing, is a process that mills wood to cut it to uniform thickness and produce a smooth surface. Planing is used to remove rough mill marks on stock that has not been surfaced at the mill. Planing is also used to *true-up* (eliminate bow in) the wood, but it cannot straighten warped stock.

Most stock purchased directly from the mill needs to be planed to produce smooth faces. Normally, planing is done in the direction of the grain, with a maximum cut of up to 1/16 in. (1.6 mm).

There are three major types of machines used for planing: the knife-blade planer, the abrasive belt planer, and the jointer. Both the knife-blade and abrasive belt planers are used to smooth the face surfaces of stock. The jointer can be used to surface the faces of narrow-width stock (boards up to six or eight inches wide). The jointer is the preferred machine for removing stock from the edges of boards.

413

Knife-Blade Planer

Figure 22-1 shows the design of a knife-blade planer, which works on the same principle as the hand plane. The knife blades shear off a uniform layer of stock as the material is passed through the machine. Knife-blade planers with cutterheads only on the top are called *single planers* or single surfacers, and are the most common type. **Double planers** have blades on the top and the bottom, so that both sides of the stock can be planed at the same time. With either type of planer, stock is pushed into the machine on a bed, and power feed rolls carry it on through the machine. Chip breakers and pressure bars help to reduce chattering and keep the stock from jumping during planing.

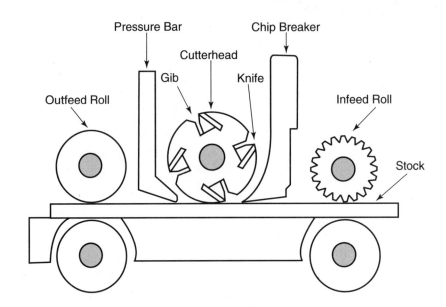

Figure 22-1
A single surfacer is a knife-blade planer with one rotating cutterhead. The knife blades, held in place by plates called gibs, shear off a thin layer of material from the top of the stock as it passes through the machine. Units with two cutterheads, called double planers, surface both top and bottom of the stock at the same time.

Planers are sized by the **bed width**, which essentially means the width of stock that can be milled in the machine. Planers range in bed width from 12 in. to 52 in. (30 cm to 132 cm). Knife-blade cutterheads are normally adjustable to suit the width of the stock to be planed. Cutterheads may carry a number of knives, or may have only one, with a counterbalancing blank on the other side of the planer head. Large planers may carry as many as 20 or 30 jointed knives. Figure 22-2 shows an operator planing a thick board on a knife-blade planer of the single-surfacer type.

One of the disadvantages of the knife-blade planer is that the knives must be removed for sharpening and changes in set-up. The practice is time-consuming, and the knives must be handled with extreme care. Another drawback of the knife-blade planer is the danger of **kickback**. This means that it is possible for the cutter to bite into the stock so that the workpiece is thrown out of the machine back toward the operator. Kickback is usually caused by such operator errors as feeding stock that is badly warped or bowed, taking too large a cut in one pass, or feeding stock that is too short to be safely planed. A safety-conscious operator always stands to one side of the machine, out of the path of the moving stock.

Figure 22-2
This worker is feeding a thick plank through a knife-blade planer. Note the vacuum system used to collect sawdust and chips. It simplifies cleanup and minimizes the amount of dust in the air of the shop. (Delta International Machinery Corp.)

Abrasive Belt Planer

The *abrasive belt planer* is a safer machine than the knife-blade planer. It eliminates kickback and the problems related to removing and sharpening knives. In addition, abrasive planers generally have lower overall maintenance costs, and are less noisy than knife-blade planers. Abrasive planing can improve lumber yields up to 20 percent by eliminating product defects caused by knife-blade planers.

The machine shown in Figure 22-3 is a high-production abrasive planer used in furniture manufacturing. It can handle parts up to 63 in. (160 cm) wide at speeds of up to 80 feet per minute. On this planer, the stock is fed into the infeed section of the first unit, which planes one side of the stock. The wood is then carried on a conveyor to the second head, which planes the bottom of the board.

Figure 22-4 shows the interior of the same machine. Note the vertical planing belt exposed on each head. The heavy sanding belt takes the place of conventional metal knife-blade cutters used in other planers. Abrasive planers

Figure 22-3
This large abrasive planer surfaces the top side of the wood in the first unit and the bottom side in the second unit. It is used for high-speed production planing in a furniture factory. (Timesavers, Inc.)

Figure 22-4
The abrasive planer shown in Figure 22-3, with covers removed to show roller drive mechanisms that carry the abrasive belts. Note that the belt is on top in unit at right, and at bottom in unit at left. This permits surfacing both sides of the stock in a single pass. (Timesavers, Inc.)

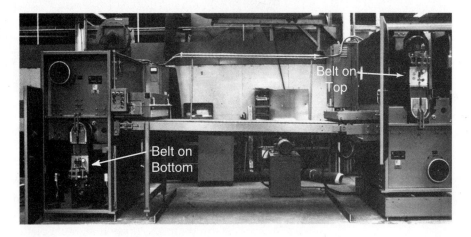

can be purchased with belts from 12 in. (30 cm) to almost 10 ft. (3 m) in width. Machines of this type can be constructed with any number of planing heads to meet high-volume production requirements.

At first glance, you might think that a sanding belt would wear out quickly in a production environment. If the machine is used properly, this is far from true: the belts are tough and will last for a long time. However, if you to try to take too large a cut with one pass, or if you attempt to plane stock with nails or other metal fasteners, the belt will tear and soon need to be replaced. Tears or holes shorten the life of the belt, leading to premature failure.

Jointing

The *jointer* is a machine that can improve overall quality of a workpiece by effectively complementing the work of the planer and circular saw. For example, even if stock is purchased with both face sides planed from the mill, it is still necessary to true-up one edge on a jointer before proceeding. A rough-edged board cannot be safely fed through the circular saw. The jointer is an important machine.

This sequence of operations is often followed when jointing. If the stock is narrower than the jointer's cutterhead, then the jointer is normally used to plane the surface and edge of the best face. The opposing face is then planed on the planer, and the board is cut to width on the circular saw.

Jointers are sized by the length of the knives in the cutterhead. A small jointer is shown in Figure 22-5. A machine such as this would have a cutterhead with three or four knives, and would run at a speed of approximately 4500 rpm.

Figure 22-5
A small jointer of the type used to true the edges of boards. (Powermatic-Houdaille, Inc.)

When a jointer is used to true an edge, the operator places the stock on the infeed table of the machine, resting against a *fence* (adjustable stop running perpendicular to the cutterhead axis), then pushes it over the rotating cutterhead. The cutterhead is located to the center of the machine's infeed table. See Figure 22-6.

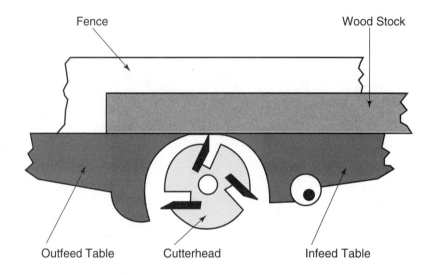

Figure 22-6
Cross-sectional view of a typical jointer. Stock is held against the fence and moved over the rotating cutterhead for material removal. The heights of the infeed and outfeed tables are independently adjustable.

After the stock passes the cutterhead, it slides onto the outfeed table and a machine guard closes the opening over the cutterhead. On a small machine such as the one shown in Figure 22-6, a *push stick*) is normally used to hold down the stock and push it on through the machine. A larger jointer is shown in Figure 22-7. With this machine, the stock is of sufficient size to allow the operator to push it through the machine without using a push stick.

Figure 22-7
This worker is facing one side of a thick workpiece, using a jointer. The dark-colored structure between the workpiece and the operator is a spring-loaded guard that swings into place to cover the rotating cutterhead as soon as the workpiece has cleared it. (Delta International Machinery Corp.)

All of the major parts of the jointer—the infeed table, outfeed table, and the fence—can be adjusted to produce the desired cut. The adjustment of the outfeed table is very important. It must be positioned so that it is level with the highest point of the knife edges. If the table is too low, the trailing edge of the board will drop down, producing a dip on that end of the stock. If the outfeed table is too high, it will push the board upward and produce a taper. In order to remove stock on the jointer, the depth of cut is established by adjusting the infeed table to a level that is lower than the outfeed table and the knives on the cutterhead.

If it is not used properly, the jointer can be a dangerous machine. Stock must be at least 12 in. (30 cm) long and at least 3/8 in. (9.5 mm) thick. End grain should never be planed unless the stock is at least 12 in. (30 cm) wide. A push stick or push block should be used when removing material from the face surfaces of small- to medium-sized workpieces. When the stock is large and heavy, it may be safer to push it through without a push stick.

Shaping

The shaper is a useful and versatile machine for producing the intricate shapes required for molding or window-framing, but it can also be a dangerous piece of equipment if it is not used properly. Shaping produces a straight line or design pattern along the length of the stock. See Figure 22-8.

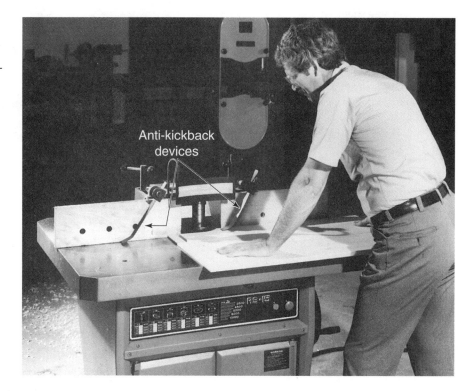

Figure 22-8
The single vertical spindle of this shaper carries a cutter that produces a desired edge-shape on stock. The worker in this photo is using the shaper to finish the edge of a door facing. The two diagonal bars in front of the fence are anti-kickback devices. (Delta International Machinery Corp.)

There are various types of shapers. The most common of these is the *single-spindle shaper*, which has a vertical *spindle* (shaft) that projects through an opening in a horizontal metal worktable. The shaper cutters are carried on the spindle. Cutters have many different shapes, ranging from grooving saws and solid cutters to slip knives. Here's how the process of shaping works:

If you are shaping material where the edge to be grooved is a straight line, the fence is used to gage the depth of the cut that will be made into the edge of the stock by the rotating cutter. After the fence has been set, the stock is placed on the table and snugged against the adjustable fence. The workpiece is then pushed carefully into the rotating shaper cutters. A push block or push stick should be used, whenever possible.

When shaping, stock must always be fed against the rotation of the cutters. The height of the cut is controlled by a handwheel that raises the height of the vertical shaft. As noted, the depth of cut is regulated by adjusting the position of the fence. The shaper cutter rotates at high speed (5000 rpm to 10,000 rpm). Figure 22-9 shows how a shaper cutter is attached to the spindle, and its relationship to the workpiece.

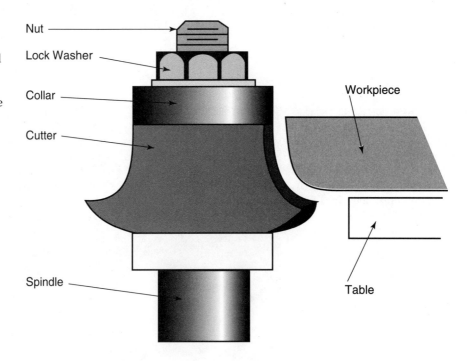

Figure 22-9
The shaper cutter is secured on the spindle with a nut and lock washer, as shown. The cutter configuration will produce a matching shape on the edge of the workpiece, as shown. When shaping the edge of an irregular workpiece, a template is often used. It rides against the collar to maintain the proper depth of cut.

Not all pieces are easy to shape. If the workpiece has been bandsawed into a curve or irregular shape, the fence cannot be used. Instead, the workpiece is pushed against the top and bottom collars, while the knives work to remove the desired amount of stock.

A third method of guiding the stock against the cutter is to use a shaped pattern or template. The stock is attached to the pattern, which is placed on the shaper table. The edge of the pattern rests against the collar that is above the cutter on the spindle (See Figure 22-9). The guiding edge of the pattern regulates the cut. One of the advantages of this method is that the entire edge of the part can be shaped, since no stock has to be left uncut to run against the collar. Another advantage is that the edge of the stock can be rough, since it is not in contact with the collar.

As in the case of the jointer, stock being processed on the shaper should be held and positioned with push sticks, guards, or hold-down devices when possible. Often, special-purpose wooden guides or fences are fastened to the shaper table. At other times, the stock is held against templates or special forming jigs.

The fence on the shaper is designed like the infeed and outfeed table of the jointer, except that it is in a vertical rather than horizontal position. In concept, the shaper is much like a vertical jointer.

In production, shaping is often performed using *multiple-spindle shapers,* which have a number of spindles arranged side by side. Automated equipment that clamps the stock to the table and feeds it into the cutters is also popular.

Routing

The router has become one of the most important machines in the furniture-making industry because of its great versatility. Routers can be mounted on a table or may be hand-held. Large **production routers** look much like vertical milling machines or drill presses, with the cutter carried on a spindle moving down into the work. Production routers may have motors capable of rotating spindles at more than 50,000 rpm.

The router is used to cut designs and small moldings, to round interior and exterior edges on stock, and to cut chair panels. It can produce intricate carvings, such as rosettes and overlays, and it can even be used to produce spiral flutings and rope moldings on lathework.

When a *hand-held router* is used, it is held securely in both hands, with one hand grasping a handle on each side of the tool. Then it is lowered down into the work. The best cut is made by using a fairly rapid feed rate and several shallow cuts. Usually a template or guide is used to produce the desired pattern. A guide on the end of the cutting tool helps to guide the cutter around the template.

Stationary routers are used to make grooves and cut irregular shapes. A pin guide mounted in the table serves as a guide. The template runs against the pin to shape pieces as desired.

Routers are popular in the furniture-making and boatbuilding industries. Computer numerical control (CNC) routers are often used to cut patterns and designs when a number of identical parts are required. Figure 22-10 shows an operator loading an uncut piece of wood onto the router table while the routing process is carried out on two other pieces. A computer controller

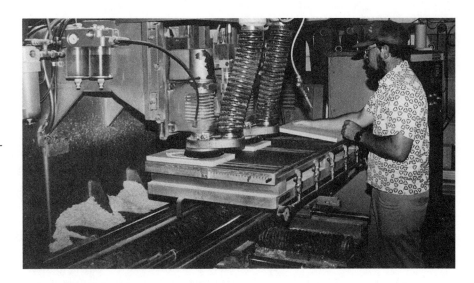

Figure 22-10
This CNC router can complete two workpieces at the same time with great precision because of the program stored in the computer controller. The operator is loading an uncut piece of wood on a transfer table in preparation for the next cycle. (Stanton Manufacturing Co., Inc. Photo by Matt Bentz.)

is located to the worker's right. The program for cutting the desired pattern is stored on a computer disk until it is needed. Then, the controller sends the electronic signals to the router to make the desired cuts.

In industry, it is common practice to use two or more routers attached to the same control system so they will perform identical movements on workpieces. Figure 22-11 shows an operator waiting for a pair of CNC routers to finish cutting parts. The vacuum hoses connected to each router head are used to draw sawdust away from the router bits, keeping the cutting paths clear.

Figure 22-11
To keep the router path clear, a vacuum dust collection system is attached to each of the router heads of this CNC tandem routing machine. The two hoses carry sawdust away to a central collection point. Note the two pieces of uncut stock on the transfer table at left, ready to move into place beneath the routers. (Stanton Manufacturing Co., Inc. Photo by Matt Bentz.)

Turning

Turning of cylindrical parts is done on a wood lathe. The workpiece is either held between centers or is attached to a faceplate mounted on the spindle of the lathe's headstock. Turning between centers is referred to as *spindle turning*, Figure 22-12.

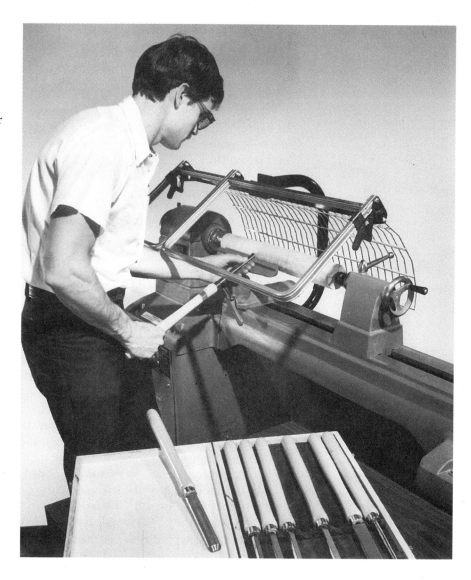

Figure 22-12
Spindle turning on a lathe is done by rotating the workpiece between centers and using a chisel-type tool to scrape material from its surface until the desired contour is achieved. (Delta International Machinery Corp.)

Spindle Turning

In the photograph, the operator is removing stock using a lathe tool called a chisel. Chisels come in six different shapes: gouge, skew, parting tool, diamond point, round nose, and square nose. The *gouge* is a round-nosed, cupped tool for roughing and making cove cuts. The *skew* is a flat tool that is used to smooth cylinders and cut shoulders. The ***parting tool*** is used to separate material and to cut off stock. The ***diamond point, round nose*** and ***square nose*** tools are used to produce specialized contours.

Lathes are sized by the swing and length of the bed. The swing is the dimension that is equal to the largest diameter piece that can be turned on the lathe. Swing is equal to twice the distance from the center of the spindle to the bed.

Major parts of the wood lathe are the headstock, tailstock, bed, dead and spur centers, and tool rest. A wooden workpiece would be secured between centers for turning as follows:

First, the center of each end of the stock to be turned must be determined by drawing intersecting diagonals from the corners. The point where the lines intersect is where the point of the tailstock center is placed to secure the part for turning. The tailstock is loosened and moved to the right to permit insertion of the workpiece, and the stock is tapped against the spur center in the headstock. The tailstock is then moved up against the stock until the point of the dead center can make contact with the point of intersection on the right end of the stock. The tailstock is then locked in place and the center turned snugly into the stock. Finally, the tool rest is moved to the area where stock removal will begin, and is raised to a height of about 1/8 in. (3 mm) above the workpiece center.

Once the workpiece is secured in the lathe, it is time to perform the initial turning (*roughing*) operation. It is advisable to use slow speeds of 1000 rpm or less to perform roughing operations. When roughing, the gouge would be held firmly on the tool holder and pushed into the work. A cut would be taken down the length of the tool holder. During turning, the tools will get hot from friction. They should be cooled in water to prevent destroying their temper and also to keep them from burning the wood.

Facing

Facing is also called ***faceplate turning***. Facing of wood is normally done on a wood lathe, with the stock screwed directly to the faceplate. Sometimes the stock is glued to a backing block, which is then screwed to the faceplate. This eliminates the problem of holes showing in the work. When the facing operation is completed, the part can be separated from the backing block with a wood chisel.

Faceplate turning can be done using the diamond point, round nose, or square nose tools. Lathe tools are positioned using the tool rest. The rest should be raised slightly above center of the workpiece, and aligned parallel to the surface to be turned. The faceplate is screwed onto the lathe spindle; turning is done using a slow speed. A scraping, rather than cutting, action is used. The gouge should never be used for faceplate turning.

Sawing

There are many different types of machine- and hand-powered sawing operations. All are chip-producing separating processes. Chips are produced when the teeth of the saw blade cut into the work. The tooth shape and the type of blades vary from saw to saw, but the principle of sawing is always the same. The teeth cut into the stock, with the blade moving (turning or being pushed or pulled) in the direction that the teeth are pointing. The action of the teeth digging into the stock helps to push the material being sawed down against the table of the saw. Sometimes, stock is fed into the blade, but in other cases the blade is pulled into the stock.

Saw cuts made across the grain are referred to as *crosscut sawing.* Cuts made in the direction of the grain are referred to as *rip sawing.* Saw blades can be purchased with either crosscut or rip teeth. Crosscut saw teeth are more "v-shaped," while teeth used for ripping are chisel-shaped. Circular power saw blades are also available with teeth that can be used for both types of cutting. They are called *combination blades.*

Circular saw blades are used for most of the machine-powered saws: portable circular saws, radial arm saws, cutoff saws, and panel saws. Blades for bandsaws are constructed as a continuous loop, with the teeth formed on the edge of the band. Other blades for power sawing come in short lengths that are held in tension vertically. These blades are used for processes such as scroll sawing.

The major sawing processes are: scroll sawing, bandsawing, circular sawing, radial arm sawing, cutoff sawing, and panel sawing. Each of these will be discussed below.

Scroll Saws

Scroll saws are often referred to as "jig saws." They are heavily used by patternmakers to produce intricate cuts within the inside dimensions of a workpiece. See Figure 22-13. Scroll saws can also produce sharp radii and fancy designs of the outside edges of the material being cut. However, this type of work can be accomplished with other machines. Cutting irregular curves inside of the stock is an operation unique to this machine. When many ornate shapes need to be cut, scroll sawing is often an ideal choice.

Blades used for scroll sawing are sized in terms of the number of *teeth per inch.* The finer-toothed blades (those with more teeth per inch) can be used to cut sharper curves. Harder material such as aluminum, copper, and plastics can also be cut with scroll saws, but fine-toothed blades and slower speeds are necessary. The blades are about 6 in. to 8 in. (15 cm to 20 cm) long, and are held in a vertical position with spring tension between upper and lower chucks. It is the up-and-down movement of the lower chuck that moves the blade.

The scroll saw can also be used to produce angled, rather than vertical, cuts. This is accomplished by tilting the table supporting the stock. Typically, tables will tilt up to 15 degrees to the left or 45 degrees to the right.

As noted, the scroll saw is ideal for making inside cuts. These cuts can be accomplished without producing an undesired saw cut across the pattern to get to the inside. To do so, drill a relief hole, slightly larger than the width of the blade, in an area where stock will be removed. Then, loosen the blade from the upper chuck of the saw. Move the lower chuck down by rolling the motor pulley by hand. The stock can then be placed on the saw table with the blade projecting through the drilled hole. Fasten the blade in the upper chuck again, and proceed with the inside cut.

Figure 22-13
To cut intricate shapes on the inside of a workpiece, such as the numerals being cut out of a thick block of wood in this photo, the scroll saw is the preferred power tool. (Delta International Machinery Corp.)

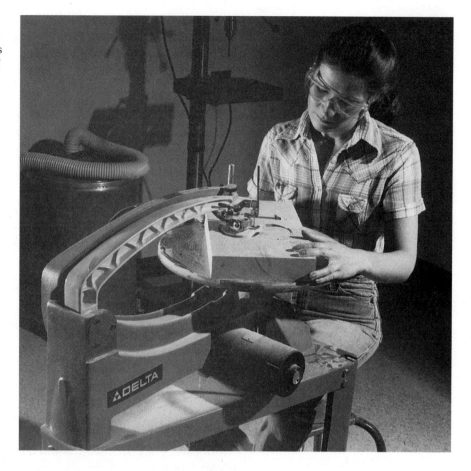

While the scroll saw is fine for making small-radius turns and cutting intricate designs, sharp turns must be made slowly or the blade will break. One disadvantage of the scroll saw is that it is not very useful for cutting long straight lines. Because of the very thin blade, it is difficult to cut long lines which are perfectly straight and do not waver from one side to the other.

Bandsaws

Bandsawing is a better process than scroll sawing for cutting long straight lines and larger-diameter arcs and circles. The *bandsaw* is used primarily for cutting curved edges. The bandsaw is unable to compete with the scroll saw, however, when sharp radii must be cut.

The blade used in a bandsaw is a *band* (continuous loop) with teeth cut into its edge. The blade will last a long time without needing to be replaced if it is operated at a speed appropriate for the material being cut. If arcs are attempted that are so sharp that they cause the blade to bind, it may break. If this occurs, the ends of the broken blade can be rejoined by welding. In some cases, a butt welding fixture is attached directly to the bandsaw.

The blade is held vertically in tension between two rotating wheels. See Figure 22-14. The upper and lower wheels can be seen in the photograph, covered by semi-circular guards, above and below the workpiece.

Figure 22-14
A bandsaw is useful for cutting circles and arcs with a fairly large radius. The continuous-loop sawblade is carried on large wheels inside the rounded safety covers at top and bottom of the machine. (Delta International Machinery Corp.)

Bandsaws are sized by the diameter of the wheels. The smallest machines are normally 10 inches, while others have wheels as large as seven or eight feet in diameter. These very large saws are used in sawmills to cut logs into timber.

As was the case with the scroll saw, angled cuts can be made by changing the angle of the table. Straight cuts are made using a fence for accuracy. On *long* straight cuts, however, the blade sometimes will pull slightly to the left or right. This is referred to as **lead,** and is usually caused by improper tracking of the blade on the wheels. Since it is difficult to avoid this problem, many operators prefer making long cuts using a circular saw. The bandsaw works best for cutting angles and large curves.

It is important to remember that the capability of every saw is limited by the shape and width of its blade. In bandsawing, the blade used is normally from 3/8 in. (9.5 mm) up to several inches in width. A circle with a 2 in. (5 cm) diameter can easily be cut with a 3/8 in. (9.5 mm) blade. However, if you try to cut a 1/4 in. (6.3 mm) radius, the blade will not be able to turn this sharply. It will bind and may break. In order to cut such a sharp radius, the cut either must be broken up into a series of tangential cuts that gradually work toward accomplishing the sharp radius desired, or a narrower width blade must be selected.

Circular Saws

When long, straight, or angular cuts are desired, the circular saw is normally the preferred woodworking machine. Circular sawing is a very versatile process. Most wood manufacturing plants have at least one machine.

However, the circular saw cannot do everything. For example, circles cannot be made with this saw without a special fixture. It is limited to making straight cuts. This includes ripping, cutting off stock, making dados and miters, and producing grooves. Despite this limitation, the circular saw is still the most important machine for separating wood material.

There are many different types of circular saws. The common table saw will be discussed first, since it will convey the basic process involved in circular sawing. Other types of circular sawing are accomplished with machines such as radial arm saws, cutoff saws, and panel saws. Portable circular saws are extensively used in the construction industry. See Figure 22-15. Another type of circular saw commonly found on construction sites is the *cutoff saw* Figure 22-16. Cutoff saws are mounted on a pivot and are popular for cutting angles on framing lumber.

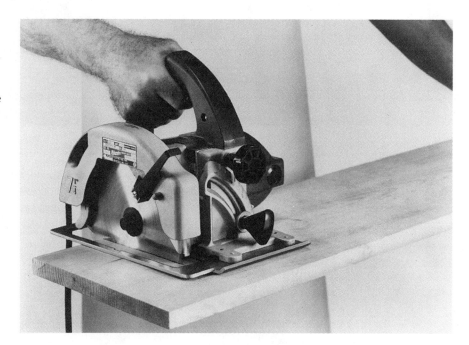

Figure 22-15
Hand-held portable electric circular saws are widely used in the construction industry because of their ease of handling on the job site. Saws are available in several different blade-size configurations, with the 7 1/4 in. size shown here among the most widely used. (Milwaukee Electric Tool)

In many woodworking operations, the table saw is a vital piece of equipment. With the table saw, the circular saw blade is carried on a horizontal shaft, called an *arbor,* located beneath the center of the table. Single-arbor saws are most common, and are often called *variety saws*, because they can perform a number of tasks. See Figure 22-17. Saws with two arbors (one with a ripping blade mounted on it, the other with a crosscut blade) are called *universal saws.*

Figure 22-16
The cutoff saw is used on many construction sites to quickly and accurately trim framing lumber to size. For rafters and similar members, precise angles can be set and cut repeatedly. (Black & Decker)

Circular saws are sized by the diameter of the blade. Common sizes for circular saws range from 7 1/4 in. (18.4 cm) portable units and 8 in. (20.3 cm) table-top models to production machines with blades 16 in. (40.6 cm) or more in diameter.

Before using the circular saw, the operator must make a decision about the type of cut. If the cut involves ripping the stock (reducing the width of the board), this is normally accomplished using a fence. In Figure 22-17, the fence can be seen to the operator's right, with the board snugged up securely against it.

When setting up the machine, the blade is raised to the level necessary to produce the desired cut by turning the hand wheel. If the cut is to be made completely through the board, the blade should be raised to a height just above the top of the stock when it is resting on the table.

The fence is adjusted by releasing a locking lever, then sliding it toward or away from the blade. If the fence is moved closer to the blade, a narrower width cut will be made; if moved away, a wider cut will result. Once the fence is positioned for the proper width of cut, the lever is used to lock it securely in place.

The stock is fed by hand into the blade. During ripping, the stock is pushed against the fence, and forward into the blade. If the stock is too narrow to be safely guided through by hand, a push stick should be used to keep the operator's hands away from the revolving blade.

When using the table saw to make cuts across the grain, a different technique is required. First, the fence is removed or moved out of the work area. A miter gage is then placed in a slot in the table. In Figure 22-17, the miter gage slot can be seen just to the left of the workpiece. The gage is used to guide the stock through the saw.

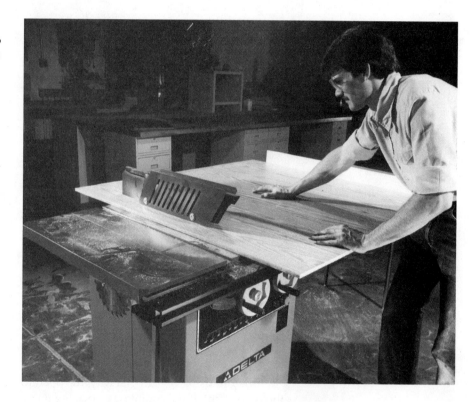

Figure 22-17
Table saws can be used to rip or crosscut stock. Many have a tilting arbor that allows making angled cuts. (Delta International Machinery Corp.).

To crosscut, the board is placed against the gage, so that it runs across the table (perpendicular to the blade). The stock is shifted left or right, so that the correct amount of material will be cut off by the blade.

Whether crosscutting or ripping, be sure to avoid standing directly behind the stock that is being removed. Sometimes *kickback* occurs, propelling loose stock violently backward. The stock being "kicked back" could cause a serious injury if it struck anyone. Kickback most often occurs when ripping. Most saws can be equipped with an ***anti-kickback device*** to eliminate this problem.

Radial Arm Saws

The radial arm saw is another type of circular sawing machine. While the table saw has its blade mounted beneath the table, the blade on the *radial arm saw* is carried on a horizontal arbor suspended on an arm above the work table. See Figure 22-18.

Figure 22-18
In addition to duplicating most of the functions of a table saw, the radial arm saw can also be used as a sander or grinder by replacing the blade with the appropriate accessory. (Black & Decker)

Radial arm saws are normally used for crosscutting, particularly trimming boards to length. Most of the cutting processes performed on the table saw can also be done on the radial arm saw. The radial arm saw is particularly handy when cutting long stock that would be unwieldy to work with on a table saw.

Another interesting feature of the radial arm saw is its capability of being used for purposes other than sawing. The blade can be removed and replaced with a sanding disc or an abrasive wheel to allow use as a sander or a grinder.

When cutting on the radial arm saw, the stock is held securely against a stop-type fence located to the rear of the saw table. The saw head is then pulled across the board toward the operator. Radial arm saws are classified by the size of the saw blade and the horsepower rating of the motor. Radial saws normally are from 8 in. to 14 in. (20.3 cm to 35.5 cm) in diameter, with 10 in. (25.4 cm) the most common size.

Miter Saws

The *miter saw* is similar to the cutoff saw, but has the capability of cutting stock (especially trim and molding) at an angle from the vertical, as well as moving through a 90-degree arc horizontally. See Figure 22-19. The miter saw is attached to a spring-loaded arm, Figure 22-20, which raises the saw off the work until it is pulled down by the operator. Most cutoff and miter saws use 10 in. (25.4 cm) blades; some use 8 1/2 in. (21.6 cm) blades.

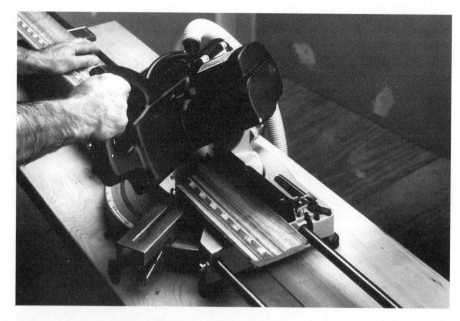

Fig 22-19
The miter saw, or power miter box, can be used to cut trim and molding stock at compound (both horizontal and vertical) angles, as shown here. (Black & Decker)

Figure 22-20
A spring-loaded arm keeps the miter saw off the work until it is pulled down by the operator. This saw is equipped with a dust collection system. (Black & Decker)

Panel Saws

Sometimes large sheets of plywood or other polymeric material must be cut into sections. While this can be done on the table saw, it is often easier to use another type of saw called a *panel saw*. See Figure 22-21. The panel saw consists of a circular saw attached to steel rails that are suspended over a rack that supports the material to be cut. The saw slides up and down easily on the rails, and can be turned to cut the sheet either vertically or horizontally.

Figure 22-21
Panel saws can quickly and accurately cut large sheets of plywood or similar materials into smaller pieces. The saw can be rotated to cut either horizontally or vertically. (Black & Decker)

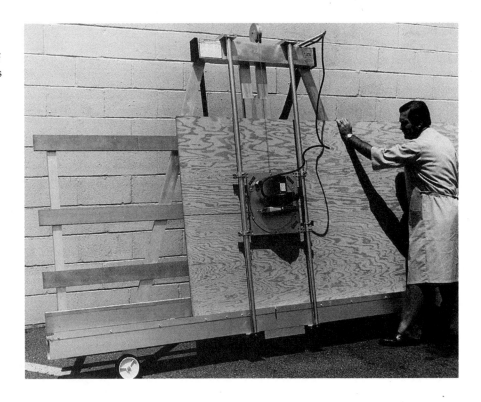

Drilling

Most readers will be familiar with the process of drilling holes using hand drills. However, there are many different ways to make holes in wood.

Machine drilling, or boring, of holes in wood is normally done with a drill press or a boring machine. The tools that are used to produce holes are called *bits*. They are held in place in the drill press by clamping them in a chuck or by inserting them in a tapered sleeve.

In addition to common drill bits (usually referred to as "twist drills"), other cutting tools are available, such as plug cutters, Forstner bits, hole saws, auger bits, and spur machine bits. Holes 1/4 in. (6.3 mm) or smaller in diameter are normally drilled with a hand-held power drill, a manually operated rotary drill, or a push drill. See Figure 22-22. Any of these tools will drill small holes quickly. Larger holes are usually made with a drill press, Figure 22-23. The operator in the photo is checking the positioning of the workpiece before starting to drill the hole. Note the large bit, which has a tapered shank that is inserted in a sleeve attached to the drill press spindle. To ensure accurate and safe operation, the workpiece is held securely in a fixture bolted to the table of the drill press.

Boring

In the furniture-making industry, holes must be bored for inserting *dowels* (wooden connecting pegs) in parts such as drawers or shelves. In such applications, perfect alignment is necessary. This is normally accomplished

Figure 22-22
Smaller holes are easily drilled with hand-held manual drills or power drills such as this one. A drill press or boring machine is normally used for larger holes. (Bosch Power Tools)

Figure 22-23
A drill press is used with larger bits. This operator is checking for proper positioning of the drill over a workpiece that is held in the fixture bolted to the drill press table. (Delta International Machinery Corp.)

using a boring machine. While boring machines look something like conventional drill presses, they are much heavier and often have several spindles. This enables more than one hole to be bored at a time.

Sometimes boring machines have ten or more spindles, each carrying a different cutting tool. Cutting bits are clamped in place using various methods, including a set screw attached to the spindle, tapered shanks, or threaded shanks. Taper-shank bits are popular.

With some single-spindle boring machines, the table moves upward toward the tool. In other instances, the table is stationary and the spindle moves toward it. Often, the travel of the tool is regulated by a foot pedal, as is the case with mortisers and punch presses.

Multi-spindle boring machines are normally available in sizes up to 6 feet long. There may be as many as twenty spindles on large production machines.

Mortising and Tenoning

Sometimes, an odd-shaped joint called a *mortise-and-tenon joint* is required to secure drawer components or other pieces of fine furniture. The vertical mortising machine cuts the rectangular opening in wood for a mortise-and-tenon joint.

Mortising machines

Only the *mortise* is cut on the mortising machine. The mortising machine looks like a large vertical drill press or boring machine. The major difference is that the stock is advanced into the chisel, rather than the chisel being advanced into the work. The hollow mortising chisel is carried in a vertical spindle, somewhat like the drill press. The outside rectangular shape of the mortise is cut with the chisel, while the inside is cleaned out with the mortising bit. The bit turns inside of the chisel, making the hole and removing chips. The chips travel up the spirals of the bit and flow away from the bit through an opening in the front of the chisel.

To cut a mortise, the mortising chisel and bit is placed in a bushing, then locked in place in the spindle with set screws. The stock is clamped to the table, and a mark showing where the first mortise is to be cut is lined up properly beneath the tool. After the depth of cut is set, the machine is turned on. When the operator depresses a foot pedal, the table moves up, and the mortising tool immediately cuts the mortise. When the pedal is released, the table retracts. The process then can be repeated.

Oscillating-chisel mortising machines are available in single or multiple head models. Some machines can operate either vertically or horizontally. Multiple head machines can be equipped with as many as 20 mortising heads.

Tenoning machines

The square mortise would be of little value without a machine to cut a tenon. The *tenon* is cut on a tenoning machine, often called a "tenoner." Single-end and double-end tenoners are used in industry to cut tenons for products such as window sash rails. They are also used to shape shoulders, and to cut corner joints for cabinets.

A typical setup with a single-end tenoning machine might consist of several tenoning heads, coping heads, a cutoff saw, and a movable carriage. When using a machine such as this, the operator would clamp the stock to the carriage. The role of the carriage is to transport the stock to the tenoning heads. The tenoning heads make the face and shoulder cuts. The stock would then be moved on to the coping heads, which would cut contours on the shoulders. The stock would then be carried on to the cutoff saw, which trims the tenon to the proper length.

Important Terms

- abrasive belt planer
- anti-kickback device
- arbor
- band
- bandsaw
- bed width
- bits
- combination blades
- crosscut sawing
- cutoff saw
- diamond point
- double planers
- dowels
- faceplate turning
- fence
- gouge
- hand-held router
- jointer
- kickback
- lead
- machine planing
- miter saw
- mortise
- mortise-and-tenon joint
- multiple-spindle shapers
- panel saw
- parting tool
- production router
- push stick
- radial arm saw
- rip sawing
- roughing
- round nose
- scroll saws
- single planers
- single-spindle shaper
- skew
- spindle
- spindle turning
- square nose
- stationary routers
- teeth per inch
- tenon
- true-up
- turning
- universal saw
- variety saw

Questions for Review and Discussion

1. What would be the advantage of using a mortise-and-tenon joint, rather than a right-angle butt joint?
2. How would you go about removing the warp in a 1 in. x 10 in. (2.5 cm by 25.4 cm) board?
3. What is the difference between a knife-blade planer and a jointer?
4. What causes *kickback* on a circular saw? How can it be avoided?

23 Fabricating Wood Materials

Key Concepts

△ Fabrication of wood materials may involve the use of interference fits, mechanical fasteners, or adhesives.

△ Nails are classified primarily by length; screws and bolts by head shape, length, diameter, and in some cases, type of thread.

△ Glued joints can fail through weakness of the adhesive itself (cohesive failure) or weakness of the substrate (substrate failure).

△ Adhesives may be thermoplastic (softening on exposure to heat) or thermoset (permanently set by a chemical reaction) in nature.

△ New wood-joining technologies include the use of high-frequency gluing and VHB (very high bond) tape.

The process of fabricating or joining wood parts is accomplished by using one of three basic methods: *interference fits* between parts designed to lock together, mechanical fasteners, and adhesives. Interference fits are not widely used in manufactured wood products; most fabrication is done with mechanical fasteners or adhesives, which lend themselves more readily to volume production.

There are numerous varieties of fasteners used in industry today. Many of these, such as screws or bolts and nuts, have threads. Other types use various mechanical methods to permit assembly (hook-and-loop materials, nails, and blind rivets).

What all fasteners have in common is the fact that they must join the materials and they must carry a load. Some types of fasteners (such as bolts and nuts) can be assembled, then later disassembled if desired. Other types of fasteners, such as blind rivets, cannot be disassembled after they are joined without damaging either the material being joined or the fastener itself. As noted, there are many different types of mechanical fasteners; entire volumes are required to do justice to all the variations. For this reason, only the major types of fasteners used with wood materials will be covered here. The same is true of adhesives: only the types most often used in specialized joints to permanently join wood parts together will be covered.

Mechanical Fasteners

Screws and nails are the most common types of fasteners used to join wood parts. Selection of the correct fastener depends on the type of material being joined, and the desired strength of the joint. Screws provide better holding power than nails, and are easier to remove if disassembly is necessary. However, in most cases, they take more time to apply than nails.

One of the first places where design engineers seem to focus their attention is on the type of threaded fasteners that are to be used to assemble a product. Threaded fasteners are preferred in cases where interchangeability and large clamping force in a small area is necessary. Disassembly is a major advantage of using threaded fasteners—many products have to be made so that they can be disassembled easily.

Wood Screws

Selection of a particular type of wood screw as a fastener sometimes depends on the type of tool that is available to drive the screw. Screws are classified according to the type of opening in their head, Figure 23-1. This opening (slot, hex key, Phillips, square, etc.) naturally dictates the type of device that can be used to drive the screw. A recessed hex key opening in a screw head has more contact with the driving tool and is desired in applications where there is heavy loading. If the equipment available is a power driver with a Phillips head, it will be of little value if the fastener has a lobular (Torx®) head design. The design of the head (flat, round, pan, hex, etc.) must also be matched to the job. Figure 23-2 shows some of the most popular types of screw head designs.

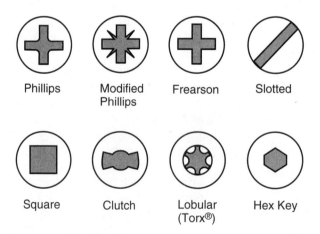

Figure 23-1
The design of the screw head selected may be influenced by the equipment available to install the fastener.

Screws may be driven (installed) with a hand or powered screw driver. Tools are also available which feed screws automatically, Figure 23-3. The screw attachment shown in the illustration turns a standard drill into a self-feeding production screwdriver.

Figure 23-2
There are a limited number of major screw head styles. Selection of the style to use is normally determined by the application.

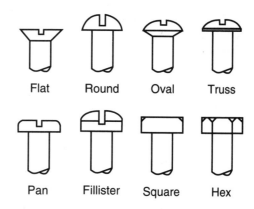

Figure 23-3
This automatic-feed screw-driving attachment can be used with a conventional portable electric drill to speed production. (Duo-Fast Corporation)

Figure 23-4 shows some of the available types of self-drilling Phillips head screws that are designed for power driving in wood and other materials. Holes do not have to be drilled in the wood for these type of screws. Note the different types of threads. The screw on the left has fine threads, all the other have coarse threads. Those intended for softer material (drywall or gypsum board) have threads the entire length of the shank; the screw for use in harder material (wood) has a shank that is only threaded for half its length. The design of the screw and screw threads influences the holding power it will exhibit when driven into the material. Fine threads, for example, would provide greater holding power, because there are more of them per inch of screw length.

Screws are also classified by their length, diameter, surface finish, and composition. Selection of the proper screw for the job starts with determining the length needed. The first consideration is the thickness of the materials to

Figure 23-4. A selection of Phillips-head screws designed to be driven by power equipment. (Duo-Fast Corporation)

Screw Specifications*

Stock No.	S-102	S-105	S-109G	S-108	S-112	S-110
Length	1 1/4"	1 1/4"	1 1/4"	1 5/8"	1 5/8"	2"
Thread Type	Fine	Coarse	Coarse	Coarse	Coarse	Coarse
Size No.	#6	#6	#6	#6	#6	#6
Head Type	Bugle	Bugle	Bugle	Bugle	Bugle	Bugle
Recess	Phillips	Phillips	Phillips	Phillips	Phillips	Phillips
Point	Pierc.	Pierc.	Pierc.	Pierc.	Pierc.	Pierc.
Materials	Steel	Steel	Steel	Steel	Steel	Steel
Finish	Black	Black	Zinc	Black	Black	Black
Per Box	6,300	6,300	6,300	5,400	5,400	4,500

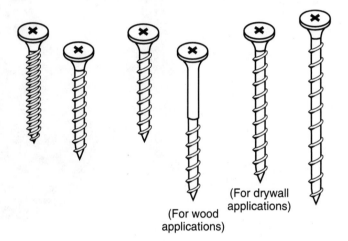

(For wood applications)

(For drywall applications)

* The standard screws listed above comply with ASTM Specification C1002.

be joined. The screw has to be shorter than this dimension. If two pieces are to be joined, a good rule of thumb would be for the screw to go through the first piece and project at least half way into the second, or base piece. To achieve maximum holding strength, the entire threaded length of the screw should enter the second piece of wood. This is not always possible with very thin stock, of course.

Wood screws are typically purchased by the box. When the desired length is known, the next consideration is the *gage number,* or diameter, of the screw. Wood screws can be purchased in lengths from 1/4 in. (6.3 mm) to 6 in. (15.2 cm) and in gages from 0 to 24. The diameter of the screw increases with larger gage numbers.

After you know the length needed and the gage number, it is time to choose the head design and material desired. Let's put it all together. If flathead screws were needed for an inside building job, and the length of the screws had to be 1 1/2 in. (38 mm), then the order might look something like the following: "(5) boxes of 1 1/2 x No. 8 flat head steel screws." If the project were outdoors, you would more likely select screws made of brass, aluminum, or stainless steel so that they would not rust.

When fastening wood with screws, it is often advisable to drill two holes. The first is called the *pilot hole* or anchor hole. It should be slightly smaller in diameter than the largest diameter of the threads. The pilot hole is drilled through the first piece of stock and about halfway into the second. Next, the pilot hole through the first piece of stock is enlarged with a drill of the same diameter as the screw shank (the section above the threads). This hole is called the *shank hole.* The two holes will enable easy insertion of the screw through the first piece of stock and secure threading into the second.

When fastening softwoods together, the pilot holes can be a bit narrower than for hardwoods. However, care must be taken to not *over-drive* (excessively tighten) the screw into the hole. This could destroy the wood in the thread area or split the stock. Proper drilling is even more critical with hardwoods, which are more likely to split.

Nails

Nailed joints are often used in industry where the appearance of nails in the joint is not a detriment to the quality of the product. Typically, nails are used for products such as packing boxes, crates, pallets, and frames. Nails with small heads (finish nails) are used as fasteners in many fine furniture applications. These nails are recessed below the surface with a *nail set,* a type of punch. The holes are hidden by filling them with a material stained to match the surrounding wood.

Nails are normally driven by hand using a hammer, or with a powered automatic nailing machine. There are five major classes of nails: common, box, casing, finish, and brad. See Figure 23-5.

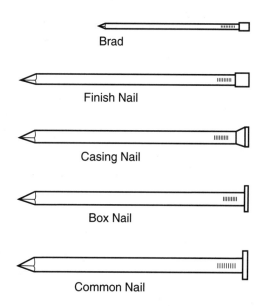

Figure 23-5
These are the major types of nails used to fasten wood. Note the differences in head shape—the brad and the casing and finish nails are designed to have their heads driven below the wood surface for a more finished appearance.

Nails are typically available in lengths from 1 in. to 6 in. (2.5 cm to 15.2 cm), but longer ones are sometimes used. Nails are most often made from mild steel or, sometimes, aluminum. Some nails have a galvanized coating for exterior use. Others are coated with zinc, cement, or resin for special applications. Hardened nails are available for driving into concrete, cement block, and brick. Nails are even made with threaded shanks for special holding applications, such as flooring underlayment or drywall.

The basic unit of classification for nails is called the *penny,* (abbreviated **d**). A 6 penny (6d) nail is 2 in. (5.8 cm) long. A 2 penny (2d) nail is 1 in. (2.5 cm) long. All types of nails are sold by the pound.

Box nails smaller than 1 in. (2.5 cm) in length, and all brads, are sized by their length and gauge number, rather than a penny designation. These small nails are available in gauges from 12 to 20, with the smallest number being the largest diameter.

A rule of thumb for selecting nails to join wood is that the nail should extend into the second sheet for a distance equal to twice the thickness of the first sheet. Larger heads should be chosen for applications in softwoods or drywall.

Common nails are driven with a claw hammer or power nailer. Brads are normally driven using a smaller tack hammer. The power nailer feeds nails in clips, and is often used for construction tasks such as nailing framing members. A similar power tool is used to drive staples when applying roof shingles, Figure 23-6.

Figure 23-6
Power nailers and staplers apply the correct amount of force to drive the fastener with minimal effort by the operator. Such devices make fastening work faster and more efficient.

Blind Rivets

Blind riveting is a unique process that can be used to fasten just about any type of materials, and is worthy of special treatment here. The *blind rivet* is also known as a *pin rivet.* It is often used in areas where there is not sufficient room to make use of a conventional rivet that must be compressed by pounding or hammering.

The blind rivet consists of a *mandrel* (shaft) and rivet head assembled as one unit. The rivet is placed in the hole drilled through the materials to be joined. The rivet is clinched by using a rivet tool to pull the mandrel, forming a head on the back side of the material. The mandrel can then be snapped off for a finished appearance. See Figure 23-7.

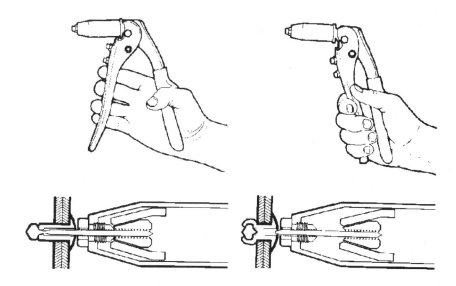

Figure 23-7
Blind rivets are useful in many fastening situations. After being inserted through the materials to be fastened, left, the mandrel is pulled outward by the rivet tool. This forms the second head on the rivet, right. The mandrel can then be snapped off. (Malco Products, Inc.)

Other Mechanical Fasteners

In addition to nails, screws, and blind rivets, other mechanical fasteners such as tee nuts and nuts and bolts are popular for fabricating wood materials. Special fasteners with threads on both ends are also used to join wood parts. This type of fastener has machine threads for accepting a nut on one end and wood screw threads on the other so that it can be screwed into the part.

Adhesives

Sometimes, it is desirable to fabricate wood materials without mechanical fasteners. With the appropriate adhesive and proper surface preparation, it is actually possible to create a stronger bond than would be possible using a mechanical fastener.

An *adhesive* is a material that adheres, or sticks to, another material. To better understand what can be accomplished with adhesives, think for a moment about the basic structure of wood. With the aid of a powerful microscope, the wood surface would appear much like a sponge cut in cross-section. (Recall the cross-sectional drawings of softwoods and hardwoods in Chapter 20.)

When you use a mechanical fastener, the raised surfaces of the wood provide secure contact when held against each other. However, the open pores provide little support to the interface area between the fabricated materials. Since adhesives are normally used in liquid form, they not only coat the raised surfaces but also flow into the depressions. This enables adhesives to provide a more complete bonding surface. However, there is much more to the process of fabrication than just bonding materials in the area where they come into contact with each other.

The linkage between surfaces in the interface area is only one type of bonding which occurs when materials are fabricated. This is actually the *poorest* type of bonding: mechanical adhesion through the film of glue accounts for only about 10 to 20 percent of the holding force of most glued materials. Most of the bonding strength occurs as a result of chemical reactions between the adhesive and the wood.

Gluing failure in the interface area is referred to as ***adhesive failure;*** if the failure occurs in the adhesive itself, it is called ***cohesive failure.*** The bond could also fail because of substrate failure. ***Substrate failure*** happens when both the cohesive strength and the strength of the adhesive bond between the adhesive and substrate exceeds the strength of the substrate. Substrate failure results in the fracture, or destruction, of the substrate.

There are many ways to classify the available types of adhesives. One of the most useful ways is to view them in terms of the type of reaction they exhibit during the bonding process. This is what makes certain types of adhesives stick to some materials and not to others. There are three types of chemical actions through which an adhesive *sets* (becomes hard): through the cooling of a molten liquid, by releasing a solvent, or by polymerization.

Hot Melt Adhesives

Adhesives that must be heated to a molten state to make them liquid, then return to a solid state as they cool, are called ***hot melt adhesives.*** These adhesives, often referred to as "hot melts," are thermoplastic solids that can be purchased in stick or pellet form. They are used with a special applicator known as a hot melt glue gun.

Hot melt applications are completed by applying the melted glue on one surface, and then placing that surface in contact with the other surface to be bonded. An example of this type of application would be to join one flap of a box to another flap.

Hot melt bonding normally takes several seconds. In some cases, the material is held in a fixture to provide proper contact during cooling.

Hot melts are popular for edge-banding laminates in furniture construction, and in all sorts of packaging applications. Hot melt adhesives are clean and easy to use, and they can be employed to fabricate almost any material.

The major disadvantage of hot melts is that they are not as strong as many other adhesives. However, for packaging and other sealing applications, this type of adhesive is normally the preferred choice.

Solvent-releasing Adhesives

Solvent-releasing adhesives give off a liquid as they cure. Sometimes, the liquid is water; in other cases, it is a solvent that is compatible with solid ingredients in the adhesive.

Solvent-releasing adhesives include polyvinyl, hide, and casein glues. Solvent-releasing adhesives are used with porous woods, which will absorb the solvent. Polyvinyl (white glue), aliphatic thermoplastic resin (yellow glue), and alpha cyanoacrylate ("super glue") adhesives are all solvent-releasing thermoplastic adhesives.

Other adhesives are designed to withstand a great deal of stretching. These are referred to as elastomers. Elastomers are solvent-releasing adhesives that are used in applications where low creep resistance is required, such as in the bonding of drywall to other materials, or in attaching flooring and tile. They are often referred to as *construction adhesives.* See Figure 23-8.

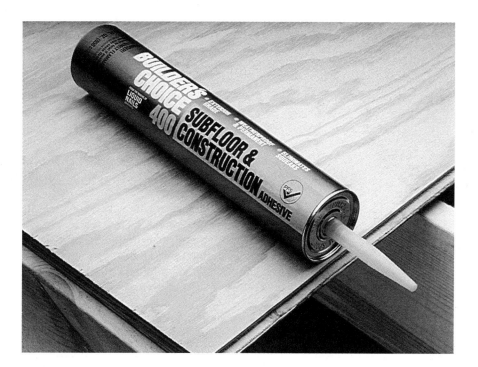

Figure 23-8
Construction adhesives are elastomer-based solvent-releasing adhesives that are widely used for fastening wood to various substrates. (Macco Adhesives)

The major component in an elastomer is natural or synthetic rubber. Rubber-based adhesives do not cure in the same way as the ureas or phenolics. With elastomers, the adhesive thickens with the loss of solvent, until it reaches a plastic state. Elastomers never become completely hard. Consequently, an elastomeric bond is not rigid and will shift under stress.

Thermoplastic adhesives can be softened by heat, and are not recommended for use in high-temperature applications. All thermoplastic adhesives acquire tack, or stickiness, when they begin to dry.

Of the three popular types of thermoplastic adhesives, aliphatic thermoplastic glue is stronger and more resistant to heat than polyvinyl glue. Alpha cyanoacrylate is stronger than either polyvinyl or aliphatic glues, and can be used to bond both porous and non-porous materials. Cyanoacrylate glue dries hard in about 24 hours, but it can be reduced to a plastic consistency through the application of acetone.

Thermoset Adhesives

One of the disadvantages of thermoplastic solvent-releasing adhesives is that they often break down in hot or wet conditions. For this reason, *thermoset adhesives* are often selected when the product application requires exposure to moisture or heat. These adhesives will not soften through exposure to heat. Curing takes place through polymerization. Once the glue has hardened, the joint is strong but brittle.

Thermoset adhesives set when a chemical reaction (polymerization) takes place between the two parts of the adhesive mixture. The ingredients vary, depending on the type of thermoset adhesive. Common types include epoxy, resorcinol formaldehyde, phenol formaldehyde, melamine formaldehyde, and urea formaldehyde.

High-frequency Electric Gluing

Thermosetting resin glues can be cured through direct contact with a heat source, or by exposing the glued material to a high frequency electric field. *Electronic gluing,* or electric welding as it is frequently called, makes it possible to create a permanent bond in seconds. Here's how it works:

When materials to be joined are placed in an electric field, a portion of the energy that is generated is changed into heat, thereby increasing the temperature of the exposed material. Increases in electrical frequency result in increased temperature in the material. Frequencies similar to that required for shortwave radio — between 10 and 40 million cycles per second — are used for high-frequency gluing. Household electricity (*alternating current*) has a frequency of 60 cycles per second.

In high-frequency gluing, most of the heat is generated in the glue area; the wood itself is not heated. Electrodes are placed parallel to the glue line for plywood, and perpendicular to the glue line for edge gluing. With high-frequency gluing, the curing time can be much shorter than would be required with contact bonding or laminating in a conventional heat press if sufficient energy is generated over a small surface area.

Many different types of fabricating can be accomplished with high-frequency gluing. A portable electronic welder is shown in Figure 23-9. Welders such as the one shown are being used daily for applications such as veneering paneling; applying laminates and banding to cabinets, furniture, sash, and doors; patternmaking; patching veneers, and bonding plastics to countertops.

A hand-held electrode gun is shown in the foreground of the photograph. This gun weighs only a little more than two pounds. All that needs to be done to produce a bond in a glue joint is to place the electrodes on the joint, and pull the trigger of the gun. It produces high-frequency energy that penetrates up to 2 inches (5 cm) of wood.

Portable machines such as the one shown are popular for small applications. In production, high-frequency batch presses are used for edge gluing. Production machines feed stock into a conveyorized glue spreader. After glue is applied and the pieces are sandwiched together, they are moved into the high-frequency welding machine, which produces the heat to set the glue permanently.

Very High Bond (VHB) Tape

One of the newest additions to the fastening field is *very high bond (VHB) tape.* The new VHB adhesives permit bonding to woods, metals, fiberglass, sealed wood, glass, stone, and plastics. VHB materials can be purchased in either conventional tape roll form, or as die-cut shapes for specialized applications.

Figure 23-9
High-frequency electronic welders generate waves that can set glues in a matter of seconds. (Workrite, Inc.)

Some VHB material can withstand temperatures up to 500°F (260°C) for limited periods of time, or 300°F (149°C) for sustained periods. VHB joining systems marketed by 3M Corporation use either adhesive transfer tape with a release liner, or an acrylic foam carrier with adhesive on both sides. Both tape and foam use an acrylic polymer adhesive that flows across the surface to which it is attached.

With VHB material, no curing period is necessary. Generally, the tape reaches maximum holding strength in about 72 hours. Residential window manufacturers are finding the VHB tape useful in producing single-unit windows with attached grilles simulating multi-pane units. The tape attaches the grilles permanently to the glass pane.

With VHB assembly, no drilling, welding or refinishing is required. Although fasteners employing VHB tape are not yet as permanent as mechanical fasteners, the technology is rapidly improving. A manufacturing test of various fasteners by a truck manufacturing firm found that VHB fasteners could actually be *more* permanent than mechanical fasteners when subjected to certain stressful conditions. A motor-driven rocker arm was set up so that it would flex a frame repeatedly at a 45-degree angle from the center line of a mounting panel. The test unit was operated outdoors with temperatures ranging from 20°F to 90°F (-7°C to 32°C). The rate of the flex stroke was 500-600 strokes per minute for 7.5 million test cycles. The test was designed to simulate the equivalent of 15 years of over-the-road driving. The test was so severe that 50 percent of the weldments, 100 percent of the rivets, and 100 percent of the industrial adhesives tested failed. The VHB adhesive passed with minimal separation, and the VHB foam tape passed with no separation.

Important Terms

adhesive
adhesive failure
blind rivet
cohesive failure
construction adhesives
electronic gluing
gage number
hot melt adhesives
interference fits
mandrel

nail set
over-drive
penny
pilot hole
shank hole
solvent-releasing adhesives
substrate failure
thermoset adhesives
very high bond tape

Questions for Review and Discussion

1. Discuss the advantages and disadvantages of each major type of mechanical fastener.
2. Why is a pilot hole recommended when drilling a hardwood?
3. Which is larger, a 6d nail or a 3d nail?
4. What are hot-melt adhesives made from? How are hot-melts applied?

Conditioning Wood Materials

Key Concepts

△ Conditioning processes are used to make wood products resistant to moisture and to temperature extremes.

△ Microwave drying (RF heating) improves the strength, impact resistance, and finished quality of wood.

△ Various processes can be used to make wood more supple and plastic-like, allowing easier forming.

△ Acetylation provides wood with increased toughness and impact resistance, while allowing it to be cut and shaped with conventional woodworking tools and techniques.

△ Injecting chemicals into wood under pressure is a method used to make wood more resistant to moisture and other conditions encountered in outdoor use.

Wood is an *anisotropic* material. This means that drastic structural changes will occur as wood absorbs moisture or as it gives up moisture (dries). The capacity to modify the internal structure of wood, through adding or removing moisture from its cells, creates an entirely new set of design requirements for products made from this material. Design factors are particularly important if the product is to be used in wet environments where it also will be exposed to hot or frigid temperatures. Often, the wood product can be conditioned to resist moisture.

In addition, conditioning and finishing processes are effective in helping to reduce the shrinkage and swelling that would occur even after conventional kiln drying.

Conditioning processes used on wood materials influence the *porosity* (moisture-absorbing capacity) of the internal structure of the material, making it more dimensionally stable. Sometimes, this is accomplished with specialized drying methods; in other cases, by modifying the structure of the wood using chemical treatment or processes such as high-pressure impregnation.

Radio-Frequency Dielectric Heating

A process that is gaining increasing use as a means of reducing the moisture content of wood is *radio-frequency (RF) dielectric heating.* This is also referred to as *microwave drying.*

In RF heating, moisture is vaporized through induced heat. One of the reasons for using RF heating is to speed the drying time of wood stock. Another is that the process improves the impact resistance, shear strength, and overall finished quality of the wood. Radio-frequency heating also prevents defects, such as end checking and distortion, that can occur with conventional drying processes. The RF process is most effective when it is used with hardwoods, such as oak, walnut, or ash.

The RF heating process uses high-frequency current (microwaves) to achieve uniform distribution of heat throughout the wood. Besides being used for drying and curing wood itself, it can be used to speed the drying of coatings, paints, and inks on wood. If excessive energy levels are used, however, RF heating can cause damage and burning of the material.

Wood Plastic Composition (WPC)

The process referred to as *wood plastic composition (WPC),* uses radioactive isotopes to transform wood into a material that is much like plastic. The material that is created is called *irradiated wood.*

To make irradiated wood, the material is placed in a vacuum chamber, and all of the air removed. A plastic monomer, normally methacrylate, is then introduced and is drawn by vacuum into the cells of the wood.

After the wood cells are filled with plastic, the workpiece is bombarded with gamma rays from radioactive isotopes. This exposure to radiation generates heat in the material, and hastens polymerization. When the workpiece cures, the wood has been transformed into a material with many of the characteristics of a plastic. See Figure 24-1.

The major advantage of WPC is that it creates workpieces that are much more resistant to moisture than conventional wood. Another important reason for selecting WPC material over conventional wood is that it has improved strength and dimensional stability.

Much of the early work on WPC was done by Dr. James Kent at West Virginia University. This research, begun in 1969, was sponsored by the Atomic Energy Commission. Researchers found that after the wood specimens were irradiated with gamma rays for twenty hours, they had greatly improved mechanical properties. After irradiation, the material was as much as eleven times harder than natural wood, seven times more abrasion-resistant, and up to six times more able to withstand compression perpendicular to the grain.

Today, there are many other types of "plastic wood." Some of the most popular variations are called impreg, compreg, and acetylated wood. These variations are discussed in the following sections.

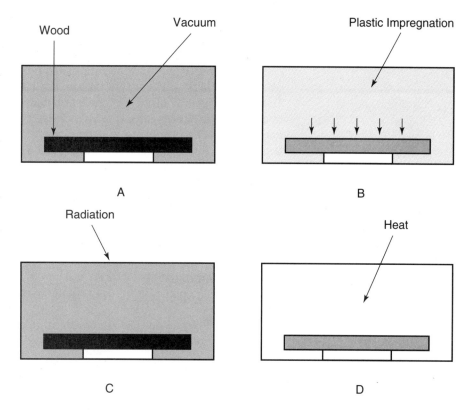

Figure 24-1
Wood is transformed into a material with many of the properties of a plastic in the wood plastic composition (WPC) process. A—A vacuum is drawn in the chamber. B—The plastic monomer is introduced and drawn into the cells of the wood. C—Gamma rays bombard the plastic-impregnated wood. D—Heat generated by the action of the gamma rays speeds up the curing process.

There are also other methods used to chemically modify wood to make it more resistant to moisture and chemical attack. For years, materials such as beeswax, tar, paraffin, or asphalt have been used to fill the pores of wood and make it more resistant to moisture and chemical attack. Even such materials as molten sulfur and inorganic salts have been used as impregnating agents. Sulfur improves the wood's resistance to chemical attack; inorganic salts reduce shrinkage as the wood dries out after being subjected to moisture.

Impreg

Another process that involves saturating the cells of wood with plastic is called *impreg*. With impreg, a fiber-penetrating thermoset resin is used. The plastic wood is then cured without compression.

The most successful use of thermosetting resins for making impreg has involved water-soluble phenol-formaldehyde resins with initially low molecular weights. Thus far, no *thermoplastic* resins have been found that can be used to saturate the cells of the wood and provide dimensional stability.

The thermoset resin penetrates the cell wall, keeping the wood in a swollen condition, then is polymerized by heat. This produces a water-insoluble resin incorporated in the cell walls of the wood stock.

Impreg is usually made using green veneers to facilitate resin pickup. The veneer is either soaked in the resin, or is impregnated with the resin under pressure. When *impregnation* is used, the workpiece is kept under pressure until the resin content equals a 25 to 35 percent gain in weight (when compared to the dry weight of the wood). If the stock is impregnated to the 35 percent level, shrinkage and swelling when exposed to moisture changes may be one-third less than wood dried by conventional methods.

When impregnation is accomplished by soaking, solution temperatures of 70°F to 100°F (21°C to 38°C) are used to facilitate resin penetration. Treatment is usually limited to stock less than 1/3 in. (8.3 mm) thick, because treating time increases rapidly with increased thicknesses. Also, thicker wood may be subject to checking and cracking when drying. Often, veneers are laminated together to make products such as handles for knives and other kitchen utensils.

Once the wood is impregnated with resin, it is dried at 175°F to 200°F (79°C to 93°C) for approximately 30 minutes. This produces a moisture content of about 10 percent. The drying process is conducted using this low temperature to keep from forcing the uncured resin to the surface of the wood.

After the treated veneer is initially dried, it is then cured by heating at high temperature, usually 310°F (154°C) for 30 minutes.

When impreg has been dried and cured, it is normally reddish-brown in color. If another color is desired, the material can be finished with most paints and varnishes.

Most adhesives can be used with impreg. However, those containing a high percentage of solvent should be allowed to evaporate somewhat before they are applied.

Compared to unmodified wood and plywood, impreg has improved properties. Swelling in water is reduced to 35 percent of that exhibited by untreated wood. This, in turn, creates a reduction in grain raising and surface checking. Impreg also exhibits improved compressive strength.

Impreg is useful when resistance to decay, rot, termites, and marine borers is required for an application. This property is probably due to the fact that the treated cell walls can't take up sufficient moisture to support biological attack.

Impreg also has a high resistance to acid and can even be used to make storage containers for batteries. It has also been used for electrical control equipment because it offers electrical resistance higher than that of untreated wood.

Despite impreg's many advantages, the product does have some limitations. While the addition of the resin increases the weight of the material, it does little to improve the wood's overall structural strength, and actually *reduces* its impact bending strength. For these reasons, it is not recommended for structural applications.

Impreg is most often used in the construction of pattern and die models, and in the manufacture of plywood paneling. A model for an automobile roof master die, made from mahogany impreg, is shown in Figure 24-2.

Compreg

Another type of plastic wood is made with a process known as ***compreg***. Compreg is similar to impreg, but the wood is compressed prior to curing. Impreg involves no compression.

Another major difference between the two methods is that while impreg is limited to use on veneers and thin stock, compreg has been successfully used to make larger solids such as water-lubricated bearings, gears,

Chapter 24 Conditioning Wood Materials

Figure 24-2
Impreg is often used to make master die models, such as this one for an automobile roof panel. (Forest Products Laboratory, USDA Forest Service).

and aircraft parts. When large solids are desired, it is often popular to sandwich treated veneers into panels and compress them without the use of bonding resins.

Water-soluble phenol-formaldehyde resin is often used for making compreg. The treated wood is compressed while the resin is curing in its cells. The percentage of resin used will vary, depending on the product application. When maximum impact strength is required, only 10-20 percent resin is used. If the product requires high dimensional stability, as much as 30 percent resin might be used.

When compreg is made from veneers, each strip of veneer is treated with water-soluble phenol-formaldehyde to a level of from 25 to 30 percent resin. The veneers are then dried at a temperature of no more than 175°F (79°C).

The next step is to compact the plies to bond the sheets together, forming the product. The pressure used depends on the resin and type of wood, but it is typically 1000 psi to 1200 psi (6895 kPa to 8274 kPa). After compression, the product is heated to produce the final cure. This is normally done at a temperature of approximately 300°F (149°C).

The compression process results in improved moldability and appearance, when compared to impreg. Compreg is used to make such products as cutlery and tool handles, model airplane propellers, and antenna masts. It is also a popular choice for constructing dies and jigs, water-lubricated bearings, pulleys, and even electrical insulators.

Case hardening, a process often used with metallic materials to produce a tough outer shell on the workpiece, can also be done with compreg. It is used to produce a tough cladding on the exterior surface of the wood by adhering compreg panels to a core of ordinary wood. An example would be compreg used as the outer ply on a plywood panel where abrasion resistance is needed. It is also used for handles of knives and other utensils.

Neither compreg nor impreg will swell and crumble like particle board or other untreated wood when used in moist environments, but both are more brittle than untreated wood.

Both compreg and impreg are hard materials because of the cured resin they contain. Consequently, conventional wood tools will dull quickly when used to cut or shape them. Metal cutting tools are recommended when machining is necessary.

Acetylated Wood

Acetylated wood was developed while searching for a modified wood that would provide the strength of compreg and impreg, but without their brittleness. Acetylation is a chemical process that is conducted in an airtight chamber.

Many different chemicals have been used to impregnate conventional wood in the acetylation process. These include the anhydrides, epoxides, isocyanates, acid chlorides, carboxylic acids, lactones, alkyl chlorides, and nitriles. The most popular chemical for making acetylated wood is acetic anhydride.

Acetylation requires the use of dry wood with less than five percent moisture content. The wood is placed in a pressure cylinder, with the temperature elevated to between 230°F and 250°F (110°C and 121°C). Acetic anhydride is introduced to the chamber to acetylate the wood. A pyridine catalyst is used to speed up the reaction. The acetylated wood process is popular for treating thin stock and veneers, and can be done on all types of hardwoods and softwoods. It reduces wood shrinkage and swelling by as much as 30 percent, compared to wood processed by conventional kiln drying.

The process also makes the wood resistant to attack by fungi, termites, and marine organisms, and causes little change in color. Once a wood has been acetylated, it is more stable in sunlight. It also will have better acoustical properties. This makes acetylated wood particularly useful for making speakers and sound systems.

Like compreg and impreg, acetylated wood creates a product with increased toughness and impact resistance. A major advantage of acetylated wood is that it is more flexible than either compreg and impreg and is much easier to machine. Unlike compreg and impreg, it can be cut and shaped with conventional woodcutting tools. Acetylated wood can be finished with the same processes used on conventional wood.

Plasticized Wood

Wood can be made supple, or *plasticized*, by exposing it to ammonia. This allows it to be more easily shaped or formed. The process is still in the early stages of commercialization, but appears to have considerable potential for the wood material processing industry.

After the material is exposed to the ammonia, it can be formed and then dried. It will keep its original properties in the new form. There is also the advantage of more rapid processing of the workpiece. The common wood-forming processes, such as laminating, soaking, or steaming, require holding the stock in a form for an extended period of time until it is dried and set. The anhydrous ammonia process requires keeping the wood in a form for only a brief period of time, and springback is minimal.

Research has demonstrated that wood can be plasticized by using anhydrous ammonia in either the liquid or gaseous state. With the liquid ammonia process, the wood stock is soaked in a tank for a period of time (normally from one to three hours for 3/4″ stock). The process makes the wood so flexible that small strips of it can even be tied in knots. There are some limitations to the process, however. The anhydrous ammonia that is used must be kept below -27°F (-33°C), so that it remains in a liquid state. The wood that is to be plasticized must be pre-cooled before treatment.

The gaseous ammonia process also begins with cooling the workpiece. It is then placed in an airtight vacuum chamber. The chamber is filled with anhydrous ammonia gas. Once the sample has been exposed long enough for it to be plasticized, the gas is removed from the chamber. The stock is then placed in a form for a brief period of time to set in the desired shape. See Figure 24-3.

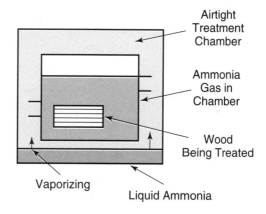

Figure 24-3
Anhydrous ammonia gas is used to plasticize wood. In the gaseous ammonia process, the wood is cooled in a soaking bath. A vacuum is then drawn in the treatment chamber, causing liquid ammonia to vaporize into a gaseous form that will penetrate the wood.

Staypack

Since compreg and impreg are both more brittle than untreated wood, the U.S. Forest Products Laboratory conducted research to find a suitable alternative. They developed another type of modified wood, called *staypack*, which doesn't use any resin. Instead, wood pre-conditioned to between 30 percent and 65 percent relative humidity is compressed and heated. The pressure (1,400 psi to 1,600 psi, or 9653 kPa to 11 032 kPa) and heat (350°F or 177°C) cause the natural lignin between the wood fibers to liquefy, and the entire mass to realign itself. This realignment process reduces internal stresses and creates a new type of wood with improved tensile and flexural properties.

Both veneers and solid wood can be used to make staypack. If solid wood is used, it must be free from knots. The resulting material can be glued and finished like untreated wood.

Like other modified wood products, staypack has increased impact strength. Its strength increases almost proportionally to the amount of compression. The greatest increase is in impact bending strength. Staypack is not as water-resistant, nor as resistant to biological attack, as compreg. However, it does absorb water at a slower rate than uncompressed wood, and is more dimensionally stable.

Typical product applications for staypack are tool handles, mallet heads, and tooling jigs and dies. Materials similar to staypack are produced in Germany (under the tradenames "Lignostone" and "Lignofol"). In England, "Jicwood" and "Jablo" are two types of compressed plywood that are similar to staypack.

Pressure-treated Wood

A great deal of research and experimentation on chemical treatment for the preservation of wood has been conducted by both government and private agencies. The Southern Pine Association classifies chemical wood preservatives into three types:

Δ   Creosote.

Δ   Oil-borne preservatives.

Δ   Water-borne preservatives.

Preservation treatments are available that can be used to make wood resistant to almost any type of harsh environment. **Preservatives** can be fire-resistant, moisture-repellent, colorless, and odorless. Some are even capable of receiving a surface finish. Most treatments impregnate the cells of the wood under pressure; such pressure-treating processes are the only methods presently used to produce a uniform deep penetration of the wood. Typically, the processes involve holding the workpiece in a closed vacuum cylinder while the preservatives are introduced into the wood.

There are two basic types of pressure treating systems used today: *empty-cell impregnation,* used to apply creosote, and *full-cell impregnation,* used to apply fire-retardant and moisture-protectant chemicals.

The empty-cell process starts by placing the wood to be impregnated in a stream of high-pressure air. The preservative is then introduced and the pressure increased to force it into the wood cells. After a given length of time, the air pressure is returned to normal. This allows the air trapped inside the wood cells to push any excess preservative out of the cells.

The advantage of the empty-cell method is that it produces more uniform penetration than the full-cell method. The empty-cell process is used to treat telephone poles, posts, and lumber for piers and bulkheads.

In the full-cell process, the wood is placed in the treatment chamber and a slight vacuum is applied. In this method, the preservative slowly penetrates the cells, rather than being forced in aggressively as it is with the empty-cell process.

The full-cell method is used when products must be able to withstand continued exposure to harsh conditions. Products treated by the full-cell method include those that must be *fire-retardant,* or must withstand continued exposure to water. See Figure 24-4.

Figure 24-4
Pressure-treated wood is widely used for decks and other outdoor structures that are exposed to harsh weather conditions. The full-cell process is used to incorporate the preservative into the wood. (Step 2 Corporation)

Other pressure-treating processes are gaining popularity. Some make it possible to treat only the end of a timber or post that will be exposed to weather or other harsh conditions. Others involve drilling holes in the product and treating only the areas around the drilled holes. Other processes use liquefied petroleum gas (LPG) to distribute the preservative.

Polyethylene Glycol

A chemical treatment that results in improved dimensional stability for wood is soaking in *polyethylene glycol (PEG).* Sometimes called *carbowax,* PEG is a white, waxy chemical that looks something like *paraffin.* Polyethylene glycol melts at 104°F (40°C), dissolves readily in warm water, and is noncorrosive, odorless, and colorless.

Polyethylene glycol is normally used to fill the cells and cell walls of green wood. No pressure is required. The process involves immersion of the wood product in a solution of PEG dissolved in warm water. Treatment is usually done at temperatures ranging from 70°F to 140°F (21°C to 60°C).

Diffusion of the material into the wood is greatly accelerated by increasing the temperature and keeping the concentration of the solution at about 30 percent. The PEG treating solution can be used many times.

PEG treatment is slow. How long it takes depends on the wood thickness and density. In many cases, it takes weeks. After parts are treated with PEG, they are *stickered* (stacked using spacers between boards) and dried in a well-ventilated heated room. Time necessary for air-drying depends on thickness, temperature, and relative humidity.

Treatment with PEG is not permanent. No curing takes place, so the chemical remains water-soluble. If the treated material is exposed to water, the polyethylene glycol will leach out. For this reason, PEG-treated wood will become sticky in high humidities (above 60 percent). Usually, treated wood is finished with two coats of a polyurethane varnish to seal in the glycol.

Polyethylene glycol is used to help in eliminating checking and cracking of wood during drying. It has even been successfully used to preserve antiques, and to prevent rare wood artifacts discovered in the ocean from cracking during drying. The material can be used in any application where the wood is wet and needs to be dried gradually without cracking.

Important Terms

- acetylated wood
- anisotropic
- compreg
- empty-cell impregnation
- fire-retardant
- full-cell impregnation
- impreg
- impregnation
- irradiated wood
- paraffin
- plasticized
- polyethylene glycol (PEG)
- porosity
- preservatives
- radio-frequency (RF) dielectric heating
- staypack
- stickered
- wood plastic composition (WPC)

Questions for Review and Discussion

1. What is the advantage of using the polyethylene glycol (PEG) process, in preference to other processes? What is the major disadvantage of the PEG process?
2. What is the greatest problem with irradiated wood?
3. What process would be most suitable for making wood impervious to chemical attack? Give reasons for your choice.
4. What are some of the methods you might select for impregnating wood to make it resistant to moisture and swelling?

25

Finishing Wood Materials

Key Concepts

△ Surface preparation is the most difficult part of finishing a wood product.

△ On open-grain woods such as oak, a filler must be used before applying the final finish.

△ Electrostatic coating and the waterfall method are types of finishing that have achieved widespread acceptance in industry.

△ Both synthetic and natural abrasives are available; synthetic types are most often used in industry. Sanding may be done by hand or machine.

When you think about finishing wood products, the first thing that may come to mind is brushing on paints, varnishes, or stains. However, there is much more to finishing wood than just applying a coating over an unfinished product. Normally, there is a great deal of surface preparation that must precede the actual application of finishing materials. Once the product is ready to accept a finish, the material is typically applied using industrial processes such as spraying or curtain coating.

As a living tree in the forest, wood often will survive for hundreds of years, despite exposure to frequently adverse conditions. This is made possible through the amazing process of photosynthesis—the foundation of growth for the tree. Growth not only occurs internally in terms of cells and passageways for nourishment, but also externally in terms of the tree's protective layer of bark.

Once the tree is harvested and processed into lumber, the green wood is exposed directly to the elements without the protection of bark. While the pattern of the grain in unfinished wood may be beautiful, that grain may discolor and develop defects if the wood is left in an unfinished state. Untreated wood will absorb dirt, oil, grease, and moisture. Discoloration from dirt and oils is difficult to remove from unfinished surfaces, especially open end grain. Excess moisture will cause unfinished wood to swell and deform. If drying is done too rapidly, the wood may warp or develop checks or cracks.

Finishing processes protect wood from exposure to contaminating influences in its environment. If finishes are properly applied they can preserve and improve the natural beauty of wood. Hasty or careless finish application, however, can destroy the quality product that has been created.

Surface Preparation

To successfully apply any finish, you must know about the structural characteristics of the wood being finished. Not all finishes work equally well on all woods. Also, the surface preparation needed for finishing hardwoods is different from the process used for softwoods.

Finishing of wood surfaces is a labor-intensive process, and often involves more planning than is necessary with most other materials. It is simpler to finish most metals and plastics than it is to finish wood and ceramics. There are many different surface preparation methods that may be used, depending on the type of stock and the desired finish. Most of the time that is consumed in final finishing wood products is devoted to surface preparation. The first step in surface preparation is usually sanding.

Sanding

Sanding can be done by machine or by hand; machine sanding is most common in industry. Machine sanding involves use of an abrasive-coated sheet, belt, or disk attached to power-driven equipment. There are many different types of natural and synthetic abrasives used to smooth wood. Abrasives are normally fastened to a backing with adhesive; individual abrasive particles work like tiny cutting tools against the surface of the wood. Manufactured abrasive materials are usually referred to in industry as ***coated abrasives***, although they are still called "sandpaper" in common usage. *Sand* is not actually used for abrasives, although the natural abrasive quartz (silicon dioxide) used to coat ***flint paper*** is chemically similar. Despite the fact that there is no sand in sandpaper, material removal by use of abrasives is still called "sanding."

Types of sanding

There are three basic types of sanding (whether done by machine or by hand): roughing, blending, and fine finishing. ***Roughing*** is done to remove the maximum amount of stock; it requires coarse-grit abrasive. If the product has surface scratches or gouges, rough sanding may be a necessary first step.

If the product is relatively free of blemishes, it may be possible to start off with blending. ***Blending*** is done with a medium-grit abrasive. Blending results in some material removal, producing a fairly smooth finish.

Fine finishing involves using a fine-grit abrasive to remove light scratches. After fine finishing, the stock is often polished or burnished with rubbing oils or rubbing compound. Most rubbing oils are made of varnish diluted in oil. They penetrate well and produce an attractive finish that is maintained by using furniture polish, wax, or mineral oil. No further finishing or surface treatment is necessary.

The most common abrasives used to remove stock from products made of wood are aluminum oxide, silicon carbide, and garnet. Garnet is the softest of the three, while silicon carbide is the hardest. Abrasive hardness, like the hardness of other minerals, is measured on Mohs scale. The scale ranges from 10 (diamond) down to 1 (talc). Silicon carbide is rated at 9.5, aluminum oxide at 9, and garnet at 7.

Abrasive grain sizes

The *grain size* of an abrasive material is used to grade the coarseness or fineness of the finishing material. A coarse abrasive has large grains, while a fine abrasive has much smaller grains.

The grain numbers used to classify abrasives refer to the number of openings in each inch of mesh through which the abrasive is sifted. As the numbers become larger, the grain size becomes smaller: a particle with a grain size of 12 (extra coarse) would be a virtual boulder next to an extra-fine particle with a grain size of 600. The standard numbering system applies to most types of abrasives, which might be designated as a "No. 80 paper" (a medium grit) or a "No. 120 paper" (a fine grit). Flint abrasives are the exception. They are graded as extra fine, fine, medium, coarse, and extra coarse.

The extra coarse, very coarse, and coarse grain types range from 12 to 50 grains per inch. These highly abrasive materials are used for heavy-duty industrial sanding machines. Medium-grit papers range from 60-100, fine from 120 to 180, very fine from 220-280, and extra fine up to 600 grains per inch.

Naturally, the grade of the paper selected will greatly influence the speed of stock removal and the number and depth of scratches generated on the surface of the material. Care should be taken to avoid making too large a "jump" from a coarser grade to a finer grade of abrasive: a No. 60 paper will produce scratches that cannot be removed with a No. 280 paper. Ideally, abrasives should not be more than two grades apart as you move to a finer paper. Thus, the No. 60 paper should be followed by a No. 100 (a No. 120 paper would eventually remove the scratches, but it would take a lot more work).

Coated abrasives are manufactured by attaching the abrasive particles to a paper or cloth backing with an adhesive. The material is then cut into the shape required to fit the desired sanding application. Sanding materials are available in sheet, disk, belt, or cylinder form.

Sanding equipment

Sanding is normally done with a belt sander, disk sander, or spindle sander. Figure 25-1 illustrates a combination belt and disc sander. Sanding also can be accomplished using a radial arm saw by removing its blade and installing a sanding disk in its place. Sanding can even be done on a bandsaw by using a special segmented blade with small abrasive disks attached to each segment.

Sanding must be done in the same direction that the grain of the stock is running. Sanding *across* the grain will produce scratches that are difficult to remove. Figure 25-2 shows an edge sander being used to surface-sand a long strip with the grain.

The abrasive belt planer is a machine used for stock removal in most furniture-making plants. On this machine, the sanding is done by a large belt held tightly on two drums. The stock is fed into the machine on a conveyor; as

Figure 25-1
This combination belt and disk sander is of a type widely used in both woodworking shops and industrial applications. Sanding is an important finishing step for most wood products. (Delta International Machinery Corporation)

Figure 25-2
Sanding is typically done with the grain of the wood. This machine is being used to edge-sand a thin strip of material. Note the ear protection worn by this worker because of the high noise level in this production area. (Robersonville Products)

much as 1/16 in. (1.6 mm) of material is removed from one face of the board with a single pass through the machine. Abrasive planers have motors heavier than those of conventional sanders. The abrasive belt, which moves at speeds up to 7500 surface feet per minute (sfpm), is normally made of tough cloth and can withstand tremendous abuse. There are many variations of the basic abrasive planer. The machine shown in Figure 25-3 is a *production molder*, a type of abrasive planer used to shape long strips of molding.

Figure 25-3
A production molder is a special-purpose abrasive machine used to create the desired cross-section on a length of molding. (Robersonville Products)

Applying Fillers

With open-grained woods such as walnut, oak, mahogany, or ash, a ***filler*** should be used to close the pores before any finish is applied. Fillers also help to produce a smoother surface and enhance the beauty of the wood by highlighting the grain pattern. Fillers are usually purchased in paste or semi-paste form, but are also available in liquid form.

Paste filler contains powdered quartz, linseed oil, turpentine, and drying agents. Its natural color is tan. It can also be purchased in other colors, or colored as desired before use. Varnish and lacquer are often used as clear fillers.

Before wood is filled, you must first clean the pores with a wash coat of shellac or lacquer. The wash coat consists of one part shellac mixed with seven parts of alcohol, or one part sealer and five parts lacquer thinner. The wash coat is applied with a brush and allowed to dry.

The filler is then applied with a stiff brush, first with the grain to cover the workpiece, then against the grain to ensure complete coverage. While the filler is still wet, any excess material is removed by wiping with a rag. Wiping should be done with the grain.

Once the filler has been applied and dried, the wood is ready for its final finish of stain, varnish, paint, enamel, or lacquer. Sometimes, further treatment is needed to change or enhance the color of the wood before finishing.

Bleaching Wood

Sometimes, it is necessary to lighten the natural wood coloring of the workpiece overall, or to lighten some areas and leave others alone in order to equalize the color of the wood. This can be done by *bleaching,* a process that involves applying a caustic liquid to the surface with a sponge or rag. In large volume industrial applications, the bleach is typically applied by spraying it onto the wood surface.

Bleaching is done to achieve honey-toned finishes. Normally, lighter-colored woods are used, since it is difficult to drastically lighten dark woods. Bleaching removes the color by oxidation. Bleaching often is done when the wood is to be stained, or to bring out the desired grain pattern.

Commercial bleaches usually consist of two solutions that are sprayed onto the surface. The two parts mix to produce the desired bleaching action. When bleach is applied in this manner no caustic residue remains, so there is no need to neutralize the surface after bleaching.

Homemade bleaches can be made by mixing oxalic acid crystals in hot water. A stronger bleach can be produced by using hydrogen peroxide mixed with caustic soda. Whenever bleaches are used, rubber gloves and eye protection should always be worn—the solutions are very caustic and require careful handling.

The bleach is applied and allowed to soak into the pores of the bare wood. The bleaching time varies according to how much the wood has to be lightened.

Once the desired degree of bleaching has taken place, lightly sponge the stock with warm water to *neutralize* (stop the action of) the bleach and remove any residue. When the surface has dried, the wood can be lightly sanded to prepare it for final finishing.

A major disadvantage of bleaching is that washing off the residue adds moisture to the wood. This sometimes causes the wood to *cup* (assume a concave shape). To keep this from occurring, water is often applied to both face surfaces of the wood to equalize the swelling.

Shading Wood

Shading or glazing is a process that is sometimes used to improve the color uniformity of the wood surface. Shading differs from bleaching, because it is carried out after the wood has been sealed with filler. Most shading is done on low-cost furniture to equalize the color produced by a previous application of stain.

The process involves spraying color-tinted shading lacquer over the area to be shaded. The lacquer adds a lighter color to the area to equalize the color throughout the product.

Shading can also be applied by mixing a color tint with commercial glazing liquid, then brushing the material on the surface. To achieve the desired look, some of the glaze is wiped off with a rag or brush. This serves to highlight certain areas, and produce an aged or antique look in the product.

Staining Wood

Manufacturing firms use stain to change the overall wood color or to beautify the grain by either adding emphasis or lightening the grain pattern. *Stain* is made of finely ground pigments held in suspension, or chemicals in solution.

Stain is applied with a brush, sponge, or rag and allowed to dry. The length of drying time depends on the type of stain. Oil stains are wiped off with a soft rag after being left on the wood long enough to achieve the desired degree of darkness. Stains can be applied to filled or unfilled wood, depending on the type of stain and effect desired.

There are four basic types of wood stain: water stain, oil stain, non-grain-raising stain, and spirit stain.

Water stains are used to create consistency of color. When a darker tone is desired, more pigment is mixed with water. The major disadvantage of a water stain is that it will raise the grain in the wood. This can be controlled by wetting the wood with a sponge prior to staining. Water stains are often sprayed.

There are two major types of *oil stains,* pigmented oil stains, and penetrating oil stains. Both types consist of pigments suspended in oil, and can be purchased in either powder or liquid form. Oil stains are applied by brushing, spraying, wiping, or dipping. As noted earlier, they are wiped off with a soft rag to control the depth of color.

Non-grain-raising (NGR) stains are widely used by furniture manufacturers. NGR stains contain dye mixed in a *methanol* (methyl alcohol) solvent, rather than water. Since alcohol evaporates fairly rapidly, these stains have a short drying time (often 15 to 20 minutes). NGR stains are normally applied by spraying or dipping.

Spirit stains have the fastest drying rate of all of the stains. They consist of *aniline dyes* (rather than pigments) suspended in alcohol. Spirit stains are the most difficult type to apply, primarily because of the rapid evaporation rate of alcohol. These stains are used mostly for repair or touch-up work.

Once a stain is applied and allowed to dry thoroughly, it should be *sealed* with shellac or a lacquer-base sanding sealer. The sealer encapsulates the stain, and permanently protects the finished surface. Sealers are normally brushed or sprayed.

After the lacquer-base sanding sealer is applied, it should be lightly sanded. The abrasive paper that is used is so fine that it produces a very fine dust and leaves no scratches in the finish.

Finish Coatings

After the wood has been filled, finished, and sealed, it may or may not be necessary to provide another protective coating. If such a *topcoat* is desired (as is the case with fine furniture), it is usually a clear coating of lacquer, oil varnish, or synthetic-resin varnish. The type of topcoat selected depends on the durability needed in the product, the extent of contamination in the environment where it will be used, and the characteristics of the material. Depending on the type of finish used, topcoats can be applied by spraying, brushing, or by wiping.

Varnishes and Lacquers

Varnishes are different from stains or paints. There are synthetic varnishes (polyurethane), oil varnishes, spirit varnishes, and acrylic varnishes. All of these different types of varnishes contain oil, resin, solvent and driers. A varnish topcoat is applied to protect the surface from water spotting, chipping, and abrasion.

Sometimes the topcoat applied is a lacquer. *Lacquer* is made of nitrocellulose, acetone, varnish resins, a thinner such as benzene, and plasticizers. Lacquer dries quickly and provides a hard finish. The preferred method for applying lacquer is by spraying, but sometimes several coats are applied by brushing. Lacquer is highly volatile, and the fumes are toxic. It should be handled and applied with care, making sure there is adequate ventilation.

Paints and Enamels

Paints and enamels are opaque finishes that hide the grain of the wood. Generally, they are applied as surface protection when it is not important to see the grain and texture of the wood. For this reason, less surface preparation is necessary when paints or enamels are to be used.

There are two basic types of paint, oil-base and latex (water-base). Naturally, the type of base determines how the paint can be thinned or removed when the painting job is completed. *Latex paints* are soluble in water. *Oil-base paints* are thinned with paint thinner, turpentine, or mineral spirits.

The type of base used is important to know when refinishing any painted product. The general rule to remember is that oil-base paints can be applied over a previous coating of latex paint. However, if latex is applied over oil-base paint, then the finish will bubble and peel.

There are many different types of oil-base and water-base paints. A *paint* consists of pigments suspended in a vehicle. *Pigment* gives the paint its opacity and color. In most cases, a greater amount of pigment would result in a duller appearance. High-gloss paints have little pigment, but more vehicle and solvents.

The *vehicle* functions as the carrier for the pigment. When the surface is painted, the vehicle forms a layer of film with the opaque pigment suspended in it. When the vehicle dries, a protective surface layer remains. Paints are normally applied by brushing, spraying, dipping, or other coating processes.

Enamel is different from paint, because it is made by adding pigment to varnish. Enamel has a gloss finish, and is more durable than latex paint. Since they are not opaque, however, they will not cover as well as paint. Enamels often require a base coat, or undercoat, to seal pores and provide a stable foundation for subsequent coats.

The first coat of enamel is applied to the base coat. Several coats may be necessary, with 24 hours allowed for drying between them. Light sanding is often done between coats to smooth the surface and aid adhesion. Enamels are thinned with mineral spirits.

Application Methods

Because of the volume of material to be finished, industrial production usually requires the use of automated application methods for finishing materials. These include several spraying methods, electrodeposition, and coating by means of roller, curtain, or dip application.

Atomized Spraying

In atomized spraying, pressure is used both to **atomize** the finishing material (break it into very fine drops) and to blow it onto the material to be coated. After the product is sprayed, the finishing material is either air-dried or heat-dried. Atomized spraying is done using air or hydraulic pressure. In Figure 25-4, a robot is shown spraying a surface coat on a display shelf. Here, the *end effector* (tool or device mounted to a robot arm) consists of a spray attachment.

Figure 25-4
Atomized spraying is a widely used industrial method of applying finishing materials. This robotic painter is spraying the finish on wooden shelf components. (Robersonville Products)

When hydraulic pressure is used to force the finishing material to the spray gun, the technique is referred to as *airless spraying.* Airless spraying is popular because it can be used with many different types of finishing materials. In addition, airless spraying saves as much as 15 percent of the material that would be lost as *overspray* with conventional spray systems.

On airless spray systems, the paint is squeezed out of the spray nozzle to be atomized, much like water coming out of a garden hose. Airless systems make possible heavier finish coats than conventional spray systems. This results in fewer coats and faster application. With airless spraying, the pump delivers finishing material to the gun under fluid pressures as high as 3000 psi (20 685 kPa).

Spraying using air pressure is a relatively straightforward process. The spraying system consists of a compressor, a regulator, an extractor, and the spray gun. The purpose of the *extractor* is to filter oil and water out of the air as it passes through the system.

There are two basic types of guns used for atomized spraying: the siphon gun, and the pressure-feed gun. The *siphon gun* is fed by suction, drawing finish upward to a fluid cap, then mixing it with air. With the *pressure-feed gun* system, the finishing material is fed to the nozzle, where it is mixed with air. The siphon type is used for smaller jobs, since a limited quantity of finishing material can be stored in the fluid cap. Most production spraying is done with the pressure-feed system.

Sometimes production spraying (atomized or airless) is done using *heated* finishing materials. Heating finishes during application reduces their viscosity for better spraying, without the addition of undesirable solvents. The process uses special guns that heat the finishing material to a temperature of 160°F to 200°F (71°C to 93°C).

Electrostatic Spraying

Electrostatic spraying is a process that was developed as a method for coating metallic materials. It can also be used to finish woods, but the wood must either be high in moisture content or coated with a conductive material. Electrostatic spraying of wood has been done in Europe, where the moisture content of the test samples is more than 12 percent. Wood in the United States seldom has more than 6 percent moisture, so precoating with conductive material is necessary.

In electrostatic spraying, the finishing material is fed to a specially designed spray gun, where it receives a negative charge. The product is given a positive charge.

When the negatively charged finishing material is sprayed out of the gun, it is drawn by attraction to the positively charged product. With this coating method, coats of even thickness can be applied with little overspray and wasted material.

There are three different techniques used to apply electrostatic coatings. These are referred to as the gun method, the disk method, and the bell method. The gun operates on conventional air pressure.

The disk method and bell methods both make use of centrifugal forces. Fluid is pumped directly to the disk and the bell. When the devices are rotated, centrifugal forces disperse the fluid to their outer edges. The material then receives a negative charge, and is pulled away by the positive charges surrounding the product to be coated.

The great advantage of electrostatic spraying is the cost—only about 40 percent of the finishing material required with conventional spraying is used with electrostatic spray coating.

Electrodeposition

Electrodeposition is similar to electroplating, and is normally used on metals. It can also be used with coated polymers. Electrodeposition is also called **electrophoresis.**

Electrophoresis was one of the first manufacturing processes to be utilized in space, as part of the Material Processing in Space (MPS) initiative undertaken by NASA. A number of successful experiments have already been conducted with electrophoresis in space.

The process is accomplished by placing the part to be coated in a vat of special paint, and then attaching it to the positive side of a charging circuit. The part becomes the anode. The paint particles contain a negative charge, and are attracted to the positive anode. After the product is coated, it is placed in an oven for drying and curing. Like electrostatic spraying, the process produces a uniform coating.

Roller Coating

Roller coating is a process that is economical for high-speed coating of flat stock such as wood, metal, plywood, and particleboard. The process consists of squeezing the stock to be coated through rubber rollers carrying a film of finishing material.

In roller coating, the rollers are partly submerged in a vat of finishing material. When the stock travels through the rollers in the same direction that they are turning, the process is referred to as *through coating*. Today, most manufacturers use what is referred to as the **reverse roller process,** in which the rollers turn in one direction and the stock moves in the other. This results in better coverage of material on the workpiece.

Roller coating is often used to apply a woodgrain finish to particle board or hardboard panels. The panels are first roller-coated, then pass through a grain printer. Next, a polyester topcoat is applied by another set of rollers. Panels are normally dried using ultraviolet (UV) light waves, rather than more intense sources of heat.

Curtain Coating

Curtain coating involves passing the stock to be coated through a continuous, sheet-like falling stream of finishing material. The curtain is made by flowing the finish over a dam, where it drops by gravity onto the workpiece to coat it. Excess material then runs off the product, is captured in a gutter, and flows to a reservoir for recycling through the system. The process is also referred to as *waterfall coating* or *flow coating*.

Many panel manufacturers finish their products this way. Films with a thickness of 0.001 in. to 0.0025 in. (0.025 mm to 0.064 mm) can be applied using curtain-coating machines.

In flow coating, the process is varied somewhat, The workpiece is placed beneath a bank of low-pressure nozzles that eject streams of nonatomized finishing materials. The process helps to reduce the amount of uncontrollable airborne contaminants, because less air is used to distribute the finishing material.

Dip Coating

In *dip coating,* the finishing material is applied through complete immersion of the product or part. The viscosity of the finishing material must be carefully controlled to prevent the lower surfaces and edges of the product from retaining an excessive amount of film.

Dip coating is fast and economical. The process is practical for coating parts that are difficult to finish by any other method. Auto bodies are often dip-coated. As noted, the critical element in dip coating is the viscosity of the finishing material. If it is too thin, it will produce inadequate coverage; if too thick, it will produce sags and runs.

Tumbling

Tumbling can be used to finish wood knobs, buttons, golf tees, beads, and other small parts. The objects to be finished are placed in a drum or barrel with a small amount of lacquer, enamel, or wax. The barrel revolves at about 25 rpm causing the parts to tumble over each other. Additional finishing material is placed in the barrel as needed. When the parts dry, they polish against each other, creating a smooth, satiny finish.

Forced Drying

Forced drying is typically accomplished by loading parts or assemblies onto conveyors that carry them through heated ovens. Figure 25-5 shows an overhead conveyor system used to transport products to drying ovens.

Forced drying is often done with ultraviolet (UV) radiation. Radiation drying is preferred for curing flat sheets. The finish coat is cured by light rays, rather than heat, converting the coating to a tough, solid film in a matter of seconds.

Some wet surfaces cannot be exposed to high temperatures immediately after finishing or they will bubble and blister. These finishes are given a *flash-off period* to allow excess thinners and solvents to evaporate. This is normally done by air drying, either at room temperature or in a warm-air enclosure.

Important Terms

airless spraying
aniline dyes
atomize
bleaching
blending
coated abrasives
cup
curtain coating
dip coating
electrodeposition
electrophoresis
electrostatic spraying

enamel
end effector
extractor
filler
fine finishing
flash-off period
flint paper
forced drying
grain size
lacquer
latex paints
methanol

Figure 25-5 This overhead conveyor system is used to convey parts into an oven for forced drying of the finishing material. (Robersonville Products)

neutralize
non-grain-raising (NGR) stains
oil-base paints
oil stains
overspray
paint
paste filler
pigment
pressure-feed gun
production molder
reverse roller process

roller coating
roughing
shading
siphon gun
spirit stains
stain
topcoat
tumbling
varnishes
vehicle
water stains

Questions for Review and Discussion

1. What is the disadvantage of atomized spraying, when compared to electrostatic spraying?
2. What finishing methods that do not produce airborne contaminants can be used on woods?
3. Why must a filler be used before applying a final finish on oak? Is a filler needed when finishing an item made from pine?
4. What method is used to remove dust from the pores of wood before applying a final finish?

Good surface preparation is vital to obtaining a durable and attractive finish on wood. Manufactured products are normally machine-sanded, but hand-sanding may be used in some situations. The abrasive being used here is a perforated steel sheet that permits rapid material removal. (Red Devil, Inc.)

The manufacturing of doors, windows, moldings, and similar products (usually referred to as "millwork"), is an important segment of the wood industry. This selection of wood windows indicates the variety of shapes available today. Millwork items are often sold in pre-finished form, with finishes ranging from paint to varnishes and synthetic coatings to vinyl cladding. (Andersen Corporation)

COMPOSITE MATERIALS

A composite is a material in which different substances are combined to create a new material with better attributes than any of those making it up. Using composites usually results in a product that is both stronger and lighter. It also makes possible combining parts to provide a significant cost savings over conventional assembly methods. While composite materials are often more costly than conventional materials, the final product is frequently less expensive. It appears likely that composites will be the material of choice for a majority of sophisticated engineering applications in the future..

Siemens

26 Introduction to Composite Materials

Key Concepts

- Δ Structural composites are built to specification from two or more composite systems.
- Δ The matrix used for a composite is typically a polymeric, metallic, ceramic, or glass material.
- Δ There are three major types of composite fillers: fibers, particulates, and fabric.
- Δ The major processes used to form composite parts are identical to those used to form plastics.

In the field of engineering materials, a *composite* is defined as a material that contains two or more distinctly different but complementary substances. In a composite material, different substances—each with unique attributes—are combined to create a new material with better attributes. A composite is an example of the phenomenon called *synergism*, in which the *whole* (in this case the composite material) is greater than the sum of the *parts* (individual substances).

Composite materials have been used to solve technological problems for a long time, but it was not until the 1960s that these materials really began to capture the attention of industry. In that decade, *polymeric-matrix composites* (PMCs) received the top billing as they came to be perceived as revolutionary new space-age materials.

In the late 1980s, the field of composites began to grow exponentially. Today, it appears that composites will be the material of choice for a majority of sophisticated engineering applications in the future.

Unique Characteristics of Composites

There are many reasons for the growth in composite use, but the driving force is the fact that using composites usually results in a product that is both stronger and lighter. However, there are a number of other advantages to be gained from substituting composites for conventional materials.

Since composite materials are made by combining two or more different materials, it is simple to consolidate parts when the composite structure is being fabricated. Often, a significant number of parts can be eliminated by the use of composites. This can provide a significant cost savings over conventional assembly methods.

The processes that are used to form the product must be taken into consideration, of course, and consolidation of parts is not always desirable. Many products have to be assembled from discrete parts, so that parts can be removed for later servicing and maintenance.

There are other reasons why composites are often chosen as the preferred material for many demanding product applications. On the bottom line, most of these relate to the fact that composite materials can provide a means of accomplishing more with less. The cost of composite materials is often greater than the cost of conventional materials, but in many instances, the final product can be less expensive.

Humans did not invent the concept of composites. While synthetic or *engineered composites* are relatively new, *natural composites* have been used throughout human history. Probably the most common is *wood,* a composite of cellulose fibers bound together with the natural glue called lignin. Wood fibers have a high tensile strength, but they are still flexible.

One of identifying features of composites is that all of them have a **matrix**. In wood, *lignin* is the matrix that binds the fibers together and provides stiffness.

Another example of a composite found in nature is *bone.* Bone consists of short collagen fibers embedded in a mineral matrix material, called apatite. It is both strong and relatively light in weight.

The development of engineered composites can be traced to ancient civilizations: more than 3000 years ago, the Incas of South America combined plant fibers with the clay slip used to make their pottery, providing increased strength. The Egyptians mixed chopped straw with their clay bricks to make a stronger and more useful composite. In the same way, the Eskimos of North America and Asia froze layers of moss in water to provide stronger blocks of ice for building.

The field of metallurgy is full of examples of experimentation and accomplishment with composites. For many years, swords of exceptional quality were produced in Asia and in Europe by *forging* (hammering) layers of different steels together. After forging, stretching, and bending the original layers, thousands of very thin layers of steel were produced. The Samurai craftsmen of Japan and the Toledo swordmakers of Spain had created a composite material that was light, strong, and very tough.

Throughout the history of civilization, discoveries were made that provided new opportunities for composite materials. Today the search for strong but lightweight materials, particularly for aerospace, defense, and medical applications, is opening many new markets for composites.

The use of composites in aircraft has been extensive. We can expect that most of the conventional metals found today in commercial aircraft will be replaced by advanced composites by the year 2000, Figure 26-1. The United States Office of Technology Assessment estimates that worldwide sales of advanced composites will grow by 15 percent annually, and that composites will become a $12 billion industry by the year 2000.

Figure 26-1
The propeller blade being set up for testing by this technician is made from an advanced composite. The blades are used on an experimental type of modified turbofan aircraft engine. (General Electric)

It is estimated that half the parts in the United States Advanced Tactical Fighter (ATF) are made from composites. Boeing estimates that the total weight of composites in one of their newer passenger aircraft will soon approach 65 percent. This will result in significant fuel savings because of the reduced weight.

Major programs sponsored by the Department of Defense (DOD) have greatly expanded research and development efforts with composites. Such programs as the National Space Aircraft Project rely heavily on the capabilities of composites.

The National Aeronautics and Space Administration (NASA) has been working toward establishing an international space station in low earth orbit. While the future development of space is uncertain at this time, the eventual completion of this project is expected to lead to rapid expansion of a whole new world of *material processing in space (MPS)*. Composites will play a major role in the MPS program. Companies involved include Rockwell International, McDonnell Douglas, Boeing Aerospace, General Electric, 3M Corporation, Fairchild Industries, GTI, and many others.

While it is difficult to find any industry today that does not take advantage of composites, the largest volume user of polymeric-matrix composites is the automobile industry. See Figure 26-2. This industry

Figure 26-2
The automotive industry uses more polymeric-matrix composites than any other type of manufacturing. Most of the body panels of this Saturn auto are formed from polymeric composite materials. (Saturn)

consumes large quantities of sheet molding compound to produce exterior body panels. The market has grown tremendously since 1984, when General Motors introduced the Pontiac Fiero—the first domestic production vehicle to use composite body panels attached to a steel space frame. This allowed GM to achieve model differentiation for limited production runs (200,000 units), while avoiding expensive tooling costs that would have been involved with steel.

Composites have also been used in limited production for such components as drive shafts and leaf springs. Ford produced 3,000 drive shafts of graphite wound with E-glass filaments for use in its 1989 and 1990 Econoline vans, while GM used glass fiber-reinforced polymeric-matrix leaf springs in the 1990 Corvette.

In the fairly near future, composite unibody frames will probably be used. This will affect the way that automobiles are assembled and also how they are serviced. Before composite unibody construction becomes economically feasible, however, much faster production methods will have to be developed. The major barrier is the lack of composite production technology capable of matching the high production rates that are possible with existing metal stamping technology.

Structure of Composite Materials

As just described, composites are playing a significant role in manufacturing today. But what really constitutes a composite? We learned that composites are formed when two or more materials are combined. However, composites are more than just a "sandwich" of materials. A composite is actually a *new structural material that has been designed to provide characteristics not found in its parent materials.*

When these materials first began to attract interest in industry, they were grouped into two distinctly different classes: structural and advanced. ***Structural composites*** were relatively simple, consisting of different types of reinforcing materials suspended in a matrix.

Advanced composites were more complex. The term was used to refer to the exotic and costly combinations of materials that could withstand exposure to harsh chemicals, temperature extremes, or even contact with space junk. Advanced composites were able to survive just about any condition.

Today, the distinction between these classifications is much less apparent. Some composites are constructed of simple and inexpensive materials; others require expensive exotic materials for specialized applications. Both structural and advanced composites can be designed to achieve increased strength and stiffness.

When advanced composites are needed, they are more expensive. Advanced composites typically contain a large percentage of high-performance continuous fibers, such as S-glass (high-strength glass), aramid, graphite, or other fibers. Advanced composites have considerably greater strength and toughness than structural composites.

Structural composites, on the other hand, are normally glass fiber-reinforced plastic (FRP). They use low-strength E-glass fibers suspended in a polyester resin. These composites have been used for years in the open molding process to lay up boat hulls, corrugated sheets, and automotive panels. Structural composites are used with other processes as well. Figure 26-3 shows an electronic hood scoop that is injection-molded from a high-temperature structural foam material.

Figure 26-3
A structural foam composite capable of withstanding high temperatures was injection-molded to form these hood scoops used to customize cars and pickup trucks. As shown beneath the left-hand scoop, the units can house instrument displays. (Martec Plastics)

Types of Composite Fillers

All composites consist of some type of *filler* suspended in a matrix binder. There are three major types of filler used to construct composites: *fibers, particulates,* and *fabric.* Figure 26-4 illustrates how these fillers are dispersed in the matrix of the composite material.

Another type of filler, called *whiskers,* is being used on an experimental basis. However, use of this material has not been refined to the extent that it receives a high degree of commercial acceptance.

Figure 26-4
There are three basic types of fillers (reinforcements) used in composites. A—Fiber-reinforced. B—Particulate-reinforced. C—Fabric-reinforced.

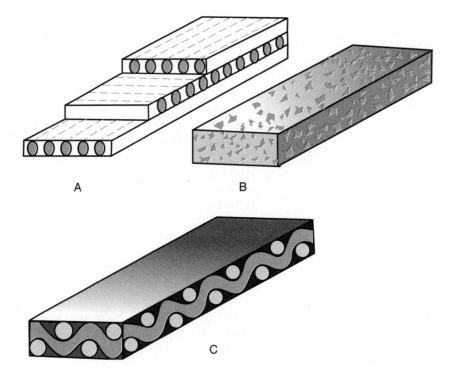

A whisker is about the same diameter as a fiber, but is usually short and stubby. The length-to-diameter ratio can be in the hundreds to one. Whiskers are more perfect than fibers, and exhibit higher properties. They are made from materials such as boron carbide, alumina, or silicon carbide. Whiskers provide increased strength, and have potential for use in situations where heavy loads must be supported.

Composite Matrices

Matrices (the plural of *matrix*) are usually made of ceramic, glass, metal, or plastic. At least 12 different types of composites can be created by combining these four basic types of matrices and the three major types of fillers (fibers, particulates, and fabric).

Each type of composite can be carried on through many other perturbations. Consider just one type, fiber-reinforced plastic composites. There are many different plastics that could be selected as the matrix, and many different types of fibers.

Most manufacturers do not make use of all of these different combinations. They select one type of matrix, then describe the composite as one of three types:

- Δ Fibrous.
- Δ Particulate.
- Δ Laminar.

This general classification is more closely aligned with the type of processing that will be involved in manufacturing products from composites.

Each of the three types of composites is different in terms of the manner in which the supportive materials (fillers) are suspended in the matrix material. Each matrix provides strengths and weaknesses that must be considered for it to be useful in a particular environment.

The matrix is the component that largely influences the nature of the composite; however, different types of composites can be produced with the same filler and matrix. It is the *manufacturing process* that really makes the difference. Different forming, fabricating, and separating processes can produce different product capabilities.

Polymeric-matrix (plastic) composites are the most popular type of composite. Ceramic-, glass-, and metallic-matrix composites are not used as extensively in industry.

Products made from polymeric-matrix materials are normally extruded, compression molded, or laminated. Polymeric-matrix composites are chosen primarily because of their impact resistance and temperature characteristics.

Polymeric-matrix composites can be of the fibrous, particulate, or laminar type. The composite that is created is influenced largely by the manufacturing process. Fiber- or particulate-reinforced composites are usually molded or cast. Metal-matrix composites are normally rolled, cast, extruded, or laminated.

The product application also weighs heavily on the type of composite matrix that is chosen. For example, to make re-entry shields for use on space vehicles, materials were chosen which could withstand high temperatures and extreme forces. A carbon matrix material was selected because of its high heat capacity. Carbon-matrix composites are made by compacting powder or a slurry of carbon in a die, then heating the compact to above 4500°F (2500°C). This turns the carbon into graphite.

Fibrous Composites

Fibers are used more than any other type of composite filler. The matrix material used with fibrous composites is relatively weak and brittle, but the reinforcing fibers give these composites tremendous tensile strength. Fibrous composites are particularly suited to product applications requiring high strength, stiffness, and light weight. Typical product applications would be fishing poles, antennas, and turbine rotor blades. The fibers used are normally filaments of boron, glass, graphite, or aramid.

Carbon-fiber composites are being used in the medical field for many purposes, from medical appliances (Figure 26-5) to actual implants in the human body. An operation that uses carbon fibers to replace ligaments in the human knee has been performed on hundreds of patients in Germany. Carbon fibers are even being used by dentists to provide reinforcements to tooth roots.

Figure 26-6 illustrates a typical fibrous composite manufacturing operation: computer-assisted filament winding around a metal mandrel. Winding is done with various filaments, including fiberglass, graphite, carbon, and aramid.

A number of types of filaments are used in making fiber reinforced composites. Besides fibers and whiskers, they may consist of reinforcing wires suspended in matrix material. The construction industry has long been familiar with the tremendous advantage of using reinforcing wires or rods in their

Figure 26-5
A specialized splint used to immobilize severe fractures has been converted from metal to high-strength carbon fiber by a medical supplies manufacturer. The composite version, shown at bottom, is lighter in weight, has greater stiffness, and does not interfere with X-rays or other diagnostic tools. (Polygon Company)

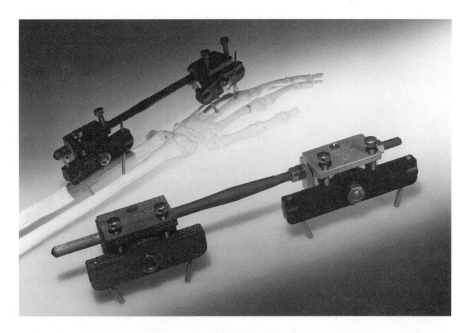

Figure 26-6
Filament winding around a metal mandrel is a typical operation in manufacturing fibrous composites. (Compositek Corporation)

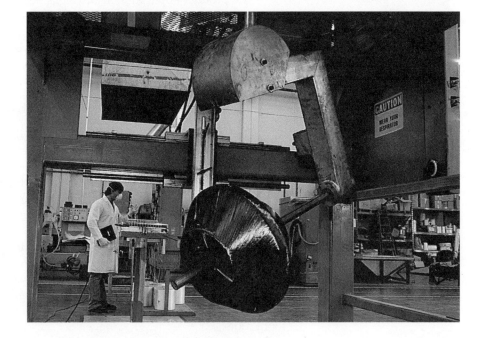

products to improve strength and the ability to withstand heavy loads. Figure 26-7 shows several applications where reinforcing rods are typically used in concrete to improve strength and durability.

An innovative approach to concrete reinforcement is shown in Figure 26-8. Pultruded small-diameter rods, composed of strands of S-glass and epoxy, are twisted into a cable and used as reinforcement in prestressed

Figure 26-7
Steel reinforcing rods are embedded in concrete to improve its ability it carry heavy loads, as shown in these examples. Sometimes wire reinforcement is used, with the wire stressed (placed in tension) either before or after the concrete sets. Prestressing and post-stressing are widely used in casting large concrete beams for bridge and building construction.

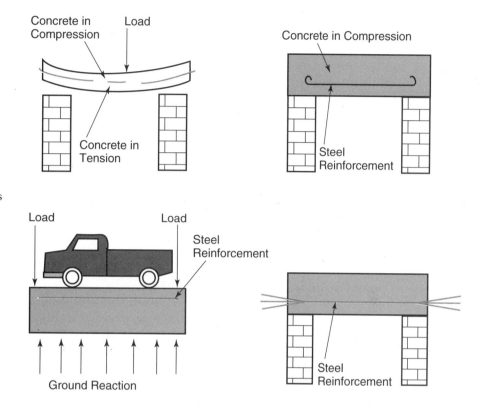

Figure 26-8
A unique reinforcement for concrete is being produced from rods of high-tensile-strength fiberglass embedded in epoxy resin. The rods are twisted into a cable and used in prestressed beams that will be exposed to corrosive salt-water environments. (Polygon Company)

concrete structures. The composite reinforcement stretches only one-third as much as steel and has a very high tensile strength. Its major advantage, however, is its resistance to corrosion, making it a desirable replacement for steel reinforcement in such concrete structures as bridges.

Particulate Composites

Layered fibrous composites are materials in which layers of reinforcement are built up so that the fiber strands are oriented in different directions. This provides greater strength and stiffness.

Particulate composites are made by suspending particles or flakes of material throughout the matrix. Particulate composites have a tough and ductile matrix, carrying a second additive in the form of hard spherical particles. Such composites provide tremendous hardness and high compressive strength. In one particulate composite, silicon carbide is embedded in a metallic matrix, producing a material that is highly abrasion-resistant.

Powder metallurgy is often used to produce metal-matrix composites. In this method, a powdered alloy is combined with a powdered metal-matrix material. Many different matrix materials can be used. Ceramic, metal, and cermet matrices are popular.

You will recall, from your reading in the metals section of this text, that *cermets* are composites made from hardened particles. Figure 26-9 shows a cermet cutting insert removing metal from a large cylindrical workpiece. Cermets are used for many different types of cutting tools, ranging from inserts for lathe and mill cutters to drill bits and friction devices used in nuclear fuel production. Figure 26-10 shows some of the geometries available in cermet inserts for lathe cutting tools.

Figure 26-9
Cermet composites are widely used as replaceable cutting inserts on tools for lathes and milling machines. This cermet cutting tool insert is being used to shape alloy steel on a lathe. (Kennametal, Inc.)

Figure 26-10
A variety of shapes is available in cermet cutting tool inserts. These inserts are for use in lathe tools. (Kennametal, Inc.)

Many different types of engineered and natural materials are used for composite matrices. Concrete, asphalt, particle board, and flake board are examples of composites that use particulate fillers. See Figure 26-11. Epoxy, polyurethane, and nylon are used for matrices as well.

Figure 26-11
The asphalt paving material being compacted by this heavy roller is a particulate composite—it consists of a filler of fine stone chips in a matrix of bituminous material. (Jack Klasey)

The particulate that is used can take a variety of different forms—powders, beads, rods, crystals, or whiskers. Sometimes the particulate is in the form of *flakes*, larger flat particles arranged parallel to each other. Metallic flakes have even been added to polymeric matrices to improve their electrical properties and act as radiation shielding.

Recent research conducted by Dr. David MacInnes, a chemist from Guilford College in North Carolina, has shown that tiny amounts of silver can be mixed with polymers to create a conductive gel. This gel can be made in any consistency from a liquid to a solid. Some firms are considering the use of the material (in a putty-like consistency) to attach electrical leads on appliances. No soldering or crimping of leads would be necessary—the wires would be pushed directly into the conductive "putty." Other types of composites are being created by filling a skeletal matrix with another, less costly material (often a particulate). In most cases, the filler is added to reduce cost, rather than to serve as a reinforcing agent. Both natural and synthetic fillers are used, depending on the type of matrix. For example, natural fibers have been used as fillers in urea and melamine plastic parts for many years.

Laminar Composites

Stock for laminar composites is available in sheet form. **Sheet stock** is manufactured by laminating two or more layers of materials together. The sheets of material, which may be unidirectional fibers or woven fabric, are called *lamina*. The alternating lamina of sheet stock are normally made of different materials.

An early example of a laminar composite was the wooden aircraft, the *Mosquito,* that was built to be flown in combat during World War II. Lamina of plywood were sandwiched with thick cores of light balsa wood.

Today, many different core materials—foam, plastics, paper, metals, and ceramics—are used. Honeycomb, waffle, and cellular sandwiches are constructed using lamina of metal foil, paper, or fibers. Panels of all shapes and configurations are made using laminated composite materials. Fiber-reinforced plastics are also made with the basic process of lamination. Forming might be accomplished using open molding or press lamination.

Coating and cladding processes are also used to produce laminar composites. Coins are made by cladding metal to metal. For products that must withstand the conditions of severe environments, inorganic laminates are often clad to metal. Inorganic to organic laminates are used as protective shields and lenses.

A closer look at how the lamina work together to create an improved product, safety glass, may be useful here to our understanding of the capabilities of composite materials. Ordinary glass is brittle. If it is broken, it will shatter into many sharp-edged pieces, making it dangerous to the user. Polyvinyl butryl plastic is a tough material that withstands strains and stresses without fracture. It is also flexible, but is soft and easily scratched. **Safety glass** is a composite created by sandwiching polyvinyl butryl between two layers of glass. The glass provides stiffness and resists scratching, while the polyvinyl butryl provides toughness and adhesive qualities.

It is doubtful that there are any manufacturing processes which are used solely with composites. Many of the processes previously discussed in conjunction with metallic, ceramic, plastic, and wood materials can also be used to shape composites.

Sheet molding processes are used to produce plastic sheet stock. Other processes, such as injection molding, resin transfer molding, prepreg tape layup, pultrusion, filament winding, and vacuum forming are also used with polymeric-matrix materials. What is important to remember is that since several different materials are being combined to make a composite, the basic process may be somewhat different. In Figure 26-12, a composite product using strands of fiberglass and roving is being produced, using the *pultrusion* process.

There are dozens of imaginative processes previously presented in this text in conjunction with metal, wood, plastic, and ceramic materials that can also be used to shape composites. In order to build on what has already been presented, the emphasis in this section of the text will be on processes for forming, separating, fabricating, conditioning, and finishing that have *not* been covered in previous sections. It should be recognized, however, that processing of composite materials is a field that is still in its infancy. There is much yet to discover about the properties of these new materials, particularly in the design of more durable tools and molds, the use of materials with faster curing times, and the reduction of waste.

Processing Composite Materials

The characteristics of the matrix determine the strength and toughness of the composite material, as well as its resistance to impact, water absorption, chemical attack, and other environmental conditions. In other words, the matrix is the *limiting factor* in composite structures.

Figure 26-12
In this pultrusion equipment, strands of fiberglass, roving fabric (on roll at top), and a polymeric matrix are combined to form the shaped rod emerging from the die block at left.
(Creative Pultrusions, Inc.)

Because of the many possible combinations of fillers and matrices, materials can be developed that will meet the requirements of just about any application. Composites are often perceived more as customized *structures* than as materials. Engineering composites can be developed to address unique product requirements, such as strength, stiffness, corrosion and wear resistance, weight, fatigue life, chemical resistance, thermal insulation, and thermal conductivity. Composite materials can be designed to last longer than conventional materials in just about any type of abusive environment.

Of the four major types of composite matrices, the polymeric-matrix composites (PMCs) are most popular. Fillers can normally be added with little difficulty during the forming process.

Polymeric matrices can be made from thermoset or thermoplastic resins. In most cases, PMCs are made from polyester resins reinforced with low-strength glass fibers. Figure 26-13 shows fiber strands being impregnated with resin before they are pultruded.

There are three steps involved in the manufacture of a fiber-reinforced polymeric composite. First, the fiber is impregnated with resin. Next, it is then placed into the matrix and formed into the desired shape. Forming is usually accomplished with such plastics processes as:

- Injection molding.
- Resin transfer molding.
- Pultrusion.
- Filament winding.
- Prepreg tape layup.

Figure 26-13
Strands of low-strength fiberglass are shown being impregnated with resin prior to being pultruded into the desired shapes. (Creative Pultrusions, Inc.)

After the composite material is formed, the part is cured and finished. Some of the newer advanced polymeric-matrix composites use high-performance fibers such as carbon/graphite, aramid, glass, or other high-strength organic fibers. Currently, these fibers are more expensive than E-glass or similar materials, but they yield the type of high-strength products required to meet the needs of the aerospace industry. Pound for pound, composites are stiffer and have greater strength than metals.

With most structural composites, the distribution of filler (fibers, particulates, or fabric) in the matrix is random. The filler is held in suspension, and never mixes completely with the matrix material. Some structural composites are made by laminating materials together in plies. Plywood, and laminated products of wood and plastic, are examples. Other materials can also be laminated together to create new composite structures.

The capability of consolidating a large number of parts into one, through processes such as lamination or encapsulation, is a major reason to use composites. This reduces assembly cost of the completed product. For example, automobiles once had some 1500 individual structural parts. Now, manufacturers have reduced this total to a few hundred parts by using composites and adopting *design for manufacturability* concepts.

Not all composites are made from solid materials. Composites can also be created from chemical reactions. What is unique about liquid composites is that the two materials are never mixed together completely, as they would be if processes such as blending or alloying were involved. With liquid composite materials, the original integrity of each material is maintained, but the matrix may be soluble enough to permit the materials to react with each other. It is this interaction of chemicals that produces new physical properties in the material.

The Nature of Composite Stock

Many different types of stock are available for manufacturing products of composite materials. This is because of the diversity of products made from composites, and the great variety of manufacturing processes that can be used to make those products.

Often, the manufacture of composite stock is accomplished during the forming or fabrication process used to develop the final product. At other times, the stock is purchased from an outside vendor.

If the final product is a laminar composite, stock is usually purchased in sheet form. Sheets are available with fiber distributed throughout the matrix. Fibers and matrix material can also be purchased separately. Fibers are available in bundled form or as continuous roving.

Prepregs

If a polymeric-matrix composite product is being developed, the manufacturer may wish to purchase *prepregs*, which are sheets of matrix material with the fibers already saturated with resin.

Prepregs are even available in tape form, with the unidirectional fibers suspended in an epoxy matrix. The tape has a removable backing that keeps the material from sticking together when it is rolled.

Depending on the type of composite desired, the raw material may also be purchased from suppliers or manufactured in-house as liquids, pellets, fibers, whiskers, ingots, masses (clay), particles, or powders. With the capability of manufacturing new materials in space, there are unlimited possibilities.

Examples of products made from composite materials can be found in just about every field. Carbon black, used to make rubber, is a composite; so is portland cement mixed with sand.

Bimetallic strips for thermostatic controls are a composite: high-expansion copper alloy is rolled together with low-expansion steel. When the material is heated, the two metals expand at different rates, causing the strip to bend and open (or close) electrical contacts.

Engineered composites have assumed growing importance since the 1960s, when organic-matrix composites (polymers) were the "state-of-the-art." Carbon fiber technology in metallic matrices was introduced late in the 1970s. Ceramic-matrix composites were born in the 1980s. What is ahead can only be left to our imagination. Materials research in space, beginning with permanent inhabitation of the space station, will undoubtedly enable thousands of new and imaginative possibilities. Composite materials are being developed by most of the countries of the world. The United States, Canada, Japan, France, and the United Kingdom are presently conducting most of the research and development work with composites.

Important Terms

advanced composites
cermets
composite
design for manufacturability
engineered composites
filler
forging
lamina
layered fibrous composites
matrix
natural composites
particulate composites
polymeric-matrix composites
prepregs
safety glass
sheet stock
structural composites
synergism
whiskers

Questions for Review and Discussion

1. What is the most common type of composite? Why do you think this is the case? Does this have anything to do with the ease of forming the material using manufacturing processes?
2. How would fillers be added to a polymeric matrix being injection-molded? Remember that this is a closed molding process.
3. Why would you want to add carbon fibers to a metallic matrix?
4. Discuss the major advantages of composites over conventional materials. If they are more expensive, why would they still be desirable in many applications?

27

Forming Composite Materials

Key Concepts

- Δ Forming processes used with composites are usually modified versions of processes used to form the material used as the matrix.
- Δ When open-molding polymeric-matrix composites (PMCs), vacuum bagging is often employed to compress the composite material.
- Δ Chemical vapor deposition is used to form large-diameter carbon fibers for reinforcement.
- Δ Alumina whiskers used to reinforce metal-matrix composites are grown in a heated atmospheric reaction chamber.

A *forming process* changes the *shape* of a material, without significantly reducing its mass or weight. Different materials exhibit different amounts of expansion and contraction during forming. In terms of the basic materials, ceramic parts are more likely to shrink during drying and curing than any of the others. This can result in some reduction in size and mass.

The selection of a forming process for use with a particular composite is controlled by the filler and matrix that are used. The operational characteristics of the manufacturing process must also be considered before the most appropriate composite is chosen.

With most material families, it is relatively easy to separate the types of processing actions used by primary and secondary firms. Primary firms are responsible for the manufacture of industrial stock; secondary firms use this stock to manufacture the final product. In the field of composites, however, it is more difficult to distinguish between primary and secondary types of processing. Some companies produce the basic composite in their plant, and then go on to make the final product. Others purchase the basic material from a supplier.

You will recall that there are four basic types of matrices used with composites: metallic, polymeric (plastic), glass, and ceramic. It might seem logical to assume that any process used to shape metals, plastics, glass, or ceramics can also be used to shape composites made from these materials. This is true, but in most cases, some modification of the process is necessary. The addition of fillers, in the form of fibers, particles, flakes, or sheets, adds a different twist to how these manufacturing processes are used.

How would the filler particles, flakes, or sheets be introduced to closed molds used in injection or resin injection molding? What about hot isostatic pressing or metal casting—when are the fibers added, or for that matter, *why* would you add fibers to molten metal anyway? These and similar questions offer many challenges and opportunities for the technologist interested in shaping composite materials.

Forming Processes

Some of the major processes used to make *polymeric-matrix composites (PMCs)* are pultrusion, injection molding, reaction injection molding, hand layup, compression molding, and laminating. Processes typically used to form *metal-matrix composites (MMCs)* are bending, rolling, and powder metallurgy. Processes such as hot isostatic pressing and casting are often used to form *ceramic-matrix composites (CMCs)*. Other processes such as centrifugal casting and lamination are often used to make *glass-matrix composites (GMCs)*. Let's take a closer look at the ways in which these processes are used to form composites.

Open Molding

The simplest way to form polymeric-matrix composites is the tried and true process of *open molding*. Materials can be applied with either hand layup or spray-up. After the resin and fiber are applied and rolled, the composite is cured using a catalyst or heat assist.

Sometimes, *woven roving* is used, but in many cases fibrous composites are formed using prepreg tapes. The use of tape simplifies the process of arranging plies according to the design requirements of the end product. After the proper plies are applied, the composite system is usually compacted using the vacuum bagging process.

Vacuum bagging

When open molding is involved, vacuum bagging is normally used to compress materials during curing. Once the desired matrix and filler materials are applied, the bagging film is placed over the mold. As described in Chapter 9, a seal is then drawn using partial vacuum. This causes the film to tighten and smooth out, and the draw is completed. The mold, composite workpiece, and bag are then placed in a curing oven to hasten curing.

Autoclave molding

Often, the conventional oven is replaced by autoclave molding. *Autoclave molding* produces a more dense part using higher pressures. Autoclave molding involves applying both pressure and an elevated temperature throughout the cure cycle.

Autoclave molding is often used to produce carbon fiber composite parts for the aerospace industry. Figure 27-1 shows workers preparing a graphite-epoxy composite tool that will be used with autoclave molding to make composite parts. The tool uses graphite/epoxy 350°F (177°C) prepreg material.

The major advantage that autoclave molding has over vacuum bagging is that autoclave molding pressurizes the part from outside of the plastic membrane. This helps to remove air inside the bag early in the cycle, then helps to consolidate the part during final curing.

Figure 27-1
These workers are using graphite/epoxy prepreg material to prepare a composite tool for use with autoclave molding. (Compositek Corporation)

Pressure Roll Bonding

Metallic matrix composite systems constructed of such material combinations as stainless steel/aluminum or tungsten/aluminum are made with the *pressure roll bonding* process. In this process, layers of sheet material (the matrix) are positioned against fibers in alternating plies. The composite sandwich is heated, then pressed or rolled to form the desired laminate. One of the limiting features of this process is that the product is cured as it is rolled. This is time-consuming, since care must be taken to avoid air bubbles and chemical reactions at the point of contact between the matrix and fibers.

Liquid Infiltration

Composite systems of steel/epoxy, boron/epoxy, steel/aluminum, glass/epoxy, copper/tungsten, and molybdenum/copper are being made using the process of *liquid infiltration.*

There are two basic approaches used with this process. The first involves adding the reinforcing fibers to a molten matrix, using either vacuum or gravity.

In the second approach, fibers are placed in the opened mold, and then the molten matrix material is poured in. As in the case of conventional open molding of plastics, careful process control is necessary. Sloppy procedures can cause defects that will show up when the fiber and matrix undergo thermal expansion.

Plating

Plasma arc spraying and electrodeposition are processes that are sometimes used to coat metal-matrix composites. Steel/aluminum, boron/titanium, and boron/aluminum composites are formed using plasma arc spraying.

Plasma arc spraying of MMCs begins with winding the reinforcing fibers around a spinning mandrel. Then, powdered matrix metal is melted in the plasma torch and sprayed directly on to the fibers. With plasma arc spraying, the coating is applied in layers. When the composite workpiece reaches the desired thickness, it is removed from the mandrel and is hot-pressed to improve density and speed up the curing process.

Composites are also plated with the *electrodeposition* technique: a mandrel acting as a cathode works in conjunction with a consumable anode of matrix material. The anode is held in an electrolytic bath.

As was the case with plasma arc spraying, continuous filament is wrapped around the spinning mandrel and matrix material is deposited on the windings. The amount of material applied depends on the type of matrix being used, and its rate of deposition.

Chemical Vapor Deposition

When large-diameter carbon fibers are to be made for use as the filler in a composite, the process known as chemical vapor deposition might be used. *Chemical vapor deposition* is similar to the electrodeposition process used to deposit coatings on metal-matrix composites.

With chemical vapor deposition, no matrix is involved. Instead, a solid material is formed through the decomposition of gaseous molecules on the heated *filament* (fiber). The filament is pulled through a cylindrical heating chamber, where reactive gases are placed in contact with its surfaces.

The reaction chamber heats the gas through either resistance or radio-frequency heating methods. The formation of coatings takes only fractions of a second, but costs for the gases, equipment, and reactor are high. Despite its high cost, chemical vapor deposition is a valued process because it can be used to create almost any type of large-diameter fiber.

Coextrusion

Coextrusion, also referred to as *continuous extrusion,* is an extension of the basic process of extrusion. Continuous extrusion is particularly useful for constructing composites, since it provides better adhesion between matrix and filler than either the open molding or liquid infiltration processes. Continuous extrusion permits the blending of different colors and types of materials.

In the coextrusion process, the matrix and filler (reinforcement) are continuously extruded together through a forming die. The advantage of continuous extrusion is that the process ensures a better distribution of matrix and filler.

The die that is used for continuous extrusion includes several different pathways or *channels.* These permit the combining of separate sources of matrix, filler, and adhesives into a single stream of composite material that is continuously extruded from the nozzle. Multilayered composite laminates are produced from broad, thin nozzle tips, while other types of products might require the use of round, square, or star-shaped tips. When coextrusion is involved, various materials are carried to the mixing nozzle, where they are blended and deposited in a common stream just before the material leaves the nozzle. Mixing tips have several inlet channels that carry the various materials through the nozzle to the point where they combine and leave the extruder.

Reinforcement or other core materials can be fed into molten polymer and extruded as bar stock. They can also be pressed into composite laminate sheet stock.

Composite products such as refrigerator liners, construction panels, and sheet stock for thermoformed containers are often manufactured using continuous extrusion.

Expansion Molding

A two-part expandable mold (tool) is used for the process known as *expansion molding.* The tool is usually made of silicone rubber.

A prepreg is placed inside the mold, and the mold is placed between two press platens. The platens provide support and keep the exterior walls of the mold from expanding. The entire assembly is then placed in an oven for curing.

The silicone rubber of the tool (mold) will expand until it reaches the curing temperature. The tool then stops expanding and provides forming pressure necessary to compress the prepreg.

Elastomeric Molding

In *elastomeric molding,* pressure is generated against a shaped preform by an expandable elastomeric bag. The elastomeric mold is made using extrusion, casting, or press molding. Silicone is the most frequently used elastomer.

The construction of composite parts using this process requires the use of an elastomeric liner and a preform. The preform is inserted in the elastomeric liner. When the elastomer is heated, it expands and applies pressure to the composite workpiece.

Fiber Spinning

Boron fibers, about the same thickness as a human hair, were the first sophisticated engineering fibers used as reinforcement filler in composite systems. Today, black carbon fibers thinner than a human hair are being produced with almost twice the fatigue resistance of steel. These fibers are often incorrectly referred to as "graphite" fibers.

Carbon fibers are 99.9 percent chemically pure carbon. They are now one of the most popular fillers for use with polymeric-matrix composites (PMCs).

The preferred polymeric-matrix material for carrying the carbon fibers is epoxy. Polyester, polyimide, and thermoplastic resins are also popular.

Carbon fibers are lighter than fiberglass, with higher stiffness. All glass fibers are strong. So are aromatic polyamide (aramid) fibers like DuPont's Kevlar®. However, carbon fibers have much greater tensile strength and elasticity than any of these fibers. When tested using Young's modulus of elasticity, some carbon fibers are four times as elastic as polyamide fibers.

When carbon fibers were first introduced in the 1960s, they were very expensive ($500 per kilogram), due to high production costs. Today, the price is significantly less, but carbon fibers are still relatively expensive. The high cost is still due to the fact that these precision fibers are difficult to manufacture, since the primary raw material—petroleum pitch—is comparatively inexpensive.

There are four major steps in the production of carbon fiber. The first step is spinning, the process that forms the basic fiber. This is done in the same way that fiber spinning is accomplished with other materials.

The second step, *stabilization,* keeps the fiber from melting throughout the various high-temperature treatments, and prepares the fiber for the third step, *carbonization.*

When the fiber is being stabilized, it is carried through an oven heated to between 400°F to 575°F (200°C to 300°C). Stabilization causes *oxidation,* changing the internal structure of the thermoplastic fiber so it will no longer melt. Additional thermal treatment will result in carbonization.

Carbonization removes most of the noncarbon elements from the fiber. Carbonization is accomplished by slowly heating the fibers to 1800°F to 2700°F (1000°C to 1500°C). As a result of the previous stabilization step, the raw material is carbonized without melting. This is an important step, since 99.9 percent of the content of carbon fiber is carbon.

The final step, *graphitization,* is an optional thermal treatment used to improve the elasticity and texture of the fiber. Graphitization is done by rapidly heating the fibers to approximately 5400°F (3000°C) and holding them at that temperature for a short length of time.

Spinning is the most costly step in the production of carbon fibers. There are three variations of the spinning process used by industry. They are referred to as melt spinning, dry spinning, and wet spinning.

Melt spinning is the process that is preferred by most polymer manufacturers. This is because melt spinning is a fast production process, and it has fewer cooling problems associated with the release of solvents. It saves time by eliminating the need to wash coagulated fibers. Melt spinning is the process that is often used to produce monofilament nylon line and glass strands.

However, melt spinning cannot be used to produce all varieties of carbon fibers. Some must be produced by either wet spinning or dry spinning.

The differences in types of carbon fibers depend on the method of dispersing the fiber and the quantity of fibers that can be drawn through the spinneret. The *spinneret* is a plate with holes drilled in it, through which the fibers are pulled.

With *dry spinning,* fibers are formed in a solution. Approximately 2000 strands are formed, using a ring-type spinneret. When *wet spinning* is done, the fibers are formed in a solution, and then allowed to precipitate in a coagulation bath. Wet spinning permits handling more fibers at one time than dry spinning. Wet spinning machines are being used with multiple spinnerets having as many as 320,000 holes. This permits manufacture of 320,000 filaments simultaneously.

Filament Winding

Filament winding can be done using either the dry or wet winding processes. Both types of winding involve the use of a forming tool called a *mandrel.*

Both types of winding processes are used to make products such as tubes, struts, automotive drive shafts, pressure vessels, and support members. Other applications include nose cones, underwater buoys, rocket engine cases, and helicopter blades.

Let's take a moment to look closer at how wet winding works. Wet winding is accomplished in the following steps:

1. Resin is mixed in a dip bath.

2. Fiber is impregnated with resin by dipping it into the bath.
3. Resin-impregnated fiber is wound around a rotating mandrel. The mandrel is coated with mold release compound.
4. The fiber is wrapped with prepreg tape.
5. The composite is permitted to gel and cure.
6. The mandrel is removed.
7. Product is finished and inspected.

The major difference between wet and dry winding is that, in *wet winding,* the filament is impregnated with resin just before it goes onto the mandrel. Wet winding is a fairly rapid process. Some manufacturers have produced tubes 150 cm x 6.5 cm (59 in. x 2.6 in.) with a wall thickness of 3 mm (0.1 in.) in approximately one minute.

Tubular products are made on winding machines that vary in construction from simple lathe-type winders to complex computer-controlled production machines. Portable equipment has also been designed for winding large storage tanks on location.

Dry winding uses prepreg tape or roving. Dry winding is normally accomplished using prepregged *fiber tows* (bundles) or tape. Dry winding is a popular process for manufacturing fishing rods and golf club shafts. The process usually involves winding the prepreg onto a warmed mandrel.

Mandrels

As noted, in both wet and dry winding, the fiber is wrapped around a mandrel. Some mandrels have open ends. Others permit the winding of parts with closed ends. Other types of mandrels are collapsible for ease of removal.

Some mandrels are made of low-melting-point metals. These types of mandrels can be melted, and the molten metal used to cast another mandrel.

Other types of mandrels are inflatable. Once they are wound and the product is cured, they are deflated and removed. Even cardboard or wooden tubes are used, in some cases, to wind parts as long as 30 or 40 feet.

Winding using a eutectic salt mandrel

Another interesting type of mandrel is made of eutectic salt. This type of mandrel is good for making large-diameter hollow parts. Here's how it is used.

Eutectic salt (a salt with a low melting point) is melted at a temperature of from 400°F to 500°F (204°C to 260°C), then poured into a two-piece ceramic mold. The salt cools and hardens to form the mandrel.

When the salt mandrel has solidified, the mandrel is removed from the mold. Laminate material is then brushed on the outside of the salt mandrel and allowed to cure. At this point, the mandrel is ready to be wound. After it is wound, the part is cured and the salt mandrel is removed by washing it out of the part with water.

Whisker Growing

The short reinforcing fibers called *whiskers* were discovered in 1952 by two researchers from Bell Laboratories. Their research revealed that the strength of their engineered metallic tin whiskers was 1,000,000 psi (6 895 0000 kPa). Since that time, research with whiskers has made slow progress.

A great deal of experimentation is taking place with whiskers used in metal-ceramic matrices. Some researchers are even investigating the use of rice hulls in MMCs. These rice hull whiskers are being used in aluminum and other low-melting-point metal matrices, with good results.

Whiskers are short fibers about 0.5 in. (12.7 mm) long. They have polygonal cross sections with large length-to-diameter ratios. Engineered whiskers are nearly perfect crystals that are grown in heated, highly temperature-controlled atmospheric reaction chambers.

Figure 27-2 illustrates the principle of growing alumina whiskers. The process is complicated; a simplified description of the basic process follows.

First, an aluminum charge is placed in a ceramic dish, or *boat*. The boat is then placed in an atmospheric reaction chamber, and the furnace is heated. Heated hydrogen forms vapors of aluminum oxide. The aluminum oxide reacts with oxygen that is present in the form of water vapor given off from the pores of the ceramic dish. The reaction forms solid, needle-like crystals of alumina that grow on the surfaces of the boat and its contents.

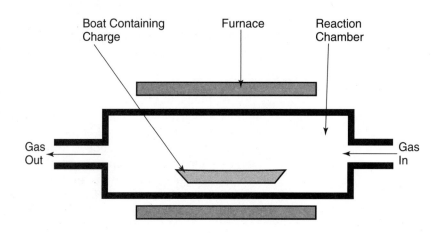

Figure 27-2
The reaction of aluminum oxide gas with oxygen inside the furnace forms needle-like crystals of alumina ("whiskers") that grow on the surfaces of the boat and the aluminum charge it contains.

Thermal Expansion Resin Transfer Molding

Thermal expansion resin transfer molding (TERTM) is a variation of the basic process of resin transfer molding. In this process, a preformed rigid cellular (foam) core is wound with reinforcement material.

Core materials are usually polyvinyl chloride, polyurethane, or polyamide rigid foams. Reinforcement materials of carbon fibers, graphite, fiberglass, polyester or Kevlar® are also popular.

Once the core is wound, it is placed inside a matched die, and is impregnated with epoxy resin. At this stage, the mold is only partially filled. Heat is applied and the foam expands more completely. This creates pressure that squeezes the reinforcement and matrix materials together.

Another interesting use for core materials in composite products is shown in Figure 27-3. Cardboard honeycomb is used as the core for a number of laminated products, including wall sections, doors, and structural supports. The application has been so popular that the manufacturer is

Figure 27-3
The hexagonal cells of cardboard honeycomb material provide strength and rigidity when laminated with sheets of facing material, as shown at bottom. The composite material permits building structures that are strong and very light. (Hexacomb Honeycomb Corporation)

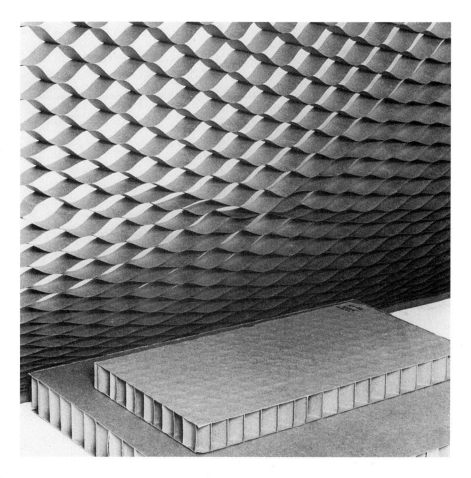

producing actual cardboard houses for installation in arid (dry) environments. The walls are made entirely of cardboard honeycomb composite panels. Refer to Figure 27-4.

Figure 27-4
Walls made of cardboard honeycomb composite panels are being used to build entire houses for installation in arid environments. (Hexacomb Honeycomb Corporation)

Important Terms

- autoclave molding
- carbonization
- ceramic-matrix composites (CMCs)
- chemical vapor deposition
- coextrusion
- dry spinning
- dry winding
- elastomeric molding
- electrodeposition
- eutectic salt
- expansion molding
- fiber tows
- filament
- filament winding
- forming process
- glass-matrix composites (GMCs)
- graphitization
- liquid infiltration
- mandrel
- melt spinning
- metal-matrix composites (MMCs)
- open molding
- oxidation
- plasma arc spraying
- polymeric-matrix composites (PMCs)
- pressure roll bonding
- spinneret
- stabilization
- thermal expansion resin transfer molding (TERTM)
- wet spinning
- wet winding
- whiskers
- woven roving

Questions for Review and Discussion

1. Explain how chemical vapor deposition can make coatings without a matrix to depend on.
2. How could carbon fibers be added to polymeric/plastic matrices? Give some examples.
3. What is melt spinning? How is it accomplished?
4. What is the difference between melt spinning and filament winding?

28 Separating Composite Materials

Key Concepts

- △ Diamond wire sawing is the preferred method for sawing expensive, extremely hard composites.
- △ Laser machining is most useful in applications where traditional processes have limitations, such as cutting and drilling superalloys.
- △ There are three basic types of lasers used in manufacturing today: CO_2, Nd:YAG, and excimer.
- △ Ultrasonic machining is used extensively with plastics, and is highly suitable for cutting composite prepregs.

There are three different types of separating processes, as described in earlier chapters of this book. One type uses mechanical forces to generate a chip *(mechanical chip-producing separating processes)*. A second type uses mechanical forces, but does not produce a chip *(mechanical non-chip-producing separating processes)*. The third type of process separates material without using mechanical force *(non-mechanical separating processes)*.

Separating Processes for Composites

Composite products are normally formed in the size and shape desired ("near-net-shape"). Because they are produced in the shape desired, the major reason for separating composite materials is to trim parts.

Conventional trimming tools such as the bandsaw, router, hacksaw, belt sander, reciprocating saber saw, and circular saw are used for many trimming applications. A less-conventional trimming tool, a 5-axis robotic router, is shown in Figure 28-1. The router is trimming a dash unit for a street sweeper to close tolerances. This chapter will be devoted to the more advanced trimming tools used with composites.

Diamond Wire Sawing

The wire that is used to saw composites is a round, high-tensile-strength copper-plated wire. The surface of the wire is impregnated with fine particles of industrial diamond that serve as the cutting "teeth." **Diamond wire** can be used, with a reciprocating motion, to cut extremely hard materials with

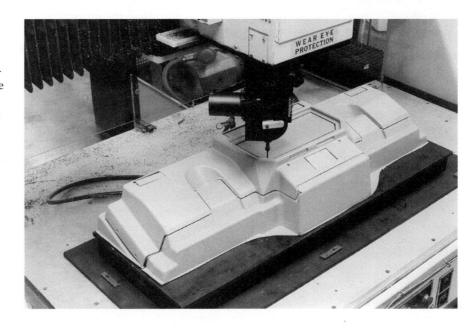

Figure 28-1
This composite part, a dash unit for a street sweeper, is being trimmed to close tolerances by a robotic router. The 5-axis machine can be programmed to precisely repeat the trimming operation on each workpiece. (PMW Products, Inc.)

virtually no scrap loss. Cuts are burr-free, and generate very little frictional heat. This makes the process particularly useful for cutting heat-sensitive semiconductor materials.

Diamond wire has been used to cut materials ranging from exotic composites, moon rocks, and medical specimens to radioactive nuclear fuel rods and iridium metal (a material so rare that it is worth nearly $500,000 for a 4 sq. in. sample). Wire sawing is a cost-effective solution for cutting expensive materials, since diamond wire produces a saw *kerf* (cut) only 0.009 in. (0.229 mm) in width.

Diamond-impregnated wire is useful for cutting fragile materials that would be cracked or broken by other cutting methods. Diamond wire saws have been successfully used to cut boron and tungsten composites, brass and bronze, carbon and graphite, carbon fibers, ceramics, chromium, crystals, fiber composites, gallium arsenide, garnet, glass, and gold. Exotic materials such as inconel (used for mufflers in jet engines), titanium, and indium antimonide (a fragile semi-conductor material) are easily cut with this process. See Figure 28-2.

A diamond wire saw that incorporates a table which floats on a cushion of air is shown in Figure 28-3. The circular cutting attachment is mounted on top of the table. With this saw, the round diamond-impregnated wire can cut around a corner, leaving a radius only half the diameter of the wire used. This makes possible machining of complex parts with a degree of accuracy not possible with most cutting tools.

Diamond wire saws typically use 100-200 feet of continuous length cutting wire wound on a drum. The wire has no joints or welds, and is designed to cut under high tension. In operation, the wire unwinds from one drum and is rewound on another. See Figure 28-4.

Figure 28-2
A diamond wire saw quickly and easily cut through the high-strength composite materials used for this helicopter rotor blade. (Laser Technology, Inc.)

Figure 28-3
This "Cloud 9" diamond wire saw can be used for extremely precise cutting of complex shapes, since the round worktable floats on a cushion of air. (Laser Technology, Inc.)

When small or fragile workpieces are to be cut, wax is used to affix the stock to a chalk-like ceramic support base. The wire is continuously drawn across the workpiece. A slurry of glycerin and silicon carbide is normally supplied at the point of contact.

Dicing

Dicing is a sawing process that is used to cut ultra-hard refractory material such as ceramics, ferrites, glass, silicon, and superconductors. It is also used to cut thin sections of such materials as strip-line microwave hybrids,

Figure 28-4
The diamond wire on this 24 in. (61 cm) wire saw is being used to slice the square composite workpiece into several thinner sections. (Laser Technology, Inc.).

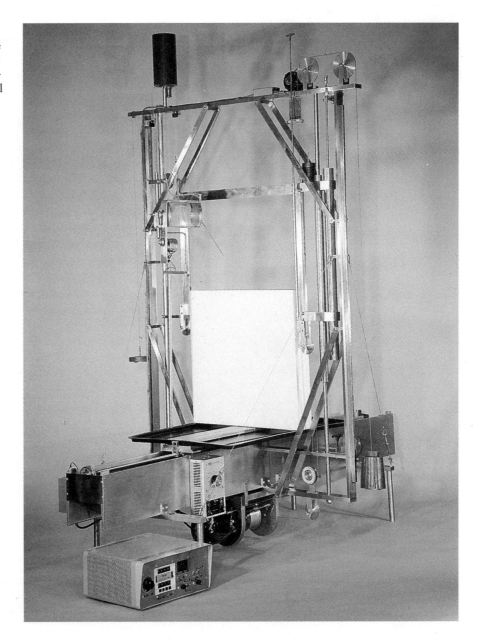

miniature chip capacitors, read-write heads, and display glass. Dicing is done on a machine called a dicing saw, Figure 28-5. Dicing saws use blades that are sintered wheels with diamond in a metal matrix bonded to the periphery of the wheel. Blades can be either metal-bonded or phenolic resin-bonded.

In dicing, both the diamond grit and diamond concentration must be considered. The *diamond grit* is the particle size. **Diamond concentration** refers to the ratio of diamond volume to the volume of the matrix material. In most cases, the higher the concentration, the faster the cutting speed that can be used for dicing.

Figure 28-5
A dicing saw uses a diamond-impregnated blade to cut very hard materials. It also can be used to cut thin sections from composites. (Aremco Products, Inc.)

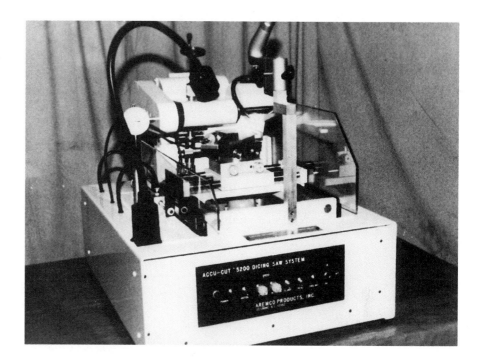

Very thin materials are diced. They are fastened to a substrate with a water-soluble adhesive for support during the dicing process. Dicing wheels are *dressed* (made sharper) by using coarse-grit dressing sticks to expose the diamond facets in the matrix. The dressing sticks also help true the blade and provide faster, cooler cutting.

Since ultra-hard materials are being cut, dicing is typically done using a coolant. Water-soluble fluids are available that permit high dicing speeds at lower temperatures, resulting in longer blade life.

Waterjet Cutting

At General Motors' Hardware Trim Plant in Detroit, Michigan, a unique type of robot was first installed in 1986. The uniqueness of these robots was that they used a stream of high-pressure water to trim thermoformed wood fiber into interior door sections. The robots were able to trim wood fiber at a speed of around 800 in. (2033 cm) per minute. The stream of water that was used to cut the material ranged from 0.004 in. to 0.014 in. (0.102 mm to 0.356 mm) in diameter, at a pressure of 55,000 psi (379 225 kPa).

The *waterjet* process provides a number of advantages when machining composites. A waterjet can be used to cut, drill, or machine. It generates no dust, requires no additional lubricant to keep the kerf clean, and does not produce heat in the cutting area.

The basic patents for waterjet technology were obtained in 1968 by Norman Franz of Ingersoll-Rand. The first commercial demonstration of the technology was made in 1971, when a waterjet was used to cut 3/8 in. (9.5 mm) pressed board for furniture forms.

Since that time, waterjet cutting has demonstrated its effectiveness in quickly and cleanly cutting materials including titanium, ceramics, glass, fiberglass, carbon fibers, stainless steel, and concrete. It is a versatile process, with applications ranging from cutting frozen meat (foods industry) to slicing bones (surgery), to extracting minerals (mining).

Waterjet cutting can be accomplished with or without abrasives. While many nonmetals can be cut without abrasives, the use of abrasives is preferred with metals and harder nonmetal materials. See Figure 28-6. The Ingersoll-Rand *hydrobrasive process* is powerful enough to cut through 3 in. (7.6 cm) tool steel at a rate of 1.5 in. (3.8 cm) per minute or 10 in. (25.4 cm) reinforced concrete at speeds exceeding 1 in. (2.54 cm) per minute. In waterjet cutting (sometimes referred to as *waterjet machining,*) material is removed by a compressive, shearing action. This makes the process faster and more cost-effective than other techniques when cutting materials such as boron/aluminum honeycomb, aluminum/boron carbide, and graphite composites.

Figure 28-6
This waterjet cutting system for sheets of flat stock up to 9 ft. by 14 ft. (2.7 m by 4.3 m) can cut at speeds of up to 500 in. (12.7 m) per minute. Computer controls are used to program movements of the cutting head in the x, y, or z axes. (Technicut)

The basic components of a typical waterjet system are a booster pump, an intensifier, and a cutting nozzle. Clean water enters the booster pump, where it passes through filters, then into the intensifier. In that component, the water velocity is increased tremendously by being hurled through tubes and swivels. The very-small-diameter jet of water exits the cutting nozzle at speeds as high as 3000 ft. (914 m) per second—approximately three times the speed of sound! Because of the very high velocity of the cutting jet, people must be kept out of the work area.

Most waterjet systems used today are computer-controlled, and many are attached to robotic manipulators (a robot arm), providing great flexibility.

There are several technical advantages of waterjet machining over conventional machining. The process is noncontacting, and because of the high pressure, cutting is accomplished with little lateral or vertical force applied to the workpiece. No airborne dust is produced because all material that is removed from the kerf is carried away by the high-pressure waterjet. A minimal amount of material is removed, since the jet's diameter is only a few thousandths of an inch.

Laser Machining

Another major advantage of waterjet machining is that there is no tool to replace. Orifices on the waterjet nozzle are made of sapphire or a similar hard material, and have an average useful life of hundreds of hours. Sometimes, nozzles can be reused after ultrasonic cleaning.

The *laser* has been used in industry for more than a generation, but is still classified as a *nontraditional machine tool*. The devices known as **programmable logic controllers (PLCs)** have been around for roughly the same length of time, and they are used everywhere in manufacturing today. This indicates that lasers are not the cure-all for every material processing application, as was once predicted. For many high-volume applications, traditional cutting, stamping, and welding processes are still more practical and cost-effective.

Lasers should be considered for applications where traditional processes have limitations. One example would be for the drilling and cutting of composites and other highly dense and ultra-hard materials.

While lasers are not appropriate for every application, they can beat many traditional manufacturing processes to the punch when flexibility, speed, or solutions to unusual problems may be needed. Lasers can be used to melt and evaporate just about any material, from honeycomb panels to engineering plastics to stainless steel, using only concentrated light energy, Figure 28-7.

Lasers are used for welding, heat-treating, and cutting. They are also used to hermetically seal electronic components, and to drill holes in superalloys. Since no tool bit is involved, lasers are even useful for drilling holes at an angle into the workpiece. With conventional drilling, if the drill is not held at a right angle to the stock, it is likely to break.

Lasers are useful for cutting and trimming composite tooling or parts to within ± 0.005 in. (0.127 mm) with a high degree of reproducibility. Lasers eliminate the presence of dust during cutting, and conserve on energy used for processing.

Lasers are often used in conjunction with computer numerical control (CNC) machine heads to provide automated cutting of laminates up to 6 in. (15 cm) or more in thickness. In most cases, the process of machining, cutting, or drilling is automated to keep operators out of the work area. See Figure 28-8.

There are three basic types of lasers used in manufacturing today. These are referred to as the CO_2, Nd:YAG, and the excimer lasers. CO_2 lasers are used for processing both metals and nonmetallics. The Nd:YAG lasers are used primarily for marking, scribing, drilling, welding, heat-treating, and cutting of metals. They are also used to transmit electrical signals through fiber optic "light pipes."

Excimer lasers are still essentially in the experimental stages of development, although they have been used in delicate eye surgery and similar applications. These lasers use high-energy gases such as argon fluoride (ArF), krypton fluoride (KrF), and xenon chloride (XeCl).

CO_2 lasers

The movies have done much to dramatize the concept of lasers, although in a context far from manufacturing. Much of what has been shown about lasers as weapons in adventure and science fiction films is theoretically impossible.

Figure 28-7
A CO_2 laser like this one can be used to cut exotic metals and many other types of materials. (Coherent General)

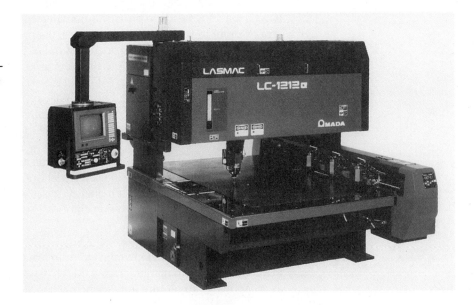

Figure 28-8
This CNC laser machine can provide high-speed cutting and piercing of metal or composite materials, with precise repeatability of actions from one workpiece to the next. It combines the flexibility of a laser cutting machine with the productivity of a traditional turret punch press. (U.S. Amada, Ltd.)

For example, it is not possible to see the beam of light projected with a CO_2 or Nd:YAG laser without special glasses or lenses. Even if such a high-energy beam were visible, it would not be exposed without protective shielding. Without protective glasses, the flash from the beam would cause blindness.

In the real world of manufacturing, lasers perform a variety of tasks. Low-powered **CO_2 lasers** in the 20 to 500 watt range are good for cutting polymers such as wood and plastic. These types of lasers are also used to cut and drill ceramics. At the low end of the power range, CO_2 lasers can be used for scribing, soldering, and trimming. More powerful units, beyond the 500 watt range, are used to weld, cut, and heat-treat metals.

It is possible to cut metal with thicknesses ranging from 1/2 in. to 5/8 in. (12.7 mm to 15.9 mm) with a 1000 watt to 1900 watt laser, but cutting material this thick can be accomplished more efficiently using conventional oxyacetylene cutting. Higher power CO_2 lasers have larger beam diameters, and are used for heat-treating and welding.

If you are interested in cutting thick metal plate using a laser, it will be a slow and painstaking process. The heat of the beam will cause burning of the metal, resulting in a cut that is fairly ragged and rough. Thus, laser cutting of thick metals is not recommended. Another reason why the laser is a poor choice for cutting hard or thick metals is that much of the light energy projected onto the workpiece is reflected away and lost. This lowers the cutting efficiency of the beam.

The CO_2 laser is well-suited to cutting and heating many types of plastics and composites. Higher-powered lasers have larger beam diameters and are used for heat-treating and welding.

Let's take a closer look at how lasers work. Remember that LASER is an acronym for *L*ight *A*mplification by *S*timulated *E*mission of *R*adiation. As you think about the process, think of the stages involved—generation of light, then amplification, then stimulation and emission of radiation.

First, electrical current is transmitted to the *lasing tube.* The tube is either a solid crystalline rod (of aluminum oxide, sapphire, or yttrium-aluminum-garnet) or a hollow tube filled with a gas (CO_2, helium, or nitrogen). When the tube is energized, it lights up and glows much like a fluorescent light tube. The tube can be energized by using alternating current (AC), direct current (DC), or radio frequency (RF). At this point, the process is very inefficient. The light that is created is very bright (several times brighter than light emitted by the sun), but is *incoherent*. This means that the light spectrum includes many different wavelengths and phases. Without some means of focusing the light and making it *coherent*, it has little or no value in manufacturing.

The second stage of the process involves passing the light energy back and forth between the cathode and anode at opposite ends of the lasing tube. Movement of the light energy through the lasing medium amplifies the light intensity. The light beam is then released to mirrors that direct it through a prism, where it is focused into a beam of the desired size. The light is now coherent, with a narrow spectrum of wavelengths that are in phase.

Once the laser beam is produced as desired it is ready to be directed toward the workpiece. The beam is normally pulsed, that is, it occurs in short on-and-off bursts. The power requirements, and the quality and size of the beam, influence the processing action: cutting, welding, or marking.

Nd:YAG lasers

The Nd:YAG is a "solid state" laser that uses a synthetic crystal lasing rod. The beam of this laser is nearly visible, and is focused with lenses to improve the output. Green safety lenses are used to view the beam of the Nd:YAG laser.

In the *Nd:YAG laser,* Nd stands for the element neodymium, an expensive metal that causes the lasing action. The lasing rod—a crystal composed of yttrium (Y), aluminum (A), and garnet (G)—is grown during the processing operation. The Nd:YAG laser is smaller than the CO_2 laser, but it is just as powerful. A 450-watt YAG crystal might be no more than 6 in. (15.2 cm) long by 3/8 in. (9.5 mm) in diameter.

There are two types of Nd:YAG laser beams, continuous and pulsed. Continuous-beam lasers rated at 25 to 50 watts are used for soldering and scribing; those rated at 50 to 100 watts are used for marking. Pulsed Nd:YAG lasers up to 450 watts are good for drilling, cutting and welding of metals.

The Nd:YAG laser is particularly useful for welding delicate components such as heart pacemakers without damaging the parts. This is accomplished because the laser emits a pulsed, highly focused beam for fractions of a second, resulting in rapid heating.

Excimer lasers

Excimer lasers use exotic gases to create a beam that pulsates on and off for periods of from 10 to 16 nanoseconds (billionths of a second). These lasers actually remove stock, molecule by molecule, rather than vaporizing it.

Excimer lasers are still in the developmental stages, and will probably not be used very extensively in manufacturing for many years. Some of the greatest successes to date with excimer lasers have been in the field of eye surgery. Laser light is used to remove unwanted tissue without generating any tissue-damaging heat.

Ultrasonic Machining

Ultrasonic machining is a process that is used extensively with plastics. It is also used to cut prepregs such as aramid fiber, graphite, boron epoxy, carbon phenolic, hybrid composites and fiberglass, as well as aluminum honeycomb core, Nomex® fabric, leathers, fabrics, films, and sealants.

With the rapid growth in the use of composites, a need developed for fast, precise, and reliable means for cutting and shaping these materials. One such method is the use of *ultrasonics* (high-frequency vibrations). The rotary ultrasonic machine tool shown in Figure 28-9 is used to process composites, ferrites, zirconium, beryllium oxide, ruby, sapphire and other difficult-to-machine materials.

This machine applies axial 20 kHz (20,000 cycles per second) ultrasonic vibrations to a rotating diamond tool. The ultrasonic vibrations reduce friction between the tool and material. This permits drilling and milling of workpieces with less pressure.

Rotary ultrasonic machining processes use a power supply to generate a high-frequency electrical signal that is transmitted to a piezoelectric *transducer* (converter). The transducer converts the electrical signal to mechanical motion. This motion is conveyed to the *horn,* which holds the rotating tool.

Figure 28-9
Vibrations at the rate of 20,000 per second, used with a rotating diamond tool, allow this equipment to perform rapid drilling and machining of composites and other materials. (Branson Ultrasonics Corporation)

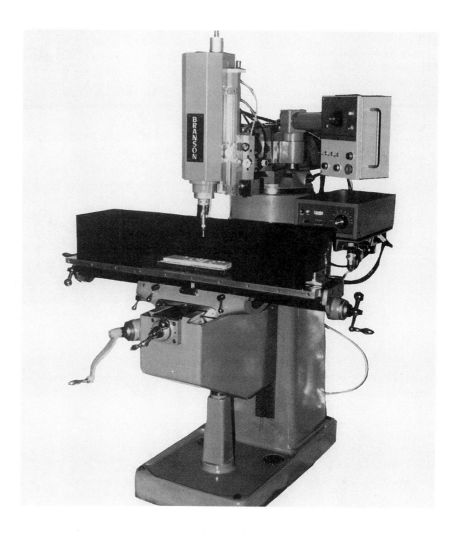

Figure 28-10 shows a close-up view of a horn extended down to the workpiece. The horn expands and contracts about 20,000 times each second. This causes the tool to vibrate in a *longitudinal* (along its length) direction, while also rotating. See Figure 28-11.

The diamond tools that are used in this process rotate at a speed of up to 4000 rpm. The combined action of the rotation and the longitudinal ultrasonic vibration provides the cutting action. Frictional heat is reduced, eliminating the need for coolant that could cause workpiece contamination. Ultrasonics enables fast and efficient cutting at lighter tool pressure than with traditional machining.

Ultrasonic drilling and milling has several advantages over conventional machining processes when used with ultra-hard materials. The process is faster, permitting continuous drilling. There is no need to withdraw a drill to flush it. Ultrasonic machine tools can drill deep, straight, small-diameter holes. As noted, minimal heat is generated in the cutting area. This makes ultrasonic machining a valuable process for use with fragile materials that are prone to thermal cracking.

Figure 28-10
The horn assembly of this ultrasonic machining system expands and contracts thousands of times each second, causing the tool to move rapidly up and down, cutting into the workpiece. (Sonic-Mill, Albuquerque Division, Rio Grande Albuquerque, Inc.)

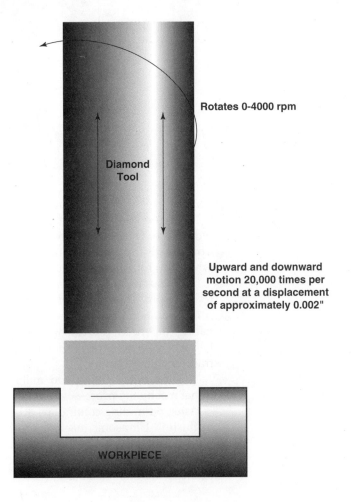

Figure 28-11
In rotary ultrasonic machining, a diamond tool is vibrated up and down, while simultaneously rotating. This provides rapid cutting action. (Sonic-Mill, Albuquerque Division, Rio Grande Albuquerque, Inc.)

Ultrasonic machining is used for a wide array of manufacturing applications, ranging from the fabrication of carbide dies and the machining of small holes in alumina substrates, to drilling of long holes through blocks for use with lasers. Ultrasonic machining can be used to drill, engrave, broach, and shape silicon, silicon carbide, silicon nitride, sapphire, glass, quartz, ferrite, alumina, alumina nitride, gallium arsenide, and many exotic composites.

Ultrasonic machining can also be done in a stationary manner, without tool rotation. This permits the cutting of holes or pockets of virtually any shape in the workpiece. The cutting tool used has a cross-section that is the size and shape of the desired hole or pocket.

As shown in Figure 28-12, a recirculating pump forces an abrasive *slurry* (abrasive suspended in a liquid medium) into the working area between the vibrating tool and the workpiece. The abrasive particles strike the workpiece at a force equivalent to 150,000 times their weight. The particles chip off microscopic flakes of material. As the tool is advanced into the workpiece, the abrasives grind out an opposite but perfect match of the toolface. The presence of the cool slurry makes this a cold cutting process.

Ultrasonic cutting

Ultrasonics can be used to drill, machine, engrave, broach, and shape materials. The principle has now been applied to hand-held ultrasonic knives that can be used to cut and trim advanced composites. See Figure 28-13.

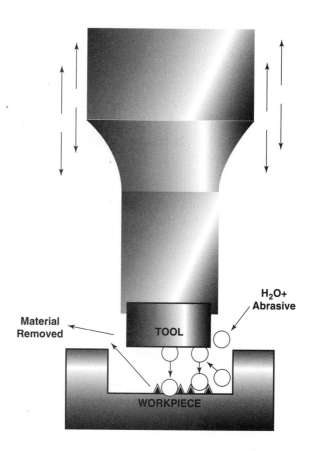

Figure 28-12
When a stationary tool is used, a slurry of abrasive suspended in water or another liquid does the actual cutting. Energy is transmitted to the abrasive particles by the rapid longitudinal vibrations of the tool. (Sonic-Mill, Albuquerque Division, Rio Grande Albuquerque, Inc.)

Figure 28-13
The blade of this hand-held ultrasonic knife vibrates 20,000 times per second, allowing it to easily cut most composites. Cutting speeds are governed by the material being cut and its thickness. (Branson Ultrasonics Corporation)

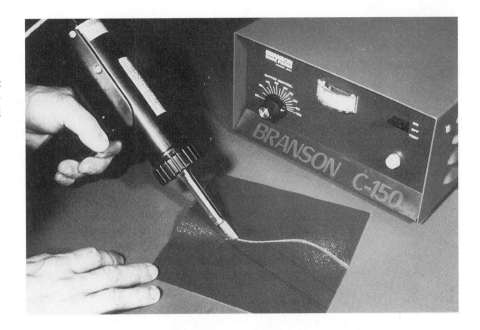

The ultrasonic trim knife has three basic components: a power supply, a transducer, and a horn/blade assembly. The power supply increases 50/60 Hz household current to 20 kHz. The high frequency energy is supplied to the transducer, which changes it into mechanical vibratory energy. This energy is transmitted to the horn and through the blade to the workpiece. The manufacturers claim that even multiple-ply composites can be cut precisely and rapidly without fiber disorientation or damage.

Important Terms

CO_2 laser
diamond concentration
diamond wire
dicing
dressed
excimer laser
horn
hydrobrasive process
kerf
laser
lasing tube
longitudinal
Nd:YAG laser
programmable logic controllers (PLCs)
slurry
transducer
ultrasonic machining
ultrasonics
waterjet

Questions for Review and Discussion

1. What are some of the advantages of diamond sawing, compared to conventional sawing, when cutting composite workpieces?
2. Explain the difference between slicing and dicing.
3. For what type of application would waterjet cutting be most suitable? Why?
4. What is the difference between ultrasonic cutting and ultrasonic machining?

29 Fabricating Composite Materials

Key Concepts

△ Pyrolysis, chemical vapor deposition, and in situ processing are important techniques used in fabricating composites.

△ Cohesive fabrication processes are typically used for custom fabrication work.

△ Various types of fibers are added to composite structures as reinforcements to improve strength and other properties.

△ Selection of the most effective fabricating process for a composite is dependent upon the type of matrix used.

△ Fabrication of superconducting composites is typically done with either powder metallurgy or in-situ processing.

Just as the separating processes used for other types of materials are used with composites, the major types of *fabrication* processes used with engineering materials are also used with composites. Of the three fabrication processes, *adhesion* and *cohesion* are used far more often with composites than is mechanical joining.

Fabrication of Composites

Cohesive bonding is a chemical process taking place between substrates. The materials are permanently joined through chemical action. We already know how this polymerization takes place in the open molding process used with plastics.

Cohesive fabrication methods used with open molding might be as simple as adding high-performance fibers such as Kevlar®, carbon, or boron during hand layup. Fibers are placed into the mold by hand and resin is sprayed or brushed over them. When a chopper gun is used, the chopped fibers are combined with the resin and sprayed together into the mold.

Most advanced composites used in the aerospace industry are fabricated by layup, rolling layers of polymer-impregnated fiber tapes onto a mold surface, and then heating the composite sandwich under pressure to cure the resin. See Figure 29-1. This type of fabrication could probably best be classified as cohesion.

Figure 29-1
Cohesion fabrication processes are used to produce the fiber-reinforced composite components of aircraft like this large military helicopter. Sophisticated composites are widely used in the aerospace industry for their combination of strength and light weight. (United States Air Force)

Other processes, such as filament winding, continuous pultrusion and vacuum molding are also used to fabricate composite parts. Processes are normally chosen depending on the type of volume that is being produced. Cohesive fabrication methods are useful when *custom* (one of a kind) parts are desired. However, when high-volume continuous production is involved, such labor-intensive cohesive fabrication methods as hand layup are typically far too slow to be cost-effective. Other fabricating processes are normally selected.

Often, an *adhesion process* will be used. Adhesion involves applying thin layers of materials to join two or more different types of composite materials. Adhesive layers are sometimes glues and cements, but many different types of materials can create adhesion. These range from frozen water and solder to waxes, hot melts, natural resins, tapes and prepreg materials. The purpose of the adhesive layer is to equalize stresses throughout the composite matrix.

There are some fabrication processes involving cohesion that are more suited to volume production. Spin welding, plastic welding, and hot gas welding are examples of such cohesive processes. Other types of cohesive processes include radio-frequency bonding and electromagnetic radiation. All these processes join the composite materials at the point of interface through melting of the surface layers.

Some composite parts are joined with friction fits, while others are joined with mechanical fasteners. One of the advantages of using composites is that mechanical fasteners often can be designed out of the product. Let's look more closely at how this might be accomplished.

Designing mechanical fasteners out of a product is possible because of the strength advantage of composites. Composites are stronger than conventional materials, so they are able to withstand greater stresses and increased loads under use. If holes for fasteners were created in a workpiece, these holes would limit the strength of the product. Since this is undesirable, the number of holes and voids in composite workpieces is kept to a minimum.

Adding Reinforcement Fibers

One of the most popular methods for fabricating composites is by adding reinforcement fibers to a matrix. The fibers that are used are normally glass, boron, carbon, ceramic, or metallic. The following sections will look more closely at the composition of these fibers.

Glass-fiber-reinforced Composites

There are three major types of glass fibers used as reinforcement in composite workpieces. These are referred to as *E glass, C glass,* and *S glass.* The type of glass that is used for electrical applications is known as E glass. It's the preferred choice for products that must serve as electrical insulators. Approximately 90 percent of all glass fibers made are of the E glass type.

Fibers of C glass are used when the product must be able to survive continued exposure to corrosive chemicals. An example of a C glass fiber use would be reinforcing containers used to handle hazardous waste and other caustic materials.

Because of their high silica content, S glass fibers added to a composite matrix are ideal for product applications where exposure to excessive temperatures is likely.

Before the advent of fiber-reinforced composite materials, most glass filaments were made by drawing the molten glass through tiny orifices, or bushings. In the late 1970s, a new process called the *sol-gel technique* became popular for making glass and ceramic fibers.

The sol-gel process combines the unique characteristics of two types of liquid, a sol and a gel. A *sol* is a colloidal suspension so thick that it shows no *sedimentation*—small particles will not settle to the bottom of the container when the solution is allowed to remain still for a prolonged length of time. In a *gel,* the liquid is so thick that it acts more like a solid.

The sol-gel process consists of three stages. First, the solution is converted into a sol. Stage two involves changing the sol into a gel. In the third stage, fibrous gels are drawn from the sol-gel solution at room temperature, then converted into glass fibers using high temperatures.

Glass fibers are available in various forms, including chopped strands, continuous yarn, roving, and fabric. Glass fibers are often used as reinforcement material with polyester, epoxy, and phenolic resins. See Figure 29-2. Glass-fiber-reinforced products are popular in the construction industry. Other products made with glass fiber reinforcement include shower and tub enclosures, boats, wall panels, tanks, window frames, and even race car bodies.

Boron-fiber-reinforced Composites

Glass fibers have low density and a high tensile strength, but are limited because they are subject to *static fatigue.* This makes them undesirable for applications where they would have to support heavy loads for continued lengths of time. When improved strength under static load is desired, advanced fibers such as boron or carbon should be utilized.

One of the unique features about boron fibers is that they are constructed as a composite. Boron fibers are brittle materials made from the chemical vapor deposition of boron onto a substrate material. Since high temperatures are required for chemical vapor deposition, tungsten wire is most often chosen as the substrate material.

Figure 29-2
These composite struts of glass fiber and epoxy are designed to be competitive with metal connecting rods for automotive and aircraft applications. A patented braiding process for the reinforcing fiber increases the strength-to-weight ratio of the parts. (Polygon Company)

The two processes normally used to deposit boron coatings are halide reduction and thermal decomposition. With *halide reduction,* fine tungsten wire is heated as it is drawn through a tightly sealed chamber. About 10 percent of the boron halide in the high-temperature reaction chamber is converted into boron as the wire is drawn through. The temperature is critical: the most significant deposition of boron occurs with the wire not moving and a chamber temperature of about 1800°F (1000°C).

Boron coating by *thermal decomposition* is done at lower temperatures. With this simpler process, carbon-coated glass fibers can be used as the substrate. The fibers that are produced are not as strong as those made using halide reduction.

Boron-fiber composites have been widely used in military aircraft such as the F-14 and F-15, as well as the space shuttles. Boron fibers are also gaining popularity for use in strengthening commercial products such as tennis rackets, golf shafts, and bicycle frames.

Carbon-fiber-reinforced Composites

Composite products reinforced with carbon fiber were first introduced in the 1950s. Since that time they have continued to find many applications in the aerospace and sporting goods industries.

Carbon fibers are usually made by carbonizing fibers of polyacrylonitrile (PAN) or rayon. This is done by exposing the fibers to a high temperature, without melting the fiber. After *carbonization*, the fiber is processed further to turn it to graphite. Graphitization was described in Chapter 27.

Matrix materials most frequently used with carbon fibers are epoxy, polyester, polysulfone polyimide, and thermoplastic resins. Polymeric manufacturing processes such as injection molding, extrusion, pultrusion, laminating, and filament winding can be used to manufacture carbon-fiber-reinforced composite products. See Figure 29-3.

Figure 29-3
These self-lubricating bearings, in a variety of sizes and types, are fabricated from carbon and polyester fibers in a high-temperature epoxy resin matrix. They are used in devices ranging from earthmovers to computer printers. (Polygon Company)

In addition to the PAN or rayon fibers used with polymeric matrices, there is another type of carbon fiber/matrix combination which is also popular. This is a carbon-fiber-reinforced carbon-matrix material, referred to as a *carbon/carbon composite (C/C)*.

Carbon/carbon composites behave more like a ceramic material than a polymeric. The C/C materials are prepared and processed differently from polymeric-matrix composites. There are three major techniques used to produce C/C materials.

The first of these techniques uses pitch from coal tar or petroleum sources to impregnate a woven preform under heat and pressure. This is then followed by *pyrolysis*, the creation of chemical changes by subjecting the material to heat. These steps are repeated as often as needed to give a product the desired density.

A second technique used to make carbon/carbon composites begins with pyrolysis to decompose the resin. This step is followed by again impregnating the preform, then repeating the pyrolysis. This serves to bond the carbon matrix to the carbon fibers.

A third technique used to produce carbon/carbon composites involves the use of chemical vapor deposition. With this process, the carbon fibers in a preform are exposed to methane, nitrogen, and hydrogen gases.

Carbon-fiber-reinforced products have found particular usefulness in harsh environments and temperature extremes, as well as with exposure to oils and fuels, acids, and hot gases. Carbon fibers can also be used with matrices other than polymers, particularly aluminum and ceramics, to provide additional unique properties.

Ceramic-fiber-reinforced Composites

Ceramic fibers have the advantage of being impervious to chemical attack; thus, they can be used in all sorts of caustic environments. Ceramics are hard, resistant to compression and abrasion, and have high thermal resistance. They are also brittle, with poor mechanical shock resistance and low tensile strength.

The most impressive metals, the **superalloys** used in jet engines, can withstand temperatures approaching 1500°F (800°C) without shielding. With the addition of special coatings, they can survive continued exposure to temperatures approaching 1800°F (1000°C).

When a material must be exposed to temperatures above these levels, ceramics offer the optimal solution. In many applications, the combination of ceramics and metallics is even better.

Often, ceramic fibers are made directly from the metal or polymeric/plastic matrix. Three major techniques are used to produce ceramic fibers: chemical vapor deposition, sol-gel processing, and polymer pyrolysis. With both sol-gel processing and polymer pryolysis, ceramic fibers can be produced from metallic and polymeric materials. By extending the pyrolysis process into ceramics, polymers with a high percentage of silicon and carbon or nitrogen can be transformed into ceramic fibers.

There are many different types and manufacturers of metallic/ceramic fibers. Alumina fiber yarns are being made by DuPont. Sumitomo Chemical Company manufactures an alumina fiber, called Saffil™, as well as other products with mixtures of alumina and silica. Under the trade name, Nextril™, 3M Company also produces a ceramic fiber, using the sol-gel process.

Metallic/ceramic fibers can be produced several ways. Many manufacturers make these fibers by developing a slurry mix of alumina particles and additives. The mix is then spun to remove water, and dry fibers are spun from the slurry.

The dry yarn is then subjected to several firings. Sometimes, a thin silica coating is applied to the yarn to cover imperfections and improve tensile strength.

Metallic-fiber-reinforced Composites

Metallic fibers such as tungsten, beryllium, and steel are used to reinforce composite materials. The advantage of using these metallic filaments is their strength. Sometimes, however, other desirable characteristics are lacking in metallic fibers. These might include the ability to withstand excessive heat or to resist chemical attack and oxidation.

Steel wire or rod is most frequently used as a reinforcement, particularly when concrete and polymeric matrices are involved. Traditional wire-drawing processes are typically used to produce fibers made of tungsten, titanium, molybdenum, and steel. Ultra-fine filaments can be produced with the ***Taylor process.***

This process starts out by encasing the metal wire in a sheath of glass. The coated wire is then heated to the point where the protective sheath softens into a plastic state. The soft-coated wire is then drawn through a tiny orifice to a fine diameter. Any remaining sheath material is removed by etching.

Ceramic and metallic fibers are also used to reinforce metallic-matrix composites. Metals are strong, but it is often necessary to modify their structural characteristics. For example, metals can be hardened, but they are generally lacking in terms of elasticity.

Another disadvantage of metals, particularly when used in many aerospace applications, is that they are heavier than composites. High-modulus fibers can be used to reinforce metals in structural applications (beams in an aircraft or a bridge, for example), and produce a more durable and lightweight product.

Fabrication of Composite Matrices

Many processes have been discussed for fabricating composite matrices. Naturally, the selection of a process depends on the type of matrix that is used. When polymeric matrices are involved, one of the most popular manufacturing processes that might be used is lamination.

With lamination, prepreg sheets are stacked together to make a composite sandwich. The sheets are separated with backing paper that is removed before lamination. Normally, an autoclave is used to speed the process of curing through heating.

When thermoplastic composite sheet materials are to be produced, the thermoforming process, followed by vacuum forming, is often used. Another process, called *reinforced reaction injection molding (RRIM)* is also gaining popularity. With the RRIM process, short reinforcing fibers are added to one of the RIM liquid components pumped to the mixing head.

Fabrication of Metallic Matrices

Metallic matrices can be fabricated as either solids or liquids. When solids are desired, this is accomplished in a vacuum, using pressure. Liquid metallic matrices involve infiltration of molten metals with other materials.

An example of a composite with a solid metallic matrix might be one that begins as a sandwich of such materials as resin binders, boron fibers, and aluminum foil. Compacting the sandwich under heat in a vacuum will cause the binder to flow around the fibers and create a bond with the aluminum matrix layer.

Sometimes, vacuum hot pressing is used to construct this type of metallic-matrix composite. Figure 29-4 shows the type of hot pressing furnace that would be used with such materials. This hot isostatic furnace has been designed for use with molybdenum. It operates at a temperature up to 5400°F (3000°C), with a vacuum of up to 60,000 psi (413 700 kPa).

Filament winding, powder metallurgy, and plasma spraying are also used to construct solid metallic-matrix composites.

Figure 29-4
Both the heat and the high vacuum needed to fabricate some types of metal-matrix composites are provided by a hot isostatic press like this one. (ISO-Spectrum, Inc.)

Sometimes, material is added to matrices which are in the liquid state. Liquid-state processes involve infiltrating fibers or preforms with liquids (in this case, molten metal). Infiltration can be accomplished under vacuum or with high pressure.

Squeeze Casting

Squeeze casting is the liquid-state process that is frequently used to fabricate metallic-matrix composites. This method has been used to cast silicon-free aluminum alloys for diesel engines, an application where high-temperature strength is critical. Ceramic-fiber-reinforced metal-matrix composite parts have also been produced for use in applications where high wear resistance is vital.

In squeeze casting of metallic-matrix composites, a fiber preform is inserted into an opened mold or die. Molten aluminum is then poured into the preheated die. When the die halves close, hydraulic pressure compacts the die halves tightly together, forcing the molten metal into the fibrous preform.

Fabrication of Ceramic Matrices

One of the most common methods for incorporating fibers into ceramic matrices is through the *slurry infiltration process.* Let's take a look at how this process works.

First, the desired type of fiber—in the form of fiber tows (bundles) —is passed through a liquid slurry made from matrix powders, a carrier liquid, and binders. After the slurry infiltrates the fibers, the tows are wound on a take-up drum and dried.

After the fibers have been prepared, they can be combined with various types of matrices. Hot isostatic pressing is a common process for consolidating ceramic-matrix composites. Melt infiltration, in situ chemical reaction, polymer pyrolysis, chemical vapor deposition, and reaction bonding are other processes that are used to prepare ceramic-matrix composites.

Ceramoplastics

Another type of composite material is made from mixing powders of lead glasses and mica. This composite is referred to as *ceramoplastic,* or glass-bonded mica.

Ceramoplastic is a moldable/machinable inorganic composite with high insulating properties. The material is used for products in many high-growth industries including communications, electronics, and space technology.

Glass-bonded mica is cold-pressed into a preform. The preform is then heated, melting the glass. The mica becomes the filler matrix for the combined materials.

After the preform is melted, it is then molded under extremely high pressures either by transfer or compression molding processes. Figure 29-5 shows a 75-ton transfer press being used to mold near-net-shape parts to tolerances as close as 0.0002 in. (0.005 mm).

There are major differences, in terms of behavior, between ceramics and plastics. Plastics are much easier to machine than ceramics. Ceramics are more stable than plastics. Plastics smoke and burn, but ceramics do not. Ceramics are able to withstand about eight times higher temperatures than plastics.

Figure 29-5
This technician is checking a small part molded from glass-bonded mica as the 75-ton transfer press behind her cycles through production of another part.
(Mykroy/Mycalex Ceramics)

The advantage of ceramoplastic is that it is able to bridge the gap between these materials. It capitalizes on the attributes of both materials. Glass-bonded mica has high dimensional stability, thermal endurance, arc resistance, and thermal conductivity. It is an excellent electrical insulator, and absorbs virtually no moisture.

Ceramoplastic is easy to machine with carbide tools, using water-soluble oil coolants. Machining is done on conventional equipment. The composite can be milled, ground, cut, drilled, filed, or sanded.

There are a number of safety precautions that must be considered when machining this material. The lead in the glass used in glass-bonded mica is a cause for concern in both environmental pollution and worker health. The lead in the final product is in a form that may leach into water. The waste from glass-bonded mica is considered to be hazardous under the Resource Conservation and Recovery Act, and must be carefully handled.

The creation of dust during manufacture and machining is to be avoided. Dust collectors must be provided so that vacuum air ducts can be placed in areas where dust can be generated. Covered or enclosed equipment should be used where possible. Water used to machine glass-bonded mica contains waste particles and should be collected, stored, and then disposed of in a manner specified by the Environmental Protection Agency.

If an excessive amount of machining is necessary, a part can be designed with inserts formed into the ceramoplastic material. This reduces the amount of hazardous airborne dust that might be generated. Insert materials such as stainless steel, brass, copper, silver, and aluminum are popular.

Even though extra care must be taken in machining ceramoplastics, the material provides significant advantages over conventional plastics and ceramics for many applications. The major advantage is that parts have practically no shrinkage during molding. Glass-bonded mica can be molded to close tolerances, even if the part is large.

Glass-bonded composites are being used to make capacitor end plates, microwave components, jet engine thermocouples, plasma engines, computer disks, memory boards, laser components, and high-temperature devices.

Production of Superconductors

Since the discovery of superconductivity by Heike Kamerlingh Onnes in 1911, approximately 27 elements and hundreds of compounds have been discovered that demonstrate this phenomenon. A *superconductor* is a material that exhibits a total absence of electrical resistance below a critical temperature. The most important characteristic of a superconductor is its ability to carry an electrical current without creating resistance.

There are three factors that control the limits of a superconductor. These are its critical temperature, the electrical current density, and the electromagnetic field that is generated around the material.

The 1972 Nobel Prize in Physics was awarded to John Bardeen, Leon Cooper, and John Schrieffer for explaining superconductivity. In simple terms, superconductivity can be described as a reversal of the usual behavior of electrons. Under normal conditions, two electrons *repel* each other. However, at very low temperatures, these electrons become *attracted* to each other. These electron pairs, referred to as the Cooper pairs, require a much greater amount of energy to create electrical resistance.

Research on superconductors made with *filaments* first began in the late 1970s. Two methods used to make the filaments are known as the "in situ" and powder metallurgy fabrication processes.

In situ methods (in situ means "in the original location") are complicated. First, a copper-rich niobium mixture is melted at about 3360°F (1850°C). The material is then cast and cold-drawn into wire. This produces fine filaments in a copper matrix. The filaments are then tin-plated.

This is a theoretical and overly simplified explanation of this process. In practice, many different processes are involved in the total production, including chill casting, continuous casting, levitation melting, and consumable-electrode arc melting.

When powder metallurgy is used, the process is much simpler. Copper and niobium powders are mixed, pressed, and then hot- or cold-extruded into fine wire. The wire is then coated with tin.

Bonding Composite Matrices

The technology for joining of wood or metal workpieces is fairly well-established. It may include using fasteners such as bolts, screws, or rivets. Other methods, such as gluing or soldering, may also be possibilities.

When joining ceramic or plastic matrices, the choices are generally limited to mechanical fastening or adhesive bonding. Adhesive bonding is preferred when the part does not have to be disassembled, because a stronger and lighter part can be made. Mechanical fasteners typically require holes, and thus require separation of the fibrous reinforcement.

Adhesive bonding techniques have been used since the beginning of time. An early use of adhesive bonding was the fabrication of mummy cases by the ancient Egyptian, using casein adhesives. Today, hundreds of different types of adhesives and similar substances are used to provide bonds between various types of composite materials.

Important Terms

adhesion process
carbon/carbon composite (C/C)
carbonization
ceramoplastic
C glass
cohesive fabrication
custom
E glass
gel
halide reduction
pyrolysis
reinforced reaction injection molding (RRIM)
sedimentation
S glass
slurry infiltration process
sol
sol-gel technique
squeeze casting
static fatigue
superalloys
superconductor
Taylor process
thermal decomposition

Questions for Review and Discussion

1. What are the advantages and disadvantages of ceramoplastics over conventional ceramics and plastics?
2. Why would one strive to eliminate the use of fasteners with composite parts?
3. What is the difference between sol-gel processing and chemical vapor deposition? Direct your explanation toward the formation of ceramic whiskers and fibers.

Conditioning and Finishing Composite Materials

Key Concepts

△ The surface energy (surface tension) of a substrate affects how well a liquid will wet its surface.

△ Controlled irradiation of plastic composites can improve their ability to withstand weathering, as well as improving other properties.

△ When cermets are made using powder metallurgy, they are conditioned by sintering to improve their density.

△ Hazardous waste composites from manufacturing plants can be detoxified by processing in a special furnace.

Conditioning processes change the internal or external properties of materials, by subjecting them to heat, shock, electrical impulses, magnetism, chemical action, or mechanical forces. Because of the exotic nature of many composite matrices, there are many unusual processes that can be used to condition and finish products made of composite materials.

Types of Conditioning Processes

Processes such as spin welding, hot gas welding, and dielectric heating also cause changes to the internal molecular structure of the composite material being joined. However, they rely on mechanical forces or on exposure to external sources of heat to permanently bond the composite materials together.

There are other processes that use nontraditional forms of energy to create internal changes in composite workpieces. Some of these processes cause the composite materials to become reactive.

One form of energy that accomplishes an internal reaction is radiation. Exposure to some forms of radiation can actually improve the behavioral characteristics of the material. Other types of radiation can destroy the workpiece and render it unusable.

Irradiation

Exposure of a material to energy from a radioactive source is called *irradiation.* Both ionizing and nonionizing methods of radiation are used to heat, dry, and cure adhesives that are used to join different materials to make composite workpieces. Ionized radiation is accomplished with electron beam accelerators, using radioactive materials such as Cobalt 60. Ionizing of atoms and molecules causes them to become highly reactive.

Prolonged exposure of polymeric materials to radiation can destroy their structure and composition. However, controlled irradiation can be used to improve the ability of plastic workpieces to withstand continued exposure to extreme temperatures. It can also be used to improve cross-linking, and can enhance the electrical resistance characteristics of wire insulation. When used with elastomers, radiation can improve vulcanization and reduce stress cracking.

Another advantage of controlled irradiation of thermoplastics is that it can improve their weatherability, impact strength, and resistance to static electricity. Controlled radiation is also used on composites to improve their resistance to abrasion and harsh chemicals.

Induction heating and dielectric heating are examples of nonionizing radiation processes. Microwave, ultraviolet, and infrared exposure sources are also used to generate nonionizing radiation. Processes such as these do not make the composite workpiece reactive, but they do have the ability to destroy molecular bonds in the materials. Normally, nonionizing radiation is used to dry or cure adhesives or layers of thin film.

Harmful effects of irradiation

Radiation is not always beneficial to polymers. It may cause damage to plastics, resulting in a reduction in their molecular weight. It can also cause cross-linking, polymerization, or degradation.

Thin film acetates, cellulosics, and the fluoroplastics are more susceptible to degradation from radiation than other plastics. A dose as small as 1 mrad (1 millirad, or 0.001 rad) can damage polytetrafluoroethylene (PTFE) or the acetals. A dose of 20 myriarads (200,000 rads) is necessary to influence the cellulosics. This should be contrasted to the much larger amount of radiation—1000 myriarads (10 Mrad, or 10 million rads)—that is required to degrade a plastic such as polystyrene.

Electrical Surface Treating

It is often necessary, when constructing composite parts, to bond the materials together. This is not always an easy task, however, because of the specific properties of the surfaces to be joined. In other instances, processes must be used to print directly on the plastic workpiece. Here, the surface characteristics of the plastic part are also critical.

The reason for this is that the liquid—adhesive or ink—must be able to wet the surface of the material. *Wettability* is dependent on the *surface energy*, (surface tension) of the material. The surface energy of the solid material affects how well the liquid wets the surface.

Figure 30-1 shows two liquid droplets (ink or adhesive), each set onto a smooth solid horizontal surface. One droplet has spread out over the substrate, the other has not. Wettability is described by the angle between a tangent line at the contact point and the horizontal line of the surface of the solid. Refer to the drawing showing "poor wettability" in Figure 30-1. Angle CBD is the angle between the tangent line on the droplet and the point of contact with the surface being wetted. If only partial wetting takes place, the contact angle will range from nearly 0° (indicating almost complete wetting) to 180°(indicating very poor wetting). The higher the surface energy of the solid material in relation to the surface tension of the ink or adhesive, the better its wettability. A smaller contact angle shows better wettability. Refer to the drawing showing "good wettability" in Fig. 30-1. Note that the angle GFH is much smaller than the angle CBD.

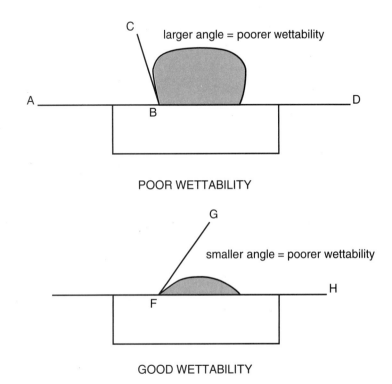

Figure 30-1
The wettability of a surface is indicated by the angle of a tangent line drawn from the point of liquid contact with the surface.

In order to determine the contact angle that is necessary for a particular material, it is necessary to know the surface energy or tension of the solid and liquids involved, which is measured in units of dyne-centimeters (dyne-cm). Some of the more common solid and liquid materials are shown in Figure 30-2.

Liquids such as thick paints and adhesives have greater surface energy than high-viscosity liquids; consequently, they have a greater contact angle. Wetting a substrate with liquid adhesive or paint is also influenced by the cleanliness of the substrate. Air bubbles, oil, dust, or moisture buildup will naturally interfere with the ability of the coating to stick to the substrate. Good wetting requires strong surface tension in the liquid or droplet that must stick to the substrate. This surface tension must be less than the adhesive force between the liquid and the substrate. This is an important concept, worth a closer look. A liquid with a high surface energy, such as an epoxy adhesive, cannot be used to wet a low-energy polyethylene solid; the epoxy would not stick well to the polyethylene. This occurs because the surface tension in the

Figure 30-2
Some dyne-cm values for typical solids and liquids. To be wettable by a given liquid, a solid must have a value at least 10 dyne-cm higher.

Surface Energy (Dyne-Cm)

Dyne-CM	Material	Form
100+	Adhesives	Highly Viscous Liquids
100+	Metals	Solid
100+	Glass & Ceramics	Solid
80-50	Most Finishes	Liquid
46	Nylon	Solid
43	Polyester	Solid
33	Polystyrene	Solid
31	Polyethylene	Solid
29	Polypropylene	Solid

epoxy material is greater than the adhesive force between the epoxy and the polyethylene. On the other hand, a low-surface-energy polyethylene liquid would adhere well to an epoxy solid, since the liquid would have a lower surface tension.

If a good bond is to occur between the solid and liquid, the substrate's surface energy should exceed the liquid's surface tension by 10 dyne-cm. The surface energy of some plastics, such as polyethylene and polypropylene, is insufficient to permit adequate bonding of printed images or adhesives. Often these plastics are desired because of their useful properties: chemical inertness, low coefficient of friction, high wear resistance, puncture, and tear resistance.

This is where the process called *electrical surface treating*, or corona discharge treating, comes in. This process is used to change the surface tension and wettability of plastic materials. This is often done when bonding plastics to metals or other plastics, or when printing directly on plastic surfaces.

Electrical surface discharge treating is based on the principle of high-voltage discharge in air. Free electrons are always present in air. In a high-voltage field, these electrons accelerate and ionize plasma gas. When the electric field is strong, the electrons collide with molecules of gas, causing an avalanche of electrons. When a plastic part is placed in the path of these electrons, the electrons impact the surface with two to three times the energy necessary to break apart molecules in most materials.

This creates free radicals in oxygen, which react to form chemical groups on the surface of the material. Typical groups would be carbonyl, carboxyl, hydroperoxide, and hydroxyl.

These chemical groups increase surface energy and enhance surface bonding to the matrix material. The high-voltage discharge used in this process modifies internal surface characteristics without visibly affecting the appearance of the material. Here's how the process works:

A plastic workpiece is placed in the machine and a high-voltage, high-frequency discharge in air takes place between the two electrodes. This *corona discharge,* at frequencies of 20 kHz to 25 kHz, accomplishes high-energy transfer as the electrons oscillate in the gap between the two electrodes.

Figure 30-3 shows a complete electrical surface treating system. The system consists of a high-frequency signal generator, high-voltage transformer, and treating electrodes. The generator produces an output signal of 175 volts which is automatically adjusted in the 20 kHz to 25 kHz range. The high-voltage transformer intensifies the output signal from the generator to the level necessary to create a discharge of the desired intensity. The part is inserted into the treating station between the two electrodes.

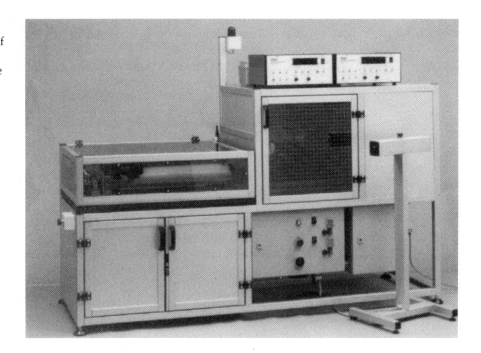

Figure 30-3
This equipment is capable of generating the 20 kHz-to-25 kHz frequency, high-voltage signal needed to create a corona discharge for electrical surface treatment. (Tantec)

Materials that are electrically surface-treated should be finished soon after treatment, since shelf life ranges from hours to years, depending on the plastic. Shelf life also is influenced by the temperature at which the material is stored. After a coating has been applied to the substrate, the bond becomes permanent.

Electrical surface treating is being used to treat polyethylenes for small pharmaceutical bottles, food containers, and baby food jars. Other product applications include treatment of video cassettes, medical vials, test tubes, and chemical containers. Corona discharge has even been used to treat electronic cable insulation to improve adhesion of inks and coatings.

Conditioning Cermets

As you read in earlier chapters of this book, *cermets* are composite materials that take advantage of the best attributes of metals and ceramics. The metallic matrix provides ductility and strength. The addition of ceramics improves hardness.

Normally, about 80 percent of a cermet composite is ceramic and 20 percent is metallic. **Cemented carbide,** the most common type of cermet, is made of hard particles of tungsten or titanium carbide, suspended in a cobalt matrix.

This type of cermet is normally made by compacting carbide and cobalt powders using powder metallurgy. After the green compact is formed, the part is sintered. Sintering recrystallizes the cobalt to improve the density of the final product.

Cermets are also made using liquid infiltration. Liquid metal is permitted to infiltrate around ceramic particles. The bond between the metal and ceramic occurs through chemical reaction, or through the solidification of metal and ceramic at the point of interface. This type of cermet is not as strong as that formed with powder metallurgy. There are a variety of methods for improving the bond between materials. Some of these involve the use of coupling agents.

Coupling Agents

We know that the interfacial bond between matrices, reinforcement materials, fillers, and laminates can be improved by adding **coupling agents,** which are frequently referred to as **promoters.** Coupling agents often are added directly to the resin along with the filler or reinforcement.

Two types of coupling agents are popular. When the composite is a laminate, particulate, or fiber-reinforced type, *titanate* is normally added to improve strength.

If a surface treatment is desired for a fiber-reinforced composite, then a *silane* agent might be preferred. This type of agent can be added to the solution, or introduced in powdered form.

Sometimes, it is necessary to keep a plastic from being degraded by exposure to ultraviolet rays. Ultraviolet stabilizers and carbon black are sometimes used with polyesters, polystyrenes, and other plastics to reduce damage from ultraviolet rays. Other additives, called **antiozonants,** are used to help prevent breakdown by ozone gas in the atmosphere.

Foaming Agents

In some applications, the addition of chemicals to make a product more resistant to a particular condition is not enough. An alternative approach may be to introduce frothed or cellular *foaming agents* into the plastic mixture to create a lighter, more cellular structure. Cellular foams are sometimes desired in applications where improved thermal insulation, light weight, and impact resistance is necessary.

There are several methods used for making cellular composites. You have previously read about the production of syntactic cellular polymers by adding microspheres or microballoons to resins. The small balloons expand when the plastic is heated during polymerization. Scrap pieces, expandable beads, and pieces of cellular foam can even be added with other material to create lighter products.

A common method of producing cellular foams is adding chemical blowing agents or volatile liquids to the resin or melt. When the chemical agent decomposes at a specified temperature, gases are given off to create a cellular structure. If the liquid that is added is volatile, it will decompose during heating. This expands the gases, causing the softened polymer to fill the mold. Methylene chloride is such a chemical. It is often used to expand flexible urethane plastics.

Temperature Degradation of Composites

The degradation of composite materials through exposure to high temperatures is quite different from what might be expected with other materials, such as glass. With glass, internal changes occur through phases of melting and crystallization. These stages are reversible.

With products made of other materials, such as wood, exposure to intense heat can char and burn the material, so that it can never be reused. This also occurs with polymeric-composite materials, where thermal degradation and crosslinking are irreversible.

Failure or destruction of parts made of composites through exposure to very high or very low (cryogenic) temperatures takes place in a different way than it does with other materials. Since fibers, particles, or other additives are used in conjunction with a matrix, these materials must each be viewed independently in terms of how well they will survive exposure to extreme conditions.

Graphite, ceramic, or boron fibers are typically resistant to high temperatures. However, these fibers are normally used in conjunction with a matrix that may be much more vulnerable. For example, graphite fiber composites are as strong as metals, and they have the advantage of being lighter. However, if graphite fibers are enclosed in an epoxy matrix, the matrix will burn and the fibers will no longer be held in suspension.

Sometimes, polymeric-matrix composite parts are needed which are more-than-usually resistant to flame and high temperatures. One method for improving their ability to survive in this type of environment is by adding boron or graphite powders to the matrix. The additives help to retard the char that begins to occur in thermal oxidation. Flame-resistant additives can also be used with matrices that have the tendency to dissipate heat.

In other end-use environments, a component may have to withstand exposure to fuels or chemical solutions. These can sometimes cause the matrix to swell or dissolve. Examples of product applications where this would be critical would be for components used on space vehicles or automobiles. Other products, such as kitchen counter or laboratory countertops, must also be constructed so that the matrix can withstand chemical attack. Sometimes, multilayered barrier films are used to strengthen the point of lamination between materials.

Destructive Detoxification

The proper disposal of hazardous wastes is a serious (and growing) problem, especially in more industrially advanced nations. In many instances, the waste material that is involved is a composite, a substance made of several different types of materials that are either partially mixed or bonded together in a laminated sandwich. A company in Seattle, Washington, Penberthy Electromelt International, Inc., has developed a special type of furnace for *destructive detoxification* of hazardous wastes. See Figure 30-4. The process is presented here because it is a new method that is being used in industry to convert nearly any type of hazardous waste material into a nonhazardous useful material.

Figure 30-4
This electric furnace is the key element in a molten glass process used to detoxify hazardous wastes. One recent project involved reducing 1300 tons (5000 drums) of radioactive hazardous wastes to 110 tons of solid glass. (Penberthy Electromelt International, Inc.)

The patented Penberthy molten glass furnace process generates high temperatures by passing electrical current through glass between immersed electrodes. The basic process of resistance heating is involved, and no arc is created.

Organic materials fed into the furnace are exposed to 2300°F (1260°C) temperatures in the presence of air and water vapor. Chlorine is reduced to hydrogen chloride and converted to calcium chloride. Carbon is oxidized to carbon dioxide, and the mineral residues that are created are melted into glass. As the waste materials are heated, they drop into the long stream of molten glass. The Penberthy furnace is not an incinerator. It does not use a flame, and does not produce ashes. The only end product is glass.

What is interesting about this process is that many different types of composite materials can be thermally destroyed and turned into glass. This includes asbestos, arc furnace dust, incinerator ashes, electroplating sludges, military wastes, paint solvents, in-plant wastes of all kinds, even radioactive wastes.

Processes Used to Finish Composite Materials

Many manufactured items require considerable time for finishing, in order to produce a high-quality final product. This is seldom the case with composites, since parts are usually designed so that no additional finishing treatments, other than trimming, are needed.

Sometimes, even the final trimming process is automated. Figure 30-5 shows an automated waterjet cutting machine being used to trim an aircraft wing section made from a complex metallic sandwich. The finished product must be capable of withstanding extreme stresses from high altitudes and supersonic speeds. The actual aircraft for which the wing was being cut is shown in Figure 30-6.

Figure 30-5
A technician monitors the operation of an automated waterjet cutter at an advanced aircraft production facility. The large part being trimmed is a metallic composite wing section. (Vought Aircraft Company)

Figure 30-6
The B-2 aircraft is not only unusual in shape and performance capabilities, but also is one of the first aircraft to be constructed almost entirely of advanced composite materials. (Vought Aircraft Company)

In some instances, finishes are applied to composites through coating processes. Many of these processes, including electrostatic coating, extrusion coating, calendar coating, fluidized bed coating, and powder coating, are popular. (Finishing can also be provided through forming processes, such as coextrusion and plasma spraying.)

Sometimes, the finished color must be different from the natural color of the stock itself. In these cases, an additive may be used to tailor the color of the product to the needs of the user.

Colorants

When a product of a particular color is needed, it is possible to accomplish this through extrusion or coextrusion. The melt gradually achieves the color of the stock material.

Another approach is to add a colorant to the base resin. ***Colorants*** are usually in pellet form and are mixed prior to forming.

Colorants may be dyes, or they may be organic or inorganic pigments. They are used to impart color, but they often afford other attributes. Colorants can be chosen to improve the product's strength and electrical properties, and to help it resist degradation from heat or ultraviolet light.

Transfer Coating

It is sometimes necessary to cover a composite material with a coating to protect it from environmental conditions, such as weathering, abrasion, corrosion, or chemical exposure. To be successful, a coating process must result in a coating that remains on the substrate.

Coating processes include extrusion, dip coating, fluidized bed coating, electrostatic coating, spray coating, brush coating, roller coating, powder coating, and plasma coating. There are many other possibilities, as well.

One type of coating process particularly useful with composites is called ***transfer coating***. With this process, a surface skin coat is produced on a coated release paper, using the basic method of roller coating. Once the coated material sets up or gels, a second coat (the "tie coat") is rolled over this skin coat. At this point, the stock is dried, and the release paper is removed.

With the transfer coating process, it is possible to join many different types of materials using a film or fabric coating. Transfer coating is also used to place printed images onto composite materials. The image is printed onto transfer paper or film and is then heat-set or ironed onto the actual workpiece. The carrier paper or film is then carefully peeled off.

Many of the finishing processes that are used with composites are used to print images onto solid substrates. Transfer printing or coating is one of these processes. While printing processes are not within the scope of this textbook, several are worth mentioning here. The most commonly used processes for imprinting images on composites in a continuous production environment are pad printing, hot stamping or marking, and embossing. Fully automated production machines can be purchased for pad printing and hot stamping.

Pad printing is accomplished using heat-set inks. The printed design in relief contacts an inked pad, then is pressed lightly against the workpiece. The wet ink that has transferred onto the product is dried by the application of heat.

Hot marking is a stamping process that can be used to impress or "brand" images onto the workpiece. It is also used with foils to transfer color to the product. An image, created in relief, is heated and then pressed into the stock. When foils are used to color an image, they are inserted between the relief form and the workpiece.

Embossing is a process that is the opposite of branding, since it produces a raised image rather than one which is sunken into the product's surface. In embossing, the design or type is constructed in sunken or female form. Often, the embossing die is made of wood, and is included as part of the die used to transfer mold, or vacuum-form, heated polymeric sheets. The embossing die presses down onto the sheet as pressure from below pushes the stock up into the recesses of the die.

Important Terms

antiozonants
cemented carbide
colorants
corona discharge
coupling agents
destructive detoxification
electrical surface treating
embossing
foaming agents
hot marking
irradiation
pad printing
promoters
surface energy
transfer coating
wettability

Questions for Review and Discussion

1. Explain how the Penberthy furnace is used to detoxify hazardous wastes.
2. What is the difference between the manufacturing processes using ionizing and nonionizing radiation to condition plastics and elastomers?
3. What is the relationship of viscosity to wettability?
4. How does the corona discharge method of conditioning improve wettability?

AUTOMATIC PRODUCTION PROCESSES

The word "automation" was introduced almost 50 years ago to describe the process of applying automatic control devices to production equipment (usually a single machine or process). While far more sophisticated methods are used today, automation still has three basic ingredients: a repeatable manufacturing or process, a control system, and a material placement system. Linking several automated processes or operations together with a management control system will provide the information necessary to achieve the level of sophistication referred to computer-integrated manufacturing (CIM).

Wes-Tech Automation Systems

Automated Manufacturing Systems

Key Concepts

- Δ An automated manufacturing system has three basic ingredients: a repeatable process, a control system, and a material placement system.
- Δ The type of automation (hard or soft) dictates the selection of hardware and control systems.
- Δ Programmable logic controllers (PLCs) have virtually replaced relay logic in manufacturing applications.
- Δ In most manufacturing operations, automation is an evolutionary process over a period of time.
- Δ Numerical control programming allows precise, repeatable operation of individual machine tools and machining cells.
- Δ Computer-integrated manufacturing is viewed by many manufacturers as an essential strategy, through reducing costs and improving product quality, for remaining competitive in the global marketplace.

The word *automation* was coined in the first half of the twentieth century by the automotive industry to describe the process of applying automatic control devices to production equipment. In those early days, automation was generally limited to controlling a single machine or process.

When we compare the methods used to automate machines and operations in the automotive industry in the 1940s and 1950s with the sophisticated automated systems used today, there seems to be little similarity. However, all forms of automation—regardless of their degree of sophistication—have three basic ingredients. These can be classified as the *building blocks of automation.* These building blocks are:

- Δ A repeatable manufacturing operation or process.
- Δ A control system.
- Δ A material placement system.

If a particular manufacturing operation or process utilizes all of these building blocks, and requires no human intervention, then it can be referred to as an *automated* operation or process. Linking several of these processes or operations together produces an **automated manufacturing system**, Figure 31-1.

Figure 31-1
This flexible manufacturing cell is a fully automated operation demonstrating all three building blocks of automation: a repeatable manufacturing process, a control system, and a material placement system. Workpieces mounted on special fixtures are transported on a pallet system (foreground) to machining centers for shaping and finishing operations. Automatic toolchanging and the ability to program the system for varied operations makes processing small lots of parts cost-effective. (Cincinnati Milacron)

One of the most difficult obstacles to overcome when creating an automated system is *variability,* (inconsistency in what takes place) when manufacturing a particular product. Variability can exist in either the manufacturing operation or process, or it can be present in the materials being used. If automation is to be successfully implemented, all sources of variability must be controlled. Sometimes this is difficult to accomplish.

Implementing concepts like statistical process control (SPC) are appropriate first steps in planning and preparing for automation. In most cases, the process of automation takes place in stages—an evolution, rather than a revolution. Once the sources of variability are reduced, other manufacturing operations and processes can be automated and integrated to create a more comprehensive automated system.

In the preceding chapters of this book, hundreds of manufacturing processes and operations have been described in detail. Virtually all of these processes or operations can be automated once they are under control.

Once variability has been brought within acceptable limits, most of the work that remains involves the design and implementation of control and material placement systems. This chapter will emphasize methods used to create these systems. It will also touch on design considerations that affect the selection of hardware and control systems used in automation.

The Nature of Automation

To develop a better understanding of the nature of automation, it will be helpful to review what *is* and *is not* required in manufacturing to achieve automation. Automation does not require large batch sizes, or a high volume of mass-produced parts. However, high-volume production is based on the concept of *interchangeability* of components and materials, a concept that is necessary for automation.

Before interchangeability can be accomplished, components and materials must be standardized. Once this is done, any part, assembly, or material in a manufactured lot can be substituted for any other one in the lot. This means that modifications or adjustments to individual parts are unnecessary: each part in a particular job lot is identical to every other part in that lot.

To achieve interchangeability, you must control variability. Controlling variability, in turn, is necessary to implement automation. Thus, reducing variability by automating an operation can provide an improvement in quality. It is important to remember that neither high volume nor mass production is required for automation.

Automation is not the same as mechanization. **Mechanization** simply means doing things with or by machines. Often, mechanization increases the rate at which work is being done by reducing variability in the characteristics and cycle times of that operation. Theoretically, mechanization also enhances the machine operator's abilities. This is why it is often viewed as one of the fastest ways to improve worker productivity. Automation relies heavily on mechanization, but it does not require an operator who functions as an integral part of the production cycle.

Automation does share the primary goal of mass production and mechanization—increasing productivity and reducing cycle times to improve throughput in a manufacturing system. *Throughput* is the amount of product that moves through a production operation, from start to finish, in a given period of time. Automation may also achieve a reduction in the cost for direct labor, since it uses the concepts of mechanization and control without human intervention.

The primary goal of automation is to improve productivity. Improvements in productivity will be realized if the manufacturing system is able to directly translate the automated system's output into throughput. This translation is dependent on manufacturing management's direction and support, as well as the reliability, availability, and maintainability of the hardware and tooling in the automated system.

Now that some of the fundamental requirements for automation have been identified, it will be useful to examine four criteria that shape the nature of automation. Two of these criteria, *replication* and *volume,* center entirely on the *product* being manufactured. The other two, **control system sophistication** and **manufacturing flexibility,** pertain to the manufacturing *process*.

Figure 31-2 is a diagram that illustrates the effects that the four criteria have on the manufacturing process. All types of manufacturing systems, ranging from those that are manual to those that are totally automated, can be portrayed in such a diagram.

Cases that would fall in the upper right-hand corner of the diagram would include processes with the most sophisticated control system and highest flexibility. This type of system would most closely approximate the artistic capability of a human being. It can be argued, of course, that machines which can mimic the capacity of the human brain cannot be created at this time. It might also be argued that a highly skilled human being will not be able to achieve the high volume and uniformity of output that can be realized from an automated system with dedicated tooling (lower left in the diagram).

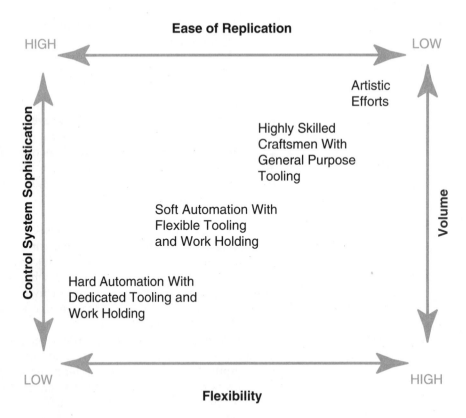

Figure 31-2
This diagram portrays the variety of manufacturing processes, ranging from hard automation—with high volume and low flexibility—to the high flexibility and sophisticated control of machine operations that rival those of a human artisan.

These two extremes illustrate points on a continuum that are defined by the four criteria of automation. This provides a framework for identifying appropriate control and material placement methods for automation in manufacturing. With the general framework for defining automation established, it is now possible to focus on the other factors that shape the control and placement systems used in automation.

Automation is important to both primary and secondary manufacturing firms. As you will recall, primary manufacturers are concerned with the production of industrial stock. In such primary industries as those that produce petroleum products, pharmaceuticals, chemicals, and food products, production is highly automated. There is a well-established technology for material control and movement, with piping, valves, pumps, and fluid processing equipment standardized and readily available.

Secondary manufacturing firms are concerned with the production of discrete products. These types of industries manufacture products that have specific characteristics and functions, such as appliances, circuit boards, or power tools. The control and material placement systems used to manufacture discrete products are generally tailored more to the specific application than the systems in process (primary) manufacturing. This means that, in primary manufacturing, automation can often be purchased as a *turnkey system* (a package that is standardized and ready to go). Secondary manufacturers often must create systems that are tailored to their unique needs.

In general, the three building blocks of automation can be found in automated systems employed by both primary and secondary manufacturing firms. The design emphasis by primary firms is generally on the process and control system. Secondary manufacturers concentrate on the control and material placement systems. However, in both cases, the design and development of an automated process must address each of the basic building blocks: repeatability, control, and material placement.

The two major types of automated systems, *hard* and *soft,* were identified in Figure 31-2. In **hard automation,** the machines and tooling are designed to produce a specific part or a family of similar parts or assemblies. Consequently, the control system must deal with only a finite sequence of logic and decision points defined by the structure of the hardware and tooling. This type of automation is dedicated to production that requires only infrequent changes in product or component configuration. Examples would be food canning, bottling operations in the beverage industry, and nail-making. Figure 31-3 shows part of a hard automation system where closures for metal beverage containers are being distributed from one operation to another.

Figure 31-3
This distribution system for metal beverage container closures is an example of hard automation. Equipment and tools are designed for efficient, high-volume production of identical products. Changes in processes or products are difficult to make.
(Ball Corporation)

Hard automation focuses on the mechanical components of the automated system. Because of this, the relative complexity and cost of automation is concentrated in the hardware and tooling components, not in the control system.

Soft automation refers to a system that can handle a wide variety of different shapes and material characteristics. Typically, the control system is very sophisticated. The hardware, which is designed to accommodate a wide range of components and materials, can appear to be deceptively simple. See Figure 31-4. This form of automation makes extensive use of modern computers and software that can accept a variety of inputs from sensors. *Sensors* are usually electrical devices that receive (take in) different types of information. This will be discussed in greater detail later in this chapter.

Figure 31-4
This unusually shaped new machine tool is claimed by its manufacturer to be five times faster and five times more rigid than conventional machining centers. The six legs, actuated by servomotors and individual ball screws, rapidly align the cutting tool at the desired angle to the workpiece. The new type of machine tool operates under conventional CNC programming, and is intended for integration into flexible machining cells and similar systems.
(Giddings & Lewis, Inc.)

The software that defines the logic of the control system can call from memory the subroutines that are appropriate for the circumstances. Generally, the control system software can be easily changed to accommodate different material or assembly requirements.

A more sophisticated adaptation of soft automation includes an element of *artificial intelligence (AI)*. As a control system concept, AI provides a great degree of flexibility. The inputs from sensors, coupled with predefined rules and relationships stored in the computer's memory, allow the control system to provide expert control in well-defined situations. This form of control (AI) gives systems of this type a limited but human-like reasoning ability. It can provide simple perception systems that are able to process information in various forms, including vision, speech, and movement. Robots are a good example of the type of hardware often used with this form of intelligent control.

Control Systems

The preceding discussion on the nature of automation made it clear that hardware and control systems are closely tied to each other. This means that one system cannot be designed without giving consideration to the other. It should also be evident that there are several types of control systems. Among them, they incorporate many forms of technology. It is important now to look more closely at the unifying concepts that form the basis for all control systems. These concepts will be useful when you must design or select control systems appropriate for specific automation situations.

Control systems address the *logic*, or principles of reasoning, that is designed into the automated system. Control systems must also deal with the *methods* that are used to make and implement decisions to operate a particular process, operation, or production system.

Manual control is the simplest type of control system. In a manual control system, an operator is required to start, stop, or adjust the process by pushing buttons, turning knobs, or engaging levers on the machine. *Automated control systems* for manufacturing also must be able to start, stop, and sequence production (advance parts). However, automated control also must simultaneously monitor the quality of the product and the functioning of the system.

Once the sequence of operations has been initiated, the control system must be able to carry out its predetermined functions, regardless of their number or complexity. Consequently, *safe operation* must be an integral part of the control system; it *must not* be left to human intervention.

The basic problem in designing a control system is determining what standards or references must be adhered to in order to satisfy the operation and complete the cycle. Consider the following example. Does a control system determine if a bottle is full on the basis of its time under the spigot, or the height of the material in the bottle? Your response to this question will define the type of control system that will be needed. There are two basic types of control systems, referred to as "open-loop" and "closed-loop."

Open-loop systems

An *open-loop control system* attempts to meet a preset standard without monitoring the output or taking corrective action. With the example just presented, an open-loop system would be appropriate if you chose time under the spigot as the standard for control. It is easy to visualize that the actual output (material dispensed to the bottle) may deviate from the desired output, since there is no feedback with this type of system. An open-loop system is composed of an input, controller, actuator, and an output. See Figure 31-5.

The *input* part of the system represents the value for a characteristic, such as time, temperature, or pressure. The *controller* in such a system provides the logic and governs the action of the actuator. The *actuator* is a device that responds to the signal from the controller. Typical *outputs* under control are electric motors, solenoid valves, or relays.

Examples of open-loop control systems can be found in such household appliances as a dishwasher or clothes dryer. Satisfactory action depends entirely on the appropriateness of the characteristic selected by the operator. The actual results may vary from the expected to the extent that other variables can influence the operation or process under control.

Figure 31-5
In an open-loop control system, the controller responds to a preset standard without monitoring the output. Thus, it might operate a motor or open a valve for a given length of time.

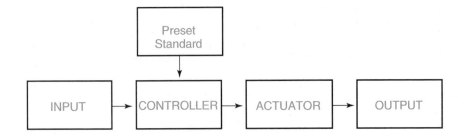

Closed-loop systems

Closed-loop control systems include the same functional components as an open-loop system, but have one significant additional feature, *feedback*. A closed-loop control system consists of an input, error detector, controller, actuator, output, and a negative-feedback circuit. See Figure 31-6.

Figure 31-6
A closed-loop control system can respond to changes in conditions or other problems by adjusting the process. Changes are based on measurements of output that are compared to the preset standard. This feedback provides the controller with information needed to make changes.

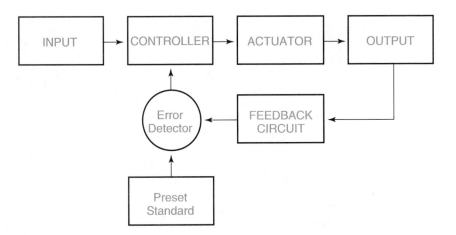

With this type of control system, the actual output is measured, then compared to a preset standard or reference. This is called *feedback.* For example, in the bottle-filling operation, the height of the material in the bottle is detected and compared to the characteristic established as a reference. Any measured deviation from the standard (detection of error) results in a compensating action from the control system. Thanks to this compensation, the bottle would be filled to the desired height.

There are four criteria used to evaluate the performance of a closed-loop control system. They are: transient response, steady-state error, stability, and sensitivity. These are shown in Figure 31-7.

In general, *response criteria* provide a means to illustrate the system's ability to achieve a desired output. A *transient response* describes how the

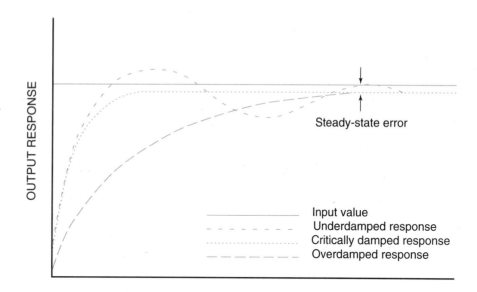

Figure 31-7
Different types of transient response by a closed-loop control system are suited to different situations. A critically damped response, for example, is needed where little deviation from the input requirement is permitted: response must be fast and accurate. Underdamped and overdamped responses are allowed under less stringent conditions.

output device under control responds to an input before steady-state conditions are achieved. Ideally, the output should immediately match the input requirement. However, as shown in Figure 31-7, there are several possible response conditions.

Examples of two different types of applications will help to make sense of this concept. In an application such as controlling the temperature in a drying oven to avoid overheating of items near the heat inlets, an overdamped (slow) response might be permissible because the temperature of the oven rises slowly. Temperature adjustment does not have to be made rapidly to match input and output.

In a different type of application, such as maintaining tension on a wire feeder in a coil winding machine, the system may require a critically damped response. In this case, tension adjustments must closely follow input requirements to avoid snarls.

Steady-state error is another characteristic that is present in some degree in every control system. *Steady-state error* is the difference between the *actual output* and the *desired output* required by the control system. Often, this error is minimized by recalibration or adjustment of the input signal.

Stability is a term that refers to how well the control system is able to maintain steady-state conditions. If the control system lacks stability, it will oscillate around the desired output level.

Sensitivity is the ratio of change in output to the change in input. This characteristic can be used to gauge the ability of the system to correct for small changes in input. In *adaptive control systems* (systems that provide control in response to changing conditions), sensitivity is a particularly important design characteristic. This will be discussed in greater detail later.

The four criteria used to evaluate the response of control systems have a direct relationship to the type of actuators that are chosen for use in the system. Actuation can be expressed in one of two ways: on/off or proportional.

On/off control

Although other factors can influence the criteria for designing control systems, the most important consideration is the mode of actuation that is chosen by the designer. In many cases, actuators are on/off devices such as ac motors, solenoid valves, and relays. Most control systems are based on the on/off mode of operation.

Relay logic, a system based on opening and closing electromechanical relays, served for many years as the basis for the majority of on/off control systems. Now, *programmable logic controllers (PLCs),* Figure 31-8, have almost completely replaced the use of relays in automation control systems. This occurred for a number of reasons. As the basis for designing relay control systems, relay (also called "ladder") logic worked so well that it also became the basis for programming PLCs. Technicians familiar with relay logic found it easy to understand PLCs, and adapt them to existing automated systems.

Figure 31-8
Special software, coupled with a touch-control display screen, allows machine operators to easily modify parts programs for a PLC like the one at right in the background. Programmable controllers are used with many types of production machines, especially in metalcutting applications.
(Giddings & Lewis, Inc.)

Another reason for using PLCs is their reduced cost. The price of a PLC is very competitive with relays for small control systems. Often, PLCs are less expensive than relay-based systems for large, complex automated operations.

However, the major reason for using PLCs does not relate to cost or to understanding the technology. When compared to relay-based systems, a PLC control system has a significantly higher level of reliability.

For this reason, PLCs can be more cost-effective than conventional relay-based systems for on/off modes of control. Consider the following example:

If a machine makes 50 parts per minute, and you allow for 50 production minutes per hour, the machine will produce 5,200,000 parts per year from a one-shift, five-day week. The life expectancy of a relay ranges from 200,000 to 2,000,000 operations, depending on the electrical load and quality of the relay. Thus, you might anticipate that each relay will fail and need to be replaced two to three times a year. In a small eight-relay system, assuming an average of 2.5 replacements per relay, this amounts to 20 relay failures per year.

The situation is complicated further by the fact that relays do not just quit. They fail *intermittently,* and then fail again. This means that there is a lot of time lost by maintenance personnel in troubleshooting and correcting problems. For this reason, lost production time to handle each relay failure might average four hours.

This boils down to 10,000 parts that are not manufactured because of each failure. With 20 failures per year, this means that 200,000 parts will be lost each year due to system "downtime."

Under the same conditions, a machine under PLC control would typically have a failure rate of less than two per year. Thanks to the built-in error checking capability of the PLC, and its light-emitting diode input/output registers, downtime might be reduced to less than two hours per failure. Lost production would be only 10,000 parts annually.

An added advantage for using PLCs is that they are easier to modify or expand than corresponding relay-based systems. Consequently, whenever an on/off mode of operation for an application is appropriate, then a PLC should be considered.

Proportional control

The proportional mode of control also has an established place in automation control systems. Often, these systems have a dedicated electronic control system with an appropriate analog interface to actuate the device. ***Proportional control*** is heavily used by primary manufacturing firms to control automated processes. In most cases, if the operation requires variation of speed, tension, or temperature, then the proportional control mode should be considered.

Control system design

Regardless of the *mode* of control selected, the design of the control system generally begins with the identification of the outputs to be controlled and their states of operation. The ***states of operation*** describe the action or condition of an output device. Examples are an AC motor with two states (on and off) or a single solenoid valve for an air cylinder, also with two states (energized and deenergized).

Let's take a moment now to apply some of the concepts that have been presented. Before you can design an automation control system, there are some preliminary steps to be taken.

- Δ First, you must identify the output devices that are to be activated or controlled.

- Δ Next, you must determine the states of the output devices.

Δ   Finally, you must identify an input that would cause the control system to activate the output device and result in the desired change in state.

After all of this has been done, a control system for the automated operation or process can be developed.

Numerical Control

One of the most important discoveries in terms of hardware used with automated systems is numerical control. The notion of numerical control of machines is often traced to the invention of punched cards used to control the weaving of complex textile patterns on a loom. The control system was developed in 1801 by Joseph M. Jacquard, a French textile weaver. It wasn't for almost 150 years, however, that a practical numerical control system was applied to industrial machinery.

The practical application of numerical control really began in 1949, when John Parsons, owner of a Michigan machining company, developed the concept of using numerical control to produce contoured parts from information stored on punched tape. At that time, his company was using tabulating equipment to generate the incremental elements for tool paths to machine patterns for helicopter blade airfoils. The process that Parsons established became the basis for ***numerical control (NC).*** Numerical control, in and of itself, is not a manufacturing process, but *a method for automating manufacturing processes.*

Parsons recognized that a tool path can be described using a series of vectors. This means that the tool moves along a path comprised of the resultants of all the axes being controlled on the machine tool. To visualize this process, it is useful to describe how Parsons tested out his approach.

He first stationed a machinist at each axis of a milling machine. Next, a person read the data defining the tool path and called it out to the operator controlling each axis of the machine. During the machining operation, one operator ran the X-axis handwheel. This controlled movement of the table back and forth (from left to right). A second operator controlled the Y-axis. This referred to the movement of the table from front to back. A third operator was used to control the Z axis, or up-and-down movement of the spindle.

The first actual machine tool to be operated by numerical control was developed in 1952 by the Massachusetts Institute of Technology (MIT). The machine was able to adapt a computer to read the encoded data and actuate motors which moved the machine the desired amount along each axis. The first commercially available NC machining center was produced by Kearney and Trecker in 1958. This machining center was able to change tools automatically, so that milling, drilling, tapping, and boring could be done on one machine.

How NC Controls a Machine Tool

Next, we will look more carefully at how the principle of numerical control can be used to control a machine tool. Focus on the lead screw that can be found on virtually all milling machines. Generally, each revolution of this screw advances the table 1/10 in. (0.10 in. or 2.5 mm). The dial on the lead screw wheel is divided in 100 equal markings so that as each mark goes by on the dial, the lead screw advances 1/100 of 1/10 in., or 0.001 in. (0.025 mm). To move the table one-fourth inch (0.250 in. or 6.35 mm), the operator rotates the lead screw wheel two full turns and then one additional half turn.

A *numerical control program* contains precise instructions for each move along each machine axis. Each instruction is sent by the machine control unit (MCU) to activate the specified output, such as an electric motor attached to the lead screw. When the motor is energized, a sensor on the lead screw will record a pulse which occurs as each division on the dial passes a reference point. Once the number of pulses representing the distance to be traveled is reached, the MCU will no longer actuate the motor turning this lead screw. The MCU to now free to act on the next instruction.

NC programming

Numerical control is a type of programmable machine tool automation. The machine is controlled by a program containing a set of instructions that can be stored on paper or mylar tape, punched cards, magnetic tape, or in computer memory. The program defines the tool-path as well as auxiliary operations.

In the transition to NC machine control, the skill and knowledge that the machinist needed to machine a part had to be included in the programmer's set of instructions. The effective operation of the NC machine tool thus became the responsibility of the programmer. The programmer determined the moves, sequences, tools, and overall motions that the machine tool used in machining the part. Under NC, the primary responsibility of the operator became monitoring of the machine tool's operation, rather than controlling it.

To support this change in function of the operator, machine tool programming languages emerged as a major component in the technology of NC. Programming languages provide a means for describing the machining operations on a component in such a way that the computer is able to translate that description into NC code. Some of the programs available to do this are: APT (Automated Programmed Tool), ADAPT, IFAPT, MINIAPT, AUTOSPOT, COMPACT II, ACTION, SPLIT, NUFORM, and UNIAPT. Of all the languages available, APT is the most commonly used, and was the first developed to use English-like statements.

Also available are higher-level programs such as PROCAD/CAM, which allow the programmer/designer to define machining paths, entry/exit points, rapid traverse paths, and other data on a computer-aided design (CAD) drawing of the part. See Figure 31-9. Once these parameters are established on the drawing and in the dialog boxes, programs of this type can generate machine code without any further involvement from the designer.

Depending on the software used, the programmer can usually view a simulation of the machining sequence on the computer screen. The ability to simulate the machining operation on a computer terminal reduces the amount of machine time and material expense needed for debugging and prototyping new parts. An additional advantage is that modifications or updates can be carried out quickly and effectively without using machine tool time.

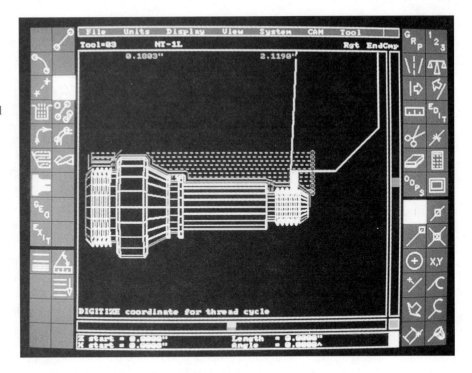

Figure 31-9
A CAD drawing on which tool paths for machining and threadcutting are displayed. This information is used by the software to generate machine code that will control a CNC machine. (Teksoft)

NC machine tool control systems

There are some fundamental differences in NC machine tool control systems that can be useful in helping to classify them. Basically, there are two types of NC control systems.

The first type of NC control is called a positioning or *point-to-point system*. This type of system moves the working spindle to specific locations on the workpiece, and then performs the programmed sequence of events. These events are typically drilling, reaming, threading, or spotfacing.

The location for the operation is specified in a two-axis coordinate system. Movement is typically confined to one axis at a time. When the table or the tool moves to a new location, it generally does so without regard to a specific path.

In point-to-point control, the programmer is required to define only the centerline of the tool. Consequently, there is no concern for specifying the tool diameter or tool offset.

The second type of NC control system is called contouring or *continuous-path control*. This system incorporates two requirements that are not present in point-to-point control. Specifically these are:

  △   The tool feed rate is controlled at all times.

  △   Cutting tool offsets or compensations are specified for every point on the tool path.

The continuous path generated by the NC part program defines the tool path and the appropriate feed rate for the entire machining operation. The constraints on feed rate input by the programmer for a specific machine tool are predicated on the size of the cut, rigidity of the machine, characteristics of the tool, and the material being machined. In application, continuous-path control runs the gamut from very simple machine control to extremely sophisticated systems that are able to generate complex curved paths.

There are two adaptations to the basic process of numerical control that have been made possible through refinements in technology. The first of these was used in the 1960s. Initially called direct numerical control, the term was later changed to *distributed numerical control (DNC)*.

In DNC, individual NC machine tools are connected through communication lines to a central command computer. Most contemporary applications of DNC utilize a small personal computer (PC) located near the machine tool and connected to a larger command computer. Such a configuration permits enhanced memory and computational capability. The advantage of DNC over conventional NC is distributed control, resulting in improved efficiency.

Today, a more popular adaptation of the basic process of NC is called Computer Numerical Control, or CNC. With CNC, a microprocessor is built into the control panel of the machine tool. This enables the CNC operator to modify programs when necessary, or to prepare programs for unique parts. See Figure 31-10.

With CNC systems, the master program is driven by a large computer at a remote site. CNC systems provide greater flexibility, accuracy, and versatility than NC or DNC systems.

Figure 31-10
This four-axis horizontal turning center features CNC operation, an automatic tool-changing system, and a pallet-changing system that permits continuous production. The CNC unit, at the right side of the machine, features a full-color graphics display for monitoring operation and modifying programs. (American SIP Corporation.)

Adaptive Control

Adaptive control enhances the capability of manufacturing processes that utilize numerical control and computer numerical control. *Adaptive control* systems can respond to the immediate conditions being encountered by a particular machining operation and make necessary changes in feed rates or other factors. Adaptive control became possible with the advent of miniaturized sensors and transducers that could measure the forces and temperatures created by the cutting tool. These input devices, used with low-cost industrial computers, have made it possible to design and implement a control system that responds to a time-varying operating environment while it provides feedback in a closed-loop system.

Adaptive control systems are able to change *input values*, arbitrary standards or references that reflected circumstances at the time they were established. This reduces the degree of error between the system's output and the real-time conditions that are impacting the system. In practice, adaptive control would assess a current condition, such as the heat generated at the cutting tool, and then make adjustments to the machine if necessary. This might result in slowing down the speed that the material is revolving, decreasing the depth of cut, or even adding more coolant.

Computer-Integrated Manufacturing

When numerical control is coupled with automated part handling and tool changing systems, it provides tremendous potential for use in many manufacturing applications. Numerical control can be viewed as a major means for integrating the manufacturing operation, since it enables the processing of data from such dissimilar functions as design engineering, manufacturing engineering, and machining. Today, numerical control is a major influence in the design of automated systems and integrated manufacturing operations.

The concept of numerical control is frequently and appropriately used as a model of integrated manufacturing. Because of this, it is often seen as an example of *computer-integrated manufacturing (CIM).* However, such a perception would be inaccurate. NC, in and of itself, is not CIM. Computer-integrated manufacturing systems can be developed with or without numerical control.

The major difference between NC and CIM is *magnitude*. In addition to the design and manufacturing operations that are served by numerical control, CIM systems generally manage data planning, sales, accounting, and management functions.

Computer-integrated manufacturing is now viewed by many manufacturing firms as an essential strategy for remaining competitive in a rapidly changing global market. The goal that many manufacturers have for implementing CIM is reducing the cost that they pay for *indirect labor* (work supportive to production) to less than the cost for shipping in a comparable product from overseas or Mexico.

Computer-integrated manufacturing involves using computers to link together the various control systems that are found in manufacturing. When properly implemented, CIM ties every aspect of a factory into a structure that makes all of its component parts and control systems visible to any user at any point in the system. This is necessary before firms can accurately understand what happens when a particular product is manufactured.

Effectively designed CIM solutions can result in significant reductions in engineering design cost and overall lead time. Firms such as General Electric,

IBM, and Ingersoll Milling have proven that CIM can result in improved product quality, engineering capability, and overall productivity.

Computer-integrated manufacturing provides significant manufacturing flexibility. Large production lots are not necessary; in fact, it is often said that, with CIM, the ideal *lot* (batch) size is 1 piece!

There are no CIM systems that can meet the needs of all applications. The type of system that a given company needs to develop depends on the types of products that it manufactures and the way that they are manufactured. Some CIM systems involve total "all-at-once" integration and automation. Others will require phasing in automation and integration over a period of time. Often, the implementation of computer-integrated manufacturing takes place in an evolutionary manner.

Despite the fact that CIM systems differ among companies and types of industries, most systems have some common elements. Normally, they utilize most of the elements of *flexible manufacturing:* NC or CNC processes, computer-aided design and manufacturing (CAD/CAM) systems, PLCs, microprocessors, industrial computers, and personal computers. See Figure 31-11. They also include production planning and control systems, collection systems for operating data, and most of the hard and soft technology systems used in automation.

Figure 31-11
Flexible machining cells like this one are extensively used in computer-integrated manufacturing operations. Such centers typically include one or more machining stations connected by a material handling system and served by automatic tool changers. (Boston Digital)

Control System Design Requirements

Control systems can be designed around a variety of different types of actuation devices. The most popular types are mechanical, hydraulic, fluidic, or electronic. A system may also combine two or more of these types of devices.

Electronic systems may have the logic and input/output functions hardwired, or they may be programmed through software. Software-based systems can involve the use of programmable logic controllers (PLCs), dedicated industrial computers, or a computer system integrated with other operations.

Regardless of the hardware and type of actuation devices used, a control system for automation must be capable of carrying out some basic operations or functions. These are:

- Δ It must be able to initiate, coordinate, cycle, and stop the various motions of the machine and its individual stations.
- Δ It must have the means to ensure that the machine does not begin or continue to operate in an unsafe condition.
- Δ It must be able to monitor and identify incomplete or defective products in a way that insures that the resultant production is acceptable.

The first two functions, addressing machine actuation and operational safety, are standard requirements for machine control systems. Standards are covered by OSHA (Occupational Safety and Health Administration) regulations and industry standards such as NMTBA-JIC (National Machine Tool Builders Association) specifications.

The third function emphasizes monitoring to assess production output. Monitoring should include techniques for incorporating the concepts of *statistical process control (SPC)* into the system. This approach provides a basis for control that is predicated on preventing unacceptable production. It also helps the control system designer focus on inputs and methods for detecting errors. Both of these elements are necessary in order to meet the final operational requirement of the control system.

Material Handling Systems

There are many different types of devices and systems that are used to provide material to automated processes and assembly machines. In general, material handling methods can be grouped into three major types or classifications—bulk systems, feeding/orienting methods, and magazine feeders. See Figure 31-12.

Bulk systems are designed to dispense liquids, gases, and granular solids. These types of handling systems need only to contain and convey the material to the dispensing mechanism. The *dispensing mechanism* distributes the desired amount of stock to the machine.

A *feeder* contains and conveys discrete components while simultaneously orienting and sorting them. Typically, parts to be handled have a specific geometry that must be oriented and sorted from other components that do not conform to the specified configuration. This *orientation* (positioning) allows the dispensing system to capture and hold the part until it is ready to be placed and then assembled or processed. A vibratory feeder with internal tooling is probably the most popular type of feeder in this class.

Magazines are material handling systems that contain and convey pre-oriented parts. This method is used when bulk or vibratory feeding methods

Figure 31-12
Three common types of material handling systems. A—Bulk systems, which feed materials that do not require orientation. B—Feeding/orienting devices, such as the vibratory feeder. These devices orient parts properly, then feed them individually. C—Magazine feeders, which dispense parts that have been loaded into them with the proper orientation.

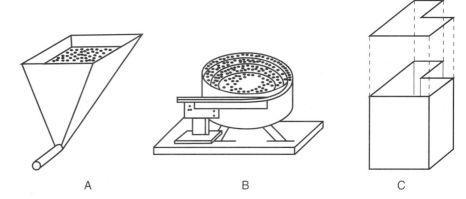

could cause the parts to become tangled or damaged on the way to the dispensing system.

A variation of magazine feeding is called reel or *strip feeding*, Figure 31-13. With this type of feeder, the parts or materials are linked together to be fed into a processing or assembly operation. Electrical terminals are frequently handled in this way, since they have the tendency to jam if fed individually.

Figure 31-13
Two types of electrical terminals are provided to this connector assembly machine, using the strip-feeding method. Note the two reels at the top of the machine, each feeding a strip of connectors. The operator is checking some connectors after assembly. (AMP Inc.)

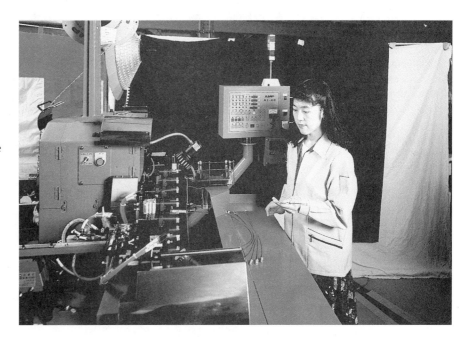

Another common example of a magazine feeding system application is supplying coiled strip steel to a die stamping press. In this case, the coil is an economical and convenient method for packaging and handling the material.

Material Feeding Techniques

Material feeding consists of orienting and providing a method for propelling components. Vibratory bowl feeders can move components, and with the help of bowl tooling, can separate and orient them. However, the act of orienting does not in itself control or place a part for acquisition by the tooling. This is the role of the dispensing system or *escapement.*

Escapements

Typically, parts coming from a feeder are fed into a track. Once in the track, the parts must come to a stopping point where an individual component can be picked and placed or stripped onto a workholding fixture or nest. See Figure 31-14. After a part is removed, the escapement immediately replaces it with the next one in line.

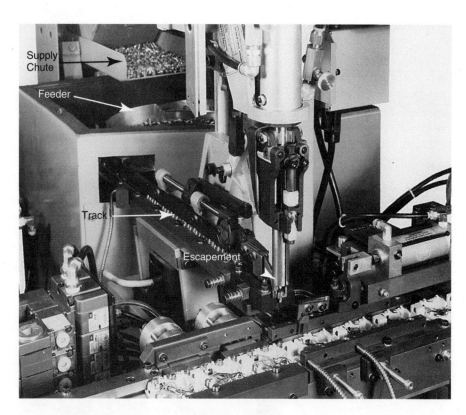

Figure 31-14
A typical escapement system. This escapement feeds screws to an automatic screwdriving device that installs them in electrical terminal block. Note the supply chute, at rear, that delivers screws to the vibratory feeder inside the enclosure. The feeder orients the screws and places them in the track leading to the escapement located just behind the vertical spindle of the screwdriving device.
(Dixon Automatic Tool, Inc.)

The component feeding system is always designed to provide more parts to the escapement than are actually required by the operation. Consequently, the escapement must be able to accept parts only when needed. It must also be able to withstand pressure from the feeding system without jeopardizing the orientation and control of the part that is being presented for selection.

Stopping parts at the pick-up point and withstanding the feeding pressure are major challenges in designing an escapement. Sometimes parts are fragile and tend to tangle. In cases such as these, the escapement must be isolated from the parts feeding system.

Once a part has moved into position at the pick-up point, there are several ways to eliminate feeding pressure. One method senses when the escapement is filled, and activates a gate or chute which causes the feeder to spill off the parts, relieving pressure on the escapement. A second method eliminates feeding pressure by using a switch to shut down the feeder.

There are many types of escapements in general use. The simplest is the *dead-end escapement.* This type of escapement tracks components to a fixed point, the dead end. This point is defined by a barrier which contains the part until it is picked.

The part contained by the barrier also acts as a barrier by blocking the next part from moving into the pick-up point. After the part is removed at the pick-up point, the remaining parts in the track move ahead until the next part encounters the barrier.

A more positive approach uses an escapement similar to the one shown in Figure 31-14. This escapement has two fingers that are able to capture and place individual parts. In this instance, the parts are machine screws that are being supplied by a vibratory feeder aligned to a track leading to the escapement. It takes the escapement two cycles to advance a part into the pick-up position. In the first cycle, it captures the part; in the second cycle, the screw is placed into position. The pick-up position in this case is the jaws of an automated screwdriver which assembles the screw into the product, a wiring block.

Figure 31-15 shows a similar assembly operation involving the placement of a machine screw in a fuse block. This photograph clearly illustrates the use of *two* escapements—one for the machine screws, the other for the fuse blocks. These escapements are able to locate and position both parts, eliminating the need for nests or other fixturing to hold the assembly. Escapements used in this manner can simplify and reduce overall costs, providing that the components lend themselves to this form of placement and control.

A third variation in escapement design provides for the tracking and control of parts in one orientation, then placing them in a different orientation at the pick-up point. This change in orientation may be a shift in the plane in which the part is located or a change in the angle of the part. A familiar example of this type of escapement is the cassette transport mechanism in front-loading VHS tape decks. This escapement moves the oriented cassette into the tape deck and then drops. This change in plane positions the cassette in alignment with the tape heads.

Despite the fact that escapements can become very complicated mechanically, their function is really quite simple. In general, escapements are mechanical devices working in tandem with feeders. Their purpose is to present oriented components at a fixed point and at a rate that exceeds the processing rate of the automated process or assembly operation.

After a part or component has been fed, oriented, and then positioned by an escapement, it is ready to be placed for assembly or processing. The means used to move and place components are termed *part-transfer mechanisms.* These mechanisms are often considered to be extensions of escape-

Figure 31-15
An assembly operation making use of two escapements. One escapement, located at the end of the diagonal track at left, feeds individual screws to the screwdriving device. The second escapement, at bottom center, regulates feeding of fuse blocks to the assembly position below the screwdriving spindle. (Dixon Automatic Tool, Inc.)

ments. However, because of the variety of methods in use today, they merit consideration as a specific technique.

Part-transfer mechanisms take the part from the escapement pick-up point and place it in a workholding fixture or nest. The methods used to make these transfers can be as simple as dropping the part into the nest, or they can be very complicated. The more complicated the transfer action, the more costly the device.

One of the simplest and most widely used transfer methods is a wiper mechanism. The part to be transferred is placed in position by an escapement, and then wiped or swept away to a nest or fixture by the transfer mechanism. A good example of this type of transfer can be found in a magazine-fed bolt-action rifle. The wiper is the bolt, which moves the cartridge from the escapement into the barrel of the rifle, the nest.

A variation on this method uses a mechanism in the escapement to hold back the part until the nest or fixture is in place to accept it. When the part is released by the escapement, it drops into the nest. In some cases, the part drops into a shuttle which carries it from the escapement to the nest. The shuttle is similar in concept to a coin slide on a gumball machine.

Shuttles

A more complicated type of shuttle is shown in Figure 31-16. In this instance, the shuttle arm is able to move up and down and also rotate. Although this shuttle is a standard design that is readily available from its manufacturer, it can be easily customized to accept tooling for specific appli-

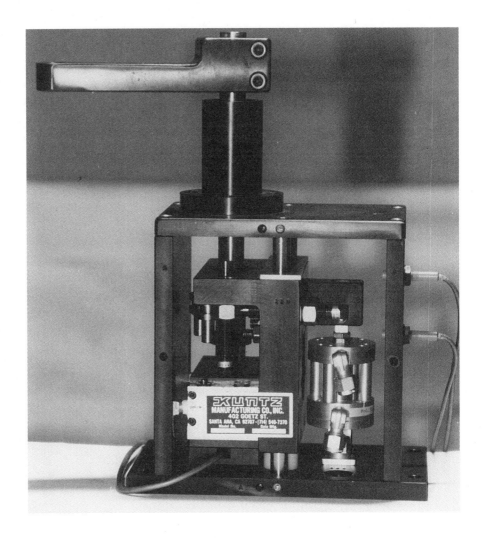

Figure 31-16
This shuttle is designed to hold back a part until it is ready for release to the nest. (Kuntz Automation Engineering)

cations. The use of standardized shuttle mechanisms is popular with machine designers.

Shuttles are widely used to link escapements to a process. Although they add complexity and cost to the material placement system, they are cost-effective in many applications. The use of a shuttle between the escapement and the operation or process being served should be considered if any of the following conditions is present:

△ The escapement's operation is jeopardized by heat, overspray, or other dangers from the processing equipment being served.

△ The part must be reoriented before it can be placed.

△ Other parts are also fed into position at this station.

△ The part being fed is an option, and thus, is not always required.

△ The escapement or operation is subject to jams or frequent maintenance. In this case, the shuttle allows the feeder and escapement to be more accessible.

- Manual feeding of components must be done at one station of an otherwise fully automated system. The shuttle provides an interface between a manual operation and an otherwise automated process.
- The type of material or the configuration of the part being fed at this station is frequently changed.

In the first six conditions outlined above, the design, operation, or safety of a specific station may be improved by incorporating a shuttle in the process. Shuttles may also simplify the service and maintenance requirements of automated systems.

In situations where the configuration of the part is frequently changed, the shuttle provides a transition point that can significantly improve the flexibility of the system. Feeders and escapements are generally dedicated to a specific part or component configuration. Therefore, if a part change is needed, the feeder and escapement mechanism can be moved out and replaced with a new unit. The shuttle tooling serves as the transition to the remaining portions of the automated system. Specifically, the shuttle provides an easily accessible alignment point. It also facilitates testing the operation of the escapement without the need to operate the entire system.

Robots

When robots are used in an automated system to transfer material, they play a role similar to shuttles. In general, *pick-and-place robots* (those that simply pick up a part and move it to another location) can carry out two important functions. They are effective in transferring material between operations. They are also able to load and unload completed components.

In order to accomplish these functions, the robot is equipped with a gripper-type end effector. In most cases, the gripper is custom designed to handle the specific part or material being moved. See Figure 31-17. In nearly all applications where robotic systems are involved, the parts must be presented to the robot in a known position or sequence, and in a known orientation. An escapement can be used to provide the needed positioning and orientation of the part for the robot.

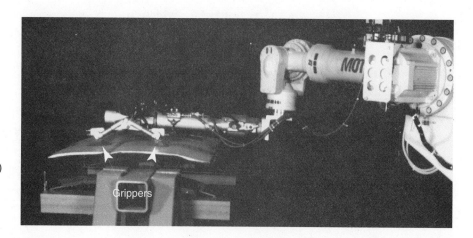

Figure 31-17
This robot uses a specially designed end-effector with vacuum grippers to handle sheetmetal parts. It is designed for high-speed transfer of parts as heavy as 110 lbs. (50 kg) between stamping machines or other devices located up to 26 ft. (8 m) apart. (Motoman, Inc.)

To successfully unload the part, the robot makes use of a nest or fixture to define the position and final orientation of the material. Using a robot to transfer parts can provide even greater flexibility than a shuttle. The robot provides the advantages of soft automation (the ability to change easily), whereas the shuttle is limited by the fact that it is a hard automation device, designed for a specific part or operation. In large complicated systems, the advantages of soft automation may be necessary.

Moving Parts Between Stations

There are various types of mechanisms used to move a part or assembly between machining operations or material placement stations. All of these mechanisms use a fixture or a nest to control the part's orientation and location, so that it is properly positioned at the next operation. This type of movement enables various operations to be performed on the part without having to re-orient and recapture that part. Two types of mechanisms are used to achieve these objectives. They are referred to as rotary or in-line transfer systems.

Rotary indexing tables

Indexing tables are the most commonly used methods for rotary transfer. An indexing table provides intermittent motion that accurately moves parts to successive operations located at the periphery of the table. The number and physical size of the operations performed is a function of the circumference of the table and the number of *dwell points* (stops) in one table revolution.

A rotary indexing table with its tooling, escapements, and vibratory feeders is shown in Figure 31-18. The view is taken from directly overhead. This particular machine assembles a health-care product. As with most automated assembly machines, functional testing is incorporated as part of the sequence of operations.

The selection process for a rotary index table is not overly complicated. First, the designer must determine the sequence and types of operations to be accomplished. Once this information is available, the remaining tasks to be completed are identifying the number of dwell points required, calculating the weight of the parts and tooling, and determining the type of mechanism needed to provide the desired rotary motion.

There are three basic driving mechanisms in general use to provide intermittent rotary motion. As shown in Figure 31-19A, the *ratchet and pawl* is the simplest and least costly of these systems. It is also the least accurate, however.

The simplicity of the ratchet and pawl driving mechanism begins with the power source used to create the intermittent motion of the table. The driver for this mechanism is frequently a single-acting air cylinder. The ratchet and pawl indexing apparatus has the advantage of rapid changeover from one index ratio to another. This allows the designer to use the same drive mechanism with several dial plates, each having a different number of points of dwell. Since the ratchet and pawl are not mechanically interlocked, the table or dial can be moved freely in one direction. This can be an advantage

Figure 31-18
An indexing table rotates precisely to accurately move parts to successive stations for assembly, testing, or finishing operations. A health-care product is being assembled on this table. Note: Guarding that would normally be in place has been removed to better show details for this photo. (Kuntz Automation Engineering)

in clearing jams or changing tooling on the dial. However, this also creates its major disadvantage. This type of drive can overrun its dwell position. This decreases its precision and limits its usefulness for many forms of automation.

A second popular method for achieving rotary motion is called the *Geneva mechanism,* Figure 31-19B. This method uses a continuously rotating driver to index the dial. The driven member has a radial slot machined in it for each point of dwell. Because of this slot, the mechanism has an inherent locking feature when in dwell. This locking feature insures good positioning and accuracy for most assembly operations.

The primary drawback for this type of mechanism is the severe acceleration and deceleration caused by the basic geometry of the Geneva drive. This motion can shake or even damage components being carried in the fixtures on the dial. These high acceleration rates also cause a great amount of stress on the mechanism. Consequently, Geneva drives must be very robust to achieve the precision and long life expected of an automated system.

The *cam drive,* Figure 31-19C, is a third type of mechanism that provides an accurate and reliable means for achieving rotary motion. Like the Geneva mechanism, this design locks the dial when in dwell. The cam drive mechanism has relatively simple construction and exhibits excellent acceleration/deceleration characteristics.

Cam-type mechanisms used on rotary machines are also used to drive in-line transfer systems (continuous conveyors). Although cam drives are expensive, they have become a very popular drive mechanism for automated assembly.

Figure 31-19
These three basic drive mechanisms are used to provide intermittent motion for indexing tables. Each of the examples shown would provide eight dwell points per revolution.

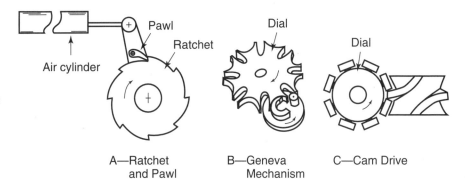

A—Ratchet and Pawl   B—Geneva Mechanism   C—Cam Drive

In-line transfer mechanisms

In-line transfer systems are generally chain or belt conveyors that convey a *pallet* (support base, frame, or tray) containing a parts-holding nest or fixture between operations. In comparison to rotary index tables, in-line mechanisms significantly expand the size and complexity that can be achieved in automated systems. Figure 31-20 provides a view of a transfer line that uses a pallet with a standard footprint to convey products between workstations. This type of conveyance is very popular for developing in-line systems and for linking together in-line automated systems.

Figure 31-20
This in-line transfer system moves assemblies between workstations for processing. The assemblies are carried on standardized pallets or trays. As shown, the system can involve switching and redirection of the pallets for proper distribution of the work. (Prodel Automation, Inc.)

There are also in-line transfer mechanisms that move intermittently, bringing the pallets into an approximate location in front of a workstation. Once the pallet is sensed as being in position, a set of locating pins accurately aligns and fixes the pallet at the station. The accuracy of the positioning is dependent on the rigidity and precision of the conveyor and the drive mechanism.

The intermittent motion that causes a transfer line to increment between stations is frequently achieved with a drive mechanism similar to the cam drive used on rotary index tables. This means that the conveyor chain and tooling accelerates and decelerates to a precise position. Once in position, the transfer line is locked during the period of dwell by virtue of the drive mechanism geometry.

There are other, less-expensive drive mechanisms that can be used. One type is a rectangular-motion device, or *walking-beam mechanism.* These mechanisms are commonly used with in-line machines where flexibility is important. In operation, the walking-beam transfer mechanism uses a transfer bar to lift the *work in process* **(WIP)** (parts) out of a workstation location and move it one position ahead to the next station. The transfer bar lowers the part into a nest, which positions it for the next operation.

An alternate approach with this method is to transfer a pallet containing the work in process. The *workholding pallet* is relocated at each station by guide pins. With this method, there is no need to place the part in a nest. Since the pallets are not attached to the transfer line, they can be changed quickly, increasing the mechanism's versatility. See Figure 31-21.

Figure 31-21
Workholding fixtures attached to pallets allow flexibility in handling work in process. The pallets can be moved onto or off a transfer line by use of a load-unload station like the turntable-equipped one in the foreground. Pallets can be moved from one transfer line to another by the use of various devices, including automated guided vehicles. The AGV in the aisle area at right is following a guide wire buried in the floor as it moves a pallet to a different transfer line. Note that the workholding fixture on that pallet is different from the fixtures used on the transfer line in the foreground. (Cincinnati Milacron)

Overhead or track conveyors are also used to carry parts or assemblies to a nest or fixture. In these instances, the work is generally oriented and has locating points that can be easily found by an escapement that captures the component at each station.

Selecting the most appropriate method for moving materials and tooling between stations or operations depends on the general characteristics of the operations and on the components being automated. The decision on which approach to take is influenced by several factors. Answers to the following questions will influence design of the system:

- How many stations or operations will have to be linked by the transfer mechanism?
- What is the weight of the components and tooling to be transferred by the mechanism?
- Is it necessary for the transfer mechanism to provide location and position accuracy?
- What is the production rate for the system?
- Do the assembly or processing stations have to operate sequentially, or will the order vary?
- Are there operations that are redundant or that can be passed by in the sequence of assembly or processing?
- Are there any operations in the manufacturing sequence that will be done manually?

In most cases, rotary methods are appropriate for relatively small, lightweight parts that require ten or fewer operations. In-line systems are able to accommodate larger components and a greater number of stations. Although variation in sequence and the inclusion of redundant stations or pass-by operations tends to favor the use of in-line transfer mechanisms, they will not be the primary determinants for any one method of transfer. The use of a shuttle can provide for flexibility or redundancy in either mechanism. In practice, shuttles can also be used to link different transfer systems together.

Conclusion

Automation today encompasses a wide variety of technologies to create hardware and control systems that have the capability of operating manufacturing machinery without human effort or intervention. There are few limitations to the application of automation. Any operation or process can be automated if the result justifies the cost of adding the necessary control and material placement systems.

Linking several automated systems together is also feasible providing that a coherent control system can be established and variability can be controlled. When this stage of automation is reached, the information necessary for carrying on automatic operations becomes part of the management control system for the manufacturing facility. At this stage of refinement, automation has truly become computer-integrated manufacturing.

Important Terms

actuator
adaptive control systems
artificial intelligence (AI)
automated control systems
automated manufacturing system
automation
bulk systems
cam drive
closed-loop control systems
computer-integrated manufacturing (CIM)
continuous-path control
controller
control system sophistication

dead-end escapement
distributed numerical control (DNC)
dwell points
escapement
feedback
feeder
flexible manufacturing
Geneva mechanism
hard automation
indirect labor
in-line transfer systems
input
input values
interchangeability
logic
magazines
manual control
manufacturing flexibility
material feeding
mechanization
numerical control (NC)
numerical control program
open-loop control system
orientation
outputs
pallet
part-transfer mechanisms
pick-and-place robots
point-to-point system
programmable logic controllers (PLCs)
proportional control
ratchet and pawl
relay logic
replication
response criteria
sensitivity
sensors
soft automation
stability
states of operation
statistical process control (SPC)
steady state error
strip feeding
throughput
transient response
turnkey system
variability
volume
walking-beam mechanism
workholding pallet
work in process (WIP)

Questions for Review and Discussion

1. Can variability be reduced by automating a manufacturing operation?
2. Describe the difference between mechanization and automation.
3. What are the distinguishing elements of automation?
4. Discuss the differences between hard and soft automation. Provide examples.
5. List several devices you are familiar with that could serve as examples of either open-loop or closed-loop control systems.

Glossary

A

Abrasive Belt Planer: A high-production wood planing machine that uses continuous abrasive-surfaced belts, rather than cutterheads with knife blades.

Abrasive Jet Machining: A grinding process that suspends tiny particles of abrasive material in a low-pressure stream of gas (dry air, carbon dioxide, or nitrogen) sprayed through a sapphire nozzle.

Abrasive Shot Method: A glass-frosting method in which a mixture of abrasives and lead shot is placed inside a frame with a sheet of glass, then agitated.

Acetylated Wood: A modified wood that provides the strength of resin-impregnated woods, without the brittleness associated with them. It can be used for treating thin stock and veneers.

Acrylonitrile-Butadiene-Styrene (ABS): A polymer made up of acrylonitrile, butadiene, and styrene monomers. ABS is widely used for plumbing pipe and fittings.

Actuator: A device in an open-loop control system that responds to the signal from the controller and performs an action.

Adaptive Control Systems: Systems that provide control in response to changing conditions.

Addition Polymerization: The linking together of monomer molecules to form long chains. The process is mechanical; there is no chemical change in the material.

Adhesion: The process of joining materials by using an agent that sticks to both pieces.

Adhesion Processes: Metal-joining methods (soldering and brazing) that form a bond between parts by filling joints with a material with a melting point lower than that of the base metal. Also, any method of producing a product that involves joining materials with a glue or other bonding agent.

Adhesive: A bonding agent that forms a film between materials to be joined.

Adhesive Failure: Failure of a glued joint that takes place in the interface area between the adhesive and the material being joined.

Advanced Composites: Term once used to describe composites consisting of more complex, exotic, and costly combinations of materials than used in structural composites.

Age-Hardening: See Precipitation Hardening.

Aging: See Precipitation Hardening.

Air Drying: Method of drying wood that reduces the moisture content to around 15 percent.

Airless Spraying: A spray method that uses hydraulic pressure, rather than compressed air, to force the finishing material into the spray gun.

Alloys: Mixtures of several different metals, sometimes with the addition of other nonmetallic elements.

Aniline Dyes: A class of dyes used in spirit stains.

Anisotropic: The property of wood that results in drastic structural changes as it absorbs moisture or gives up moisture (dries).

Annealing: A process for softening metal and relieving stresses by holding it at high temperature for an extended length of time, then slowly cooling it. Also used to relieve stresses in ceramic or glass products.

Annealing Lehr: An oven through which molten glass is pulled to relieve internal stresses.

Annular Ring: The layers of cells appearing in concentric circles and representing each year's new growth of a tree.

569

Anode: A positive electrode.

Anodizing: An electrochemical process, often used on aluminum, that converts the surface of the metal into a hard oxide layer that is resistant to corrosion. Also called anodic oxidation.

Anti-Kickback Device: A safety mechanism attached to power saws to eliminate the problem of kickback.

Antiozonants: Additives used to help prevent breakdown of plastics by ozone gas in the atmosphere.

Arbor: In a machine tool, the rotating spindle upon which a cutting tool is mounted.

Arbor-Type Cutters: Cutters that are mounted directly on the milling machine arbor or spindle.

Artificial Aging: A hardening process for aluminum alloys conducted at heat levels above room temperature.

Artificial Intelligence (AI): A control system concept in which inputs from sensors, coupled with predefined rules and relationships stored in the computer's memory, allow the control system to provide expert control in well-defined situations.

Ashing: The use of a very fine wet abrasive to reduce overheating of plastic and improve the cutting action.

Assembly: The process through which machined parts, components, and sub-assemblies are joined to assume the final shape of the product.

Assembly Line: An arrangement of machines, equipment, and workers in which work passes from operation to operation in a direct line until the product is assembled.

Atomize: To break a liquid material into very fine drops.

Attrition Milling: A batch operation in which the materials to be ground are placed in a stationary chamber, along with the grinding media. A rotating shaft provides the agitation needed for grinding action.

Austempering: A form of tempering in which metal is cooled in a salt bath to increase ductility and toughness.

Austenite: The most dense form of low carbon steel.

Autoclave Molding: A process involving applying both pressure and heat during the cure cycle to produce a more dense composite part.

Automated Control: A control system that simultaneously monitors the quality of the product and the functioning of the system; it must be able to start, stop, and sequence production.

Automated Manufacturing System: Production system that results from linking together several processes or operations that require no human intervention.

Automatic Feed: The capability of a lathe to mechanically advance the tool into the workpiece and move it to the left or right as it cuts.

Automation: The process of applying automatic control devices to production equipment.

B

Bagasse: A cellulosic byproduct of the pressing of sugar cane. It provides the fiber for manufacture of insulation board.

Ball Mill: A large steel cylinder that is partially filled with spheres of heavy steel and tumbled or rotated to grind materials. Also, a small porcelain jar mill used to mix clay and additives.

Band: The continuous loop blade used in a bandsaw.

Band Casting: A continuous production process used to produce thin strips of ceramics for use as electrical substrates and heat exchanger devices. Also called tape casting.

Bandsaw: A saw with a continuous loop blade, used primarily for cutting curved shapes.

Batch: A number of essentially identical parts or units manufactured at the same time. Also called a "job lot."

Batch Furnace: A type of furnace used to heat-treat small quantities of parts carried on a rack or flat car that is loaded into the furnace.

Bed Width: The width of stock that can be milled in a planer or similar machine.

Bed-Type Machine: A type of milling machine in which a worktable is mounted directly on the machine bed.

Bending: A process used for producing curved shapes in either solid or laminated wood materials.

Beneficiated: The term used to describe the practice of processing or refining raw material at plants located near the mine in order to reduce transportation costs.

Beveling: The process of changing the sharp (90 degree) angle where a vertical and horizontal surface join into a less-sharp angle, usually 45 degrees.

Binders: Phenolic resins used to bond together wood chips or flakes when producing various types of composition board. Also, chemicals that are added to dry ceramic powders to hold the mixture together when it is pressed.

Bioceramics: Term used to refer to ceramic devices used in the medical field.

Bisque Firing: The first firing of a ceramic material,

done without glaze; it solidifies and fuses the body so it can be handled.

Bisque Ware: The ceramic product that is fired without a glaze (coating).

Bits: Tools used in a drill press or boring machine to produce round holes.

Blank: The stock on which a flat die is used to roll threads. The term is often used for any metal piece being shaped by forging or stamping processes.

Blanking: A process used to punch a flat part blank from sheet metal. It produces very little waste.

Blanks: Slices cut from a mass of highly viscous ceramic stock, and used in the jiggering and jollying processes of production. Also called pugs.

Bleaching: A process that involves applying a caustic liquid to the surface of a workpiece to lighten the natural wood coloring.

Blending: A type of sanding that results in some material removal, producing a fairly smooth finish; it requires a medium-grit abrasive.

Blind Holes: Holes that don't run through the workpiece.

Blind Rivet: A fastener, consisting of a mandrel (shaft) and rivet head assembled as one unit, that permits riveting when only one side of the workpiece is available.

Blow and Blow Process: A bottle-forming process using a blast of air to blow the molten glass into a desired shape.

Blown Film: Film that is produced by extruding a tube of plastic material which is then expanded in diameter and reduced in thickness by internal air pressure, and finally slit to form a flat sheet.

Board Foot: The basic unit of measurement for dimension lumber, 1 inch thick by 12 inches by 12 inches.

Bonding: A forming process that involves the use of heat and pressure to compact particles or chips of wood into wood composition board.

Boothroyd and Dewhurst Method: A method used to obtain numerical data which can be used to assess design efficiency, and to evaluate the need for redesign of the product.

Boring: A technique that is used to improve the accuracy of a drilled hole through the use of a boring bar.

Brazing: A joining method that holds base metals together by adhesion of a melted filler metal, rather than by melting the base metal.

Brittleness: The tendency of a material to break without warning.

Broach: Cutting tool that creates a planing action when pushed or pulled across or through a workpiece.

Broaching: A process used for the internal machining of keyways, splines, and irregularly shaped openings.

B-Stage: Term used to refer to a partly cured preform that is dry, but still slightly tacky.

Buffing: Polishing process that uses muslin disks on a wheel to provide a smooth glossy finish.

Bulging: One type of process that can be used to expand tubular metal shapes.

Bulk Density: The type of density that is often measured on ceramic parts. It is based on the bulk volume of the part.

Bulk Molding Compounds (BMC): Puttylike materials used as stock in some closed molding processes.

Bulk Systems: Handling systems designed to convey liquids, gases, and granular solids to a dispensing mechanism.

Burrs: Sharp edges produced when metal is deformed by shearing, trimming, stamping, or machining.

Butt Welding: A form of arc welding in which heat is generated from an arc produced as the ends of two parts to be welded make contact. Also called flash butt welding.

C

C Glass: The type of glass used when a product must survive continued exposure to corrosive chemicals.

Calcination: A process in which material is first heated, then milled, to produce particles that are sufficiently fine to be mixed with the clay for ceramic production.

Calendering: A method of progressive squeezing between sets of rollers to produce rigid but pliable PVC film.

Cam Drive: A type of mechanism that provides an accurate and reliable means for achieving rotary motion.

Captive Suppliers: Those for whom a manufacturing company is the principal, or sometimes only, customer.

Carbon Steel: Steel that includes carbon as an alloying element.

Carbon/Carbon Composite (C/C): A carbon-fiber-reinforced carbon matrix material that behaves more like a ceramic than a polymeric.

Carbonitriding: A process similar to nitriding that is used for hardening low-carbon steel parts.

Carbonization: The heating step in the production of carbon fiber, during which heat is applied to burn off most of the non-carbon elements of the fiber.

Carburizing: A method of casehardening in which the heated workpiece is covered with a carbon-hardening powder that melts and forms a thin coating.

Carriage: A device that can slide left or right along the ways of a lathe, supporting the cross slide, compound rest, and tool post.

Case Hardening: Process that is used to produce a hard shell or case on a workpiece.

Cast Iron: One of the two basic forms of iron. Cast iron contains between 1.7 percent and 4 percent carbon. It is used when strength and durability are critical.

Casting: A process in which a liquid resin is poured into a mold and solidifies.

Catalyst: An agent that enables a chemical reaction to proceed at a faster rate.

Cathode: A negative electrode.

Cell Casting: A molding process in which acrylic sheets are cast and allowed to cure between sheets of plate glass.

Cellulose: One of the two major ingredients in wood, making up about 70 percent of volume.

Cellulosics: A group of commercial plastics made from a natural polymer, cotton cellulose.

Cemented Carbide: A cutting tool tip material (tungsten or titanium carbide, suspended in a cobalt matrix) brazed onto the tool blank to allow it to cut harder material at higher speeds.

Cementite: The phase that is reached when steel is slowly cooled after being heated to above 1341°F (727°C). Cementite has a carbon content of 2.0 percent.

Centerdrilling: Drilling holes in each end of the stock so a workpiece can be secured between lathe centers.

Centerless Grinding: A special type of cylindrical grinding that can be used for external or internal applications.

Ceramic: Term used to describe products made from clay, an inorganic, nonmetallic solid material derived from naturally decomposed granite.

Ceramic Engine: An internal combustion engine with components that are either entirely ceramic or ceramic-coated metal.

Ceramic-Matrix Composites (CMCs): Composite materials in which the matrix is a ceramic.

Ceramoplastic: Glass-bonded mica, moldable/machinable inorganic composite with high insulating properties.

Cermet: A ceramic composite material, made from carbide and sintered oxides, that is used to make disposable tip inserts for cutting tools.

Charged: Term used to describe the addition of a small amount of fine abrasive to the wheel while buffing.

Chemical Blanking: An etching process that forms a part by etching completely through the workpiece.

Chemical Etching: The process by which glass is frosted, or made opaque, with the use of an acid.

Chemical Milling: Process used to etch or form metallic parts. Also called chemical machining or photoforming.

Chemical Vapor Deposition: A conditioning process in which a gas reacts and contacts the heated part, resulting in the production of a very fine-grained coating.

Climb Milling: A process in which the cutter is positioned on top of the workpiece, with the work moving in the same direction as the cutter's rotation.

Closed Molding: A forming process in which stock is compressed between the halves of a two-piece mold.

Closed-Die Forging: A method of forging in which a preheated metal billet is placed between dies that completely encompass the billet and restrict its flow.

Closed-Grain Woods: Hardwoods that have small pores, such as maple.

Closed-Loop Control Systems: A basic control system in which the actual output is measured, compared to a preset standard or reference, and adjusted if necessary.

Coarsen: The process of loosening particles to create aggregates that function like larger particles, so the powder can be more easily compressed.

Coated Abrasives: Finishing materials consisting of an abrasive adhered to a backing material, such as paper or cloth.

Coated Preform: A shape made by spraying chopped fiberglass and a binder material onto a form.

Coefficient of Thermal Expansion a: The amount of increase in volume of solid, liquid, or gas for a rise of temperature of 1° at constant pressure.

Coextrusion: A process in which the matrix and filler are continuously extruded together through a forming die; this is particularly useful for constructing composites and blending different colors and types of materials.

Cohesive Fabrication: A method of producing a product that involves the physical blending together of materials, such as the process of hand layup of fiberglass roving and a resin.

Cohesive Failure: A bonding failure that occurs in the adhesive itself.

Coining: A precision closed-die forging process that causes metal to move, as the die halves are closing, from thinner to thicker areas.

Coinjection Molding: A molding process in which two or more materials are injected into the mold to make an individual part.

Cold Compression Molding: A molding process in which the material to be formed is compressed between unheated male and female molds. It is also referred to as cold press molding.

Cold Isostatic Pressing: A dry powder pressing process used in ceramics manufacture. Higher pressure is used than required for dry pressing.

Cold-Rolled Steel: Steel that is rolled while cold, and has a shinier surface finish than hot-rolled material.

Collet: A socket, equipped with threads on one end and split jaws on the other, which is used in a lathe to automatically center stock that is round, square, octagonal, or hexagonal.

Collimated: Term used to describe the focused, parallel light rays of a laser beam.

Colorants: Dyes or pigments that are added to plastic to impart color, and often, to improve other properties of the product.

Column-and-Knee-Type Machine: A type of milling machine in which a worktable is mounted on a knee that moves up and down the column.

Combination Blades: Circular saw blades with teeth that can be used for both crosscut and rip sawing.

Commodity Resins: Standard-grade resins such as low-density polyethylene (LDPE), polypropylene homopolymer (PP), crystal polystyrene (PS), and rigid polyvinyl chloride (PVC).

Common Lumber: Wood that is graded by number, ranging from 1 to 4, with Number 1 presenting the best grade (no knots or knotholes).

Community of Interest: Term used in Germany and other European countries to describe an arrangement in which two competitors join together to produce a common product. In North America, the arrangement is often called a joint venture or cross-licensing agreement.

Compact: The formed part that results from green machining. Also, the action of squeezing metal molecules more closely together through forging or embossing in a press.

Components: Individual parts that are assembled to make up a product.

Composites: Material family that consists of stock that results when two or more distinctly different but complementary substances are physically combined.

Composition Board: A sheet product made from wood that has been broken down into particles or fibers, then reconstituted. Types of composition board include insulation board, hardboard, waferboard, particleboard, and oriented strand board.

Compound Rest: A device mounted on the lathe carriage, that can be adjusted to hold the cutting tool at different angles to the workpiece.

Compreg Process: A process that involves impregnating a thin sheet of wood with liquid phenolic resin, then sandwiching it with other pieces in a compression molding press.

Compression Molding: A closed molding process that can be done with either thermosetting or thermoplastic polymers. Thermosets are popular.

Compression Seal: Bond in which parts are pressed into position in the completed assembly.

Computer Numerical Control (CNC): A system in which a program is used to position tools and/or the workpiece and carry out the sequence of operations needed to produce a part.

Computer-Integrated Manufacturing (CIM): A plant-wide or company-wide management philosophy for planning, integration, and implementation of automated production. CIM is an integration of separate computer systems, that allows data to be shared, transferred, and modified with ease.

Condensation Polymerization: A chemical reaction in which polymer chains become cross-linked. Usually, a byproduct such as water is formed and given off.

Conditioning: Applying processes that change the internal or external properties of materials.

Conductive Glass Solder: A material that is used to provide both a seal and an electrical path.

Conifers: Cone-bearing trees that have needles which remain green all year long. Softwood lumber is produced from conifers.

Construction Adhesives: Elastomers that are solvent-releasing adhesives, typically used in applications such as the bonding of drywall to other materials, or attaching flooring and tile.

Consumable Products: Substances that are "used up" soon after manufacture, such as foods and many types of paper products.

Continuous Casting: A molding process, used for acrylic sheet, in which a highly viscous syrup is slowly cured between two continuously moving stainless steel belts.

Continuous Die Cutting: A process that allows a material in roll form to be cut as it passes through either a flatbed press or a rotary machine.

Continuous Extrusion Blow Molding: A process in which a parison is continuously formed and molded in a series of molds.

Continuous Fibers: Unbroken, long glass reinforcing fibers. They are sometimes sprayed with binders to hold them together in a continuous mat.

Continuous Furnace: A type of reheating furnace in which the charge introduced at one end moves continuously through the furnace and is discharged at the other end.

Continuous Improvement: An approach to value analysis that involves analyzing the costs involved in each step of production.

Continuous Path Control: A type of NC control system in which the tool feed rate is controlled at all times, and cutting tool offsets or compensations are specified for every point on the tool path. Also known as contouring.

Contour Machining: A process that selectively etches a desired area to some specified depth.

Contouring: A method used to remove metal from surfaces of irregularly shaped parts by selectively etching to the desired depth.

Control System Sophistication: The degree of complexity of the system controlling production machinery. It is one of the four criteria that shape the nature of automation.

Controller: That part of an open-loop control system which provides the logic and governs the action of the actuator.

Cope: One-half the frame that makes up the flask used in sand casting. See Drag.

Copolymer: A molecule that results when two or more kinds of monomers are combined to form the long-chain molecule.

Core: A molded sand shape used to provide a cavity in a part produced by the sand molding process.

Corona Discharge: A high-voltage, high-frequency discharge in air. See Electrical Surface Treating.

Coupling Agents: Substances added to the resin component of a composite to improve the strength of the bond between the resin and reinforcement materials or fillers. Also called promoters.

CO^2 Laser: A device that produces a narrow beam of coherent (all one wavelength) light that can be used to cut, scribe, solder, weld, or heat-treat many materials.

CPM: Abbreviation for the Critical Path Method; a technique which is used in many areas of planning, particularly in the manufacturing and construction industries.

Cracked Ammonia Atmosphere: A mixture of hydrogen plus nitrogen used in a kiln to improve the quality of the metallizing coating on a ceramic part.

Crazing: A defect in plastics consisting of a network of fine cracks on or just under the surface.

Creativity: The second of the three major factors emphasized in most value analysis programs. This includes an analysis of how well the competitor addressed factors such as ease of use, aftermarket service, and maintenance requirements.

Creep: The slow and continuous increase in length, over time, when a material is placed under a steady load and constant temperature.

Creep Failure: A change in the dimensions of a material due to creep (stretching) caused by stress from a constant load.

Critical Path: The longest route in a process flow model. It is called critical because, if any task or operation is delayed in this path, there is a corresponding delay in the time needed to build the product.

Crocus Cloth: A coated abrasive product that is made from purple iron oxide and is used to polish and buff metals.

Cross Slide: The device, supported by the lathe carriage, that is used to adjust the advance of the cutting tool into the workpiece.

Crosscut Sawing: Saw cuts made across the grain of the wood.

Cryochemical Drying: See Freeze Drying.

Cryogenic: Term used to describe ultracold temperatures as low as -300° F (-185° C).

Crystal Pulling: A crystal-growing method in which a small, single-crystal seed is brought into contact with molten semiconductor materials which influence growth; the crystal is then attached to a rotating pulling bar and slowly pulled away.

Crystallites: Long-chain cellulose molecules that are arranged in a nearly-parallel orientation and help to form the cell wall in woods.

Cup: A concave shape that may result from using too much water to wash the bleach off a wooden workpiece.

Cupola: A high-temperature furnace used to produce cast iron.

Curing: A chemical reaction in which cross-linking of polymer chains takes place.

Curtain Coating: A process in which stock to be coated is passed through a continuous, sheet-like falling stream of finishing material.

Custom: In manufacturing terms, the production of a one-of-a-kind item or a small number of items made to meet a specific requirement.

Customized Products: Goods that are manufactured to fit the needs of consumers in different markets.

Cutoff Saw: A circular saw, mounted on a pivot, that is often used on construction sites to cut angles on framing lumber.

Cutting Speed: The distance in feet per minute that the cutting edge of the drill actually travels. It may also be called peripheral speed.

Cyaniding: A case hardening technique used to create a high nitrogen-low carbon case.

Cylindrical Grinding: Grinding process used on workpieces with curved surfaces or cylindrical shapes.

Czochralski Method: A crystal growing process in which the seed is dipped into a crucible containing melted silicon, then withdrawn. The molten material freezes to the crystal.

D

Danner Process: A method used to make glass tubes by flowing molten glass down over a hollow mandrel. Air is blown through the mandrel to keep the walls of the tubing from collapsing.

Dead Center: A nonrotating device used to hold one end of a long workpiece in a lathe.

Dead-End Escapement: A dispensing device that aligns and holds a part at a barrier (dead end) until it is picked up by a tool or workholding fixture for use.

De-Air: To remove the air from highly viscous ceramic stock, and produce blanks or pugs.

De-Airing Pug Mill: A machine that consists of knives on a rotating shaft that cut and fold clay stock to remove bubbles. The process increases the density and strength of parts produced from the clay.

Deburring: Removing burrs through various finishing processes.

Decals: A printed paper transfer with ceramic powders adhered to a varnish binder. It is applied as an overglaze decoration.

Deciduous: Trees that produce hardwood lumber. They are broad-leafed species, typically shedding their leaves each fall.

Deep Drawing: Process in which a female die is pressed into thin sheet stock, stretching it over a male forming punch.

Defiberize: The process of drying out wood particles to be used for hardboard, so that most of the chemicals are removed.

Deflocculant: A water-absorbing chemical that is used to help make slip more fluid or less fluid.

Deform: To stretch metal as a result of stamping.

Dehydrate: To remove moisture.

Deming Plan for Total Quality Control: The theory developed by Dr. W. Edwards Deming, who believed that quality control should check the process, not the product.

Densification: The process used to increase the density of a ceramic part. It is usually done by sintering.

i The weight of the material per unit volume.

Depth of Cut: The amount of material being removed from the workpiece in a single pass of the cutting tool.

Design: The process that determines a product's structure, method of assembly, number of parts, type of materials, tolerances, and surface finishes.

Design For Assembly (DFA): A process that simulates the actual assembly of a product, allowing it to be evaluated in terms of its suitability to modern methods of assembly.

Design for Manufacturability: A manufacturing philosophy in which the product design process, from its inception, involves manufacturing personnel.

Design Simplicity: The third of the three major factors that are emphasized in most value analysis programs. It includes an analysis of how critical the parts of the product are, how the material was used, the ease with which the product can be assembled and serviced, and other factors.

Destructive Detoxification: A patented process used to convert nearly any type of hazardous waste material into a nonhazardous useful material.

Devitreous Glass Solder: A material similar to a thermoset, in that it can be reheated without becoming highly liquid and flowing freely.

Diamond Concentration: The ratio of diamond volume to the volume of the matrix material in a dicing saw blade.

Diamond Point Tool: A chisel used to produce specialized contours on a lathe.

Diamond Wire: A high-tensile-strength copper-plated

wire impregnated with fine particles of industrial diamond to serve as cutting teeth; used to saw composites.

Dicing: A sawing process used to cut ultra-hard refractory material, such as ceramics, ferrites, glass, silicon, and superconductors.

Die Block: A closed die consisting of a hardened cylinder with a hole running through its center.

Die Casting: The process of forcing molten metal under high pressure into the cavity of a die to form a part.

Die Cavity: The open area inside a mold that will be filled with the material being formed.

Die Cutting: A process that is used to cut paper, plastic sheets, and similar materials by forcing sharp-edged cutting rules through the stock.

Die Set: The combination of the die block assembly and one or more punches.

Diffusion: A process used with ceramics to move particles of material into the void spaces between them, improving the density of the part. Diffusion also helps decrease shrinkage during firing.

Dimension Lumber: Wood that is normally purchased in standard-length boards up to one inch thick and twelve inches wide; it is ready to use, without any additional sanding or surface preparation.

Dip Casting: See Dip Molding.

Dip Coating: Applying finishing material through complete immersion of the product or part. Also, a method used to apply a plastisol to the gripping surfaces of metal products, such as tool handles.

Dip Molding: An application in which a heated metal tool is immersed in a dip tank to acquire a coating of plastisol. Products such as surgical gloves are made using this method.

Distributed Numerical Control (DNC): A control system in which individual NC machine tools are connected through communication lines to a central command computer.

Doctor Blade Casting Process: A process that uses a metal squeegee ("doctor blade") to regulate the amount of ceramic slurry that is allowed to flow onto a metal belt.

Dopants: In semiconductor production, a form of ion implantation in which alloying elements are added to liquid metal to produce unique properties.

Doping: An ion implantation process used in producing microelectronic devices. It permits alloying with small amounts of different metals. In ceramics, the process of adding organic binders and lubricants to the clay mix to reduce it to a semi-liquid plastic state.

Double Planers: Knife-blade planers with cutterheads on both top and bottom, so that both sides of the stock can be planed at the same time.

Dowels: Wooden connecting pegs often used in the furniture-making industry.

Drag: One-half the frame that makes up the flask used in sand casting. See Cope.

Drape Forming: A forming process in which a thermoplastic sheet is clamped, heated, then drawn down over the mold.

Drawing: A metal-forming action involving both stretching and compressing to form a three-dimensional part from flat stock.

Dressed: Term used to describe the process of exposing new abrasive surfaces on a grinding wheel or diamond-impregnated saw blade.

Drilling: The process of producing a round hole in a workpiece using a fluted cutting tool called a drill.

Dry Bag Isopressing: A process using a male die punch with a liquid-filled rubber bag in an open-cavity lower die block. The filled bag exerts pressure, forcing the ceramic powder against the punch to form a part.

Dry Pressing: A process of forming ceramic parts by placing the dry powder, binders, and lubricants in a mold and applying pressure. After pressing, the part will be fired in a kiln.

Dry Spinning: A process through which fibers are formed in a solution. Approximately 2000 strands are formed at one time, using a ring-type spinneret.

Dry Winding: A process that involves winding a prepreg tape or roving onto a warmed mandrel.

Ductility: A material's ability to be formed plastically, without breaking.

Dust Pressing: A process very similar to dry pressing, but with a higher percentage of water in the mix.

Dwell Points: Stops in one revolution of an indexing table.

Dynamic Compaction: A new field of study that has evolved from explosive forming. It permits firing of ceramics at lower temperatures, thus decreasing the amount of contaminants in the part.

E

E Glass: The type of glass that is used in electrical applications because of its insulating properties.

Elastomeric Molding: A molding process that uses an expandable elastomeric bag to apply pressure to a shaped preform.

Glossary

Elastomers: A class of highly resilient plastics that are more like rubber than plastic.

Electrical Surface Treating: A high-voltage discharge process used to change the surface tension and wettability of plastic materials. Also known as corona discharge treating.

Electrochemical Machining (ECM): A method of material removal that shapes a workpiece by removing electrons from its surface atoms. In effect, ECM is the opposite of electroplating.

Electrodeposition: A finishing process, similar to electroplating, that is used on metals and coated polymers. Also known as electrophoresis.

Electrodischarge Machining (EDM): A process that uses electrical energy in the form of sparks to erode stock from a metal workpiece.

Electroless Plating: A process that uses chemical rather than electrolytic action to produce the desired metallic coating on a substrate.

Electromagnetic Forming: A process that forms a workpiece by using intense pulsating magnetic forces. Sometimes called magnetic pulse forming.

Electron Beam Machining (EBM): A thermoelectric process that focuses a high-speed beam of electrons on the workpiece.

Electron Beam Welding: A fastening process that generates heat by bombarding the weld site with a narrowly-concentrated beam of high-velocity electrons.

Electronic Gluing: Curing of thermosetting resin glues by exposing them to heat generated by a high frequency electric field.

Electrophoresis: See Electrodeposition.

Electroplating: A means of depositing a metal or alloy from an anode through an electrolyte solution onto an article to be plated (connected as the cathode).

Electropolishing: Smoothing and enhancing the appearance of a metal surface by electrochemically removing an extremely thin layer of material.

Electrostatic Powder Coating: A finishing method in which negatively charged particles of plastic powder are attracted to a positively charged part, resulting in an evenly distributed coating.

Electrostatic Spraying: A process that was developed as a method for coating metallic materials. It uses negative and positive charges to evenly coat the workpiece with finishing material.

Embossing: A stamping process that produces a raised image on the product's surface.

Emery Paper: A coated abrasive product made from aluminum oxide and iron oxide. It is used to polish and buff metals.

Empty-Cell Impregnation: A type of pressure treating that uses high-pressure air to force a preservative (creosote) into the cells of the wood.

Enamel: An oil-based paint that is made by adding finely ground pigment to varnish; it provides a hard, glossy, very durable finish.

Encapsulating: A protective method similar to potting, except that the mold is removed from the product after casting, rather than becoming part of the product.

End Effector: The tool or device mounted on the end of the arm of an industrial robot.

End Mill: A type of cutter, normally used in vertical milling, that is fed directly down into the workpiece.

Engineered Composites: Human-made composite materials.

Engineered Materials: Materials that are designed and constructed by people, rather than occurring naturally.

Engineering Resins: High-performance resins, such as those formulated to be exceptionally resistant to chemical attack, extreme heat, or impact.

Equilibrium Diagram: See Phase Diagram.

Escapement: A device for orienting and dispensing individual components to a workholding fixture or tool.

Etchant: The agent that is used to etch (chemically machine) metals, glass, or other materials.

Eutectic Salt: A salt with a low melting point, used to form a mandrel by pouring it into a two-piece ceramic mold.

Eutectoid: Term that is used to describe an alloy that changes immediately from solid to liquid at a specific temperature.

Evaporative Vacuum Method: A process in which the metal to be deposited on the plastic is evaporated from disk-shaped targets placed in a vacuum chamber.

Excess Inventory: A form of waste that increases the cost of a part because of additional handling, facility costs, and paperwork.

Excimer Lasers: A type of laser, still in the developmental stage, that removes stock, molecule by molecule, rather than vaporizing it. Since they do not generate heat, excimer lasers can be used for procedures such as eye surgery.

Exotherm Brazing: A joining method, that uses high-frequency current through an AC electrical coil to heat the joint area.

Expandable Bead Foam: A type of molding in which hollow beads containing a tiny amount of compressed gas or volatile liquid are heated inside the mold, causing them to foam to 50 times their original size.

Expanding: A process used to increase the size of tubular parts that have been formed by drawing.

Expansion Molding: A molding process in which a prepreg is placed inside a silicone rubber mold. In a curing oven, the mold expands to provide the forming pressure necessary to compress the prepreg.

Explosive Forming: The process that uses the shock wave from an explosion to force sheet stock against a forming die.

Extractives: Materials found in small quantities in wood, such as starches, oils, tannins, coloring agents, fats, and waxes.

Extractor: A device used to filter oil and water out of the air as it passes through an air-pressure spray gun system.

Extruder: A machine that converts thermoplastic powder, pellets, or granules into a continuous melt, then forces the material through a die to produce the desired shape.

Extrusion: A process in which a hot or cold semisoft solid material is forced through the orifice of a die to produce a continuously formed piece in the desired shape.

Extrusion Blow Molding: A process in which a melted plastic is forced through an extrusion die to form a tubular parison. The parison is then clamped in a mold and expanded by air pressure to fill the cavity.

Extrusion Cutter: A machine that uses a high velocity rotating knife to cut the extrudate.

F

Fabricate: To assemble or make from component parts.

Fabricating: The processes used to join and fasten materials together: adhesion, cohesion, and mechanical joining.

Faceplate: A disk fixed with its face at a right angle to the live spindle of a lathe for the attachment of the work.

Faceplate Turning: A turning operation done with the stock mounted on the faceplate of the lathe. Also known as facing.

Facing: A technique that is used to square the end of a workpiece to be turned and reduce the workpiece to the desired length.

Facing-Type Cutters: Cutters that are bolted directly on the nose of the machine spindle.

Factory and Shop Lumber: Wood that is used primarily for remanufacturing purposes in mills that produce fabricated doors, windows, cabinets, moldings, and trim items.

FAS: A hardwood grading term, meaning firsts and seconds. The best grade of hardwood stock, thus the highest quality grade, is firsts; the next best grade, seconds.

Fatigue: The tendency of a material to crack and fail due to repeated stress.

Feedback: In a control system, the process of measuring the actual output, then comparing it to a preset standard or reference.

Feeder: A device that contains discrete components and conveys them to the point of use while simultaneously orienting and sorting them.

Fence: An adjustable stop running perpendicular to the cutterhead axis on a saw, jointer, or similar woodworking machine.

Ferrite: The softest and most ductile form of low-carbon steel found on an equilibrium diagram.

Ferrous Metals: Metals that are produced from iron ore.

Fiber Tows: Bundles of fiber material used as prepregs in some composite production processes.

Fibrils: Bundles of cellulose molecules that form the wall of wood cells. Also known as lamella.

Filament: Flexible fiber that is drawn very fine.

Filament Winding: A process for fabricating a composite structure by winding a continuous fiber reinforcement over a rotating core or mandrel. The fiber may be impregnated with a matrix material before or during winding.

Filler: A material that is suspended in a matrix binder to form a composite.

Filler Metal: In welding, a metal (such as brass, solder, or iron), that is used to help fill any voids in a joint.

Films: Thin, usually flexible sheets of material that are produced through continuous casting or other methods.

Filter Cake: The squares or thin slabs of material that result when a filter press dewaters slip.

Filter Presses: An iron frame with nylon filters that is closed with hydraulic pressure to press water out of slip.

Fine Finishing: A type of sanding in which the stock is often polished or burnished with rubbing oils or rubbing compound; it requires a fine-grit abrasive.

Finishing: Actions that improve the outward appearance or protect the exterior surface of a part.

Finishing Cut: A shallow cut, usually made at a fairly high speed with a moderately slow feed rate, that is used to bring the workpiece to final dimensions.

Fire Clay: A material used to make refractory brick that is produced by dry pressing.

Fire-Retardant: A material that inhibits the spread of fire.

Firing: Heating greenware to an elevated temperature in a kiln or furnace to fuse the materials together, providing strength and permanence to the product.

Firing Range: The span of temperatures between the point of sufficient vitrification and too much vitrification.

First Wave: Term used to describe the first of a series of basic changes in manufacturing productivity. Generally assigned to the advent of mass production, beginning in the mid-1920s.

Fixture: A special workholding device.

Flame Hardened: Term used to describe metallic workpieces that have been heated and quenched. Flame hardening can be used for *zone or spot hardening* (treating only part of a workpiece).

Flame Polishing: A process used to reduce the size and quantity of surface defects in small-diameter rods and filaments by rotating the material in a helium/oxygen flame.

Flame Spray Coating: A process in which a fine plastic powder is dispersed through a specially designed spray gun burner nozzle, so it is melted quickly to form a smooth, uniform, permanent layer.

Flame Spray Gun: A finishing process in which molten particles of ceramic material are sprayed onto a substrate.

Flame-Spraying: The process of melting metal in a special gun, then spraying it as a thin coating on a substrate. Also called wire metallizing.

Flash: The excess material that is forced out between the halves of forging dies. It must be removed as a finishing step.

Flash-Off Period: Time allowed after finishing to allow excess thinners and solvents to evaporate before the finish is subjected to forced drying.

Flask: The rectangular box used to hold the mold in sand casting.

Flat-Platen-Pressed: Designation of a type of particleboard that is produced when pressure is applied in a direction perpendicular to the faces.

Flexible Manufacturing: An automated production system that permits programming to be easily altered and tooling to be changed automatically to produce different components in small lots.

Flint Paper: A type of coated abrasive consisting of grains of quartz (silicon dioxide) adhered to a paper backing.

Float Glass Process: A process used to produce sheet glass by floating it on top of a bath of molten tin.

Floating Zone Method: A crystal growing process in which a rod of polycrystalline silicon is brought into contact with a single crystal seed. A molten zone around the interface between the seed and rod will solidify with the same crystalline structure as the seed.

Fluidized Bed Coating: A finishing process in which coating powder is mixed with jets of air. The air holds the powder in suspension so that it behaves like a fluid.

Fluidized Bed Furnace: A type of furnace that suspends heated aluminum oxide particles in a chamber with hot gases.

Flutes: Channels on a drill bit through which chips are carried out of the hole being drilled.

Flux: A material used to prevent the formation of oxides or other surface impurities in a weld, or to dissolve them as they form.

Fluxing: In the calendering process, the mixing of dry ingredients prior to adding liquids.

Foam-Frothing: Process in which a urethane chemical mixture in a partially expanded state is mixed with chemical blowing agents that give off gases to create a cellular structure.

Foaming Agents: Chemicals added to plastics to create a lighter, more cellular structure.

Foil: A printed, melamine-impregnated paper that is introduced into a mold in the in-mold decorating process.

Follower: A freewheeling spindle attached to the tailstock of a lathe in the spinning process. It helps hold the stock in place.

Forced Drying: A process usually accomplished by loading parts or assemblies on conveyors that carry them through heated ovens. Forced drying is often done with ultraviolet (UV) radiation.

Forestry: The management of forest lands for timber, wildlife habitat, and recreational uses.

Forging: To shape hot or cold metal with compressive force by means of successive hammering.

Forming: Act of applying pressure to industrial stock in order to change its size and/or shape.

Forming Block: The mandrel or mold used to shape stock in the spinning process.

Forming Process: A manufacturing activity that changes the shape of a material, without significantly reducing its mass or weight.

Forming Tool: The device used to apply pressure against the face of the workpiece in a spinning operation.

Foundational Technical Development: Event or discovery that played a major role in shaping the development of human technological culture, and led to the development of many other forming processes.

Franklin Fiber: A microscopic crystalline fiber, derived from gypsum, that can be used as a reinforcing filler in plastics.

Free Blowing: A process in which a heated plastic sheet is clamped between an upper forming ring and lower metal platen and air is blown through a hole in the platen to force the plastic upward into a bubble.

Free Foam Molding: A molding process in which granulated polymer, a chemical blowing agent, and other additives are fed into a mold cavity. When the mold is heated, foam forms and expands to fill it.

Freeze Drying: As used in ceramic production, a process that rapidly freezes the water in clay particles, encasing drops of salt solution that hold the desired metal ions. Also known as cryochemical drying.

Frit: Particles of glass that have been milled to the desired size. When added to a glaze, frit lowers the necessary firing temperature.

Full-Cell Impregnation: A type of pressure treating used to apply fire retardant and moisture protectant chemicals. In this system, the preservative slowly flows into the cell.

Function: The first of the three major factors emphasized in most value analysis programs. This includes a careful analysis of all parts and components of the product in terms of cost.

Fusion Sealing Method: A process that uses special glass compositions to make direct glass-to-metal seals in an oxidizing environment.

G

Gage Number: A unit of measure used to determine thickness of metallic stock. The smaller the gage number, the thicker the metal. Also, designation identifying the diameter of a wood screw or rod stock.

Gang Cutting: A form of traveling wire electrodischarge machining that uses multiple heads to simultaneously cut single or multiple workpieces.

Gang Drilling: A process that uses multiple spindle drill presses in order to drill many holes at one time.

Gas Metal Arc Welding (GMAW): An electric arc welding method that uses an inert shielding gas, supplied through a nozzle, to prevent oxidation of the weld. Also referred to as MIG welding.

Gas Tungsten Arc Welding (GTAW): An electric arc welding method that uses a nonconsumable tungsten electrode. Like GMAW, it makes use of an inert shielding gas supplied through a nozzle.

Gel: A liquid that is so thick it acts more like a solid.

Gel-Coat: A specially pigmented resin mixed with a catalyst and sprayed into an open mold to form the smooth finished surface of the product.

Geneva Mechanism: A popular method for achieving rotary motion by using a continuously rotating driver to index the dial.

Gilt: Decorative gold material applied to fine china after glazing.

Glass Loading: A term describing the percentage of resin encapsulated in the fiberglass reinforcing mat in a product. Higher glass loading results in greater weight and adds strength to the final product.

Glass Solder: A bonding agent that allows glass parts to be joined and later separated without destroying them.

Glass-Matrix Composites (GMCs): Composite materials in which the matrix is glass.

Glass-to-Ceramic Seals: A bond used to join unlike materials in products such as spark plugs.

Glaze: A specially formulated glass that melts on the surface of a ceramic workpiece, and adheres to the body after cooling.

Glost Firing: The second firing of a ceramic that is done to sinter the clay body and develop the glaze.

Gouge: A round-nosed, cupped chisel used for roughing and making cove cuts on a lathe.

Grade: The classification of wood according to its quality, based on appearance, strength, and lack of defects in the stock.

Grafting: A condition that occurs under radiation processing, when two or more elastomers attach to a molecular chain without the use of a catalyst.

Grain: The appearance of annual rings and fibers in wood when viewed longitudinally.

Grain Size: The unit of measure used to grade the coarseness or fineness of an abrasive finishing material.

Granulator: A type of grinding machine that is used when the manufacturing process is automated or semi-automated to recycle scrap material back into the process.

Graphitization: An optional final step in producing a carbon fiber; it involves rapidly heating the fiber to high temperature. The process improves the elasticity and texture of the fiber.

Green Compact: In powder metallurgy, the compressed part that ejected from the die but not yet fired. The part can be handled, but is extremely fragile at this stage.

Green Machining: The process of machining a dry pressed ceramic part while it is still green (before it has been sintered or fired).

Green Wood: Stock that has just been cut from a log and contains a great deal of moisture, or sap.

Greenware: Term used to describe the delicate state of a molded clay product before firing.

Grinding: A cutting process that uses abrasive particles to perform the cutting action.

Grinding Media: Small pieces of rock, rubber, ceramic, or other material used in a ball mill to pulverize the material being ground.

Grog Mix: Clay that is mixed with fine particles of previously fired clay and used to make a mold for glass produced by the sagging process.

Gun Drilling: A process which is used to drill deep, accurate holes. It was originally used to drill gun barrels.

H

Halide Reduction: A chemical vapor deposition process used to place a boron halide coating on heated tungsten wire as the wire is drawn through a tightly sealed chamber.

Hand Forging: A metal forming process dating back to about 4000 B.C. Forging was used to shape primitive metal tools.

Hand Layup: A fabricating process that essentially consists of applying a layer of roving to a mold, then saturating the roving with resin until desired thickness and strength is reached.

Hand-Held Router: A tool with a spindle-mounted cutter that is held securely against a template or guide to produce the desired pattern.

Hard Automation: A system in which the machines and tooling are designed to produce a specific part or a family of similar parts or assemblies.

Hard Good Consumer Products: Nonconsumable goods made through manufacturing processes and intended for eventual purchase and use by individuals.

Hardboard: A continuous mat of wood pulp that is pressed into a sheet of strong, hard material.

Hardinge Conical Mill: A grinding device that turns around a horizontal axis, while the mix is continuously fed in at one end and automatically discharged from the other.

Hardwood: A wood produced from a deciduous tree.

Headstock: The driven end of the lathe that holds a chuck or faceplate for mounting stock to be turned.

Heat Treating: Term used to describe processes that use controlled heating and cooling of materials in their solid state to change mechanical or physical properties without altering their chemical composition.

Heat/Quench/Temper Process: A way to harden steel by heating the workpiece above its upper transformation temperature, then quenching it in water, warm oil, or air. Finally, it is tempered by reheating to improve its tensile strength.

Heavy Gage Stock: A classification for sheet metal that is 0.125 in. (3.17 mm) or greater in thickness.

Hermetic Seal: A method of closure that prevents movement of gases or liquids into or out of a product.

High-Energy-Rate Forming (HERF): Forming process that uses a spark-generated shock wave to force the sheet stock against a forming die.

High-Pressure Lamination: The process used to adhere plastic coatings to particleboard for making office and household furniture.

High-Temperature Superalloys: Metal alloys created to withstand abuse from extreme forces and severely oxidizing high-temperature environments.

Horizontal Milling Machine: This type of milling machine has the cutting tool carried on an arbor or spindle that travels along an axis parallel to the worktable.

Horn: In ultrasonic machining, the structure that holds the tool and conveys mechanical motion to the workpiece.

Hot Forging: A variation of dry pressing that involves heating a pressed part to high temperature, then pressing it in a cold steel die.

Hot Isostatic Pressing: A ceramic pressing process involving the use of both heat and pressure to form the product. Also called gas pressure bonding.

Hot Isostatic Pressing: A compacting process that uses both heat and pressure to achieve maximum density of powdered metal or ceramic parts.

Hot Marking: A stamping process used to impress a heated, raised image into the surface of workpiece.

Hot Melt Adhesives: Thermoplastics used as adhesives. They become fluid when heated, then solidify upon cooling. They will not soften upon reheating.

Hot Melts: See Hot Melt Adhesives.

Hot Press Compression Molding: A molding process in which a plastic mixture of resin, reinforcement, filler, and additives is compressed between heated die halves.

Hot Wire Cutting: A simple process that utilizes a thin wire heated by high electrical resistance. The hot wire slices through foam plastic by melting the material.

Hot Wire Welding: A simple process in which a wire is placed between two thermoplastic surfaces to be joined and heated by electrical resistance.

Hot-Rolled Steel: Steel that is squeezed between rollers while it is hot, and can be identified by its bluish surface coating.

Hydrobrasive Process: A waterjet process that includes abrasive particles in the water.

Hydroforming: A draw forming process that employs hydraulic fluid in place of the rubber pad used in marforming.

Hygroscopic: Term describing the property of being water-absorbing.

I

Impreg: A process that involves saturating the cells of wood with a fiber-penetrating thermoset resin, then curing it without using compression. A 25 to 35 percent gain in weight (when compared to the dry weight of the wood) is achievable.

Impregnation: The process of immersing porous P/M parts in heated oil or resin after they are removed from the sintering oven. Also see Impreg.

Independent Chuck: An adjustable workholding device used on a lathe for irregularly-shaped workpieces. Its jaws are turned in toward the work one at a time.

Indirect Labor: Work supportive to production.

Indirect Two-Sheet Blow Molding: A process in which an air tube is placed between two heated sheets of stock. When the mold closes, air is blown through the tube, expanding the sheets against the walls of the mold.

Induction Bonding: A process of welding that uses induced electric currents to generate heat within the work.

Induction Hardening: A fast-acting hardening process that uses electromagnetic currents to generate the necessary heat in the part.

Industrial Ceramics: The manufacturing segment that produces products from engineered (human-made) materials.

Industrial Stock: See Raw Materials.

Industry: Term used to group all the companies that manufacture a particular product, such as "the automobile industry."

Inefficiency: A form of waste that adds unnecessarily to the cost of producing the part, often by requiring additional processing.

Infiltration: In powder metallurgy, the practice of placing a lower-melting-point metal on or below the green compact before sintering. The molten metal will be drawn into the porous compact to improve its mechanical properties.

Ingot: The large brick of stock (iron or other metal) that is melted into a liquid state for casting at a foundry.

Injection Blow Molding: A two-step molding process in which a heated parison (plastic tube) is formed in the mold cavity, then forced against the mold's walls by injecting air.

Injection Molding: A process in which a material (usually a resin or ceramic) is heated until it reaches a plastic state, then forced under pressure into a closed mold.

In-Line Transfer Systems: Chain or belt conveyors that convey a pallet containing a parts-holding nest or fixture between operations.

In-Mold Decorating: Finishing process accomplished by applying foils to compression-molded and transfer-molded products.

Input: The part of an open-loop control system representing a value for a characteristic, such as time, temperature, or pressure.

Input Values: In a control system, arbitrary standards or references that reflected circumstances at the time they were established.

Inserting: An application of ultrasonics in which a metal component is embedded into a preformed hole in a plastic part.

Insulation Board: A type of composition board that has high resistance to heat transfer, is low in density, and is often used as an acoustical barrier.

Integrators: Firms that know how to utilize all of the principles of effective technology management.

Intercellular Layer: The thin adhesive material found between the cell units of wood.

Interchangeability: A concept that makes automation possible by allowing the substitution of standardized parts, assemblies, or materials in a manufactured lot. It makes modifications or adjustments to individual parts unnecessary.

Interference Fits: One of the methods used in joining wood parts, in which the parts are designed to lock together.

Interlayer: A polyvinyl butyrl plastic sheet that is laminated between two pieces of glass to form automotive windshields. It holds the glass together and keeps it from shattering upon impact.

Intermittent Extrusion Blow Molding: A process in which the parison, immediately after being ejected from the mold, is quickly formed into a product.

Internal Threads: Threads that are on the inside of a hole.

International Joint Venture: A manufacturing operation that is jointly owned by companies from different nations.

International Standards Organization (ISO): The agency responsible for establishing and publishing manufacturing and other types of standards worldwide.

Investment Casting: A precision casting method in which the cavity created by the pattern is destroyed when the casting is removed. Also known as "lost wax casting."

Ion Implantation: A metal hardening process that improves the material's ability to withstand friction and corrosion.

Ionizing Radiation Processes: Those that produce high-level radiation, either gamma rays generated from radioactive materials such as Cobalt 60, or high-energy electron rays created by electron accelerators.

Ions: Electrons that have been drawn free of their atoms in the electrochemical machining or electroplating processes.

Irradiated Wood: See Wood Plastic Composition (WPC).

Irradiation: Exposure of a material to energy from a radioactive source.

ISO 9000 Certification: A rigorous procedure that verifies a company's ability to control all processes affecting the quality of its product or service.

J

Jacobs Chuck: A three-jawed device used to hold a workpiece in the headstock of a lathe. It is used for smaller stock than the similar Universal Chuck.

Jiggering: A ceramic production process that uses a template or forming tool to force a slice of clay over a revolving convex mold.

Job Lot: See Batch.

John Mill: A grinding machine in which agitator pegs of tungsten carbide are set into the walls of the grinding cylinder. The cylinder is turned at high speed, producing a uniformly ground product.

Joint: The point of interface that is formed when two materials are joined together.

Jointer: A machine that is used to true the edges of stock. Usually used to complement the work of the planer and circular saw.

Jollying: A ceramic production process in which a plaster mold forms the outside surface of the product, and a forming template shapes the inside.

Just-in-Time Manufacturing: A philosophy that eliminates excess inventory by placing the burden directly on suppliers to deliver parts or materials of an agreed-upon quality at a specified time and in the desired quantity.

K

Kaizen: Japanese term, meaning "incremental movement." It is a synonym for the value analysis technique called "continuous improvement."

Kanban: A term used by the Japanese to describe the Just-in-Time Manufacturing philosophy.

Kaolin Clay: A clay material with a high melting point, used for making refractory products.

Kaolinite: A white mineral, primarily aluminum oxide and silica dioxide, that is found in clay.

Kasenit: One type of carbon-hardening powder used in carburizing.

Kerf: A cut made in metal or other material by a saw or cutting torch.

Kickback: Machine hazard encountered in woodworking when the workpiece is thrown out of the machine back toward the operator.

Kiln: A furnace that fires ceramic materials at high temperatures to fuse them.

Kiln Drying: A method of drying wood, in temperature- and humidity-controlled ovens, to reduce its moisture content to about seven percent.

Kiln Furniture: Struts or other devices used to support a part with an unusual shape in the kiln.

Kiss Cutting: A diecutting method that is used for cutting self-adhesive materials without penetrating the paper or film backing.

Knurling: A technique used to press diamond-shaped indentations on the circumference of a workpiece. Knurling is used to provide a gripping surface on a part or to expand the diameter of a part.

L

Lacquer: A finishing material made of nitrocellulose, acetone, varnish resins, a thinner, and plasticizers. Lacquer provides a smooth, hard finish.

Lamina: The sheets of material, which may be unidirectional fibers or woven fabric.

Laminating: A process that involves sandwiching layers of plastics or composite materials into sheets or parts and applying pressure to bond the layers together.

Lancing: A punching process used to make a tab without removing any material.

Laser: An active electron device that converts input power into a very narrow, intense beam of coherent visible or infrared light; used for cutting, drilling, welding, heat treating, soldering, and wire stripping.

Laser Welding: Micro-spot welding with a laser beam.

Lasing Medium: A mixture of carbon dioxide and nitrogen used in a laser.

Lasing Tube: A cylinder filled with a gas, such as CO_2 or argon, that is energized by an electric current and releases energy in the form of coherent light. The output of the tube is focused into a beam to do work.

Latex Paint: A water-based type of paint that is widely used for finishing.

Lathe: The most common machine used to perform machining operations on round parts.

Lathe Dog: A stock-holding device used on a lathe. It clamps to the stock and engages the faceplate to rotate the workpiece.

Layered Fibrous Composites: Materials in which layers of reinforcement are built up so that the fiber strands are oriented in different directions, providing greater strength and stiffness.

Lead: On a bandsaw, the tendency to pull slightly to the left or right of the desired cutting line, usually caused by improper tracking of the blade on the wheels.

Lean Production: A process of production in which a company goes through continual improvement to the extent necessary to assure that better quality products are being consistently produced at less cost.

Leather-Hard: The state of clay after a ceramic product has been fired.

Left-Hand Tools: Lathe tools that cut from left to right.

Light Gage Stock: A sheet metal classification for material up to 0.0031 in. (0.079 mm) in thickness.

Lignin: A natural adhesive that bonds adjacent layers of cells together. It makes up about 25 percent of the total volume of wood.

Line Balancing: The act of eliminating slack time when a process flow model has been completed and the critical path has been identified. Various techniques are used.

Liquid Infiltration: A process used to produce metal-matrix composites, either by adding reinforcing fibers to a molten matrix, or by placing fibers in an opened mold and pouring in the molten matrix material.

Liquid-Phase Sintering: Wetting ceramic particles with a viscous liquid before heating the part to the sintering temperature.

Liquidus Line: On a temperature graph, the point at which a molten alloy begins to change from a liquid to a mushy solid.

Live Center: A rotating device used, along with a faceplate and lathe dog, to hold long pieces of stock in place on a lathe.

Logic: The principles of reasoning that are designed into an automated system.

Longitudinal: Relating to the dimension along the length of an object.

Longitudinally: Term used to describe shrinkage of wood lengthwise, or in the direction of the grain.

Lost Motion: The expending of energy without adding any value to the product.

Lower Transformation Temperature: On an equilibrium chart, the temperature where a material begins to change from one phase to another.

Low-Pressure Lamination: A lamination process that involves applying a thermoplastic film as a protective covering for photographs or documents.

Lubricants: Materials mixed with clay to improve the shaping characteristics of the material and to aid in removing the product from the mold.

M

Machine Planing: A process that mills wood to cut it to uniform thickness and produce a smooth surface.

Magazines: Material handling systems that contain and convey pre-oriented parts.

Major Material Families (MMFs): The types of materials

(metals, ceramics, polymers, and composites) most often processed to manufacture hard good consumer products.

Mandrel: A bar, rod, or similar shape that serves as a core around which other materials are wound, cast, forged, or extruded.

Manual Control: The simplest type of control system, in which an operator is required to start, stop, or adjust the process by operating control devices on the machine.

Manufacturing: The use of tools, processes, and machines to transform or change materials or substances into new products.

Manufacturing Flexibility: One of the four criteria that shape the nature of automation: the degree to which the system can change and the ease with which that change can be made.

Manufacturing Processes: The different activities or operations carried out in transforming industrial stock into products.

Manufacturing Productivity: The output of product per unit of effort; a measure of how effectively a nation's manufacturers use labor and equipment.

Maquiladoras: Industrial parks along the U.S.-Mexico border that pay no duty on materials imported from the U.S., and pay only duty on value added when finished goods are shipped back into the U.S.

Marforming: A flexible draw forming process that is often a cost-effective alternative to stretch draw forming.

Martensite: The metallic structure that is created when steel is rapidly quenched or cooled, usually by plunging it into a liquid bath.

Maskant: A type of resist that is not photo-sensitive. It is applied to a surface, then mechanically removed from areas that are to be etched.

Mass Production: A system involving continuous production of one type of product, such as an automobile, typically in large quantity.

Master Unit Die Method: This proprietary quick-change die system uses inserts in a frame or fixture that remains in the machine.

Matched Mold Forming: A molding process involving a two-part metal die used to compact a softened thermoplastic sheet.

Material Feeding: General term for the process of conveying components to the point of use. Feeding may consist strictly of component movement, but also can include orienting and sorting.

Matrix: In a composite, the material that binds fibers together and provides stiffness.

Mechanical Etching: Processes, such as sandblasting or bombarding with an abrasive and lead shot mixture, that are used to frost glass, or make it opaque.

Mechanical Joining: A type of fabrication that involves the use of fasteners to physically attach two or more materials to each other.

Mechanical Stretch Forming: A process similar to plug-assist thermoforming, but without the use of a vacuum. The mechanical stretch forming process uses only a plug to depress and stretch the stock into the female mold.

Mechanization: Doing things with or by machines.

Medium Gage Stock: A classification that consists of sheet metal from 0.031 in. to 0.109 in. (0.079 mm to 2.77 mm) in thickness.

Melt Spinning: A process often used to produce monofilament nylon line and glass strands.

Memory: The capability of a material to return to its original shape after it is bent, stretched, or otherwise distorted.

Mers: The individual repeating units in the polymer.

Metallics: Material family that includes elements and compounds classified as metals, such as copper, steel, or aluminum alloys.

Metallizing: A process used to apply metal coatings to ceramic, glass, or other types of surfaces. A typical application is to provide a transmission path for electrical current on electrical devices and components.

Metal-Matrix Composites (MMCs): Composite materials in which the matrix is a metallic substance.

Methanol: Methyl alcohol. A solvent used as a vehicle in non-grain-raising stains.

Microspheres: Hollow glass bubbles that are mixed with many different types of plastics to improve the physical properties of parts.

MIG: Abbreviation for "metal inert gas" welding. See Gas Metal Arc Welding.

Milling: A machining process that uses a multitoothed cutter to produce slots, grooves, contoured surfaces, threads, spirals, and many other configurations. Also, a grinding process in which hard clay is broken down into fine particles by passing through a series of rollers.

Miter Saw: A saw that has the capability of cutting stock at an angle from the vertical, as well as moving through a 90° arc horizontally. It is used primarily when doing interior trimwork.

Mix Muller: A pulverizing machine used for clay. It consists of a large circular pan in which two large steel-tired wheels revolve.

Moisture Content: The quantity of water in a mass of wood, expressed in percentage by weight of water in the mass.

Molding: The process in which a material, usually in a softened (plastic) but not liquid state, is formed into the desired shape using pressure and sometimes heat.

Molinex Mill: A grinding device in which eccentric grinding disks, mounted on a rotating shaft, are staggered to function as an auger. This mill makes it possible to obtain a superfine grind.

Monomers: Simple molecules that are linked together by the thousands to make up the long chain that is a polymer.

Mortise: A groove or slot that is cut into a piece of wood to accept a tenon.

Mortise-and-Tenon Joint: A joint that is used to secure drawer components or other pieces of fine furniture.

Multiple-Spindle Shapers: Wood-shaping machines that have a number of spindles arranged side by side.

N

Nail Set: A small cylindrical steel tool, usually tapered at one end, that is used to drive a nail head below or flush with a wooden surface.

Natural Aging: The process by which some aluminum alloys harden and become stronger over a period of time at room temperature.

Natural Composites: Composites (materials containing two or more different, but complementary, substances) that occur in nature.

Nd:YAG Laser: A type of laser that emits either continuous or pulsed beams. It is often used for welding delicate components.

Near-Net-Shape Manufacturing Process: Any process that produces parts in their desired final size, with little or no need for additional production steps.

Near-Net-Shape Part: A product that is made in the size desired, without significant shrinkage taking place during the manufacturing process.

Necking: A die-reduction process that stretches the metal part at the same time that it reduces its cross-sectional area.

Nesting: The positioning of blanks so that multiple parts can be cut from one workpiece with a minimum amount of waste.

Neutralize: To stop the action of bleaching once the desired degree of lightening of wood color has taken place.

Nibbling: A process that uses a small round or triangular punch to rapidly take small "bites" out of sheet metal, allowing the cutting out of limited numbers of flat parts with complex shapes.

Nitriding: A case hardening method that uses heated gaseous ammonia to form a very hard shell on the metal.

Nonconsumable Products: Goods that are durable in nature and intended for long-term use, such as plastic containers, vehicles, machinery, or furniture.

Nonferrous Metal: Any metal other than iron and its alloys. Nonferrous metals include copper, aluminum, lead, nickel, zinc, tin, and brass.

Non-Grain-Raising (NGR) Stains: A stain consisting of a dye in a methanol solvent, widely used by furniture manufacturers, because they have a short drying time and do not cause swelling (raising) of the wood grain.

Nonionizing Radiation: Process that generates low-level radiation, such as that produced by microwave, infrared, or ultraviolet energy source.

Normalizing: Annealing process that involves cooling the heated metal to room temperature in still air to produce a more uniform and fine-grained structure.

North American Free Trade Agreement (NAFTA): An agreement made between the U.S., Mexico, and Canada in the 1990s. The agreement was intended to encourage the development of new manufacturing partnerships involving companies in the three member countries.

Nosing: A die-reduction process that is used to taper or round the end of tubing.

Notching: This process, which is similar to nibbling, is typically done using a punch press.

Nuclear Fuel Rods: The product resulting from combining finely crushed radioactive ceramic powders with binders and an oil lubricant.

Numerical Control (NC): A method of automating manufacturing processes by describing tool paths and machine movements in terms of numerical increments. This movement might be along a machine axis, or a vector that is comprised of the resultants of all the axes being controlled on the machine tool.

Numerical Control Program: The method used to store and convey precise instructions for each move along each machine axis; each instruction is sent by the machine control unit to activate the specified output.

O

Oil Stains: Wood stains with pigments suspended in oil. After being applied, they are wiped off after varying periods of time to achieve the desired depth of color.

Oil-Base Paint: A type of paint that uses a petroleum-based vehicle. It must be thinned or cleaned up with a petroleum-based solvent.

Oligomers: Acrylic polymers made up of two, three, or four monomer units that become solid polymers when exposed to light. One basic use of these materials is stereolithography.

Open Molding: A process used to form many polymeric-matrix composite products.

Open-Die Forging: A method of forging in which the workpiece is formed between flat dies that compact, but do not completely enclose, the heated metal part.

Open-Grain Woods: Hardwoods that have large pores, such as oak.

Open-Loop Control System: A basic control system that attempts to meet a preset standard without monitoring the output or taking corrective action.

Organosols: A suspension of polyvinyl chloride powder in a volatile solvent with a small amount of plasticizer. They are less viscous than plastisols.

Orientation: Properly positioning a component for assembly or processing by a machine.

Oriented Strandboard: A type of particleboard that is made from large, irregularly shaped wood fibers, rather than particles or flakes.

Outputs: The operations of devices (such as electric motors) that are actuated by an open-loop control system.

Overaging: A problem that can occur when too high a heat level is used in artificially aging aluminum alloys. Weakening of the part results.

Over-Drive: To excessively tighten a screw into a hole.

Overglaze: A decoration applied over the glaze on a ceramic product. It must melt at a temperature lower than the melting point of the glaze.

Overproduction: A problem that occurs when market demand slows and there is no control over manufacturing, resulting in unsold products being carried as excess inventory.

Overspray: Term used to describe sprayed finishing material that does not settle on the workpiece.

Oxidation: Process that takes place during the production of carbon fiber. It changes the internal structure of the thermoplastic fiber, so it will no longer melt.

Oxyacetylene Rod Gun: A finishing tool that operates in a manner similar to that of the flame spray gun, except that a rod of coating material is fed into the oxyacetylene flame, instead of ceramic powder.

Oxyacetylene Welding: A joining method that uses heat from mixing and burning two gases, oxygen and acetylene, in a welding torch. The heat generated is sufficient to melt and weld many soft metal alloys.

P

Pad Printing: A stamping process that uses a design in relief to transfer ink from a pad to the surface of the workpiece.

Paint: A mixture of pigment suspended in a vehicle, such as water or oil.

Pallet: The support base, frame, or tray used on an in-line material transfer system.

Panel Saw: A circular saw attached to steel rails so that it slides up and down easily to cut large sheets of plywood or other panels. The saw can be rotated to cut across panels, as well.

Paraffin: A white, solid wax derived from petroleum.

Parison: A hollow tube of heat-softened resin that serves as the stock in blow molding..

Particleboard: A panel material composed of small discrete pieces of wood bonded together with a resin in the presence of heat and pressure. It is widely used for underlayment and as core material for furniture.

Particulate Composites: Materials that are made by suspending particles or flakes of material throughout a tough, ductile matrix.

Parting: A turning technique used to cut off a section of the end of a workpiece or to produce a groove around the circumference of the part.

Parting Line: The seam between closed forging dies, where a thin wing of flash is generated.

Parting Tool: A lathe tool (chisel) that is used to separate material and to cut off stock.

Part-Transfer Mechanisms: Devices that take a part from the escapement pick-up point and place it in a workholding fixture or nest.

Pass: A new layer of metal that is deposited in one trip along the axis of a weld.

Paste Filler: A thickened form of wood filler, containing powdered quartz, linseed oil, turpentine, and drying agents, used to close pores and enhance the beauty of open-grain hardwoods before any finish is applied.

Pearlite: The soft and relatively ductile state that occurs when steel reaches 0.8 percent carbon.

Penny: The basic unit of size classification for nails.

Perforating: A punching process that produces closely and regularly-spaced holes in a straight line across sheet metal, usually to facilitate bending.

Peripheral Speed: See Cutting Speed.

Peripheral Turning: See Straight Turning.

Permanent Mold Casting: A casting method that uses a two part mold again and again. The mold halves open after the part solidifies, allowing part removal.

PERT: Abbreviation for Program Evaluation and Review Technique, a type of process flow model that was first used in the late 1950s. It is used in conjunction with CPM (the Critical Path Method).

Phase Diagram: A graphic representation of the phase changes that take place in a metal at various temperatures. Also called an equilibrium diagram.

Photoinitiator: A component of photopolymeric resin that absorbs laser energy and initiates polymerization.

Photoresist: A light-sensitive coating that is applied to a substrate, then exposed and developed to provide a mask during chemical etching.

Physical Frothing: A process accomplished by aggressively whipping air into the resin, and then rapidly cooling the polymer.

Physical Properties: Attributes used to distinguish one material from another, such as weight, color, electrical conductivity, and reaction to heat.

Pick-and-Place Robots: Programmable mechanical devices that simply pick up a part and move it to another location. These robots are used primarily for material transfer, but also are used to do some simple assembly tasks.

Pickling: A chemical process using acid for cleaning.

Pigment: The finely ground material that gives opacity and color to paint.

Pilot Hole: One of the two holes drilled when fastening wood with a screw. It extends through one piece of stock and into the other, and is slightly smaller in diameter than the screw.

Planer: A machine used to shape long, flat, or flat contoured surfaces by moving the workpiece under a stationary single-point tool or tools.

Planing: An operation that is used to remove large amounts of material from horizontal, vertical, or angular flat surfaces.

Plasma-Arc Gun: See Plasma Arc Spraying.

Plasma Arc Spraying: A metallizing process in which an electric arc heats a high-speed stream of argon or another inert gas. A powdered metal coating material is fed into the stream to be melted and blown onto the part.

Plasticity: The ability of the material to change shape or size as a result of force being applied.

Plasticized: Term describing wood that has been made supple by exposing it to ammonia so it can be more easily shaped or formed.

Plasticizers: Additives used to provide plastic materials with increased flexibility.

Plastics: Synthetic materials that are capable of being formed and molded to produce finished products.

Plastics Welding: A cohesion process using plastic filler rod and a low temperature air stream to join plastic components of relatively heavy thickness.

Plastisol: A pourable, viscous liquid consisting of a vinyl resin dissolved in a plasticizer.

Plating: The deposition of metal on a cathode (negative electrode).

Plies: Thin layers of wood are glued together to form plywood.

Plug-Assist Thermoforming: A forming process in which a vacuum, in conjunction with a cylinder-activated plug, is used to depress a heated thermoplastic sheet into the mold.

Plywood: A laminated wood sheet made by gluing a number of layers, or plies, together at right angles to each other.

PLZT: A unique type of ferroelectric material (lead zirconate-titanate doped with lanthanum) that shifts polarity under the influence of an electric field, acting as a solid-state ceramic switch.

P/M Injection Molding: In this process, very finely ground powders are coated with thermoplastic resin, then parts are injection molded in the conventional manner. After molding, the part is sintered.

Point-to-Point System: A type of NC control system that moves the working spindle to a specific location on the workpiece, and then performs a programmed sequence of events. Also known as positioning.

Polishing: A process that uses both a fine abrasive and a wax compound to produce a highly polished surface on plastic.

Polyethylene: A polymer that consists entirely of ethylene monomers. It is used for many types of containers.

Polyethylene Glycol (PEG): A white, waxy chemical treatment that results in improved dimensional stability for wood. Also called carbowax.

Polymer: A long chain molecule made up of thousands of smaller molecules linked together.

Polymeric-Matrix Composites (PMCs): Materials, first developed in the 1960s, that combined polymers and such reinforcing materials as fibers of glass or carbon.

Polymerics: Material family that includes synthetic polymers (plastics) and natural polymers (woods).

Polymerization: The chemical reaction, triggered by heat, pressure, or a catalyst, that causes monomers to blend chemically, linking their molecular chains to form polymers.

Polystyrene: A synthetic resin used for injection molding, extrusion, or casting for molding plastic objects.

Polyvinyl Chloride (PVC): A polymer of vinyl chloride that is widely used for both molded rigid product, such as plumbing pipe, and pliable films.

Pores: The open ends of vessels in a hardwood. They appear as holes on cut ends of logs.

Porosity: The moisture-absorbing capacity of a material.

Pot Broaching: A type of broaching used to economically produce items such as precision external spur gears and automotive front-wheel-drive transmission gears.

Potting: Encapsulating method in which polyester thermosetting resin is poured into the housing of a product. The resin and housing form an integral part of the product.

Powder Metallurgy (P/M): A process that involves compacting metallic powders in a permanent (reusable) mold. Heat is applied to the molded powder to melt and fuse the powder into a finished metal part.

Precipitates: In precipitation hardening of nonferrous metals, small particles of a different phase that are added to the original matrix to achieve hardening.

Precipitation Hardening: The process in which small particles of a different phase are added into the original matrix to harden a nonferrous metal. Also called age-hardening, or aging.

Precision Grinding: Process often used on materials that are too hard to cut with conventional tools.

Preforms: Pellets or tablets compressed from powdered polymers, used as stock in some closed molding processes.

Prepregs: Cores of partially cured resin with reinforcement, used as stock in some closed molding processes. Also, sheets of polymeric matrix material having fibers that are already saturated with resin.

Preservatives: Treatments that impregnate the cells of the wood under pressure to make it fire-resistant, moisture-repellent, colorless, and odorless.

Presintering: A heating process that causes drying, decomposes organic binders, vaporizes water from the surface of particles, changes some ions, and then decomposes additives.

Press and Blow Machine: A machine used primarily for container production. The neck of the container is made by pressing or blowing, then the rest of the container is formed by blowing.

Pressure-Feed Gun: A type of gun used for atomized spraying of finishing materials. In this gun, finishing material is fed into the nozzle under pressure, then mixed with air.

Pressure Roll Bonding: A process used to produce metal-matrix composites by laminating layers of sheet material with fiber reinforcement in alternating plies, using heat and pressure.

Primary Processing: Activity carried out by firms engaged in the preparation of standard industrial stock.

Process: The action or activity by which a product is made.

Process Action: The description of what happens when a process changes the internal structure or the outward appearance of a material.

Process Flow Model: A graphic model that shows the sequence of operations required to manufacture a given product.

Production Molder: A type of abrasive planer used to shape long strips of molding.

Production Routers: Vertical-spindle machines widely used in the furniture industry to cut designs and small moldings, to round interior and exterior edges on stock, and to cut chair panels.

Programmable Logic Controllers (PLCs): Control devices widely used in industry to guide machine tools through a programmed sequence of operations. Programs can easily be changed to produce different components or reflect changes in dimensions.

Promoters: See Coupling Agents.

Proportional Control: A system that is extensively used to control automated processes when an operation requires variation in such factors as speed, tension, or temperature.

Pugging: A shredding and vacuum process used to remove air bubbles from dewatered clay filter cake.

Pugs: See Blanks.

Pull Broach: A tapered cutting tool used to shape holes in a workpiece.

Pultrusion: A process that involves pulling continuous roving impregnated with resin through a forming die. The resulting shaped material is cured in an oven.

Punch: The part of a closed die arrangement that automatically pushes wire stock in precut lengths through the hole in the die block.

Punch Press: The machinery used in stamping sheet metal parts. Also known as a stamping press.

Punch-and-Die Shearing: A production shearing process using a shaped punch and matching die.

Punching: A process similar to blanking, except that material stamped out of the sheet metal is usually scrap. The part is what is left.

Pure Metal: A metal that contains only one kind of atoms, such as iron or copper. No other metallic elements are mixed in.

Push Stick: A safety device used to hold down the wood stock and push it on through a saw or similar machine.

Pyrolysis: The creation of chemical changes by subjecting a material to heat.

Pyrometric Cones: A set of wedge-shaped pieces of clay, each with a specific melting point, used to determine the proper firing range for ceramics.

Q

Quenching: Rapid cooling of metal, usually by plunging into a liquid such as water or oil.

R

Radial Arm Saw: A saw with its circular blade mounted on a horizontal arbor suspended from an arm above the work table. This type of saw is used primarily for crosscutting.

Radially: In a direction across the rings of the tree.

Radiation Processing: A type of polymerization process that produces internal change through exposing the stock to high-energy ionizing radiation.

Radio-Frequency (RF) Dielectric Drying: A process for drying hardwood that makes use of frictional heat from the vibration of radio-frequency waves. It is faster than kiln or air drying, and results in improved lumber quality.

Radio-Frequency (RF) Heat Sealing: The process of using high-frequency radio waves, combined with the energy and pressure exerted by pressing the parts together, to cause the molecules to oscillate and create heat.

Rake Angle: The angle on the end of each row of teeth of a tap. Wider flutes at the bottom of the tap help to remove chips when the tap is backed out of the hole.

Ram Pressing: A form of wet pressing in which material is formed under pressure in a die.

Ratchet and Pawl: The simplest, but least accurate, of the basic driving mechanisms used to provide intermittent rotary motion.

Raw Materials: Crude or partially processed material that can be converted by manufacture, processing, or combination into a new and useful product. Also known as *industrial stock*.

Reaction Bonding: A process that uses a nitrogen-rich atmosphere during firing. It yields a part with less shrinkage than conventional sintering processes.

Reaction Injection Molding (RIM): A molding process involving combining two or more reactive liquids which quickly polymerize once they enter the mold.

Reactive-Phase Sintering: A variation of liquid-phase sintering except, that the liquid breaks down or disappears, rather than remaining present during sintering.

Reaming: A final finishing process that improves the dimensional accuracy and surface finish of a drilled hole.

Reconstituted Wood: A type of wood-based, engineered material made by cutting and reassembling very thin strips of wood. It is often used as a veneer.

Refractories: Ceramics that can withstand continued exposure to high temperatures.

Refractory: Nonmetallic ceramic-like materials that remain solid and intact when they are subjected to high temperatures.

Refractory Metals: High-temperature metals that can withstand heat and maintain strength. Refractory metals include niobium (Nb), tungsten (W), and molybdenum (Mo).

Refractory Slurry: A substance sprayed on or used as mortar in joints between bricks in a furnace lining.

Reinforced Reaction Injection Molding (RRIM): A process in which short reinforcing fibers are added to one of the RIM liquid components as it is pumped to the mixing head prior to injection.

Relay Logic: An older type of machine control system, based on the opening and closing of electromechanical relays.

Replication: One of the four criteria that shape the nature of automation: the ability to exactly reproduce the product as many times as necessary.

Resin: Term used in the plastics industry for all processed material, up to the point where industrial stock is created.

Resin Transfer Molding (RTM): A molding process in which catalyzed resin is transferred from a separate chamber into the cavity of a two-part matched mold.

Resistance Welding: A welding method in which the materials to be joined are clamped between two opposing electrodes and a current supplied. The resistance that is produced creates heat, bonding the metal layers together.

Resonance Point: The specific frequency at which an ultrasonic cutting tool must vibrate to be effective.

Response Criteria: Methods used to evaluate a closed-loop system's ability to achieve a desired output.

Reverse Roller Process: A finishing process in which the rollers turn in one direction and the stock moves in the other. It provides better coverage of finishing material on the workpiece.

Rework: The process of repairing defective parts, assemblies, or entire products. Since it does not add value to the product, rework is a form of waste.

Riddle: A screened sifter that is used in the sand casting process to shake sand over the pattern until it is well-covered.

Right-Hand Tools: Lathe tools that cut from right to left.

Ring Broach: A cutting tool, consisting of a series of high speed steel rings, that is used for pot broaching.

Ring-and-Stick Broaches: Cutting tools that are used to broach teeth on gear blanks at the same time that other operations are being performed.

Rip Sawing: Saw cuts made in the direction of the grain of the wood.

Roll Bending: A process used to bend circular, curved, and cylindrical shapes from bar, rod, tube, angle, or channel stock.

Rolled Glass: A translucent flat glass that is poured and rolled to the desired thickness.

Roller Coating: A high-speed coating process in which stock is squeezed between rubber rollers carrying a film of finishing material.

Roller Forming: A metal-forming method that uses rollers to progressively squeeze continuous strips of metal to form straight lengths of various cross-sections.

Rotary Shearing: A hand process that is similar to straight shearing, except that the cutting blades are rotary wheels used to cut either straight or circular shapes.

Rotary Swaging: A process that takes a solid rod, wire, or tube and progressively reduces its cross-sectional shape through repeated impacts from two or four opposing dies. It is also referred to as radial forging.

Rotational Molding: A method used to make hollow plastic articles from plastisols, using a rotating hollow mold. The hot mold fuses the plastisol into a layer of gel, which is then chilled and the product stripped out of the mold.

Rough Grinding: A process used for rapid material removal on castings, forgings, and welded parts.

Roughing: The initial turning operation performed on a workpiece in a lathe. Also, a type of sanding that is done to remove the maximum amount of stock; it requires coarse-grit abrasive.

Roughing Cut: A fairly deep initial cut done to bring a workpiece close to finished size.

Rough-Sawn Lumber: Wood that requires subsequent planing or surfacing to smooth boards prior to use.

Round Nose Tool: A chisel used to produce specialized contours on a lathe.

Roving: A woven fiberglass mat used for reinforcement in plastics.

Runners: In injection molding, the pathways that carry plastic to the mold cavity.

S

S Glass: The type of glass added to a composite matrix when the product will undergo exposure to excessive temperatures.

Safety Glass: A composite that is created by sandwiching a layer of polyvinyl butryl plastic between sheets of glass.

Saggers: Refractory boxes that protect the workpieces from contamination in the firing operation after the glaze is applied.

Sagging: A technique used for shaping glass by placing sheet glass over a mold, and then applying heat until it softens and sags to conform to the shape of the mold.

Sandblasting: A mechanical etching process in which alumina oxide abrasive is propelled against glass by compressed air. It is used to create designs on the glass.

Sand Casting: In this process, the mold is expendable—it is destroyed when the part is removed after casting.

Scrap: Excess material that is generated during the material removal process.

Screw Threads: A helical groove machined or formed onto a round piece of stock.

Scribing: Incising the surface of the material by use of a laser.

Scroll Saws: Saws used to produce intricate cuts within the inside dimensions of a workpiece. Also referred to as jig saws.

Sealing: A cohesion process that is typically used for joining plastic films.

Seam Welding: A longitudinal weld used to join sheet metal parts or to make tubing.

Second Wave: Term used to describe the second of a series of basic changes in manufacturing productivity. In this wave, beginning in Japan after World War II, emphasis shifted from a focus on quantity to one on quality.

Secondary Processing: Activities engaged in by firms that use industrial stock to manufacture hard good consumer products.

Sedimentation: The settling to the bottom of a container of small particles when a solution is allowed to remain still for a prolonged length of time.

Select Hardwood: Grade of hardwood that is below FAS and above Commons.

Select Lumber: Wood that is graded from A to D, with A presenting the best quality surface appearance.

Sensitivity: In a control system, the ratio of change in output to the change in input.

Sensors: Electrical devices that receive different types of information and transmit it (provide feedback) to a computer or other control device.

Separating: Removing material or volume from a workpiece. Some separating processes are chip-producing; others are not.

Setters: Solid masses of clay placed under certain areas of a ceramic workpiece to reduce warpage during firing.

Shading: A process that involves spraying a tinted lacquer on wood to improve its color uniformity after bleaching.

Shank Hole: One of the two holes drilled when fastening wood with a screw. It enlarges the pilot hole to the same diameter as the screw shank.

Shank-Type Cutters: Cutters that are fastened directly into the spindle on the milling machine.

Shape Memory: A property of some metal and plastic materials that allows them to be heated and formed into a desired shape, then return to their original shape when heated again.

Shaper: A machine tool for cutting flat-on-flat, contoured surfaces by reciprocating a single-point tool across the workpiece.

Shearing: A mechanical separating process often used to cut flat stock from sheet or plate.

Sheet Molding Compounds (SMC): Sheets of resin that are formed by closed molding..

Sheet Stock: Material that is manufactured by laminating together two or more layers of materials.

Shell Broach: A type of internal cutting tool that is often used to cut helical splines, such as the rifling used inside gun barrels.

Shielded Metal Arc Welding (SMAW): An electric arc welding method in which vaporization of a coating on the electrode provides a gas that shields the weld joint from atmospheric contamination.

Shot Peening: A cold working process that is accomplished by bombarding the surface of a part with small spheres of cast steel, glass, or ceramic particles called shot.

Shrinkage: A reduction in overall size that can occur during curing of plastics, the firing of ceramics, the cooling of a metal casting, or the drying of wood.

Silicon Carbide: The compound of silicon and carbon, widely used for manufacturing grinding wheels. Also, a nonporous composite that is used to form complex, near-net shapes.

Silicon Nitride: A material, produced by use of reaction bonding, that is resistant to thermal shock and to chemical reagents. It has considerable manufacturing potential.

Simultaneous Engineering: Bringing a manufacturing engineer into the design process to determine how the new product will be manufactured, thus avoiding problems that could result in waste.

Single Planers: Knife-blade planers with cutterheads only on the top.

Single-Plane Die Cutting: A process involving a machine that uses a flat die set with multiple patterns to press into the stock.

Single-Spindle Shaper: A wood-shaping machine with a cutter mounted on a vertical spindle (shaft) that projects through an opening in a horizontal metal worktable.

Sintering: The process of holding parts for a short length of time at a temperature just below the base metal's melting point in order to create an internal metallurgical bond between the particles of powder.

Siphon Gun: A type of gun used for atomized spraying of finishing materials. In this gun, material is drawn upward to a fluid cap by suction, then mixed with air.

Sizing: A method of coining that is used with forged parts to improve surface finish.

Skew: A flat chisel that is used to smooth cylinders and cut shoulders on a lathe.

Slab Broach: A tool with cutting teeth on its flat face, used to shape the outside of a workpiece.

Slack Time: The difference, if any, between the critical path time and the target completion time. Elimination of slack time is desirable for efficient manufacturing.

Slag: A hard coating, resulting from the interaction of flux and impurities, that is formed on the surface of a weld joint.

Slip: A very liquid mixture of clay and water.

Slip Casting: A process that involves depositing a layer of thin water-clay mixture ("slip") on the inner surfaces of a mold. The process results in a hollow product, such as a figurine. Also called drain casting.

Slitting: A variation of shearing that creates a slit in metal where another part or device can be inserted.

Slotter: A machine tool used for making a mortise or shaping the sides of an aperture.

Slurry: A liquefied ceramic material. Also, an abrasive suspended in a liquid medium.

Slurry Infiltration Process: A method for incorporating fibers into ceramic matrices by passing the fibers through a liquid slurry. They can then be dried and combined with various types of matrices.

Slush Casting: A process that involves inverting the mold when the part has just begun to solidify, so that metal which is still molten can be poured out. The result is a hollow shell casting.

Soft Automation: A system in which the machines and tooling are designed to produce a wide variety of different shapes and material characteristics.

Softwood: A wood produced from a coniferous tree.

Sol: A colloidal suspension so thick that it shows no sedimentation.

Soldering: An adhesion process for joining metals that uses a low-melting-point filler metal.

Sol-Gel Technique: A process used to make glass and ceramic fibers by combining the unique characteristics of two types of liquid, a sol and a gel.

Solid-Phase Sintering: Using diffusion to move surface material and fusing it where the particles come in contact with each other.

Solidus Line: On a temperature graph, the point at which a molten alloy changes from a mushy solid to a hardened solid.

Solvent-Releasing Adhesives: Adhesives that give off a liquid as they cure.

Solvents: Chemicals that soften thermoplastic materials, allowing them to be joined by cohesive bonding.

Specific Heat: The amount of energy necessary to produce a one-degree change in the temperature of the material.

Spin Welding: A welding process in which one of the parts is held stationary, while the mating part is spun rapidly against it. Friction created at the point of contact generates heat and brings the two pieces to their melting point.

Spindle: Term for a rotating shaft used on many production machines. Usually, a cutting tool is mounted on the spindle.

Spindle Turning: The process of turning a workpiece that is held between centers on a lathe.

Spinneret: A plate with holes drilled in it, through which polymeric fibers are pulled during the spinning process.

Spinning: A process that involves stretching sheet stock over a rotating male or female mold.

Spirit Stains: Stains consisting of an aniline dye suspended in alcohol. Because they are very fast-drying, these stains are used primarily for repair or touch-up work.

Split Pattern: A two-piece pattern used in the sand casting process.

Spray Drying: A process used to produce uniformly sized particles of clay powder by spraying slip into a column of heated air.

Sprayup: A process that uses a chopper gun to spray strands of roving and catalyzed resin into an open mold.

Springback: An increase in the dimensions of the compacted part upon ejection, produced by stored elastic energy.

Sputtercoating: A process using a high level of kinetic energy and a cathode which emits electrons to bombard argon gas molecules and ionize them. The sputtering action this creates transfers metal to the part.

Sputtering: See Sputtercoating.

Square Nose Tool: A chisel used to produce specialized contours on a lathe.

Squeeze Casting: A fabrication process in which a fiber preform is placed in an open two-part mold and molten metal is added. When the die halves close, the pressure forces the metal into the preform.

Stability: The ability of the control system to maintain steady-state conditions.

Stabilization: A step in the production of carbon fiber that causes the material to oxidize, changing its internal structure to prevent melting.

Stain: A finishing material consisting of finely ground pigments, suspended in a vehicle, used to change the overall wood color or to emphasize its grain pattern.

Staking: Process that uses ultrasonic vibrations to melt thermoplastic studs and mechanically lock dissimilar materials in place.

Stamping: A chipless process that produces a sheet metal part with a single downward stroke of the ram in a stamping press.

Stamping: A cold-forming process that uses a set of matched molds in a stamping press to compact stock under pressure.

Standard Industrial Classification (SIC): A U.S. Government system of assigning standard numerical codes to industries to separate them into classifications and sub-classifications.

Staple Fiber: A basic type of glass reinforcing fiber, usually between 6 in. and 15 in. in length.

States of Operation: Term used to describe the action or condition of an output device, such as "on" or "off."

Static Fatigue: A condition in which a material (such as a glass fiber) is not able to support heavy loads for continued lengths of time.

Stationary Routers: Tools that use a pin guide mounted in the table and a template that runs against the pin to make grooves and cut irregular shapes.

Statistical Process Control (SPC): A system that provides a basis for measuring a process to predict, as quickly as possible, any trends away from the expected output.

Staypack: A modified wood that is less brittle than those made by resin impregnation. Staypack is made by using pressure and heat to cause the natural lignin between the wood fibers to liquefy, and the entire mass to realign itself.

Steady-State Error: The difference between the actual output and the desired output required by the control system.

Steatite: A type of ceramic material made from the natural crystalline form of magnesium silicate or talc, mixed with clay and barium carbonate flux.

Steel: An alloy of iron and carbon, or of iron and other alloying elements.

Stereolithography: Prototyping process that builds an object, layer by layer, by exposing a liquid photopolymer to ultraviolet laser light. The photopolymer is cured and hardened as the light strikes it.

Stick Broach: A cutting tool used for pot broaching.

Stickered: Term describing the stacking of lumber using spacers between boards.

Stock: Raw material that can be processed by secondary industries into manufactured products.

Straight Shearing: The action of cutting sheet metal to size in rectangular pieces.

Straight Turning: The simplest form of turning operation, in which the tool is fed into the stock, then moved to the left or right to remove stock and reduce the workpiece diameter. Also called peripheral turning.

Stretch Draw Forming: This process uses mating dies that must match perfectly.

Strip Feeding: A feeding system that is a variation of magazine feeding. It consists of parts that are formed into, or held in, a long strip to be fed to the point of use.

Structural Composites: Term once used to describe composites formed from relatively simple materials.

Structural Foams: Strong, lightweight foams that have a surface layer or skin with a density much like unfoamed plastic and a porous core with uniformly sized bubbles.

Structural Lumber: Material purchased by the construction industry for uses ranging from light wall framing members to beams, stringers, posts, and timbers for heavy structural applications.

Stud Welding: A form of arc welding that is used to fasten a threaded metal stud to another part.

Sublimation: A process in which a substance passes directly from the solid state to the gaseous state, without becoming liquid.

Substrate Failure: A bonding failure that results in the destruction of the substrate (material being joined).

Successive Transformation: Term that describes the process of production in stages, which may extend across a number of industries.

Superalloy: A thermally resistant alloy for use at elevated temperatures where high stresses and oxidation are encountered.

Superconductor: A material that exhibits a total absence of electrical resistance below a critical temperature.

Surface Energy: The surface tension of a liquid material, which affects how well that liquid will wet a surface.

Surface Grinding: A form of precision grinding done on flat workpieces.

Sweat Soldering: A type of soldering that uses capillary action to draw melted solder into a close-fitting joint.

Swing: Term used to specify the maximum diameter of a workpiece that can be turned on a lathe.

Synergism: A phenomenon in which the whole is greater than the sum of the parts.

Synthetic: Term for anything that is human-made.

Synthetic Metal Research: A field of research concerning conductive polymers and related topics.

T

Tablet Pressing: A variation of dry pressing.

Tailstock: The fixture at the end of a lathe opposite the headstock. The tailstock is used to mount a live (rotating) center or dead (nonrotating) center that holds one end of the workpiece.

Tangentially: Term used to describe shrinkage of wood around the tree (in the direction of the annual rings).

Tap: A shanked tool with rows of cutting teeth, separated by flutes, that are used to cut threads inside a hole.

Tape Casting: See Band Casting.

Taper: A gradual diminution of thickness, diameter, or width in an elongated workpiece.

Tapping: Process that is used to cut threads inside a hole.

Tapping a Heat: The process used to release molten metal from a furnace into a heavy metal container called a ladle.

Target: A flat plate of coating material that is bombarded by a beam of electrons during sputtering.

Taylor Process: A process used to produce ultra-fine filaments by encasing metal wire in a sheath of glass, heating the coated wire until the protective sheath softens into a plastic state, then drawing the soft-coated wire through a tiny orifice to a fine diameter.

Technical Strategist: A new type of technical manager trained to solve problems and to be able to work effectively with both people and technology.

Teeth Per Inch: The measuring unit used to identify many types of saw blades: the greater the number of teeth per inch, the finer the cut.

Tempered Hardboard: A type of hardboard impregnated with a resin/oil blend to give it better strength, stiffness, and resistance to water and abrasion.

Tempering: The process of reheating a hardened metal, then cooling it to improve toughness and make it less brittle. Also, a heat-treating process used with glass products to increase their annealed strength. In ceramic production, the process of mixing and kneading liquids into a dry clay material to produce stock that is pliable enough for forming.

Tenon: A tonguelike projection from the end of a framing member that is made to fit into a mortise.

Tensile Strength: The limit of stress and strain a material is capable of bearing.

Tensile Stress: The amount of stress that must be applied to a metal to cause deformation and lengthening.

Thermal Conductivity: The amount of heat that is carried or stored in a material.

Thermal Decomposition: A chemical vapor deposition process used to deposit a boron halide coating on a carbon-coated glass fiber by drawing the fiber through a tightly sealed chamber.

Thermal Energy: The high-velocity stream of electrons that is generated by the electron beam gun.

Thermal Expansion: The dimensional changes that take place as a metal is heated or cooled.

Thermal Expansion Resin Transfer Molding (TERTM): A process in which a preformed foam core is wound with reinforcement material and impregnated with epoxy resin. When the assembly is heated inside a mold, the foam expands to squeeze the reinforcement and matrix materials together.

Thermal Heat Sealing: Using a heated tool or die at a constant temperature to join layers of plastic film. Widely used for packaging applications.

Thermal Impulse Heat Sealing: A variation of Thermal Heat Sealing used to provide better joint appearance with thin films.

Thermal Shock Resistance: The ability of a material to withstand sudden changes of temperature.

Thermochemical: Term used to describe a process involving both heat and a chemical reaction.

Thermoplastics: Plastics that can be melted and reformed as desired into new products. They are produced by addition polymerization.

Thermoset Adhesives: Adhesives that, once cured, will not soften through exposure to heat.

Thermosets: Plastics that are formed by condensation polymerization, a chemical reaction that causes the polymer chains to become cross-linked. Thermosets cannot be melted and reformed like thermoplastics.

Third Wave: Term used to describe the latest of a series of basic changes in manufacturing productivity. It refers to the aggressive implementation of a new set of management techniques in manufacturing, which first became noticeable in the early 1980s.

Thread Rolling: A chipless cold-forming process that can be used to produce either straight or tapered threads.

Threading: The process of cutting internal or external threads on a part, using a boring bar or a cutting tool.

Throughput: The amount of product that moves through a production operation in a given period of time.

Throwing: A foundational process in ceramics that involves hurling onto a revolving potter's wheel a body of clay that is in a plastic state. The clay is then shaped by hand into a symmetrical form.

TIG: Abbreviation for "tungsten inert gas" welding. See Gas Tungsten Arc Welding.

Tin Plate: A mild steel that is coated with tin.

Tip Angle: The included angle formed when measuring from one side of the tip of a drill to the other.

Tool Holder: The device that holds a lathe cutting tool. It is mounted on the tool post.

Tool Post: The device upon which a lathe cutting tool holder is mounted.

Tool Steel: A high-carbon steel that is hard and difficult to bend. It is used to make tools, such as forging dies, screwdriver shafts, chisels, and milling cutters.

Topcoat: A final clear coating of lacquer, oil varnish, or synthetic-resin varnish that is applied to a wood workpiece to provide a protective layer.

Torch Cutting: The use of an oxyacetylene torch to separate metallic materials.

Total Quality Approach: One that places emphasis not only on quality improvement, but also on reducing inventories through the adoption of a just-in-time (JIT) supply and production system. Also referred to as Total Quality Control (TQC) or Total Quality Management (TQM).

Toughness: Resistance to breaking; a physical property used to describe a material.

Tracheids: The cells through which sap is transferred vertically in softwoods. In addition to transporting sap, the trachieds provide strength to the wood.

Traditional Ceramics: Term for products made from glass and clay.

Transducer: A device used to convert one type of energy to another, such as changing an electrical signal into a mechanical motion.

Transfer Coating: A process involving deposit of a surface skin coat on coated release paper, then transferring it to the product being finished.

Transformation Toughening: A process used to increase the strength and toughness of ceramics in products that must exhibit high wear resistance.

Transient Response: The response of the output device under control in comparison to an input before steady-state conditions are achieved.

Transporting: The action of moving parts from one operation to another, or materials from storage to a workstation. This action often contributes to a form of waste called waiting.

Traveling Head Press: A machine that is designed to make multiple die cuts in wide sheets or roll goods.

Travelling Wire EDM: A type of EDM in which the cutting is done by using a round wire that travels through the workpiece.

Trickledown Effect: The spreading of technological improvements and process improvements from the firms that pioneered them to other (usually smaller) firms in the same and related fields.

Trost Fluid Energy Mill: A dry milling process in which grinding is achieved by collision between the particles being ground and the energy supplied by a compressed fluid entering the grinding chamber at high speed. Also known as the jet mill.

True: The action of making both ends of the stock perpendicular before turning it on a lathe.

True-Up: Term used to describe eliminating bow in a piece of wood by planing.

Tumbling: A finishing method in which plastic, wood, or metal parts are placed inside a rotating drum to rub against each other to remove flash and achieve a smooth finish.

Tumbling Compounds: Polishing materials, such as abrasive particles, waxes, sawdust, or wood plugs, that are rotated with parts in a drum.

Turning: Separating method used with a rotating workpiece on a lathe to reduce its diameter, change its profile, or produce a taper on it.

Turning Processes: Processes that are used to machine rotating parts.

Turnkey System: An automated system package that is standardized and ready to be put into operation.

Twist Drill: The most common type of drill for metals, normally made of high-speed steel or carbon steel.

U

Ultimate Tensile Strength: Point at which the stress and strain on the material becomes greater than its overall loadbearing capacity.

Ultrasonic: Term used to describe very high frequency vibrations that can generate heat for many kinds of joining tasks involving metals or plastics.

Ultrasonic Machining: A process that removes material by erosion, using vibrations generated by high-frequency sound waves.

Ultrasonic Welding: A welding process that works on the principle of changing sound energy to mechanical movement to generate heat for joining metals or plastics.

Underglaze: A decoration that is protected from wear by the overlying glaze, but does not provide the extensive range of colors available with overglaze decoration..

Underlayment: Particleboard laid down as a firm base for flooring materials.

Universal Chuck: A three-jawed device used to hold a workpiece in the headstock of a lathe. The three jaws make it self-centering.

Universal Drilling Machine: A machine with an adjustable head for drilling holes at an angle.

Universal Saws: Saws with two arbors; one with a ripping blade mounted on it; the other with a crosscut blade.

Up Milling: A process in which the cutter is positioned on top of the workpiece, with the workpiece fed in the direction opposite the rotation of the cutter.

Upper Transformation Temperature: On an equilibrium chart, the temperature where the changing of a material from one phase to another is complete.

Upsetting: A type of forging that thickens or bulges the workpiece while also shortening it by compression. Upsetting is actually a combination of forging and extrusion.

Uranium Carbide: A ceramic form of uranium frequently used for reactor fuel rods because of its excellent thermal conductivity.

Uranium Dioxide: A highly heat-resistant oxide of uranium in a ceramic form that makes it ideal for use as a reactor fuel.

V

Vacuum Bagging: A process in which a vacuum acts to press a plastic bag against resin-impregnated roving to force out excess air.

Vacuum Metallizing: A process used to apply very thin metal coatings to a plastic substrate by vaporizing the metal in a vacuum.

Value Analysis: One of the primary methods that firms use to gather information on the design of competitive products.

Vapor-Phase Sintering: Using vapor pressure to cause material to be shifted from the surface of the particles to the point where the particles contact each other to reduce both size and the pore space.

Variability: The inconsistency that occurs in manufacturing a particular product.

Variety Saws: Single-arbor saws that are capable of performing a number of tasks.

Varnishes: Transparent surface coatings consisting of oil, resin, solvent, and driers that are used to protect the surface of a wood workpiece from water spotting, chipping, and abrasion.

Vehicle: The fluid component of paint that acts as the carrier for pigment.

Veneer: Thin sheets of hardwood for laminated surfaces.

Vertical Milling Machine: A type of milling machine in which the cutter is positioned perpendicularly to the work table.

Very High Bond (VHB) Tape: An adhesive, in either conventional tape roll form, or as die-cut shapes for specialized applications, that forms extremely strong bonds between dissimilar surfaces.

Vessels: Cellular structures found in hardwoods that carry sap vertically.

Vibration Welding: A quick method for joining most thermoplastics. It uses vibration to produce frictional heat to make the joints.

Vibratory Deflashing: A process that uses high frequency vibrations to improve the abrasive (rubbing) action between the parts.

Vibratory Finishing: A deburring method in which parts and an abrasive finishing medium are placed in a rotating closed container.

Viscosity: The thickness or ability of a liquid material to flow.

Vitreous Solder: A low-melting-point glass that can be used to seal glass with a higher melting point.

Vitrification: The process of firing materials at a particular temperature to create a glassy bond and close pores, making the material denser and stronger.

Volume: The quantity of product that is to be produced; one of the four criteria that shape the nature of automation.

W

Waferboard: A type of particleboard that is made using high-quality wood flakes that are about 1.5 inches square.

Waiting: A form of waste that involves the expenditure of time without adding value to the product.

Walking-Beam Mechanism: A rectangular-motion mechanism that uses a transfer bar to lift work in progress out of a workstation location and move it one position ahead to the next station.

Waste: Scrap that isn't recycled for use in the production process. One manufacturer defines waste as "Anything other than the minimum amount of equipment, materials, parts, space, and worker's time, which are absolutely essential to add value to the product."

Water Stains: A wood stain that uses water as a vehicle in which to suspend color pigments.

Waterfall Method: Application of glaze by moving the workpiece through a curtain (waterfall) of continuously-flowing material.

Waterjet: A tool that produces a very fine, high-pressure jet of water to cut, drill, or machine materials.

Waterjet Machining: A process that uses a high-velocity stream of water to cut materials ranging from paper to stone or metals.

Ways: Guides machined on the top of the heavy cast iron bed of a lathe.

Wear Parts: Term for moving parts exposed to extensive friction.

Weir: A refractory barrier over which molten glass flows out of the furnace.

Welding Processes: A group of processes that are used to join materials through the application of heat and/or pressure.

Wet Bending: A wood bending technique that uses either steaming or soaking to soften (plasticize) the wood so that it can be more easily formed. Laminated wood can also be bent by forming with adhesives in a two-piece mold.

Wet Pressing: A ceramic forming process that uses higher moisture content and less pressure than dry pressing. However, it can result in dimensional tolerances varying by as much as two percent.

Wet Spinning: A process through which fibers are formed in a solution, and then allowed to precipitate in a coagulation bath.

Wet Winding: A process in which the filament is impregnated with resin just before it is wound onto the mandrel.

Wettability: A measurement of a liquid's ability to wet the surface of a solid material.

Wheel Electrodes: Rolling electrical contacts used to resistance-weld a longitudinal seam.

Whiskers: A type of experimental filler used in composites. Whiskers are more physically uniform than fibers, exhibit higher properties, and provide increased strength.

Wiping: Polishing operation performed on plastic using an uncharged wheel.

Wire Metallizing: See Flame Spraying.

Wood Filler: A paste used to close the pores of open-grained wood before finish is applied.

Wood Plastic Composition (WPC): A process that uses radioactive isotopes to transform wood into a material that is much like plastic. The material is referred to as irradiated wood.

Work Cell: A manufacturing environment in which several machines are operated by one employee.

Work in Process (WIP): Parts or components that are moving through a manufacturing operation. WIP is the stage between raw materials and finished goods.

Workholding Pallet: A transfer base containing work in process. It is moved from station-to-station by conveyor or other means, then relocated at each station by guide pins.

Workpiece: The individual unit of material that is being processed or changed into a product. Also called *stock*.

World Class Manufacturing (WCM): A term sometimes used as a synonym for lean production. It is applied to companies that have gone through continual improvement to the extent necessary to assure that better quality products are being consistently produced at less cost.

Woven Roving: A fibrous-glass material used as reinforcement in the hand lay-up method of open molding.

Wrought Iron: One of the two basic forms of iron. Wrought iron is tough, because it contains very little carbon.

Z

Zinc-Arc Spray Metallizing: A process in which zinc wires are melted by an electric arc. The resulting molten zinc is deposited on the plastic substrate by a stream of air.

Zirconia: The principal ceramic material in zirconium oxide. It is made more wear-resistant by transformation toughening.

Index

A

Abrasive belt planer, 415-417
Abrasive grain sizes, 461
Abrasive jet machining, 147-149
Abrasive shot method, 357
Abrasives, deburring, 200
Acetylated wood, 454
Acrylonitrile-butadiene-styrene, 211
Actuator, 545
Adaptive control, 553-554
Adaptive control systems, 547
Adding reinforcement fibers
   boron-fiber-reinforced composites, 517-518
   carbon-fiber-reinforced composites, 519-520
   ceramic-fiber-reinforced composites, 520-521
   glass-fiber-reinforced composites, 517
   metallic-fiber-reinforced composites, 521
Addition polymerization, 214
Adhesion, 278
Adhesion processes, 183, 516
Adhesives, 278, 443-448
   high-frequency electric gluing, 446
   hot melt adhesives, 444
   solvent-releasing adhesives, 444-445
   thermoset adhesives, 446
   very high bond (VHB) tape, 446-448
Adhesive failure, 444
Advanced composites, 479
Age-hardening, 192
Aging, 192
Air drying, 396
Airless spraying, 468

Alloys, 54
Anisotropic, 449
Annealing, 195, 284, 376
Annealing lehr, 337
Annular ring, 391
Anode, 158
Anodic oxidation, 201
Anodizing, 201
Anti-kickback device, 430
Antiozonants, 532
Applying fillers, 463
Arbor, 133, 428
Arbor-type cutters, 135
Arc welding, 169-171
Artificial aging, 192
Artificial intelligence (AI), 544
Ashing, 294
Assembly, 39
Assembly line, 15
Atomized spraying, 467-468
Attrition mill, 350
Austempering, 196
Austenite, 187
Autoclave molding, 492
Automated control systems, 545
Automated manufacturing systems
   adaptive control, 553-554
   computer-integrated manufacturing (CIM), 554-555
   control system design requirements, 555-556
   material handling systems, 556-567
   nature of automation, 540-550
   numerical control (NC), 550-553
Automatic feed, 126
Automatic production processes, 538
Automation
   control systems, 545-550
   nature of, 540-550

B

Bagasse, 406
Ball mill, 349
Bandsaws, 426-427
Band/tape casting, 333-335
Batch, 15
Batch furnaces, 189
Bed width, 414
Bed-type machine, 133
Bending, 411-412
Beneficated, 347
Binders, 317
Bioceramics, 311-312
Bisque firing, 370
Bisque ware, 370
Blank, 74
Blanking, 117-119
Bleaching, 464
Blending, 460
Blind holes, 140
Blind rivets, 442-443
Blow and blow machine, 343
Blow molding, 227-230
Blown film, 240
Board foot, 401
Bonding
   advantage of lamination, 410-411
   cementing, 277-279
   composite matrices, 526
   composition boards, 403-409
   hardboard, 404-405
   insulation board, 405-406
   lamination, 409-411
   manufacture of particleboard, 408
   particleboard, 407-408
   plywood, 409-410
   reconstituted wood, 409
   waferboard and oriented strandboard, 408-409

Boothroyd and Dewhurst method, 42-45
Boring, 120, 433-435
Boron-fiber-reinforced composites, 517-518
Brazing and soldering, 181-184
Brittleness, 308
Broaching, 127-132
B-stage, 222
Buffing, 294
Bulging, 90-91
Bulk molding compounds (BMC), 221
Bulk systems, 556
Burrs, 198
Butt welding, 174-175

C

C glass, 517
Calcination, 348, 369
Calendering, 253-254, 335
Cam drive, 564
Captive suppliers, 38
Carbon steel, 58
Carbon/carbon composite (C/C), 520
Carbon-fiber-reinforced composites, 519-520
Carbonitriding, 194
Carbonization, 496, 519
Carburizing, 193
Case hardening
  carbonitriding, 194
  cyaniding, 193
  flame hardening, 194
  induction hardening, 194
  nitriding, 193-194
Cast iron, 58, 101
Casting, 220, 250-251, 337
  centrifugal casting, 110
  die casting, 105-106
  investment casting, 109
  nylon casting, 251
  permanent mold casting, 106-107
  powder metallurgy, 102-105
  processes, 101-110, 334-335
  sand casting, 107-109
  sheet casting methods, 250-251
Catalyst, 213
Cathode, 158
Cell casting, 250
Cellulose, 391
Cellulosics, 212
Cemented carbide, 122, 532
Cementing and bonding, 277-279
Cementite, 187
Centerdrilling, 120, 126
Centerless grinding, 147

Centrifugal casting, 110, 254, 340-342
Ceramic engine, 310
Ceramic-fiber-reinforced composites, 520-521
Ceramic-matrix composites (CMCs), 308, 492
Ceramic materials
  annealing, 376
  calcination, 369
  conditioning, 367-378
  densification, 367-369
  firing, 370
  freeze drying, 369-370
  ion implantation, 375-376
  tempering, 377
  transformation toughening, 373-374
  vapor deposition coating processes, 374-375
  vitrification, 372-373
Ceramics
  in space, 306-307
  in the automotive industry, 310
  traditional, 308-311
Ceramic materials, fabricating
  encapsulation, 362
  fusion sealing, 361-362
  laminating, 362
  metallizing, 359-361
  refractory materials, 364-365
Ceramic materials, finishing
  flame and plasma spraying, 380-381
  flame polishing, 380
  glazing, 382-386
  grinding, 379-380
  laser processing, 381-382
Ceramic materials, forming
  dry forming processes, 319-325
  glass forming processes, 337-345
  preparing for processing, 318-319
  wet forming processes, 326-337
Ceramic materials, separating
  etching processes, 356-357
  green machining, 356
  grinding, 347-356
Ceramoplastics, 523-525
Cermets, 123, 310, 484, 531
Charged, 294
Chemical blanking, 154
Chemical etching, 356-357
Chemical milling, 154-156
Chemical vapor deposition, 374
Circular saws, 427-430
Climb milling, 134
Closed molding
  blow, 227-230
  compression, 222-224

  injection, 224-226
  reaction injection, 226-227
  resin transfer, 230
Closed-die forging, 67-69
Closed-loop control systems, 546
CO_2 lasers, 507-509
Coated abrasives, 460
Coated preform, 222
Coating processes
  coating storage vessels in the field, 296
  conductive spray coating, 299
  dip-coating, 295-296
  electroless plating, 300
  electroplating, 300
  electrostatic powder coating, 296-297
  flame spray coating, 299
  fluidized bed coating, 298
  vacuum metallizing, 299-300
Coextrusion, 494-495
Cohesion processes
  dielectric sealing, 276
  hot gas welding, 269-271
  hot wire welding, 270-271
  induction welding, 271-272
  radio-frequency heat sealing, 274-276
  spin welding (friction welding), 272
  thermal heat sealing, 276
  ultrasonic welding, 272-274
  vibration welding, 271
Cohesive fabrication, 515
Cohesive failure, 444
Coining, 67
Coinjection molding, 226
Cold compression molding, 222
Cold forming, 79
  advantages, 82
Cold heading, 79
Cold isostatic pressing, 324
Cold-rolled steel, 58
Collets, 122
Collimated light, 153
Colorants, 288, 536
Coloring, 301-302
Column-and-knee-type machine, 133
Combination blades, 425
Commodity resins, 210
Common lumber, 400
Community of interest, 19
Compact, 356
Compacts, 87
Competition, global marketplace, 19-20
Composite materials
  conditioning cermets, 531-532
  coupling agents, 532

Index

destructive detoxification, 533-534
electrical surface treating, 528-531
foaming agents, 532
irradiation, 527-528
processes used to finish, 535-537
separating, 501-514
temperature degradation of composites, 533
types of, 527-533
Composite materials, fabricating
adding reinforcement fibers, 517-521
bonding composite matrices, 526
ceramic matrices, 523-525
composite matrices, 521
metallic matrices, 522-523
production of superconductors, 525-526
Composite materials, separating processes
diamond wire sawing, 501-503
dicing, 503-505
laser machining, 507-510
ultrasonic machining, 510-514
waterjet cutting, 505-507
Composite matrices, 480-481
Composite stock, nature of, 489
Composition boards, 403-409
Compound rest, 124
Compreg, 452-454
Compreg process, 253
Compression molding, 222-224
Computer numerical control (CNC), 100
Computer-integrated manufacturing (CIM), 15, 554-555
Condensation polymerization, 214
Conditioning, 30, 283
Conductive glass solder, 364
Conductive plastics, 286-288
Conductive spray coating, 299
Conifers, 390
Construction adhesives, 445
Consumable products, 27
Continuous casting, 250
Continuous die cutting, 262
Continuous extrusion blow molding, 243
Continuous fibers, 340
Continuous furnaces, 189
Continuous improvement, 39
Continuous-path control, 552
Contour machining, 154
Contouring, 95, 155
Control systems
closed loop systems, 546-548
control system design, 549-550
design requirements, 555-556
on/off control, 548-549

open loop systems, 545-548
proportional control, 549
Controller, 545
Conventional milling, 349
Cope, 108
Copolymer, 212
Core, 108
Corona discharge, 530
Coupling agents, 532
Cracked ammonia atmosphere, 360
Creativity, 39
Creep, 56, 218
Creep failure, 218
Critical path, 47
Crocus cloth, 200
Cross slide, 124
Crosscut sawing, 425
Cryochemical drying, 370
Cryogenic conditioning, 197
Crystal pulling, 375
Crystalline structure of steel, 185-188
Crystallites, 391
Cupola, 101
Curtain coating, 469-470
Customized products, 19
Cutoff saw, 428
Cutting speed, 135-136, 258-260
Cyaniding, 193
Cylindrical grinding, 146-147
Czochralski method, 376

D

Danner process, 340-341
Dead center, 122
Dead-end escapement, 559
De-airing pug mill, 348
Deburring, 198-200
Deciduous, 390
Defiberize, 404
Deflashing and deburring
tumbling, 265
vibratory deflashing, 265
Deflocculant, 336
Deform, 87
Degating, 264
Dehydrate, 369
Densification, 367-369
Depth of cut, 124
Design
in manufacturing, 39-40
simultaneous engineering, 40
value analysis, 39-40
Design for assembly (DFA), 40-45
Destructive detoxification, 533-534
Devitreous solder, 362
Diamond concentration, 504
Diamond point tools, 424

Diamond wire sawing, 501-503
Dicing, 503-505
Die block, 80
Die casting, 105-106
Die cavity, 220
Die cutting, 260-263
Dielectric sealing, 276
Diffusion, 325
Dimension lumber, 398
Dip coating, 295-296, 470
Dip casting, 295
Dip molding, 295
Distributed numerical control (DNC), 553
Doctor blade casting process, 333
Dopants, 375
Doping methods, 201, 318, 375-376
Double planers, 414
Drag, 108
Drape forming, 236
Drawing, 87-90, 337
Drilling
boring, 433-435
drill sizing systems, 138-139
mortising, and tenoning, 435-436
Dry bag isopressing, 324
Dry forming processes
cold isostatic pressing, 324
dry pressing, 319
dust pressing, 323-324
explosive forming, 325
hot isostatic pressing, 324-325
Dry milling, 351-353
Dry pressing
refractory materials, 322
steps in, 321
Dry spinning, 496
Drying methods, 396-397
Ductility, 56
Dust pressing, 323-324
Dwell points, 563
Dynamic compaction, 325

E

E glass, 517
Elastomeric molding, 495
Elastomers, 213
Electrical surface treating, 528-531
Electrochemical machining (ECM), 156-159
Electrodeposition, 469, 495
Electrodischarge machining, 160-161
Electroless plating, 201, 300
Electromagnetic forming, 94-95
Electron beam machining (EBM), 159-160
Electron beam welding, 178-180
Electronic gluing, 446

Electrophoresis, 469
Electroplating, 200, 300
Electropolishing, 199
Electrostatic powder coating, 296-297
Electrostatic spraying, 468-469
Embossing, 86-87, 537
Emery paper, 200
EMI/RFI shielding, 287-288
Enamel, 466
Encapsulating, 252, 362
End effector, 150, 467
End mill, 135
Engineered composites, 476
Engineered materials, 29, 283
Engineering resins, 210
Equilibrium diagram, 186
Escapements, 558-560
Etchants, 155
Etching processes
   chemical etching, 356-357
   mechanical etching, 357
Eutectic salt, 497
Eutectoid, 188
Evaporative vacuum method, 299
Excess inventory, 37
Excimer lasers, 510
Exotherm brazing, 183
Expandable bead foam, 245
   molding, 245-246
Expanding, 90
Expansion molding, 495
Explosive forming, 97-98, 325
Extractives, 391
Extractor, 468
Extrusion, 78-79, 240, 329-333
Extrusion blow molding, 243-244
Extrusion cutter, 260

F

Faceplate, 122
Faceplate turning, 424
Facing, 120, 424
Facing-type cutters, 135
Factory and shop lumber, 399
Fatigue, 56, 308
Feedback, 546
Feeder, 556
Fence, 417
Ferrite, 186
Ferrous metals, 57-60
Fiber drawing, 339-340
Fiber spinning, 495-496
Fiber tows, 497
Fibrils, 391
Fibrous composites, 481-484
Filament, 494
Filament winding, 279-280, 496-497

Filler metal, 169
Film blowing, 240-241
Films, 250
Filter pressing, 353-354
Finish coatings
   paints and enamels, 466-467
   varnishes and lacquers, 466
Finishing cuts, 124
Fire clay brick, 364
Fire-retardant, 457
Firing
   presintering, 370
   range, 372
   reaction bonding, 372
   sintering, 370-372
First-wave manufacturing, 34-35
Fixture, 145, 220
Flame and plasma spraying, 380-381
Flame polishing, 380
Flame polishing and solvent-dip polishing, 295
Flame spray coating, 299
Flash, 67, 243, 264
Flash butt welding, 174
Flash-off period, 470
Flask, 107
Flat-platen-pressed, 408
Flexible manufacturing, 555
Float glass process, 337-339
Floating zone method, 376
Fluidized bed coating, 298
Fluidized bed furnaces, 189
Flutes, 136
Flux, 170, 183
Fluxing, 253
Foam cutting, 263-264
Foam frothing, 290
Foaming agents, 532
Follower, 99
Forced drying, 470
Forestry, 397
Forging
   closed-die, 67-69
   open-die, 66-67
Forming block, 98
Forming processes
   calendering, 253-254
   casting, 250-251
   centrifugal casting, 254
   ceramic materials, 317-346
   chemical vapor deposition, 494
   closed molding, 220-230
   coextrusion, 494-495
   composite materials, 492-499
   elastomeric molding, 495
   expandable bead foam molding, 245-246
   expansion molding, 495
   extrusion, 240

   extrusion blow molding, 243-244
   fiber spinning, 495-496
   filament winding, 496-497
   film blowing, 240-241
   glass, 337-345
   injection blow molding, 244-245
   integral foam molding, 246-248
   laminating, 252-253
   liquid infiltration, 493
   metallic materials, 66-112
   open molding, 237-240, 492-493
   plastic materials, 220-254
   plating, 493-494
   potting and encapsulating, 251-252
   pressure roll binding, 493
   pultrusion, 241-243
   rotational molding, 248
   stamping, 230-231
   thermal expansion resin transfer molding (TERTM), 498
   thermoforming, 231-237
   vacuum bagging, 249-250
   whisker growing, 497-498
   wire coating, 241
   wood materials, 403-412
Forming tool, 100
Franklin fiber, 288-289
Free blowing, 228
Free foam molding, 248
Freeze drying, 369-370
Friction welding, 272
Frit, 384
Full hardening
   case hardening, 192-194
   conditioning processes used, 190-192
   heat/quench/temper hardening, 190-191
   precipitation hardening, 191-192
Full-cell impregnation, 456
Function, 39
Fusion sealing, 361-362

G

Gage number, 62, 440
Gang cutting, 163
Gang drilling, 136
Gas metal arc welding (GMAW), 171-173
Gas tungsten arc welding (GTAW), 171
Gas welding
   arc welding, 169-171
   butt welding, 174-175
   electron beaming welding, 178-180
   gas metal arc welding, 171-173

Index

inertia or friction welding, 175-178
laser welding, 180-181
resistance welding, 173-174
ultrasonic welding, 181
Gel, 517
Gel-coat, 238
Geneva mechanism, 564
Gilt, 384
Glass flakes, 289-290
Glass forming processes
centrifugal casting, 340-342
drawing, 337
fiber drawing, 339-340
float glass process, 337-339
glass blowing, 342-344
glass pressing, 345
glass rolling, 344-345
sagging, 345
tube drawing, 340
Glass soldering, 362-364
Glass to ceramic seals, 360
Glass-fiber-reinforced composites, 517
Glass-matrix composites (GMCs), 492
Glaze
additives, 384-386
characteristics, 383-386
Glazing, 382-386, 464
Global marketplace, competition, 19-20
Glost firing, 370
Gouge, 423
Grade, 400
Grafting, 285
Grain size, 461
Granulator, 265
Graphitization, 496
Green compact, 103
Green machining, 356
Green wood, 396
Greenware, 319
Grinding
cylindrical grinding, 146-147
filter pressing, 353-354
grinding wheels, 141-144
media, 349
milling, 348-352
spray drying, 354
surface grinding, 144-146
wheels, 141-144
Gun drilling, 137

H

Halide reduction, 518
Hand forging, 66
Hand layup, 237-239
Hard automation, 543
Hardboard, 404-405
Hardinge conical mill, 349
Hardwoods, cellular structure of, 391-392
Headstock, 121
Heat treating, 189
Heating and cooling times (chart), 285
Heat/quench/temper hardening, 190-191
Heavy gage stock, 119
HERF, 98
Hermetic seal, 360
High-energy-rate forming (HERF), 98
High-frequency electric gluing, 446
High-pressure lamination, 253
High-temperature superalloys, 60
Horizontal milling machine, 133
Horn, 510
Hot forging, 321
Hot isostatic pressing, 105, 324-325
Hot marking, 537
Hot melts, 278-279, 444
Hot press compression molding, 223
Hot pressing, 326-328
Hot wire cutting, 263
Hot wire welding, 270-271
Hot-rolled steel, 58
Hydroabrasive process, 506
Hydroforming, 90

I

Impreg, 451-452
Independent chuck, 121
Indirect labor, 554
Indirect-two-sheet blow molding, 228
Induction bonding, 271
Induction hardening, 194
Induction welding, 271-272
Industrial ceramics, 308-311
Industrial stock
determining steel type, 62-63
making, 209-211
nature of, 61-63, 313-315
Industry, 20
Inefficiency, 37
Inertia or friction welding, 175-178
Infiltration, 105
Injection blow molding, 244-245
Injection molding, 224-226, 333

In-line transfer systems, 565
In-mold decorating with foils, 301
Input values, 554
Inserting, 273
Insulation board, 405-406
Integral foam molding, 246-248
Integrators, 38
Intercellular layer, 394
Interchangeability, 541
Interference fits, 437
Interlayer, 362
Intermittent extrusion blow molding, 244
Internal threads, 140
International joint ventures, 15
International Standards Organization, 19
Investment casting, 109
Ion implantation, 201-202, 375-376
Ionizing radiation processing, 284
Ions, 159
Irradiated wood, 450
Irradiation, 527-528
ISO 9000 certification, 19

J

Jacobs chuck, 121
Jiggering, 328
Job lot, 15, 250
Joint, 169
Jointing, 417-419
Jollying, 328
Just-in-time manufacturing, 37

K

Kaizen, 39
Kaolin clay, 364
Kasenit, 193
Kerf, 153, 502
Kickback, 414
Kiln, 308
Kiln drying, 396
Kiln furniture, 356
Kiss cutting, 263
Knurling, 74, 120

L

Lacquer, 466
Laminar composites, 485-486
Laminating, 252-253, 362-364
Lamination, advantage of, 409-411
Lancing, 116
Laser cutting, 152-154
Laser machining
CO_2 lasers, 507-509

excimer lasers, 510
Nd:YAG lasers, 510
Laser processing, 381-382
Laser welding, 180-181
Lasing tube, 152
Latex paints, 466
Lathe
   components, 124-125
   dog, 125
   types, 120-121
Layered fibrous composites, 484
Lead, 427
Lean production vs. mass production, 15-16
Leather-hard, 319
Left-hand tools, 122
Light gage stock, 119
Lignin, 391
Line balancing, 49
Liquid infiltration, 493
Liquid-phase sintering, 371-372
Liquidus line, 185
Live center, 122
Longitudinal, 395, 511
Lost motion, 37
Lower transformation temperature, 187
Low-pressure lamination, 253
Lubricants, 317

M

Machine planing, 413
Magazines, 556
Major material families (MMFs), 26
Mandrel, 295, 443, 497
Manual controls, 545
Manufacture of particleboard, 408
Manufacturing
   design, 39-40
   flexibility, 541
   introduction, 13-32
   major industries, 21-23
   processes, 29
   productivity, 17
   superiority, competition, 16-20
Maquiladoras, 17
Marforming, 88
Martensite, 187
Maskant, 155
Mass production, vs. lean production, 15-16
Master unit die method, 106
Matched mold forming, 236-237
Material feeding techniques
   escapements, 558-560
   handling systems, 556-567
   manufacturing, 24-28
   robots, 562-563

shuttles, 560-562
Material-removal processes, 293-295
Materials, properties, 54-55
Matrix, 476
Mechanical etching, 357
Mechanical fasteners, 438-443
   blind rivets, 442-443
   nails, 441-442
   wood screws, 438-441
Mechanical joining, 167-168
Mechanical properties, 55-57
Mechanization, 541
Medium gage stock, 119
Melt spinning, 496
Memory, 217
Metal classification
   ferrous metals, 58-60
   high-temperature superalloys, 60
   refractory metals, 60
Metallic-fiber-reinforced composites, 521
Metallic materials, conditioning, 189-197
   cryogenic conditioning, 197
   normalizing, 195
   softening, 195
   stress relieving, 195-196
   tempering, 196-197
Metallic materials, fabricating
   brazing, 182-183
   brazing and soldering, 181-184
   gas welding, 169-181
   mechanical joining, 167-168
   soldering, 183-184
   welding processes, 168-169
Metallic materials, finishing
   anodizing, 201
   deburring, 198-200
   electroless plating, 201
   electroplating, 200
   ion implantation, 201-202
   metallizing, 202-203
   surface preparation, 198
Metallic materials, forming
   bulging, 90-91
   casting and molding, 101-110
   drawing, 87-90
   electromagnetic forming, 94-95
   embossing, 86-87
   expanding, 90
   explosive forming, 97-98
   extrusion, 78-79
   forging, 66-69
   necking, 91-93
   nosing, 94
   peen forming, 95-97
   rotary forming processes, 70-78
   spinning, 98-100
   stamping, 82-86

upsetting, 79-82
Metallic materials, separating, 113-166
Metallizing
   powder coating, 203
   spray metallizing, 360
Metal-matrix composites (MMCs), 492
Metals, unique characteristics, 53-57
Microspheres, 289
MIG, 171
Milling
   conventional milling, 349
   cutting speed, 135-136
   dry milling, 351-353
   milling cutters, 135
   wet milling processes, 349-351
Miter saws, 431-432
Moisture content, 394
Moisture, effects on wood, 394-396
Molding, 220
Molding, and casting, 101-110
Molinex mill, 350
Monomers, 211
Mortise, 435
Mortise-and-tenon-joints, 435
Mortising machines, 435
Multiple-spindle shaper, 421

N

Nail set, 441
Nails, 441-442
Natural composites, 476
Nd:YAG lasers, 510
Near-net-shape manufacturing process, 102
Necking, 91-93
Nesting, 117
Neutralize, 464
Nibbling, 119
Nitriding, 193-194
Nonconsumable products, 27
Nonferrous metals, 59, 191
Non-grain-raising (NGR) stains, 465
Nonionizing radiation processes, 285
Normalizing, 195
North American Free Trade Agreement (NAFTA), 17
Nosing, 94
Notching, 119
Nuclear fuel rods, 323
Numerical control (NC), 550-553
Numerical control (NC) programming, 551-552

Index

O

Oil-base paints, 466
Oligomers, 281
On/off control, 548-549
Open loop systems, 545-548
Open molding
    autoclave molding, 492
    hand layup, 237-239
    sprayup, 239-240
    vacuum bagging, 492
Open-die forging, 66-67
Open-grain woods, 392
Organsols, 296
Oriented strandboard, 408-409
Outputs, 545
Overglazes, 385
Overproduction, 36
Overspray, 468
Oxidation, 496
Oxy-acetylene rod gun, 381
Oxy-acetylene welding, 169

P

Pad printing, 537
Paints and enamels, 466-467
Pallet, 565
Panel saws, 432-433
Particleboard, 407-408
Particulate composites, 484-485
Parting, 120
Parting line, 67
Parting tool, 424
Part-transfer mechanisms, 560
Pass, 170
Paste filler, 463
Pearlite, 187
PEEK (polyether ether ketone), 216
Peen forming, 95-97
Penny, 442
Perforating, 116
Peripheral speed, 260
Peripheral turning, 124
Permanent mold casting, 106
PERT, 46
Phase diagram, 186
Photoinitiator, 281
Photoresist, 155
Physical frothing, 290
Physical properties, 54, 55
Pick-and-place robots, 562
Pickling, 199
Piezoelectric effect, 55
Pigment, 466
Pilot hole, 441
Planing
    abrasive belt planer, 415-417
    broaching, 127-132
    slotting, 127
Planning for production
    Boothroyd and Dewhurst method, 42-45
    design for assembly (DFA), 40-45
    design in manufacturing, 39-40
    planning, 45-50
    waves of change, 33-38
Plasma arc gun, 381
Plasma arc spraying, 202, 360, 494
Plastic additives
    foam-frothing, 290
    Franklin fiber, 288-289
    glass flakes, 289-290
    microspheres, 289
    preservatives, 288
Plastic materials
    annealing, 284
    characteristics, 207-208
    conductive plastics, 286-288
    creep, 218
    development and introduction, 207-218
    major classes, 215-218
    plastic additives, 288-290
    radiation processing, 284-286
    separating, 257-268
    structure, 211-215
    thermoplastics, 215
    thermosets, 215-217
    unique characteristics, 207-208
Plastic materials, fabricating
    cementing and bonding, 277-279
    cohesion processes, 269-277
    filament winding, 279-281
    sterolithography, 280-281
Plastic materials, finishing
    coating processes, 295-301
    coloring, 301-302
    deburring, 198-200
    in-mold decorating with foils, 301
    material-removal processes, 293-295
    other finishing processes, 301-302
Plastic memory, 217-218
Plasticity, 56
Plasticized wood, 454-455
Plasticizers, 288
Plastics, 207, 209
Plastics, machining
    cutting extruded lengths, 260
    cutting speeds, 258-260
    deflashing and deburring, 264-265
    degating, 264
    die cutting, 260-263
    foam cutting, 263-264
    grinding, 265-266
Plastics welding, 270
Plastisol, 248, 295
Plating, 159, 493-494
Plies, 409
Plug-assist thermoforming, 233
Plywood, 409-410
P/M injection molding, 105
Polishing, 294
Polyether ether ketone (PEEK), 216
Polyethylene, 211
Polyethylene glycol (PEG), 457-458
Polymeric-matrix composites (PMCs), 492
Polymerization, 213-215
Polymers, natural and synthetic, 212-213
Polystyrene, 207
Polyvinyl chloride (PVC), 253
Pores, 391
Porosity, 449
Pot broaching, 131
Potting, 252
Powder coating, 203
Powder metallurgy (P/M), 102-105
Precipitation hardening, 191-192
Precision grinding, 141
Preforms, 220
Prepegs, 220
Preservatives, 288
Presintering, 370
Press and blow machine, 342-343
Pressure bubble plug-assist vacuum forming, 236
Pressure roll binding, 493
Pressure-feed gun, 468
Pressure-treated wood, 456-457
Primary processing, 25
Process action, 29
Process flow model, 46
Production molder, 463
Production of superconductors, 525-526
Production routers, 421
Programmable logic controllers (PLCs), 507, 548
Promoters, 532
Proportional control, 549
Pugging, 353-354
Pull broach, 128
Pultrusion, 241-243
Punch, 80
Punch press, 83
Punch-and-die shearing, 115
Punching, 117
Pure alloy, 54
Pyrolysis, 520
Pyrometric cones, 372

Q

Quenching, 191

R

Radial arm saws, 431
Radial forging, 76
Radiation processing, 284-286
Radio-frequency dielectric drying, 397
Radio-frequency dielectric heating, 450
Radio-frequency heat sealing, 274-276
Rake angle, 140
Ram pressing, 326
Ratchet and pawl, 563
Raw materials, 22
Reaction bonding, 372
Reaction injection molding (RIM), 226-227
Reaming, 139
Reconstituted wood, 409
Refractory materials
  metals, 60-61
  slurry, 364
  spraying, 364-365
Reinforced reaction injection molding (RRIM), 521
Relay logic, 548
Replication, 541
Resin, 209
Resin transfer molding (RTM), 230
Resistance welding, 173-174
Resonance point, 156
Response criteria, 546
Reverse electroplating, 201
Reverse roller process, 469
Rework, 37
Ribbon blow molding machine, 343-344
Riddle, 108
Right-hand tools, 122
Ring broach, 131
Ring-and-stick broaches, 132
Rip sawing, 425
Robots, 562-563
Roll bending, 70-72
Rolled glass, 344
Roller forming, 70
Rolling coating, 469
Rotary forming processes
  roll bending, 70-72
  roller forming, 70
  swaging, 76-78
  thread rolling, 72-76
Rotary indexing tables, 563-567
Rotary shearing, 115
Rotary swaging, 76-78
Rotational molding, 248
Rough grinding, 141
Roughing, 460
Roughing cuts, 124
Rough-sawn lumber, 399
Round nose tools, 424
Routing, 421-422
Roving, 220
Runners, 219

S

S glass, 517
Safety glass, 486
Saggers, 383
Sagging, 345
Sand, 460
Sand casting, 107-109
Sandblasting, 357
Sanding
  abrasive grain sizes, 461
  sanding equipment, 461-463
  types, 460-461
Sawing
  bandsaws, 426-427
  circular saws, 427-430
  miter saws, 431-432
  panel saws, 432-433
  radial arm saws, 431
  scroll saws, 425-426
Scrap, 113
Screw threads, 126
Scribing, 381
Scroll saws, 425-426
Sealing, 269
Seam welding, 173
Second wave, 35
Secondary processing, 25
Second-wave manufacturing, 35-37
Sedimentation, 517
Select hardwoods, 401
Select lumber, 400
Sensitivity, 547
Sensors, 544
Separating, 30
Separating processes
  abrasive jet machining, 147-149
  blanking, 117-119
  chemical milling, 154-156
  drilling, 136-139
  electrochemical machining, 156-159
  electrodischarge machining, 160-161
  electron beam machining, 159-160
  grinding, 140-147
  laser cutting, 152-154
  milling, 132-136
  nibbling, 119
  planing, 126-132
  reaming, 139
  shearing, 114-117
  tapping, 140
  travelling wire EDM, 161-163
  turning, 120-126
  types, 113-163
  ultrasonic machining, 156
  waterless machining, 149-154
Shading wood, 464-465
Shank hole, 441
Shank-type cutters, 135
Shape memory, 217
Shaper, 127
Shaping, 419-421
Shearing, 114-117
Sheet molding compounds (SMC), 221
Sheet stock, 485
Shell broach, 130
Shielded metal arc welding (SMAW), 170
Shot peening, 95
Shrinkage, 250
Shuttles, 560-562
Silica, 383
Silicon carbide, 364, 372
Silicon nitride, 372
Simultaneous engineering, 40
Single plane die cutting, 261
Single planers, 414
Single-spindle shaper, 420
Sintering, 104, 325, 367, 370-372
Siphon gun, 468
Skew, 424
Slab broach, 130
Slack time, reduction, 49-50
Slag, 170
Slip casting, 335
Slitting, 116
Slotting, 127
Slurry, 317, 513
Slurry infiltration process, 523
Slush casting, 107
Smoothing and polishing, 294
Soft automation, 543
Softening processes, 195
Softwoods, cellular structure of, 393-394
Sol, 517
Soldering, 183-184
Sol-gel technique, 517
Solid-phase sintering, 371
Solidus line, 185
Solvent-releasing adhesives, 444-445
Solvents, 277
Spark erosion machining, 160

Specific heat, 55
Spin welding (friction welding), 272
Spindle turning, 423-424
Spinneret, 496
Spinning, 98-100
Spirit stains, 465
Split pattern, 108
Spray drying, 354
Spray metallizing, 360
Sprayup, 239-240
Springback, 321
Sputtercoating, 300
Sputtering, 375
Square nose tools, 424
Squeeze casting, 523
Stability, 547
Stabilization, 496
Staining wood, 465-466
Staking, 273
Stamping, 82-86, 230-231
Standard Industrial Classification (SIC), 20
Staple fibers, 340
States of operation, 549
Static fatigue, 517
Stationary router, 421
Statistical process control (SPC), 556
Staypack, 455-456
Steady state error, 547
Steatite, 320
Steel
   crystalline structure of, 185-188
   determining type, 62-63
Stereolithography, 280-281
Stick broach, 131
Stickered, 458
Stock, 398
Straight shearing, 114
Stress relieving process, 195-196
Stretch draw forming, 88
Strip feeding, 557
Structural composites, 478
Structural foams, 246
Structural lumber, 399
Stud welding, 174
Sublimation, 370
Substrate failure, 444
Successive transformation, 24
Superalloys, 520
Superconductor, 525
Surface energy, 528
Surface grinding, 144-146
Surface preparation of wood
   applying fillers, 463
   bleaching, 464
   finishing processes, 198
   sanding, 460
   shading, 464-465
   staining, 465-466

Swaging, 76-78
Sweat soldering, 184
Swing, 120
Synthetic metal research, 286

T

Tablet pressing, 320
Tailstock, 122
Tap, 140
Tapers, 126
Tapping, 140
Tapping a heat, 101
Taylor process, 521
Technical strategists, 18
Teeth per inch, 425
Temperature degradation of composites, 533
Tempered hardboard, 405
Tempering, 196, 348, 377
Tenon, 435
Tensile strength, 93
Thermal
   decomposition, 518
   energy, 160
   expansion, 55
   expansion resin transfer molding (TERTM), 498
   heat sealing, 276
   impulse heat sealing, 276
   shock resistance, 372
Thermoforming
   drape forming, 236
   matched mold forming, 236-237
   pressure bubble plug-assist vacuum forming, 236
   vacuum forming, 231-236
Thermoplastics, 215
Thermoset adhesives, 446
Thermosets, 215-217
Third wave, 37
Thread rolling, 72-76
Threading, 120
Throughput, 541
Throwing, 328
TIG, 171
Tin plate, 62
Tip angle, 137
Tool holder, 125
Tool post, 124
Tool steel, 59
Topcoat, 466
Torch cutting, 169
Total Quality Approach, 37
Tracheids, 393
Transducer, 510
Transfer coating, 536-537
Transformation toughening, 373-374
Transient response, 546

Travelling wire EDM, 161-163
Trickledown effect, 16
Trost fluid energy mill, 351
Tube drawing, 340
Tumbling, 265, 293-295, 470
Turning
   depth of cut, 124
   facing, 424
   lathe components, 124-125
   operations, 126
   processes, 120
   spindle turning, 423-424
   tools, 122-124
   types of lathes, 120-121
   workholding devices and methods, 121-124
Turnkey system, 542
Twist drill, 136

U

Ultimate tensile strength (UTS), 93
Ultrasonic
   cutting, 513-514
   machining, 156, 510-514
   welding, 181, 272-274
Underglazes, 385
Universal chuck, 121
Universal drilling machines, 139
Universal saws, 428
Up milling, 134
Upper transformation temperature, 188
Upsetting, 79-82
   cold forming advantages, 82

V

Vacuum bagging, 249-250, 492
Vacuum forming, 231-236
Vacuum metallizing, 202, 299
Value analysis, 39-40
Vapor deposition coating processes, 374-375
Vapor-phase sintering, 371
Variability, 540
Variety saws, 428
Varnishes and lacquers, 466
Vehicle, 466
Veneer, 397
Vertical milling machine, 133
Very high bond (VHB) tape, 446-448
Vessels, 391
Vibration welding, 271
Vibratory deflashing, 265
Vibratory finishing, 198
Viscosity, 317
Vitreous solder, 362
Vitrification, 372-373

W

Waferboard, 408
Walking-beam mechanism, 566
Waste, 113
Water stain, 465
Waterfall method, 382
Waterjet
  cutting, 505-507
  machining, 149-154
Waves of change in manufacturing, 34-38
Ways, 124
Wear parts, 310
Weir, 345
Welding processes, 168-169
Wet bending, 411-412
Wet forming processes
  band/tape casting, 333-335
  calendering, 335
  casting, 337
  extrusion, 329-333
  hot pressing, 326-328
  injection molding, 333
  jiggering, 328
  jollying, 328
  slip casting, 335-337
  throwing, 328
  wet pressing, 326
Wet milling processes, 349-351
Wet pressing, 326
Wettability, 528
Wheel electrodes, 174
Whisker growing, 497-498
Whiskers, 479
Wiping, 294
Wire coating, 241
Wire metallizing, 203
Wood, classification, 390-391
Wood grades, 400-401
Wood materials
  conditioning, 449-450
  plasticized wood, 454-455
  polyethylene glycol (PEG), 457-458
  pressure-treated wood, 456-457
  radio-frequency dielectric heating, 450
  staypack, 455-456
  wood plastic composition (WPC), 450-454
Wood materials, fabricating
  adhesives, 443-448
  mechanical fasteners, 438-443
Wood materials, finishing
  application methods, 467-470
  finish coatings, 466-467
  surface preparation, 460-466
Wood materials, forming
  bending, 411-412
  bonding, 403-411
Wood materials, separating
  drilling, 433-436
  jointing, 417-419
  planing, 413-416
  routing, 421-422
  sawing, 424-433
  shaping, 419-421
  turning, 423-424
Wood plastic composition (WPC)
  acetylated wood, 454
  compreg, 452-454
  impreg, 451-452
Wood screws, 438-441
Wood structure
  cellular structure of hardwoods, 391-396
  cellular structure of softwoods, 393-394
Work in process (WIP), 15
Workcell, 37
Workholding devices and methods, 121-124
Workholding pallet, 566
World-class manufacturing (WCM), 38
Woven roving, 492
Wrought iron, 58

Z

Zinc-arc spray metallizing, 299
Zirconia, 373